Analog Electron

Analog Electronics

An Integrated PSpice Approach

T. E. PRICE

PRENTICE HALL

LONDON   NEW YORK   TORONTO   SYDNEY   TOKYO   SINGAPORE
MADRID   MEXICO CITY   MUNICH   PARIS

First published 1997 by
Prentice Hall Europe
Campus 400, Maylands Avenue
Hemel Hempstead
Hertfordshire, HP2 7EZ
A division of
Simon & Schuster International Group

MicroSim and PSpice are registered trademarks of
MicroSim Corporation

Typeset in 10/12pt Times
by Mathematical Composition Setters Ltd, Salisbury, UK

Printed and bound in Great Britain by
T. J. International Ltd

Library of Congress Cataloging-in-Publication Data

Price, T. E.
    Analog electronics : an integrated PSpice
approach / T.E. Price
      p. cm.
    Includes bibliographical references and index.
    ISBN 0-13-242843-1
    1. Electronics.   2. Analog electronic systems.   3. PSpice.
I. Title.
TK7816.P69   1996
   621.3815—dc20
                                 96-29174
                                       CIP

British Library Cataloguing in Publication Data

A catalogue record for this book is available from
the British Library

ISBN: 0-13-242843-1

1 2 3 4 5   01 00 99 98 97

Contents

APPENDICES

Preface

This text is devoted to the use of electronic devices in analog or linear circuits which are used extensively for amplification and signal processing, rather than digital or logic circuits which form the basis of microprocessors and computers. The subject is very large and it would be difficult to cover everything in one book. This book concentrates on the introductory material associated with the subject by starting with a description of the basic active devices of diodes, transistors, both bipolar and field effect, and integrated circuits. For each device there are examples of simple circuits which form the basis of more advanced systems of amplification or signal processing. There are also chapters on active filters, power amplifiers, power supplies and the interface between analog signals and digital signals. It is anticipated that students taking this course are also studying mathematics and circuit theory in parallel courses. The mathematical requirements are not excessive, but there is an obvious need to be familiar with algebra, complex numbers involving magnitude and phase, and at least the basic concepts of differentiation and integration. It is assumed that students are familiar with Kirchhoff's voltage laws, Thévenin and Norton theorems and superpositions. These are used extensively and form the basis of the analysis of circuits comprising resistors, capacitors, inductors and voltage and current sources. However, because of the difficulty some students experience when analyzing the frequency response of circuits which include both resistors and capacitors, a chapter has been included on the frequency response of *RC* networks, and an appendix included covering the simple basics of the magnitude and phase of such networks.

The academic level of this text has been set above technician level, but slightly below the more traditional degree level, which often tends to be very analytical. For technician-based courses it may be necessary to be selective with some of the analytical material, while for degree-level courses the lecturer may wish to further develop some of the material to more accurately reflect his particular interest, or to better satisfy the aims and objectives of the particular course.

Analog circuits are difficult to understand and small variations in component values or changes to the configuration of components can completely alter the way in which the circuit operates. The process is not helped by the fact that the solid-state devices are so simple in appearance, and yet so complex in terms of electrical operation. The inadvertent application of the wrong voltage polarity or too large a voltage can damage them beyond repair, but more importantly the effect that they have on a signal depends on its amplitude, polarity and rate of change. It is the author's view that a basic understanding of the operating principles of the active devices can help in understanding the way in which the devices operate when in a circuit. For this reason the chapters associated with the diode, bipolar and field effect transistors include a brief description of the physics of operation of each device. However, the author appreciates that many circuit designers feel that this is unnecessary, and that it is sufficient to treat the devices as 'black boxes'. For this reason the sections devoted to this material are marked as optional and it is left to the lecturer to decide whether to include this material, to use it simply as review material or to ignore it completely. In a similar vein it is important to be familiar with the concept of an equivalent circuit for active devices. These are essential for circuit analysis and are one of the reasons why students have difficulty with analyzing analog circuits. Each active device can be represented by a number of different equivalent circuits. The author has described a number of them, but only uses the simplest for hand-based calculations. However, some of the more complex versions are described to enable the student to gain some appreciation of the type of models which are used in modern-day computer-aided design software which is used for circuit simulation. This material is again marked as optional and may be ignored without any loss of understanding of the main text.

The wide availability of powerful PCs and software for circuit simulation means that no course on electronics should ignore the use of circuit simulation. The most widely used simulation software for analog electronics is based on SPICE. There are now many proprietary versions of SPICE and one which has gained in popularity over recent years is MicroSim PSpice®. The attraction for the author is the availability to teaching staff of a free demonstration version which is capable of handling about 20 components. This is more than adequate for the circuits used in an introductory course on analog electronics. Numerous examples are provided in most chapters to give a more realistic demonstration of the operation of a particular device or circuit. The examples may be regarded as 'virtual laboratory exercises' which can provide dc values of voltage and current, frequency response and voltage traces similar to those observed on an oscilloscope. Circuit simulators are ideally suited to examining the effect of changes in component-values or the effect of different frequencies. However, this does not spell the end of the need to understand how to analyze a circuit. Circuit analysis, and a knowledge of the different configurations of components and active devices, is still important. The circuit simulator is the last step in the process of design, to provide confirmation of the analysis, to examine the effect of component tolerance, and to confirm the bandwidth or the large-signal response, before the design is committed to manufacture. Students should be encouraged to use and to become familiar with the operation of a circuit simulator, but should not be allowed to use simulation in a trial-and-error approach to design. Apart from being very time consuming such an approach provides no understanding of how the circuit operates.

It is possible to generate explicit formulae for circuits, but the author has avoided the temptation of providing a summary of important formulae for each chapter, because of the pressure this places on students to learn these formulae parrot fashion rather than getting to grips with the operation of the circuit. Also a slight rearrangement of the components can result in a completely different formula, even though the operation of

---

®MicroSim and PSpice are registered trademarks.

the new configuration may be very similar to the original. It is much more important that the student understands how the circuit operates, and how it can be analyzed, than to learn a formula. It is also important that students are encouraged to evaluate properties (currents, voltages, impedances) as they proceed through a network, rather than constructing a complex algebraic equation for the whole circuit. There is always the danger of losing a term in rewriting the equation many times during its construction.

Each chapter is provided with numerous worked examples, when it is hoped that this approach of evaluating values as the analysis proceeds is demonstrated, but also the examples provide a means for introducing actual values of gain, resistance, frequency, etc., to illustrate a particular circuit or method of analysis. There are also numerous questions at the end of each chapter to provide further exercises for students on each topic. Answers are provided for all questions and if they cannot be obtained then a student should be encouraged to seek help, rather than simply noting the answer and ignoring the method by which it is achieved (a solutions manual is to be made available for teaching staff). The questions are there to provide assistance with the use of the different analytical methods and circuits used in each chapter.

It is assumed that a practical course is running in parallel to the theoretical course, since electronics is very much a 'hands-on' design, build and test type of subject. Laboratory exercises which can be used to demonstrate the theoretical aspects of a circuit and which allow the student to examine the effect that the circuit has on a signal, and to measure its gain, frequency response or impedance levels, are an important means for assisting in the understanding of the theory. The author has always favoured the prototype-board approach for practical classes, where the student assembles the circuit from a collection of discrete components, rather than use a pre-assembled circuit, although it is appreciated that this is not always practical for more complex circuits. This approach does ensure that the students are acquiring a skill in selecting and handling components, as well as constructing circuits. Practical exercises need not be complex, and the emphasis should be on creating an exercise which clearly demonstrates some aspect of circuit theory, and provides an opportunity for the student to make use of theoretical concepts for calculating gain or frequency response or impedance level, which can then be verified by measurement. Access to computers also allows the results to be confirmed by simulation. Laboratory exercises should be used to demonstrate some aspect of theory and also to allow the student to 'design' a circuit to satisfy a design specification. The design need not involve anything more complex than a single-stage amplifier. A design which specifies the supply voltage, gain, frequency response and impedance levels, to name but a few parameters, and which must use preferred component values, can present a budding engineer with a complex task of applying theory, using simulation and finally of building and testing.

A problem which is likely to arise with a text which relies so extensively on the use of computer simulation to both illustrate circuit operation and to form the basis for numerous problems, is that of assessment. It is an unfortunate fact that any course leading to the award of a qualification must involve assessments. Computer simulation does not lend itself to the more traditional form of engineering examination. Such assessments do usually include a laboratory-based element, but this often represents only a small percentage of the complete assessment, primarily because of the difficulty in making an accurate assessment of laboratory work. Those universities and colleges which have adopted competence-based teaching and assessment are better placed to incorporate computer simulation into the assessment by producing detailed competencies for different aspects of simulation and analysis. Such competencies could be loosely based on the three types of analysis which can be performed with a circuit simulator, namely dc, ac and transient. These modes of operation correspond quite well to the stages in which analog circuits are presented – dc biasing, small-signal analysis, large-signal wave shaping – so that assessment exercises could be devised which test the students' understanding of the circuit theory and also the operation of the simulator, and assessments would be delivered at the appropriate times during the period allotted for the course.

Acknowledgements

The author would like to thank the many students at the University of Paisley who, over a number of years, have inadvertently acted as guinea pigs to the development of this electronics course on analog electronics. He would also like to thank the many colleagues at the University for their helpful suggestions during the development of the course.

CHAPTER 1

Introduction

This chapter provides an overview of linear circuits which use semiconductor devices and passive electronic components, and some circuits are introduced as examples, but without explanation of how they operate. The explanations are left to later chapters. There is also a brief description and explanation of the circuit simulator PSpice. A brief users' guide and tutorial is provided in Appendix 7. Circuit simulators are now used extensively in colleges, universities and industry and in the author's opinion it is appropriate for circuit simulators to form an integral part of an undergraduate course on analog electronics. However, one cannot rely entirely on a simulator and it is important that future electronic engineers have a sound understanding of the basic principles of analog electronics and are able to derive design equations, at least to a first order of accuracy. Simulation is used to provide accuracy and detail which is otherwise often difficult to achieve with hand calculations based on approximate design equations.

1.1   Linear Circuits

An electronic circuit is formed from a collection of passive components and semiconductor devices which in some way modify an input signal. The electronic engineer must be able to analyze such an array of components, and must also be able to select suitable components, in terms of value, tolerance, voltage and power rating, and finally cost, in order to design and construct the circuit. For the designer of analog circuits the choice is very large and failure to make the correct decision can affect the overall performance of the circuit. The engineer is assisted in the design process with the use of circuit simulators, which are design aids which can be used to verify hand-based calculations and to investigate very quickly the effect of changing component values. It is particularly important to use a circuit simulator for circuits which are to be mass produced to investigate the effect of the variation in component values resulting from component tolerance.

Analog Signals

Analog or linear signals are continuously varying, unlike digital signals which only have two levels, ON and OFF, or high and low. When the continuously varying signal is applied to one or more active devices (diode, transistor, operational amplifier, etc.) which form a circuit, the relationship between the input and output is often complex because of the non-linear behaviour of active devices. Thus, not only is the signal continuously varying, but the

1

variation of the signal with respect to the original signal may change as the signal progresses through the circuit. A further complication arises from the rate at which the variations occur which may influence the output; that is, the output is affected by the rate of variation, or the frequency of the signal. Analog circuits are generally frequency dependent. All of these factors combine to make the analysis of analog circuits difficult, and, for the inexperienced, frustrating. Relatively minor modifications of the arrangement of components, or simply component values, produce marked changes in the relationship between the output and the input signals, or in the transfer function or frequency response.

Computer Simulation

A computer-aided design (CAD) software program for circuit simulation can greatly simplify the task of circuit analysis, but it is still necessary to have a detailed understanding of the basic principles which apply to the division of currents and the formation of voltages for particular combinations of components. The circuit designer must always know what to expect from the simulator, and should always be able to predict a dc or a small-signal ac response. The simulator is then used very rapidly, and accurately, to investigate the operation in more detail for different component values or different input frequencies. Many circuit simulators can provide a *worst-case analysis* of a circuit where all of the component values, or a selected group, are allowed to vary by an amount determined by their tolerance value. A table of results, or a set of curves, is generated which shows the effect of the changes on the performance of the circuit. This type of analysis is now regularly performed on circuits which are to be mass produced to ensure that the variation of component values within a batch from a manufacturer will not affect the overall performance.

Computer simulation is only as good as the *models* which are used to represent the components. For passive components this is not a major problem, although temperature effects and stray capacitance can be difficult to model, but for active components, such as transistors, the correct choice of model is important if the program is to produce accurate results.

1.2   Passive Components

Passive components usually refer to resistors, capacitors and inductors, but in addition there are transformers, radio frequency chokes, crystal filters, ceramic filters, transmission lines, fuses, devices for preventing power surges, thermistors, photoconductors and many others. Many are very specialized with a very limited range of application, unlike the first three which form the basic circuit elements of any linear circuit. It is assumed that anyone studying analog electronics would become acquainted with handling components through a practical, skills-based course which involves circuit construction and testing with electrical instruments.

Resistors, Capacitors and Inductors

A small selection of the most commonly used components, namely resistors, capacitors and inductors, is shown in Figure 1.1. The most immediate visible difference between components is in the variation in physical size, both between different components and also for a given component type. Resistors vary in size depending on their power rating rather than their value, and low-power devices (1/10th or 1/8th watt) are small, while high-power devices (1 watt and above) are much larger.

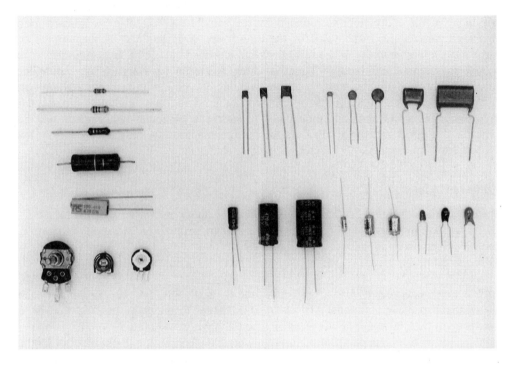

FIGURE 1.1   Selection of passive components.

There are considerable variations in the size of capacitors which are a function of the value of capacitance, the voltage rating and the type of construction. Because capacitance is a function of electrode area, large-value capacitors are generally physically large. Electrolytic decoupling capacitors fall into this category with the size increasing with capacitor value. The voltage rating is also an important factor because it determines the minimum thickness of the dielectric between the capacitor electrodes. Thus an emitter bypass capacitor for a transistor amplifier may only be required to support a dc voltage of 1–2 V across the emitter-bias resistor, while if the same value capacitor were to be used to decouple the dc supply rail of a transistor circuit, it may be required to support 10–20 V and will be physically larger. The voltage rating will affect the size and the cost. Precision, high-quality (low-loss-factor) capacitors are usually based on glass, plastic or mica dielectrics which have relatively low dielectric constants and consequently, for a given size, it is difficult to achieve large values of capacitance with these materials. However, they are generally only required for high-frequency circuits where large values are not required. A wide range of capacitor values and voltage ratings may be obtained with ceramic dielectrics. These materials can exhibit very large values of permittivity (dielectric constant), but this is often accompanied by large temperature coefficients.

Inductors are generally associated with telecommunication circuits where they provide frequency selectivity in tuned circuits, or form part of an *LC* circuit for an oscillator. The inductor is based on a coil of wire, usually wrapped around a ferroelectric material, or completely surrounded by the same material, which exhibits a high magnetic permeability. The value of the inductance is determined by the number of turns and the permeability of the ferroelectric material. Values of a few microhenrys to a few millihenrys are typical. There is often a means for varying the values by mechanically changing the position of the magnetic material, with respect to the coil, in order to alter the magnetic coupling. This can be used to 'tune' a filter to a particular frequency. Inductors are usually designed to satisfy a particular requirement and the type of wire used and the type of magnetic core material are selected to satisfy the specification for frequency, *Q* factor, size, cost, etc.

Preferred Values

Resistors and capacitors can have many different values, but for ease of manufacture, the values follow clearly defined progressions known as *preferred values*. For example, the 10% tolerance range is:

$$10, 12, 15, 18, 22, 27, 33, 39, 47, 56, 68, 82$$

This range is repeated with multiplying factors of 10, 10^2, 10^3, and so on. For each value the particular component is assumed to have a spread of $\pm 10\%$; that is, a component with a nominal value of 10 could have a value of 9 or 11, while a component with a value of 12 could have values of 10.8 or 13.2, and likewise for 15 the values could be 13.5 or 16.5, and so on. For a 5% tolerance the number of values in a range is increased, and similarly for 2% and 1%. Naturally the higher the tolerance (less spread of values) the greater the cost. Resistor values range from 1 ohm to 1×10^6 ohm (1 MΩ) or greater, while capacitors range from 1×10^{-12} F (1 pF) to 1000×10^{-6} F (1000 μF) or greater; the inductor may typically range from 1×10^{-6} H (1 μH) to 100×10^{-3} H (100 mH) or greater.

The preferred values have implications for the circuit designer because it is likely that the value obtained from a particular design calculation will not coincide with a preferred value. The designer could use a higher-tolerance component in order to obtain the required value, or use a variable component, but this would add to the production cost. Alternatively the nearest preferred value could be used, but then it is necessary to verify that the design specification can still be satisfied. The circuit simulator can play an important role in this process by enabling the designer quickly to assess the effect of alternative component values on the circuit performance. It is important to note, however, that the actual value of a component can lie anywhere within the tolerance band. The simulator can be used to determine the effect, on the circuit performance, of each component having either the lowest or highest value within the tolerance band. This *worst-case analysis* can be used to identify any potential problems which may arise as a result of component values departing from their nominal values, for example if a transistor amplifier is to be biased to a very precise value will changes in resistor values affect the operation of the amplifier? It is very important to perform worst-case analysis for any circuit which is to be manufactured in large quantities to ensure that the design specification is satisfied for all values of each of the components.

▶ **EXAMPLE 1.1** ─────────────────────────────────

A potential divider with nominal resistor values of 1 kΩ and 10 kΩ placed across a 9 V supply. Determine all possible values of the voltage across the 1 kΩ resistor assuming 10% resistor tolerances.

▶ *Solution*

Nominal value

$$V_{nominal} = \frac{9\ V}{1\ k\Omega + 10\ k\Omega}\ 1\ k\Omega = 0.818\ V$$

Maximum resistor values:

$$V^{++} = \frac{9\ V}{1.1\ k\Omega + 11\ k\Omega}\ 1.1\ k\Omega = 0.818\ V$$

Minimum resistor values:

$$V^{--} = \frac{9\ V}{0.9\ k\Omega + 9\ k\Omega}\ 0.9\ k\Omega = 0.818\ V$$

Minimum for 1 kΩ, maximum for 10 kΩ:

$$V^{-+} = \frac{9\ V}{0.9\ k\Omega + 11\ k\Omega}\ 0.9\ k\Omega = 0.681\ V$$

Maximum for 1 kΩ, maximum for 10 kΩ:

$$V^{+-} = \frac{9\ V}{1.1\ k\Omega + 9\ k\Omega}\ 1.1\ k\Omega = 0.980\ V$$

Notice that if the tolerance for both resistors moves in the same direction then the voltage is unchanged, but when their values move in opposite directions then the effect can be quite significant with the voltage ranging from 0.681 V to 0.980 V, that is −16.7% to +19.8% with respect to the nominal value. The effect can be much more complex for circuits which contain many more components. The circuit simulator can be used to perform worst-case analysis, although some care needs to be exercised in the selection of components because of the time required to complete such an analysis for a complex circuit. Some components may be more critical than others and it is for the designer to be able to identifycript the most likely components rather than relying on the simulator to find the critical ones as the result of simulating change in all of them.

1.3 ▷ Semiconductor Devices

Semiconducting materials have revolutionized the development of electronic circuits by enabling components to be fabricated that have the ability to amplify signals. The most important device is the transistor which as a *bipolar transistor* is a low-impedance current-operated device, while for the *field effect transistor* it is a high-impedance voltage-operated device. The transistor has three terminals, although some have an additional terminal which is attached to the metal case to provide electrical screening. Some field effect transistors also have an additional input electrode. With a three-terminal device it is possible to have one common terminal and separate input and output terminals. This arrangement means that the input and output stages are isolated from each other in whichever manner the transistor is configured, that is *common-emitter*, *common-collector* or *common-base*. However, in practice there is always some interaction between input and output resulting from stray capacitance, and feedback within the transistor itself.

An electronic circuit comprising resistors, capacitors and transistors can change signals in some way. These signals may have originated from transducers which respond to some physical effect, such as temperature or pressure, or from radio or a television aerial. The circuit may consist of discrete components, but it is also possible to fabricate complete circuits within a single small piece of semiconductor material to produce the *silicon chips*, as is the case for digital integrated circuits (ICs). Analog as well as digital ICs form a significant part of modern computers, video recorders, satellites and many, many other complex electronic products which are currently marketed.

A small selection of semiconductor devices is shown in Figure 1.2. The devices with two leads are diodes, those with three leads are transistors, while the multilead packages are integrated circuits. The variety of devices is very large and new ones are continually being added. Their physical size is usually related to the power dissipation when in use. The transistor for the output stage of an audio amplifier may be required to reproduce several hundred watts in the loudspeakers and thus needs to be large, because a fraction of the power is dissipated in the transistor as heat, while the transistor in a television amplifier will only be required to handle a few microwatts and can be manufactured in a relatively small package. The actual piece of semiconductor material from which the transistor is made in both cases is usually much smaller than the final packages shown in Figure 1.2.

Diodes

Diodes are formed from a piece of semiconductor such as silicon so that one part of the material is n-type and the other part is p-type (this will be considered in more detail in Chapter 2). When a dc voltage is applied current will flow when the p-type region is positively biased. However, if the p-region is negatively biased no current will flow.

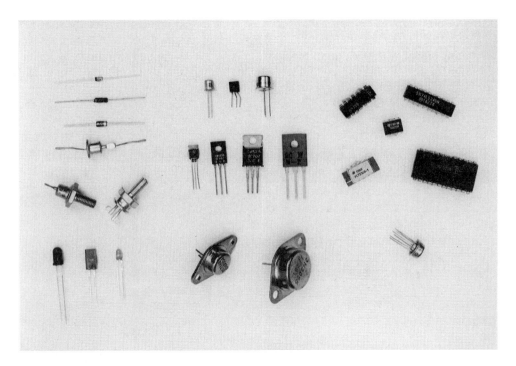

FIGURE 1.2   Selection of diodes, transistors and integrated circuits.

This ability to differentiate between positive and negative voltages is an important circuit function.

A simplified schematic of the n-type and p-type regions of a diode and the associated electrical symbol for several different types of diode is shown in Figure 1.3. The simplest is the small-signal or rectifier diode, but by appropriate control at manufacture of the semiconductor material it is a simple matter to produce a voltage regulator or Zener diode. With certain semiconductors the pn junction emits light when it conducts current (light-emitting diodes or LEDs). Other diodes can be made to respond to light and to act as a photo-sensitive switch or a detector of optical signals (photodiodes).

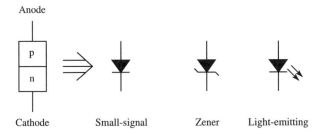

FIGURE 1.3   Schematic of pn diode and diode symbols for a small-signal diode, a Zener diode and a light-emitting diode.

One of the simplest and probably the most widespread applications of the diode is as a rectifier as shown in Figure 1.4. This circuit is described in Chapter 3. An alternating signal with peak amplitudes of $+V$ and $-V$ is applied to a diode by means of a transformer. The diode conducts when the voltage on its anode is positive, and current flows through the load resistor. The diode does not conduct when the voltage applied to its anode is negative. The result is that the current only flows through the resistor for the positive half cycles of the input wave, and a unidirectional waveform with an amplitude of V/n is produced across the resistor, where n is the step-down ratio of the transformer. This simple circuit provides the means for converting alternating voltages to direct voltages. Since all electronic circuits and electronic equipment require direct current (a dc supply), the diode rectifier is an important circuit element.

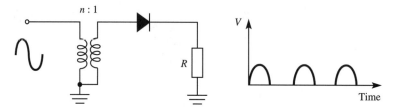

FIGURE 1.4   Simple half-wave rectifier circuit.

▷ **EXAMPLE 1.2** _____

Determine the voltage across the 100 Ω resistor (v_{out}) in Figure 1.5 if the forward voltage drop across the diode is 0.7 V and the reverse leakage current is 1 µA.

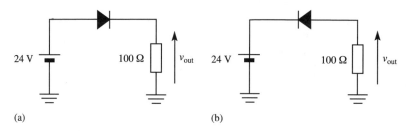

FIGURE 1.5   Forward- and reverse-biased diodes.

▷ *Solution*

Forward biased, circuit (a):

$$24\ V = 0.7\ V + v_{out}$$

and

$$v_{out} = 23.7\ V$$

and

$$I = \frac{23.7 \text{ V}}{100 \ \Omega} = 237 \text{ mA}$$

Reverse biased, circuit (b), the only current flowing in the resistor is the leakage current, and:

$$v_{\text{out}} = 1 \ \mu\text{A} \times 100 \ \Omega = 0.1 \text{ mV}$$

Four-Layer Diodes

A further development of the pn junction is the four-layer structure comprising pnpn layers of semiconductor material. It may be used as a two terminal device – a DIAC – or with an additional terminal as a silicon-controlled rectifier (SCR). A further development is the TRIAC, which is basically a DIAC with an additional terminal. The symbols for these devices are shown in Figure 1.6. They all perform the function of an electronic switch, and can be used in place of an electromagnetic relay.

SCR   DIAC   TRIAC

FIGURE 1.6   Symbols for four-layer diodes.

The SCR can be made to conduct when a positive voltage is applied to the anode with respect to the cathode, and when a short trigger pulse is applied to the gate electrode. The device will continue to conduct until the anode voltage drops below the level required to sustain conduction (1–2 V). The device is ideally suited to the control of large voltages and currents associated with electric furnaces or electric motor speed control and can cope with very large currents and voltages.

The DIAC conducts in either direction when the voltage exceeds the breakdown voltage for the device, which can range from tens of volts to several hundreds. It is often used in circuits which use SCRs.

The TRIAC is a DIAC with a third electrode which can be used to turn the device ON for either voltage polarity. It is a relatively low-power device, unlike the SCR, and an example of its use is in domestic electric light dimmers. A detailed analysis of the devices, and their circuits, is not included in this text, because these devices normally form part of a power electronics or power systems course. However, the TRIAC is an interesting device and its operation as a light dimmer can be investigated with the aid of computer simulation using PSpice which contains a model for the 2N5444 device. Computer-aided design is considered in Section 1.4. The following example demonstrates the use of computer simulation and shows how a complex device like a TRIAC can be used to control ac power.

▷ **EXAMPLE 1.3** _____

Use PSpice to simulate the waveforms produced in the 'dimmer' control circuit shown in Figure 1.7 using a 2N5444 device when the input is 110 V and 60 Hz.

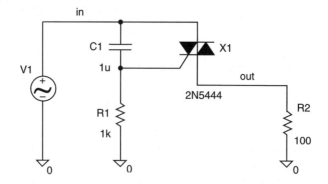

FIGURE 1.7   Dimmer control circuit based on a TRIAC type 2N5444.

The control electrode of the TRIAC is supplied with a voltage from the junction of C_1 and R_1. The resistor R_1 would be variable in an actual light dimmer, while the 100 Ω resistor represents a load, such as a light bulb.

From **Analysis** in PSpice, select **Setup...** and **Transient...** and enter 200 µs for the **Print Step:** and 34 ms for the **Final Time**. The simulation takes a few minutes. The exact time depends on the processor being used and whether a maths co-processor is present.

▷ *Solution*

The output from Probe is shown in Figure 1.8.

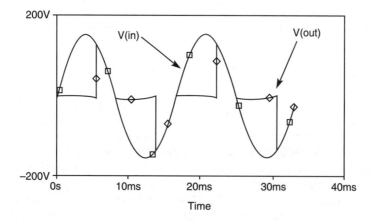

FIGURE 1.8   Output for the TRIAC circuit from Probe.

It can be seen from Figure 1.8 that the current only flows in the load for a small fraction of each positive and negative half cycle and as a result the average power dissipated in the load is small. The amount of power is varied by changing the value of the resistor R_1. Increasing the value increases the power, decreasing it decreases the power. This simple circuit forms the basis for the domestic light dimmer, where the resistor is replaced by a potentiometer.

This circuit forms the basis of a PSpice tutorial in Appendix 7.

This example demonstrates how a simulator can provide results without knowledge of how the circuit or the TRIAC operates. However, the simulator alone does not provide any guidance as to how the circuit may be optimized for a particular application.

Transistors

There are two types of transistor, the low-impedance current-operated *bipolar* transistor and the high-impedance voltage-operated *field effect* transistor. Within these two categories there are further subdivisions. There are two types of bipolar transistor, *npn* and *pnp*. For the field effect devices there are the *junction* field effect transistors (JFETs), *n-channel* and *p-channel*, and the *metal–oxide–semiconductor* field effect transistors (MOSFETs), again with n-channel and p-channel devices, but in addition there is a further subdivision of the MOSFET into *depletion* and *enhancement* devices. The symbols for the various transistors are shown in Figure 1.9.

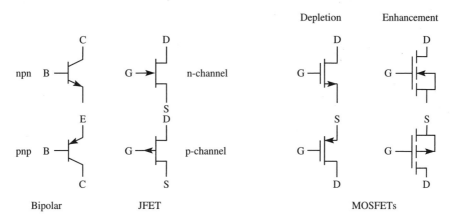

FIGURE 1.9   A selection of transistor symbols for bipolar and field effect transistors.

An important linear application for the transistor is amplification. All of the different types of transistor can be used for this purpose and a simple single-stage amplifier using an npn bipolar transistor is shown in Figure 1.10. The signal at the output (v_{out}) is an amplified version of the input voltage. The other transistor types can be used with variations of the circuit shown. This and other more complex circuits are considered in later chapters.

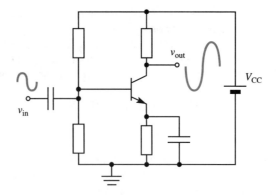

FIGURE 1.10   Simple transistor amplifier.

Linear Integrated Circuits

Integrated circuits were originally developed to perform the Boolean operations required for computers. These circuits replaced discrete component circuits which used diodes, resistors and transistors. It was quickly realized, however, that certain discrete component linear circuits, notably the common-emitter differential gain stage, could also be fabricated as an integrated circuit to operate as an amplifier with a differential input and a single-ended output. These *operational amplifiers* are now the most commonly used linear integrated circuit. Many other linear or analog functions have now been integrated and there is a large selection of circuits for processing analog signals, from dc to very high frequencies.

Common features of the operational amplifier are the very large values of voltage gain ($>10^5$), high input impedance and the low output impedance. Operational amplifiers can be used for a wide variety of functions by the addition of a few external components, for example the addition of two external resistors as shown in Figure 1.11 result in the inverting and non-inverting amplifiers (note the polarity of the output

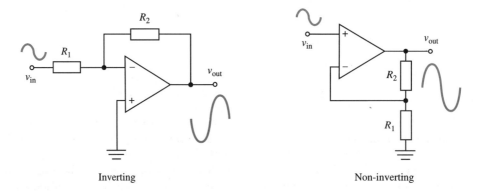

Inverting   Non-inverting

FIGURE 1.11   Inverting and non-inverting amplifiers.

waveform for each amplifier) with gains determined by the value of the resistors. These circuits are described in Chapter 13.

For telecommunication purposes there are video amplifiers, modulators and demodulators, phase-locked loops, switched capacitor filters and many more specialist circuits.

Many analog functions are now being performed with digital circuits, and an important stage in the process is the conversion of an analog signal to a digital signal. For this there are a selection of analog-to-digital (A/D) and digital-to-analog (D/A) converters which will be considered in Chapter 17.

1.4   Computer-Aided Design

The dc or ac performance of a network may be examined by application of Kirchhoff's voltage and current laws to determine the currents in the various branches and the voltages on the various nodes in the circuit. This type of analysis normally forms part of an electrical principles or electrical theory course, and it is assumed that such a course is available to students studying analog electronics. A detailed understanding of what follows is not required to use a circuit simulator, and is simply presented to provide some appreciation of how simulators work.

For the simple network composed of passive components shown in Figure 1.12 it is a simple matter to consider the mesh currents and to sum the voltages around the different loops.

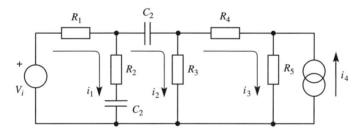

FIGURE 1.12   Application of Kirchhoff's laws to a simple network.

Consider the loop comprising V_1, R_1, R_2, and C_1. There are two currents associated with this loop, i_1 and i_2. Summing the voltages around the loop gives:

$$V_i = i_1(R_1 + R_2 + X_{C1}) - i_2(R_2 + X_{C1})$$

Additional equations are obtained by considering the other loops of the network and a complete set of equations are given as:

$$V_i = i_1(R_1 + R_2 + X_{C1}) - i_2(R_2 + X_{C1})$$

$$0 = -i_1(R_2 + X_{C1}) + i_2(R_2 + R_3 + X_{C1} + X_{C2}) - i_3R_3$$

$$0 = -i_2R_3 + i_3(R_3 + R_4 + R_5) + i_4R_5$$

$$V_i = i_1R_1 + i_2X_{C2} + i_3(R_4 + R_5) + i_4R_5$$

For those of you who are familiar with matrix algebra these equations can be written in the form of a matrix as follows:

$$
\begin{bmatrix} V_i \\ 0 \\ 0 \\ V_i \end{bmatrix} = \begin{bmatrix} i_1 \\ i_2 \\ i_3 \\ i_4 \end{bmatrix} \begin{bmatrix} (R_1 + R_2 + X_{C1}) & -(R_2 + X_{C2}) & 0 & 0 \\ -(R_2 + X_{C2}) & (R_2 + R_3 + X_{C1} + X_{C2}) & -R_3 & 0 \\ 0 & -R_3 & (R_3 + R_4 + R_5) & R_5 \\ R_1 & X_{C2} & (R_4 + R_5) & R_5 \end{bmatrix}
$$

This matrix equation is unique to the above network, but it could have been expressed as a general impedance matrix as follows:

$$
\begin{bmatrix} V_1 \\ V_2 \\ \vdots \\ \vdots \\ V_n \end{bmatrix} = \begin{bmatrix} i_1 \\ i_2 \\ \vdots \\ \vdots \\ i_n \end{bmatrix} \begin{bmatrix} Z_{11} & Z_{12} & \cdots & \cdots & Z_{1n} \\ Z_{21} & Z_{22} & \cdots & \cdots & Z_{2n} \\ \vdots & \vdots & \vdots & \vdots & \vdots \\ \vdots & \vdots & \vdots & \vdots & \vdots \\ Z_{n1} & Z_{n2} & \cdots & \cdots & Z_{nn} \end{bmatrix}
$$

A matrix of the form shown above can be constructed for any network by simply following a set of rules. These rules can be written as a piece of computer software which can interpret information about the way a network is connected together and can create a description of the circuit in the form of a matrix. The voltages and currents at different parts of the network are then obtained by the application of matrix algebra. In practice it is more efficient to write the matrix in the form of admittances $(Y = I/V)$ rather than impedances, since the admittance array usually results in fewer mathematical operations to obtain the solution, and therefore improves the speed of analysis.

The benefit to be gained by using computer software for circuit analysis is the speed and accuracy of the results once the admittance matrix has been formed. For example, to determine the frequency response of a network it is necessary to perform the same calculation for each frequency. This repetitive calculation is ideally suited to a computer, and a complete and accurate frequency response for a complex network can be obtained in a matter of minutes.

PSpice

Many CAD programs for circuit simulation owe their origin to SPICE which was developed at the University of Berkeley in the early 1970s. The program was developed in response to the potential complexity of the then new integrated circuits which were being designed for use in computers. During the years since its development it has undergone many changes and improvements and it has also spawned many variations. Originally intended to operate on mainframe computers it has now been ported to workstations and more recently to the PC. To promote their version of SPICE, MicroSim Corporation provide an evaluation copy of PSpice, and in particular there is now a version of PSpice which runs under Windows 3.1 (or higher). All engineering departments in universities and colleges have PCs, as do many students, and it would seem to be appropriate to include the use of computer simulation in a course on analog electronics.

This book is based on PSpice version 5.4 (version 6 is now available). The evaluation version has full schematic capture and a graphical post-processor to examine the data produced during the simulations. It is available free to members of engineering departments in universities and colleges and may be obtained from:

MicroSim Corporation
20 Fairbanks
Irvine
CA 92718
USA

It may also be found on the Internet. The evaluation version is limited to 20 components, but for many undergraduate problems this is perfectly adequate. The full version is capable of simulating circuits with many hundreds of components, the main limitation being the amount of memory available. The minimum computer requirement for the Windows version of PSpice is 640 kilobytes of DOS memory and at least 4 megabytes of extended memory, MS-DOS 3.0 (or later), Microsoft Windows 3.0 (or later), VGA, EGA colour or monochrome screen, a mouse, serial port for printing and an 80387 or 80487 maths co-processor. The co-processor does make a significant difference to the speed of operation. Installation is simple and is a matter of following the 'install' instructions. When correctly installed a new Windows application group is formed with the following program icons:

Schematics	The graphical circuit editor and design manager for PSpice.
PSpice	The analog, digital and mixed analog/digital circuit simulator.
Probe	The graphical waveform analyzer used to view and manipulate PSpice simulation results.
Stimulus Editor	The analog and digital stimulus generation tools.
Parts	The utility for creating semiconductor device models and subcircuit definitions.

MicroSim provide detailed instructions for using each of the above programs in a 'User's Guide' and 'Reference Manual' and there is no intention in this text to repeat these instructions apart from a simplified tutorial (see Appendix 7) to ensure that the prospective reader can understand the examples provided at different places in this text.

PSpice Basic Principles

All of the SPICE derivatives offer three basic forms of analysis: dc, ac and transient. The dc mode is fundamental to all three and is performed at the start of any analysis to determine the dc operating conditions. This analysis establishes all of the node voltages, current provided by any source and current in any active devices. It is also possible to sweep a dc voltage and to record the results. Thus the $I-V$ characteristics of a transistor may be obtained or the performance of a voltage regulator may be examined as the unregulated input voltage varies. During ac analysis the frequency of a source is varied to examine the frequency response of a network. For transient analysis the simulator acts in a manner similar to that of an oscilloscope in allowing the voltage or current waveforms to be examined at different parts of the circuit. The input signal can be a simple sine wave, a pulse waveform or an arbitrary-shaped waveform which is created from a piece-wise construction of voltages and times or with the Stimulus Editor.

With PSpice the input data for the simulator is provided in the form of a circuit schematic which is drawn on the screen with the schematic drawing program. The symbols for the schematic are contained in various libraries; there are libraries for resistors, capacitors, etc., libraries for transistors, diodes and operational amplifiers, libraries for voltage and current sources and some specialist libraries for behavioural models which will be described in more detail in a later chapter. The original version of SPICE was intended to be used on a mainframe computer which only had alphanumeric displays. It was not possible to use schematic capture with these displays and the input data was in the form of a *netlist* which comprised a list of components, their node connections and their values. The present-day SPICE variants also generate a netlist, but only after the schematic has been entered. However, the netlist can be generated manually and PSpice run without resource to the schematic capture program. It is in fact sometimes more convenient to edit the netlist to observe the effect of component changes, rather than resort to the schematic capture program (see the tutorial in Appendix 7).

Component values are modified with schematic capture by double clicking on the default value and then entering the value by means of the keyboard. The values of dc and ac voltage sources may be changed, again by double clicking on the symbol and entering the appropriate values in the menu box which appears on the screen. After selecting the type of analysis and the conditions of the analysis, the simulation is performed and the results observed.

When entering component values, PSpice recognizes the commonly used letters which are used to identify small and large exponent values, that is:

10^{-15} f
10^{-12} p
10^{-9} n
10^{-6} u
10^{-3} m
10^{3} k
10^{6} meg

NB Note the difference between 10^{-3} and 10^{6}.

When drawing a schematic it is important to include a ground connection symbol. For convenience, when using Probe, it is useful to name the nodes which are to be observed; alternatively *markers* can be used to mark nodes or branches. This is not necessary for the success of the simulation, but simplifies the selection of the signals when using Probe. If the nodes are not named then the simulator creates its own names and these are not always easy to identify.

All of the commands required to operate PSpice can be performed with the mouse acting as a pointer and left clicking on the menu option in common with all Microsoft Window operations.

A simplified user's guide is included in Appendix 7 for those unfamiliar with PSpice together with a tutorial. Those unfamiliar with PSpice should read through the appendix and work through the tutorial.

▶ **EXAMPLE 1.4** ———————————————————————————————

Use PSpice to investigate the gain of an inverting amplifier based on a 741 operational amplifier and having a gain of 100. Use a 1 kΩ resistor for the input

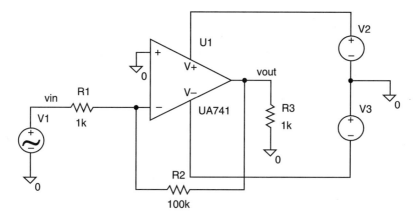

FIGURE 1.13   Schematic of circuit from PSpice.

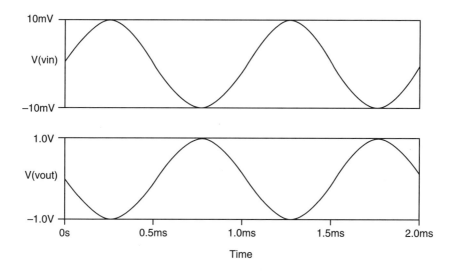

FIGURE 1.14   The 1 V output from Probe for the 10 mV input signal.

and 100 kΩ for the feedback resistor, as shown in Figure 1.13, and apply power supplies of ±15 V. Apply a 1 kHz waveform with a peak amplitude of 10 mV and 1 V.

▷ *Solution*

The circuit schematic from PSpice is shown in Figure 1.13. To perform the simulation select **Analysis, Setup…** and **Transient…**. For **Print Step:** enter 10 μs and for **Final Time:** enter 2 ms. Enable **Transient…** and run the simulation.

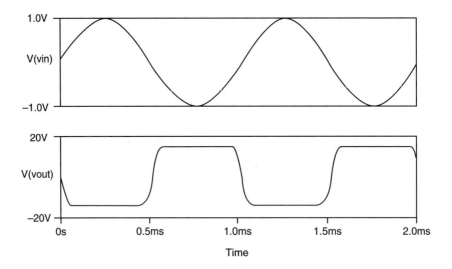

FIGURE 1.15   Waveform for a 1 V input which shows the output is clamped to ±14.6 V.

From Probe select **Trace** and plot v_{out}, Select **Plot, Add Plot**, select **Trace** again and plot v_{in}. The output from Probe is shown in Figure 1.14

The output when the input is increased to 1 V is shown in Figure 1.15.

▷ **EXAMPLE 1.5** ───────────────────────────────────────

For the same circuit as used in Example 1.4 increase the frequency of the input sine wave to 100 kHz and apply the 1 V peak signal to the inverting amplifier.

▷ *Solution*

From **Analysis, Setup…** and **Transient…** set the **Print Step:** to 20 ns and the **Final Time:** to 1000 ns. The output waveform from Probe is shown in Figure 1.16.

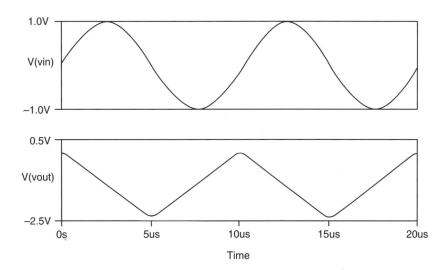

These exercises illustrate the ease with which a circuit simulator can be used to investigate the effect of applying a signal to a circuit, in this case an operational amplifier with negative feedback. The output is quite different in each case and it should be apparent that it is not enough to be able to operate the simulator, it is also necessary to understand what is happening. Why is the output a sine wave when the input is 10 mV, a square wave when it is 1 V and 1 kHz, but a triangular wave when it is 1 V and 100 kHz? The answer to these questions will become apparent in subsequent chapters.

▶ PROBLEMS

1.1 The resonant frequency for a parallel tuned circuit is:

$$f = \frac{1}{2\pi\sqrt{LC}}$$

If the tolerance on the inductor is ±5% and for the capacitor it is ±10%, determine the maximum percentage change in the resonant frequency with respect to the nominal

frequency if the full range of components are used.

[+8%, −7%]

1.2   Determine the maximum and minimum values for v_{out} for the potential divider shown.

[-16%, $+19.5\%$]

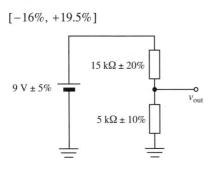

1.3   Determine a suitable power rating for the 250 Ω series resistor from 1/8 W, 1/4 W, 1/2 W, 1 W and 10 W resistors in the Zener diode circuit shown.

[10 W]

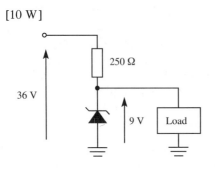

1.4   A sine wave with a peak amplitude of ±10 V is applied to two diodes which are placed in parallel as shown. If the forward voltage drop for the diodes is 0.7 V sketch the waveform of the voltage produced across the 1 kΩ resistor.

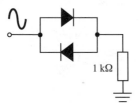

1.5   Locate the library within PSpice which contains the active devices shown below:

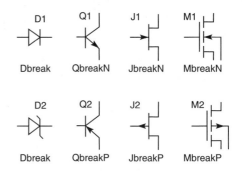

and confirm the identity of each device.

1.6   Locate the PSpice library which contains the symbols shown below:

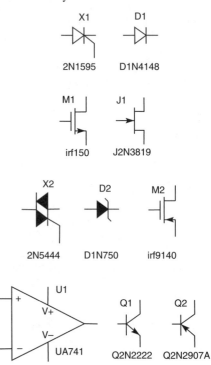

1.7 Use PSpice to investigate the effect of applying a 10 mV peak, 1 kHz sine wave to a non-inverting amplifier constructed with the 741 operational amplifier as shown in Figure 1.12. Let $R_1 = 1$ kΩ, $R_2 = 9$ kΩ, the dc supply be ±15 V and the load resistor 1 kΩ. Verify that the gain is 10. Increase the amplitude of the input signal and estimate the maximum amplitude which can be applied before the output signal becomes distorted.

1.8 For the circuit of Problem 1.7 increase the input to 1 V peak and estimate the maximum frequency which can be applied before the output is distorted.

CHAPTER 2

Semiconductors and PN Junctions

The group of materials which are classified as semiconductors has revolutionized the electronics industry. They are used to manufacture a wide range of electronic devices from discrete diodes and transistors to integrated circuits. To understand how these materials are used to make these devices it is necessary to investigate the conduction process in intrinsic and extrinsic semiconductors, the classification of p-type and n-type material and the formation of pn junctions. The simplest device is a pn junction diode and in this chapter a variety of diode characteristics and diodes are considered. It is also important to consider the electrical circuit models for diodes which are used for circuit analysis, and in particular for use in circuit simulators.

For those who only wish to pursue a circuit theory course it is possible to miss Sections 2.1 and 2.2, which deal with the physics of pn junctions, and jump to Section 2.3 which treats the diode as a circuit element.

2.1 Conduction (optional)

For the purposes of understanding materials which are used in electronics it is sufficient to divide them into conductors, insulators and semiconductors. The distinguishing feature is the ease with which the materials conduct electricity. Metals are generally good conductors and current flows freely in copper wires which are used to interconnect components in an electrical circuit, or in the tracks of a printed circuit board. On the other hand most plastics, ceramics and glass-like materials are very poor conductors and do not conduct electricity. Thus the plastic coating on connecting wires acts as a very good insulator as does the fiber-glass of a printed circuit board.

Semiconductors are neither good conductors nor good insulators, but there is an added difference which is related to the way in which charge is transported through the material. The semiconductor is also a crystalline material in the same way that natural diamond and sapphire are crystals, and this property together with the nature of the charge carriers makes the material important for the fabrication of electronic devices.

Intrinsic Semiconductor

The most commonly used semiconductor is silicon. Other semiconductors are germanium, gallium arsenide and gallium phosphide to name but a few. During the manufacture of silicon all impurities are removed and a single-crystal ingot is grown some 100 mm to 150 mm in diameter. This *intrinsic silicon* is a very poor conductor but it can be made to conduct by adding to it, or *doping* it with suitable impurities during the growth of the ingot. The type of impurity and the amount determine the conductivity of the silicon and also the way in which electric current is transported through the silicon. To understand how the impurity affects the intrinsic silicon it is necessary to know something about the way individual atoms are arranged in a single crystal.

An engineering concept of an atom is of a positively charged core or nucleus, surrounded by a cloud of electrons. The electrons are in a constant state of motion and constantly move around the core at discrete distances from the core. They sweep out an annular volume around the central core in the form of shells, rather like the skins of an onion. These shells are spaced out from the core at well-defined distances which can be specified in terms of *energy levels*. The positive charge of the nucleus is balanced by an equal negative charge of the electrons. The electrons in the outermost shell are only loosely bound to the nucleus, because of the screening effect of the innermost shells, and are continually swapping places with the electrons of adjacent silicon atoms. In *crystalline* material the atoms are precisely aligned with respect to each other in a well-ordered pattern and the electrostatic forces which exist between the atoms are the same for all atoms throughout the *crystal lattice*. Electrons in the outermost shells swap places with electrons on adjacent atoms, but cannot leave the vicinity of the atoms. Silicon has a *quadravalent* atom with four electrons in its outermost shell and they form *covalent bonds* with adjacent atoms. This is shown in diagrammatic form in Figure 2.1 with the bonds being represented by curved lines joining together the atoms which are represented by circles. The bonds represent the electrostatic forces which bind the atoms together. The regular arrangement of circles and curved lines is intended to provide a two-dimensional representation of the lattice structure of a crystalline material. The structure of the real crystal is three dimensional and is much more complex.

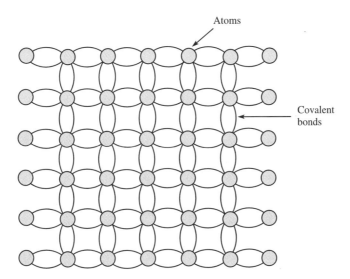

FIGURE 2.1   Idealized picture of the silicon lattice with atoms bound together by means of covalent bonds.

At 0 K all of these electrons are bound to the atoms and none is free for conduction. Thus at 0 K intrinsic silicon is an insulator. For conduction to take place some of the electrons must break free from the atom. In terms of the atomic structure the electrons must acquire energy in order to break free from the binding forces of the atoms. One such form of energy is heat and at room temperature (300 K) there is sufficient energy to allow a significant number of electrons to break free and move between the atoms. When an external electric field is applied these electrons move in the direction of the field and the silicon conducts. However, the resistivity is very large and at room temperature is approximately 230 kΩ cm (2.3 kΩ m). By contrast the resistivity of copper is 2×10^{-6} ohm cm (2×10^{-8} ohm m).

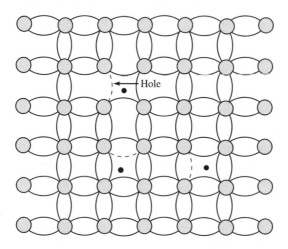

FIGURE 2.2   Crystal lattice with some broken bonds (holes) and free electrons.

When an electron breaks free from the silicon atom a covalent bond with the adjacent atom is broken and this is illustrated in Figure 2.2 by the dotted lines. These incomplete bonds represent the absence of an electron – there is a *hole*. The holes can move through the lattice in the same way that the free electrons can move, but when an electric field is applied they move in the opposite direction to that of the free electrons; they appear to have a positive charge equal to the negative charge of the electrons. In intrinsic silicon at room temperature there are an equal number of negatively charged electrons and positively charged holes. In Figure 2.3 the electrons are represented by the black dots, the hole by the absence of a dot and the atoms by circles.

When an electric field is applied the electrons will tend to move towards the positive terminal, but if a broken covalent bond or hole exists in the direction in which they are moving then there is a high probability that the electron will reform the covalent bond. Other bonds will then break to satisfy the thermal energy equilibrium requirements of the system. There is a continuous process of bonds breaking and reforming with electrons tending to move in one direction and the holes (broken bonds) moving in the other direction.

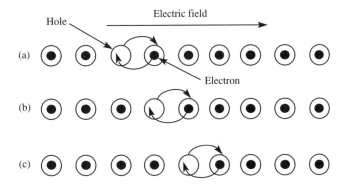

FIGURE 2.3   Simplistic view of the movement of a hole in one direction and an electron in the other, where (a), (b) and (c) represent three separate instants in time.

The number of electrons and holes, and thus the resistivity of silicon, can be altered significantly by adding certain impurities. When such impurities have been added the silicon is often referred to as *extrinsic silicon*.

Extrinsic Semiconductor

The most commonly used impurities are boron and phosphorus. The atoms of these materials are very similar in size to the silicon atoms and they are easily incorporated into the silicon lattice thus replacing some of the silicon atoms. More importantly the electrons in the outermost shells of the atoms of these two elements have similar energy values to those of the electrons in the outermost shells of the silicon atom and they can change places with the electrons in the silicon atom. In the case of phosphorus, which has a pentavalent atom with five electrons in its outermost shell as compared with the quadravalent silicon atom, the interaction results in the surplus fifth electron of the phosphorus atom being made available for conduction. There is one additional electron for every phosphorus atom. This is shown in Figure 2.4 where the phosphorus atoms replace a small number of silicon atoms. Four of the five outermost phosphorus electrons are used to complete the covalent bonds with the silicon, while the fifth electron is left free to move throughout the lattice. Thus as phosphorus is added, more and more electrons become available for conduction.

There is a limit to the number of phosphorus atoms which can be added because they cannot exceed the number of silicon atoms. For silicon there are 5×10^{22} atoms cm^{-3} (5×10^{28} atoms m^{-3}). Typically the maximum concentration for phosphorus is 1×10^{20} atoms cm^{-3} (1×10^{26} m^{-1}).

In the case of boron, the behaviour is a little more complicated. In intrinsic silicon all of the outermost electrons attached to each silicon atom are bound to the atoms by electrostatic forces associated with the covalent bonds, apart from a small number which at room temperature are able to break free and take part in the conduction process. When boron is added some of the silicon atoms are replaced by the boron atoms, but because boron is trivalent one of the covalent bonds in the lattice structure is

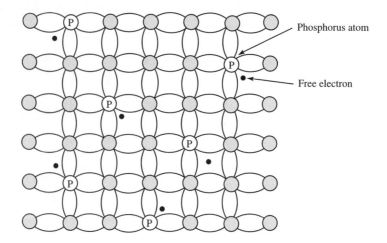

FIGURE 2.4   Extrinsic silicon with some silicon atoms replaced with phosphorus atoms.

incomplete, and thus a hole is present. This hole acts as a positive charge in the same way as the hole which is created in intrinsic silicon when covalent bonds break as a result of thermal energy. For every boron atom present, there is a hole (broken bond) as shown in Figure 2.5.

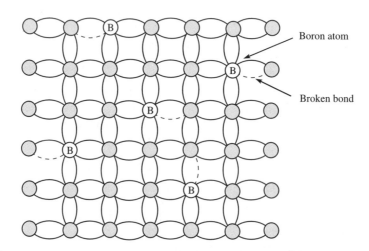

FIGURE 2.5   Extrinsic silicon with some of the silicon atoms replaced with boron atoms.

When phosphorus is added to silicon it is said to be *n-type* because of the additional negative electrons which are generated. With boron added, the silicon is said to be *p-type* because of the additional positively charged holes which are generated. If both impurities are present then the resultant conductivity type is determined by the impurity which is in the majority. The resistivity is proportional to the number of impurities. The resistivity for n-type silicon with 1×10^{19} atoms cm^{-3} of phosphorus is about 160 ohm cm (cf. the 230 kΩ cm for intrinsic silicon).

Conductors, Semiconductors and Insulators

The ability of conductors to conduct electricity easily (and of insulators not to conduct) can now be explained in terms of the energy required to free the electrons associated with the individual atoms. For a conductor this energy is very small and there are large numbers of electrons available for conducting electricity. For the insulator the energy required is large so that at room temperature there are few, if any, free electrons and no conduction takes place. However, if the temperature is sufficient (~1000°C) then an insulator will conduct. For a semiconductor, some electrons are available at room temperature in intrinsic silicon, but many more can be made available with the addition of a suitable impurity (boron or phosphorus). The resistivity is controlled by the amount of impurity which is added. In addition the current in a semiconductor can be carried by either electrons or holes. The ability to manufacture semiconductors with different conductivity types (n- or p-type) and different resistivities provides the means for producing a wide range of different types of solid-state devices, from the simple diode to complex integrated circuits.

Resistivity

When a voltage is applied across a bar of silicon current will flow. The current is carried by either electrons or holes depending on whether the silicon is n-type or p-type. The current density is proportional to the rate of flow of the carriers, that is:

J = rate of flow of charge

$$= (q)(n)(v) \tag{2.1}$$

where q is the charge of an electron (1.6×10^{-19} C), n represents the number of electrons (or 'p' for holes) and v is the drift velocity of the electron (or hole).

The velocity is proportional to the electric field, that is:

velocity $\propto$ electric field

$$v = \mu E \tag{2.2}$$

where μ is the constant of proportionality and is known as the mobility (cm^2 V^{-1} s^{-1}). The mobility is a measure of how easy it is for carriers to move through a crystal lattice, and values are available for a large number of semiconductors.

Thus, the rate of flow of charge, or the current density is:

$$J = qn\mu_n E \text{ for n-type} \tag{2.3}$$

or

$$J = qp\mu_p E \text{ for p-type} \tag{2.4}$$

where μ_n, μ_p are the mobilities for n-type and p-type silicon respectively.

Equations 2.2 and 2.3 can be rewritten as:

$$J = \sigma E \tag{2.5}$$

where σ is the conductivity of either the n-type or p-type material. Equation 2.5 is simply Ohm's law expressed in terms of current density, conductivity and electric field. The resistivity is defined as the reciprocal of the conductivity, that is:

$$\rho = \frac{1}{qn\mu_n} \left(\text{or } \frac{1}{qp\mu_p} \text{ for p-type} \right) \tag{2.6}$$

For silicon the mobility of electrons in lightly doped material is approximately 1300 cm^2 V^{-1}s^{-1} while the mobility for holes is approximately 550 cm^2 V^{-1}s^{-1}. The mobility decreases as the impurity concentration increases, and there is published data of this variation in the form of graphs and tables. However, for the purposes of simple estimates of the resistivity of a silicon sample with a certain impurity concentration, it is sufficient to assume that the mobility is constant.

▷ **EXAMPLE 2.1** ────────────────────────────────────

A bar of n-type silicon is 1 mm × 1 mm × 10 mm long and has an impurity concentration of 5×10^{16} impurities cm^{-3}. If the mobility of the electrons is 1300 cm^2 V^{-1}s^{-1} determine the resistance of the bar.

▷ *Solution*

The resistivity is

$$\rho = \frac{1}{qn\mu_n} = \frac{1}{(1.6 \times 10^{-19})(5 \times 10^{16})(1300)} \approx 0.1 \ \Omega \, \text{cm}$$

and the resistance is:

$$R = \frac{(\rho)(\text{length})}{\text{area}} = \frac{(0.1)(1)}{0.01} = 10 \ \Omega$$

──

2.2 PN Junctions (optional)

For a simple rectangular bar of silicon the current is carried by either electrons or holes, depending on whether it is n-type or p-type, and the bar behaves as a simple resistor with current flow being proportional to the voltage. However, if a bar of silicon is n-type at one end and p-type at the other, then current flow is changed. The bar no longer behaves as a simple resistor, and the current is a non-linear function of the applied voltage.

To understand what happens when there is a junction between n-type and p-type silicon it is necessary to reconsider what happens when impurities are added to silicon.

The impurities become locked into the silicon lattice. For the pentavalent phosphorus atom one of its electrons is released to take part in conduction, but this means that the charge on the nucleus of the phosphorus atom is no longer balanced and there is a resultant positive charge around the phosphorus atom. The trivalent boron atom gains an electron when it becomes locked into the lattice of the quadravalent silicon and the resultant charge around the boron atom is negative. For a single piece of silicon these charges on the impurity atoms are cancelled by the presence of free electrons or holes and there is no net surplus charge. This is illustrated in Figure 2.6(a). However, the situation changes when an np junction is formed, as shown in Figure 2.6(b).

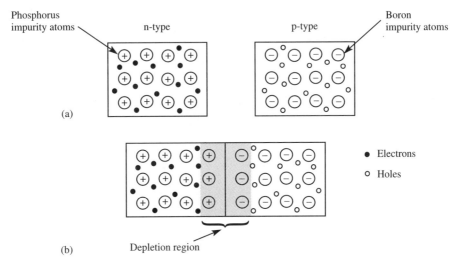

FIGURE 2.6   Diagrammatic representation of (a) individual pieces of n-type and p-type semiconductor and (b) the two pieces joined to form a pn junction.

In the n-type material the phosphorus atoms which have contributed an electron for the conduction process are shown with a positive charge. For the single piece of silicon there are equal numbers of positively charged atoms and negatively charged electrons. In the same way for the p-type material there are an equal number of negatively charged atoms and positively charged holes. The electrons and holes are free to move, but the atoms are fixed in the crystal lattice. When the two pieces of material are brought together, as in Figure 2.6(b), some of the electrons in the n-type material will cross over into the p-type material, and some holes in the p-type region will cross over to the n-type. If a hole and an electron meet they will combine and cancel each other's charge. When this happens there is no longer charge balance throughout the block of semiconductor because some of the electrons and holes have disappeared, and as a result there is a charge from the fixed impurity atoms. On the n-type side there is a net positive charge, while on the p-type side there is a net negative charge. These charges only exist near the junction, because as more holes and electrons recombine this excess charge increases and eventually it increases to such an extent that it repels holes attempting to cross over from the p-type side, and electrons attempting to cross from the n-type side. This process only occurs during a very small fraction of a second during the formation of the junction, and equilibrium is established resulting in a region either side of the junction which is empty of holes and electrons, as shown in blue in Figure 2.6(b). This is known as a *depletion region*. The charge either side of the junction produces a potential, which prevents further movement of electrons and holes. This potential barrier can be increased or decreased by application of an external voltage.

Forward Bias

When a negative voltage is applied to the n-type region of a pn junction and a positive voltage to the p-type region then the electrons and holes are pushed towards the junction. Some of these carriers will overcome the potential barrier and current will flow. As the voltage increases the current also increases, but it is not a linear increase of current with voltage. It is affected by the presence of the built-in potential barrier. The forward-biased junction is illustrated in Figure 2.7, where the number of battery symbols represents the increase in voltage and the current is represented by the vertical bar graph alongside each representation of the junction.

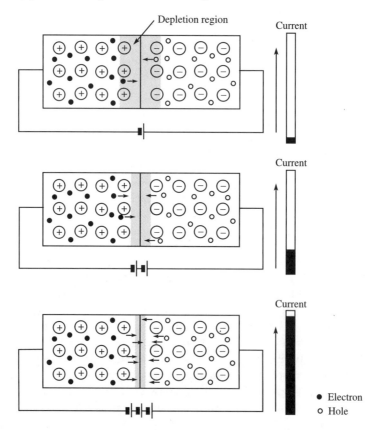

FIGURE 2.7   The forward-biased pn junction for different values of voltage.

As the bias voltage increases the width of the barrier, or the depletion region, decreases and the current increases.. The current of the forward-biased diode consists of electrons flowing from left to right across the junction and holes flowing in the opposite direction.

Reverse Bias

When the voltage polarity is reversed the free electrons and holes in the bulk of the material are pulled back from the junction so that the number of exposed impurity atoms is increased and the depletion region increases in width as shown in Figure 2.8. An alternative explanation is that potential barrier across the junction is increased.

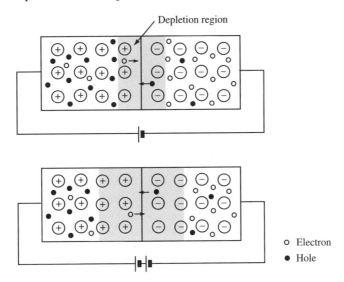

FIGURE 2.8   PN junction with reverse bias.

In Figures 2.6 and 2.7 the n-type and p-type materials are shown to contain only electrons or holes respectively. However, it was noted that for intrinsic material above 0 K there are equal numbers of electrons and holes which are produced as a result of thermal energy causing some of the covalent bonds to break, so producing an electron and a hole. Extrinsic silicon is affected in the same way with some of the covalent bonds being broken by thermal energy. The number of holes and electrons produced in this way is very small in comparison with those produced by the addition of impurities. However, n-type silicon does contain a small number of thermally generated holes, and p-type silicon contains a small number of thermally generated electrons. There is a relationship between the number of thermally generated carriers and those resulting from the addition of impurities in a single piece of material which is given by:

$$n_n p_n = n_i^2 \text{ for n-type silicon} \tag{2.7}$$

where n_n is the concentration of electrons in n-type silicon produced by the addition of impurities, and p_n is the number of thermally generated holes. The electrons are referred to as *majority carriers*, because they are in the majority, while the holes are referred to as *minority carriers*. The product is always equal to a constant n_i^2 which is the concentration of holes and electrons for intrinsic silicon ($n_i = 1.45 \times 10^{10}$ atoms cm^{-3}).

For p-type silicon the relationship is:

$$p_p n_p = n_i^2 \text{ for p-type silicon} \tag{2.8}$$

where p_p is the majority carrier concentration and n_p is the concentration of the minority carriers in p-type material.

In Figure 2.6 the electric field, which prevents the majority carriers on either side of the junction from crossing the junction, enables the minority carriers to cross. In Figure 2.8 it is the minority carriers which are seen to be flowing across the junction with arrows attached to them (cf. Figure 2.7 in which the majority carriers have arrows attached). However, the number of minority carriers is very small and the current quickly saturates. This reverse bias current, or *leakage current*, is typically a few picoamps or nanoamps depending on the area and the impurity concentrations.

▷ **EXAMPLE 2.2** ────────────────────────────────────

Phosphorus is added to silicon to give an impurity concentration of 2×10^{18} atoms cm^{-3}. Determine the number of majority and minority carriers.

▷ *Solution*

Phosphorus is a pentavalent atom and is an n-type dopant for silicon. Therefore the majority carriers are electrons and the concentration is:

$$n_n = 2 \times 10^{18} \text{ cm}^{-3}$$

The number of minority carriers is:

$$p_n = \frac{n_i^2}{n_n} = \frac{(1.45 \times 10^{10})^2}{2 \times 10^{18}} = 105$$

2.3 Diode Equation

The pn junction described in Section 2.2 forms a two-terminal device known as a junction diode. Theoretical analysis of the electric field and charge in the vicinity of the depletion layer results in a relationship between the current and applied voltage described by the following equation:

$$I = I_S \left[\exp\left(\frac{V}{V_T} \right) - 1 \right] \tag{2.9}$$

where I_S is the *saturation current*, or leakage current, V is the junction voltage and V_T

is the thermal voltage given by

$$V_T = \frac{kT}{q} \approx \frac{T}{11600} \tag{2.10}$$

where k is Boltzmann's constant and q the electronic charge. At room temperature ($T = 300$ K) V_T is 0.026 V, or 26 mV.

For forward bias when V is positive, then for values of V much larger than 26 mV, the exponential term is much larger than unity and the equation is given approximately by:

$$I_F = I_S \exp\left(\frac{V_F}{V_T}\right) \tag{2.11}$$

From equation 2.11 it can be seen that the forward current increases exponentially with applied voltage.

For a reverse bias $(-V)$ which is much larger than V_T the exponential term is very small compared with unity and the current is approximately:

$$I_R \approx I_S \tag{2.12}$$

The reverse current remains constant with increasing bias. The current–voltage relationship is shown in Figure 2.9. The current increases rapidly in the forward voltage direction, but remains constant at a very small value in the reverse direction.

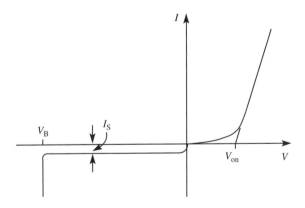

FIGURE 2.9 Current–voltage characteristics for a pn junction diode.

The saturation current (I_S) is very temperature sensitive. As the temperature increases, more covalent bonds are broken and the number of minority carriers on either side of the junction increases. Since the saturation current is largely due to the minority carriers, an increase in the temperature causes the saturation current to increase. The saturation current approximately doubles for every 10°C rise in temperature for a silicon diode.

▷ **EXAMPLE 2.3** ─────────────────────────────

Determine the forward and reverse currents for a diode for which $I_S = 2 \times 10^{-13}$ A at a forward voltage of 0.65 V and a reverse voltages of 0.1 V and 1 V at 27°C and 77°C.

▷ *Solution*

27°C

Forward voltage:

$$I_F = 2 \times 10^{-13}\left[\exp\left(\frac{0.65}{0.026}\right) - 1\right] \approx 14.4 \text{ mA}$$

Reverse voltage of 0.1 V:

$$I_R = 2 \times 10^{-13}\left[\exp\left(\frac{-0.1}{0.026}\right) - 1\right] = -1.95 \times 10^{-13} \text{ A}$$

or a reverse voltage of 1.0 V:

$$I_R = 2 \times 10^{-13}\left[\exp\left(\frac{-1}{0.026}\right) - 1\right] = -2 \times 10^{-13} \text{ A}$$

77°C

At 77°C there is a change of temperature of 50°C which results in a five-fold doubling of the saturation current to 64×10^{-13} A. The thermal voltage (V_T) also changes.

$$V_T = \frac{kT}{q} = \frac{(1.38 \times 10^{-23})(273 + 77)}{1.6 \times 10^{-19}} = 0.03 \text{ V}$$

Forward voltage:

$$I_F = 64 \times 10^{-13}\left[\exp\left(\frac{0.65}{0.03}\right) - 1\right] \approx 16.4 \text{ mA}$$

Reverse voltage of 0.1 V:

$$I_R = 64 \times 10^{-13}\left[\exp\left(\frac{-0.1}{0.03}\right) - 1\right] \approx -62 \times 10^{-13} \text{ mA}$$

or for a reverse voltage of 1.0 V:

$$I_R = 64 \times 10^{-13}\left[\exp\left(\frac{-1}{0.03}\right) - 1\right] \approx -64 \times 10^{-13} \text{ mA}$$

Turn-on Voltage

The current in the forward direction is determined by the applied voltage through an exponential relationship, as shown in equation 2.10. Typical values may be obtained by assuming a value of 1×10^{-14} A for I_S as shown in Table 2.1.

TABLE 2.1   Variation of current with voltage for a junction diode.

V	I
0.5	0.002 mA
0.55	0.01 mA
0.6	0.1 mA
0.65	0.72 mA
0.7	5 mA

It can be seen that while current flows for all values of forward voltage, below 0.6 V the current is relatively small. It is only when the forward voltage reaches 0.6 V that an appreciable current flows. This *turn-on* voltage (V_{on} in Figure 2.9) depends on what is regarded as a significant current, but since the diode is often used as a form of switch, then any voltage exceeding 0.6 V is usually sufficient to turn the diode ON. The value of V_{on} is also dependent on the value of I_S. The value used in the above example is typical for a small-signal or small rectifier diode. However, for a large-power diode or a diode formed from a different semiconductor material it may be significantly different.

Reverse Breakdown Voltage

As described above the reverse current quickly saturates at I_S. However, with increasing voltage a point is reached when the current starts to increase very rapidly and the diode may be damaged if there is no external resistance to limit the current. The diode is said to have broken down and the *breakdown voltage* can range from a few volts to several hundreds of volts depending on the type of diode. There are two main mechanisms for breakdown, *avalanche breakdown* and *Zener breakdown*.

Avalanche breakdown occurs for voltages greater than about 10 V. When electrons cross the depletion region as depicted in Figure 2.8 they travel at very high velocity in the large electric field that exists in this region (about 1×10^7 V cm^{-1} for silicon). At these velocities if the electron strikes an atom there is a good chance that as a result of impact an additional electron will be dislodged from the atom. There are now two electrons, and the process can be repeated to produce an avalanche of carriers within the depletion region. Holes are also created which travel in the opposite direction. Since the current in the external circuit depends on the number of carriers, this increase in electrons and holes represents a very rapid increase in the external current which is only limited by the external resistance.

The value of the breakdown voltage is dependent on the impurity concentrations. For practical diodes one side of the junction usually has a very high concentration of impurities, while the other side is more lightly doped. Thus a diode may be n^+p or p^+n. It is the concentration of the lightly doped region which determines the breakdown voltage. Some typical values for large-area silicon diodes are shown in Table 2.2.

TABLE 2.2   Variation of breakdown voltage with substrate impurity concentration.

Concentration (atoms cm^{-3})	Breakdown (V)
1×10^{15}	300
1×10^{16}	60
1×10^{17}	15

Diodes can be tailored to satisfy many different needs by varying the impurity concentration. For handling high voltages, as may be the case with rectifier diodes, then low impurity concentration is necessary, but for small-signal applications minimizing the series resistance may be more important, in which case high impurity concentrations are more appropriate.

For voltages of less than 10 V the breakdown mechanism changes. Zener breakdown occurs when both sides of the junction have high impurity concentrations. It can be seen from the tabulated figures above that as the concentration increases the breakdown voltage decreases, but below about 10 V the method of breakdown changes. With a large number of impurity carriers on either side of the junction ($n^{++}p^{++}$ diodes), the depletion layer is very narrow. With a narrow depletion layer the electric field is large (field = voltage/distance), so large that it is able to break the covalent bonds either side of the junction to release large numbers of carriers which then form the breakdown current. The curve of current versus voltage, in the reverse-bias direction, is similar for both types of breakdown, but the temperature coefficient for the breakdown voltage is different − it is positive for avalanche and negative for Zener.

Small-Signal Resistance

When a diode is forward biased it exhibits a *small-signal resistance* which varies with the value of the dc current. The value of the resistance is obtained as the slope of the current−voltage graph (Figure 2.9) in the forward direction and is given by:

$$\frac{1}{r_e} = \frac{dI_F}{dV_F} = \frac{d}{dV_F} I_S \exp\left(\frac{V_F}{V_T}\right)$$

$$\frac{1}{r_e} = \frac{I_S}{V_T} \exp\left(\frac{V_F}{V_T}\right)$$

$$\frac{1}{r_e} = \frac{I_F}{V_T}$$

The small-signal resistance is:

$$r_e = \frac{V_T}{I_F} \text{ ohms} \tag{2.13}$$

With $V_T = 0.026$ V at room temperature, then r_e is 26 ohms at 1 mA.

Junction Capacitance

For a reverse-biased diode the current is very small and for small-signal purposes it appears as a very large resistance, typically several megohms. However, there is another important factor which can make the diode appear as a very low impedance at high frequencies. In Figure 2.8 the majority of the electrons and holes are drawn away from the vicinity of the junction, so that the regions immediately adjacent to the junction are depleted of mobile electrons and holes. The regions between the contacts and the edges of the depletion layer are relatively good conductors, so that the diode appears as two conducting regions separated by an insulating layer (the depletion region). This structure is equivalent to a capacitor and the capacitance is given by:

$$C = \frac{C_0 A}{x_d} \tag{2.14}$$

where C_0 is a constant dependent on the impurity concentration, A is the cross-sectional area and x_d is the width of the depletion layer. The value of the capacitance is increased by increasing the impurity concentration. The width of the depletion layer increases with increasing reverse voltage (as illustrated in Figure 2.8). Thus the junction diode behaves as a voltage-dependent capacitor and is extensively used as a tuning element in radio and television circuits. The width of the depletion layer is proportional to the applied voltage as:

$$x_d \propto V^n$$

where n is a value between $1/2$ and $1/3$. Thus the capacitance can be described by:

$$C = \frac{C_0 A}{(\phi + V)^n} \tag{2.15}$$

where ϕ is the built-in voltage associated with the potential barrier across the pn junction. Typically the value of ϕ is 0.3 V.

▶ **EXAMPLE 2.4** ──────────────────────────

A pn junction diode has an area of 100 μm^2 and the zero-voltage capacitance is 0.01 pF per μm^2. Determine the capacitance for reverse bias of 5 V and 10 V.

▶ *Solution*

Reverse bias of 5 V:

$$C = \frac{(0.01 \times 10^{-12})(100)}{(0.3 + 5)^{1/2}} = 0.43 \text{ pF}$$

Reverse bias of 10 V:

$$C = \frac{(0.01 \times 10^{-12})(100)}{(0.3 + 10)^{1/2}} = 0.31 \text{ pF}$$

2.4 ▷ Other Types of Diode

The basic pn junction diode is widely used for small-signal detection and also as a rectifier for converting alternating voltage into direct voltage. For rectifiers the area of the diode may have to be large in order to conduct large currents when forward biased. For small-signal applications the diode is usually made as small as possible to minimize the effect of junction capacitance. There are, however, a number of other applications where the structure and materials are modified to provide a number of other uses for the simple pn junction.

Zener Diode

The general shape of the current–voltage characteristic for a pn junction diode is shown in Figure 2.9. In the reverse-bias condition the junction breaks down at a voltage which is related to the impurity concentration in the more lightly doped region of an n^+p or p^+n diode. For the Zener diode the breakdown is more precisely defined than for the simple pn junction diode. For the Zener diode the breakdown voltage is specified as V_Z. Provided the current is limited then the breakdown voltage (V_Z) can be used as a voltage reference as illustrated in Figure 2.10. Notice the different symbol used for a Zener diode.

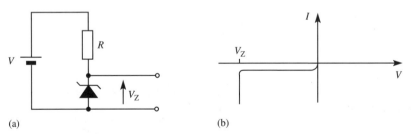

(a)   (b)

FIGURE 2.10   (a) Zener diode being used as a simple voltage reference and (b) the reverse bias characteristic.

The dc supply V is assumed to be greater than the breakdown voltage (V_Z) of the diode. The resistor R limits the current to a safe value. For avalanche breakdown the voltage is not well defined and it has a large positive temperature coefficient, so that it is not ideally suited for use as a voltage reference. But for the Zener diode the voltage can be tailored during the manufacturing process to a precise value. Additionally, the knee of the curve at breakdown depicted in Figure 2.10(b) is sharper for Zener breakdown than for avalanche. The temperature coefficient of breakdown is negative for Zener and positive for avalanche. In reality the change from Zener to avalanche is not sudden, but gradual. For voltages of <3 V it is predominantly Zener, while for voltages of >10 V it is predominantly avalanche. Between 3 V and 10 V both effects occur together. This fact is important, because the two effects have temperature coefficients of opposite polarity, and for voltages of about 3–5 V the two coefficients cancel to give a voltage reference diode which has a temperature coefficient which is almost independent of temperature.

Schottky Diode

In a Schottky diode the p^+ or n^+ side of the p^+n (or n^+p) junction is replaced by a metal electrode. The current–voltage characteristics are the same as those for the pn junction, with the exception that the turn-on voltage is lower, typically 0.2 V to 0.4 V depending on the type of metal used. The advantage of the Schottky diode is its improved switching speed and it is used extensively in microwave circuits for high-speed switching. The diode equation is similar to that for the pn junction except that the saturation current is determined by thermionic emission. The resultant current is larger than the saturation current associated with a pn junction diode of similar area which results in the lower turn-on voltage.

▷ **EXAMPLE 2.5** ───────────────────────────────────

If the turn-on voltage for a diode is defined as the voltage required to produce a forward current of 0.1 mA determine the value of I_S necessary for the turn-on voltage to be 0.4 V.

▷ *Solution*

Conventional diode $V_F = 0.6$ V:

$$0.1 \times 10^{-3} = I_S \left[\exp\left(\frac{0.6}{0.026} \right) - 1 \right]$$

$$I_S = \frac{1 \times 10^{-4}}{1.0524 \times 10^{10}} = 0.95 \times 10^{-14} \text{ A}$$

Schottky diode $V_F = 0.4$ V:

$$I_S = \frac{1 \times 10^{-4}}{\exp(0.4/0.026) - 1} = 2.08 \times 10^{-11} \text{ A}$$

Light-Emitting Diode

Light-emitting diodes are formed from compound semiconductors, such as gallium arsenide and gallium phosphide. When a forward bias is applied to pn junctions formed in these materials the holes and electrons which cross the junction change energy levels, and in doing so emit light at a frequency determined by the difference in energy level. Typically light-emitting diodes (LEDs) emit light in the infra-red (invisible), red, yellow, green and blue regions of the energy spectrum; different semiconductors are used to produce the different wavelengths The visible emitters are used for displays, with the diodes being arranged to reproduce the seven segments for alphanumeric displays. The infra-red diodes are used for optical communications, either over very short distances as in optical isolators, or over very long distances in optical fiber telecommunication systems.

Photodiode

The photodiode is a pn junction diode which is packaged in a transparent plastic material in order that light can reach the junction. Silicon is partly transparent to optical radiation. Since light is a form of energy, when it is absorbed by the silicon, the energy of the photons is transferred to the silicon where it may break covalent bonds and form electron–hole pairs. If these holes and electrons are near the depletion region then the electrons are attracted by the electric field in one direction and the holes are attracted in the other direction across the junction. These carriers are in addition to those produced by thermal energy.

The photodetector can be used with zero bias, but for many applications a reverse bias is applied to create a large depletion layer so that as many electron-hole pairs as possible are captured (any produced some distance from the depletion layer simply recombine and do not contribute to the external current). The resultant variation of the saturation current with light intensity is illustrated in Figure 2.11.

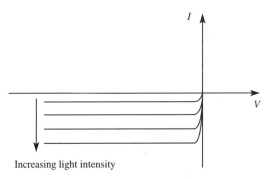

Increasing light intensity

FIGURE 2.11   Variation of the saturation current in a photodiode.

The photodetector can be used to measure light intensity by simply measuring the current, or it can be used to detect a signal from an optical fiber in an optical communication system.

A simple detector circuit is shown in Figure 2.12. The light is directed onto the diode by means of a lens and the current flowing in the resistor (R) is a function of the light intensity. The intensity can be measured as a voltage drop across the resistor. Alternatively if the light is modulated, then the modulated voltage appears across the resistor and can be coupled to a suitable amplifier.

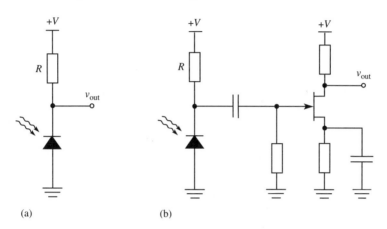

FIGURE 2.12   (a) Simple photodetector circuit and (b) detector and amplifier.

Four-Layer Diodes

The addition of two further layers of n- and p-type material to the conventional diode completely changes the $I-V$ characteristics, as illustrated in Figure 2.13.

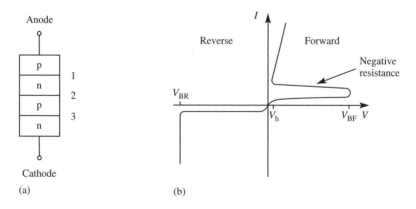

FIGURE 2.13   (a) Simplified drawing of the four-layer diode and (b) the $I-V$ curves.

Under reverse-bias conditions with a negative voltage applied to the anode and a positive voltage applied to the cathode, junctions 1 and 3 are reverse-biased, while junction 2 is forward biased. With two of the three junctions reverse biased the current is initially limited to the leakage current of a reverse biased junction. As the voltage is increased junction 1 will break down because of the higher impurity concentrations in the vicinity of this junction, but the current is still restricted to the leakage current of junction 3. With further increase of voltage junction 3 breaks down. The overall $I-V$ characteristic in the reverse-bias condition is similar to that of a conventional pn junction diode, with a breakdown voltage (V_{BR}) which can vary from a few tens of volts to several hundred volts, depending on the impurity concentrations and fabrication conditions.

The forward-bias characteristics are very different from those of the conventional diode. With a positive voltage applied to the anode, junctions 1 and 3 are forward biased while junction 2 is reverse biased. The current is limited to the leakage current of junction 2. At some voltage (V_{BF}) junction 2 breaks down and the current increases. However, unlike the reverse-bias breakdown the voltage across the device decreases to a small (~2 V) holding voltage (V_h). Between V_{BF} and V_h the device exhibits a negative resistance, with the voltage decreasing as the current increases, as shown in Figure 2.12(b).

The current in both forward and reverse directions beyond the breakdown voltages is limited by the resistance in the external circuit.

A much more useful device is obtained with the addition of a third electrode to the p-region between junctions 2 and 3 as shown in Figure 2.14.

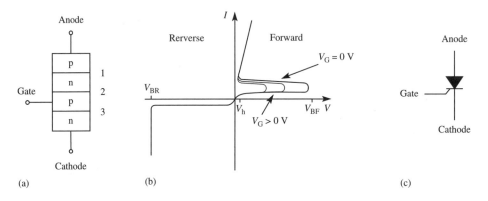

FIGURE 2.14   Diagram of the silicon-controlled rectifier (a), its $I-V$ characteristics (b) and its symbol (c).

The additional *gate* electrode controls the breakdown voltage in the forward direction, with different values of voltage affecting the value of the anode–cathode voltage at which the device reverts to a low-voltage high current mode of operation. The device is known as a *silicon-controlled rectifier* (SCR) or *thyristor*.

The SCR is the solid-state equivalent of an electromagnetic relay. The coil voltage of the relay is replaced by the gate voltage of the SCR. However, unlike the relay, the removal of the gate voltage does not switch the device OFF. In order to turn the device OFF it is necessary to reduce the voltage across the anode and cathode to below the holding voltage ($V_h \approx 2$ V). This happens automatically every half cycle when an alternating voltage is applied to the anode, and the SCR is, therefore, widely used for ac power control of furnaces and motors.

▶ **EXAMPLE 2.6** ─────────────────────────────────

Use PSpice to create the $I-V$ curves for the SCR under forward-bias conditions. Select the 2N1595 SCR from the **eval.slb** library and draw the schematic as shown in Figure 2.15. From **Analysis** and **Setup...** select **DC Sweep...**. Set the sweep conditions for the current source for 0 to 10 mA with 10 μA steps. Set the value of V_1 to 1 V.

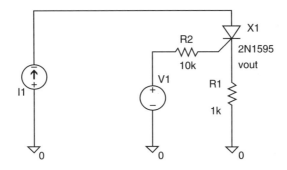

FIGURE 2.15   PSpice schematic of the circuit to obtain the $I-V$ characteristics for an SCR.

NB Notice the use of the current source rather than a voltage source for the anode. This is necessary to reproduce the negative resistance region of the $I-V$ curves.

▶ *Solution*

Depending on the type of computer this simulation can take from 1 hour to several hours. Increasing the step size is likely to result in a poorly defined breakdown voltage. The

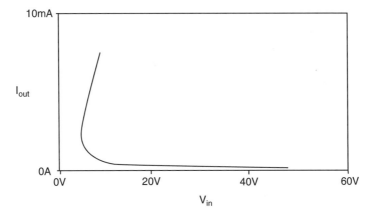

FIGURE 2.16 $I-V$ curves produced with Probe.

resultant curve is shown in Figure 2.16 which displays the voltage across the current source (V_{in}) and the current through R_1 (I_{out}).

It can be seen that breakdown occurs at about 50 V and that the holding voltage is about 5 V. Increasing the value of V_1 will reduce the value of the breakdown voltage.

The DIAC and TRIAC are variations of the four-layer structure. The DIAC is equivalent to two four-layer diodes being placed in anti-parallel as shown in Figure 2.17.

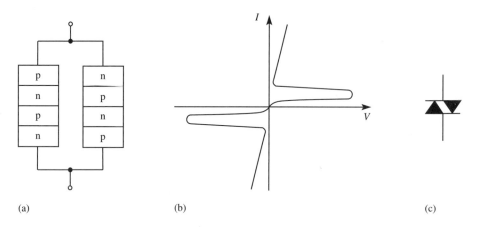

(a) (b) (c)

FIGURE 2.17 Two four-layer diodes placed in anti-parallel (a), the characteristics (b) and the symbol (c).

The DIAC is used in power electronics and often forms part of the control circuitry for an SCR.

A more useful device is created with an additional electrode to form the TRIAC. Like the gate electrode of the SCR, the gate electrode of the TRIAC controls the point at which breakdown occurs, but for both polarities, forward and reverse. The TRIAC is widely used for low-power ac control, such as light dimmers. It provides control for both positive and negative cycles of the ac

Gate

waveform. An example of a PSpice simulation involving the use of a TRIAC in a light dimmer circuit is shown in Example 1.3 in Chapter 1 (Figure 1.7).

2.5 ▷ Equivalent Circuit for Practical Diodes

For a circuit simulator such as PSpice it is necessary to describe the operation of the diode in mathematical terms which can be used by the simulator to enable it to reproduce the dc current–voltage relationships and the small-signal effects observed in real diodes. The dc characteristics of the ideal diode are described by the diode equation (equation 2.9), but in practice real diodes depart from the ideal.

A graph of $\ln I_F$ versus V is shown in Figure 2.18. The ideal saturation current (I_S) is obtained by extrapolating to the vertical axis intercept. For the ideal diode the slope of the graph in the linear region is $1/V_T$, but in practice the measured value is less than this. The difference can be explained theoretically and can be accounted for by introducing a factor η so that the slope becomes $1/\eta V_T$. A further departure of the measured characteristic from the ideal occurs at high currents when the curve departs from a straight line, as shown in Figure 2.18. The actual value of the current at which this occurs differs for different types of diode. For a small-signal diode it may be a few milliamps, while for a rectifier diode it may be several amps. The departure can be explained by taking account of the series resistance of the semiconductor material which exists between the external contact and the edge of the depletion layer. The voltage which appears across the depletion layer is less than that applied to the external terminals by the amount that is dropped across the series resistance (IR_S).

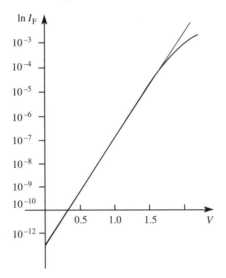

FIGURE 2.18   Graph of $\ln I_F$ versus V in the forward direction.

The characteristics of the real diode can be described by including η and R_S in the diode equation. The modified equation is:

$$I = I_S\left[\exp\left(\frac{V - IR_S}{\eta V_T}\right) - 1\right] \tag{2.16}$$

For silicon η is approximately 2. The exact value is obtained by plotting $\ln I$ versus V and measuring the slope for the linear region $(1/\eta V_T)$. The series resistance may be determined by trial and error as the value required to account for the departure of the measured characteristic from the ideal.

In addition to η and R_S there is a capacitor (C) to account for the capacitance associated with the depletion layer. The capacitance is voltage dependent. The final circuit is shown in Figure 2.19.

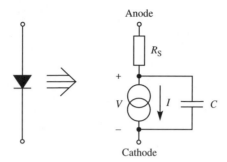

FIGURE 2.19   Equivalent circuit for a diode.

Some of the important SPICE parameters for the equivalent circuit of the default generic diode (from the breakout library in PSpice) are listed in Table 2.3. These parameters are accessible for editing within PSpice.

TABLE 2.3   SPICE parameters for a generic diode.

Parameter	Default value
IS	1×10^{-14}
RS	0
CJO	0
N	1
BV	∞

For an actual diode, such as 1N4148, the above parameters would have specific values which are obtained from careful measurement of both the dc and small-signal diode characteristics. The saturation current parameter I_S is important because it establishes the value of current for a particular forward voltage. However, current can also be increased by including a factor for the area, if a general purpose diode is being used instead of a named diode, or if a particular diode were not available, then the nearest equivalent could be used and the match achieved by changing the area factor. The area factor scales parameters such as saturation current, series resistance and capacitance. The default value of the series resistance is set to zero for the generic diode, as is the junction capacitance. Similarly the reverse breakdown voltage is assumed to be infinite.

2.6   Practical Examples using PSpice

There are many applications for diodes in electronics, but for the purposes of this chapter they are restricted to an investigation of some of the properties of the diode which are related to the material and physical properties, and which affect the equivalent circuit. The use of PSpice provides a simple means for making a detailed examination of the dc characteristics, temperature effects and junction capacitance.

DC Characteristics

The dc characteristics for both forward and reverse bias can be usefully examined with PSpice.

▷ **EXAMPLE 2.7** ————————————————————————————

Use PSpice to create the circuit schematic shown in Figure 2.20. The diode used is an unnamed generic diode, and the model parameters as identified in the above table are added to the default model. From **DC Sweep...** in **Analysis** the voltage sweep is set to −22 V to +1 V in 0.1 V steps. From **Edit** select **Model...** and add the following in the open menu window:

is = 1e − 14 rs = 50 bv = 20 cjo = 2e − 12

where the saturation current is 1×10^{-14} A, the series resistance is 50 Ω, the breakdown voltage is 20 V and the zero-voltage junction capacitance is 2 pF.

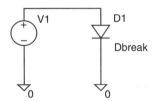

FIGURE 2.20   Simple diode circuit used to perform a DC sweep.

▷ *Solution*

The resultant I–V curve is shown in Figure 2.21.

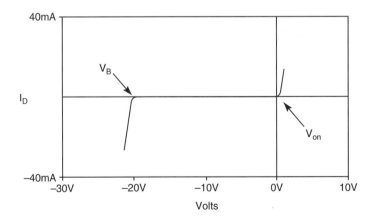

FIGURE 2.21   Current–voltage variation for the diode with the breakdown voltage set to 20 V.

Temperature Effect

In Section 2.1 on intrinsic semiconductors it was noted that at any temperature above absolute zero a large number of the covalent bonds in the crystal lattice break to release equal numbers of holes and electrons. The number increases as the temperature rises. At room temperature (300 K) the number is 1.45×10^{10} cm^{-3}. This number increases as the temperature rises and consequently the number of minority carriers as given by equations 2.4 and 2.5 also increases. This increase will cause the leakage current (I_S in equation 2.6) in the reverse-biased diode to increase. For a silicon diode the leakage current approximately doubles for every 10°C rise in temperature.

Under forward-bias conditions the diode current is inversely proportional to the exponential of the thermal voltage ($\exp(V/V_T)$), where V is the applied voltage. From equation 2.10 the thermal voltage is given by:

$$
\begin{aligned}
V_T &= \frac{kT}{q} \\
&= \frac{(1.38 \times 10^{-23}\ \text{J K}^{-1})(300\ \text{K})}{1.6 \times 10^{-19}\text{°C}} \\
&= 0.0258\ \text{V at } 27\text{°C}
\end{aligned}
$$
(2.17)

From equation 2.17 it can be seen that as the temperature increases V_T also increases, which reduces the contribution from the exponential term. The combined effect of an increase in the value of the saturation current and the increase in the thermal voltage results in an overall reduction in the turn-on voltage of approximately 2 mV per degree centigrade for a silicon diode. The change in the turn-on voltage can be used as a temperature sensor, although more accurate commercial sensors are based on the base–emitter voltage of a transistor.

▶ **EXAMPLE 2.8** ─────────────────────────────────────

Use PSpice to observe the effect of temperature on the forward *I–V* curves of a silicon junction diode shown in Figure 2.20 over the temperature range −50°C to 150°C and verify that the turn-on voltage decreases at 2 mV per degree centigrade.

▶ *Solution*

Temperature (°C)	Voltage (mV)
−50	702
0	596
50	488
100	378
150	265

Use a nested **DC Sweep...** to vary the forward bias for a number of different temperatures. The curves are shown in Figure 2.22.

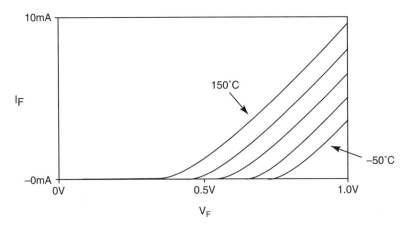

FIGURE 2.22   Forward voltage characteristics showing the variation of the turn-on voltage for different temperatures.

With the aid of the cursor within Probe it is possible to measure the forward voltage at a given current, say 10 μA, for each temperature. The results are on page 48.

From the table it can be seen that for a temperature change of 200°C the voltage changes from 702 mV to 265 mV or 437 mV, that is, -2.18 mV °C^{-1}.

Schottky diode

The most significant difference between a Schottky diode and a conventional pn junction diode is the lower turn-on voltage. If a Schottky diode is not available as a library device for a circuit simulator then it can be obtained by using a generic diode, but with I_S increased to lower the voltage at which a significant current flows, that is to reduce the turn-on voltage.

▷ **EXAMPLE 2.9** ——————————————————————————

If a conventional silicon diode has a saturation current of 1×10^{-14} A and a turn-on voltage defined as 0.6 V, determine the saturation current required to simulate a Schottky diode which is required to have a turn-on voltage of 0.4 V for the same forward current as the conventional diode. Use PSpice to verify the result.

▷ *Solution*

From the diode equation (equation 2.6) the forward current for a voltage of 0.6 V at room temperature and with a saturation current of 1×10^{-14} A is:

$$I = 1 \times 10^{-14} \exp\left(\frac{0.6}{0.026}\right) = 105 \,\mu A$$

In order to obtain the same value of current for the Schottky diode with a forward voltage of 0.4 V the saturation current must be:

$$I_S = I \exp\left(\frac{-V}{V_T}\right) = 105 \times 10^6 \exp\left(\frac{-0.4}{0.026}\right) \approx 2.2 \times 10^{-11} \text{ A}$$

To verify the result use the schematic shown in Figure 2.20 and from **Analysis** and **Setup...** select the **DC Sweep...** to sweep the dc voltage from 0 V to 1 V with a **nested sweep** to vary the model parameter I_S from 1e − 14 to 2.2e − 11. The resultant Probe plot is shown in Figure 2.23.

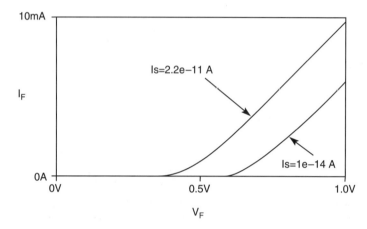

FIGURE 2.23   Simulation of a Schottky diode achieved by changing the value of I_S.

It can be seen from the forward characteristics that the turn-on voltage for the Schottky diode is approximately 0.4 V, compared with the 0.6 V for the conventional diode.

Junction Capacitance

A very simple method of observing the variation of the junction capacitance is with a parallel *LC* circuit in which a reverse-biased diode is used for the capacitor as shown in Figure 2.24. This also represents a practical application of the variable capacitance of a diode being used for automatic tuning of radio frequency stages of fm radios and televisions.

The resonant frequency for the *LC* circuit is:

$$f = \frac{1}{2\pi\sqrt{LC}}$$

where C is the capacitance of the reverse-biased diode. The large-value capacitor (C_1) prevents the dc supply from being shorted to ground through the inductor. The 100 Ω resistor lowers the Q of the circuit and results in a more realistic frequency response, and the 1 MΩ resistor isolates the dc supply from the tuned circuit. As the dc voltage

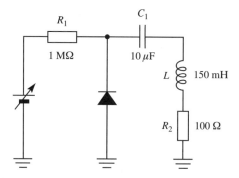

FIGURE 2.24   Circuit schematic to demonstrate the effect of variation of the diode capacitance.

increases, the width of the depletion layer increases and the capacitance of the diode decreases, which causes the resonant frequency to increase. The most effective way of examining the circuit is with PSpice.

▷ **EXAMPLE 2.10** ——————————————————————————————

Use PSpice to examine the circuit shown in Figure 2.24 for dc voltages of 2 V, 10 V and 18 V for the frequency range 10 MHz to 25 MHz and with the zero-voltage junction capacitance (cjo) set to 2 pF for the diode from the breakout library of PSpice devices. Determine the value of diode capacitance for each of the three voltage steps. The PSpice schematic is shown in Figure 2.25.

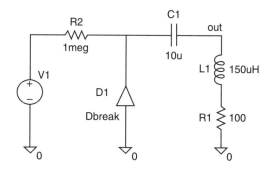

FIGURE 2.25   PSpice schematic of the circuit shown in Figure 2.24.

▷ *Solution*

From **Analysis** select **AC Sweep...** and set the frequency range to start at 10 MHz and end at 25 MHz. Select **Parametric...** option, select **Voltage Source**, enter **V1** for the name of the source, from **Value List** select **Values** and enter in **Values:** 2 10 18 to change the value of the dc source and to automatically repeat the frequency sweep. The resultant frequency curves from Probe are shown in Figure 2.26.

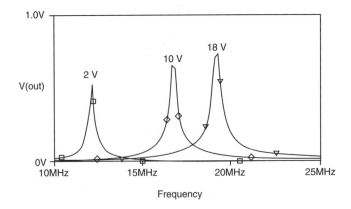

FIGURE 2.26   Variation of the voltage with frequency for the tuned circuit for different values of dc voltage applied to the diode.

Values of diode capacitance for a particular voltage may be obtained from the equation for the resonant frequency by using the cursor measuring facility within Probe to measure the frequency at resonance. Thus the following values of frequency and the corresponding capacitance may be obtained:

Voltage (V)	Frequency (MHz)	Capacitance (pF)
2	12.1	1.15
10	16.7	0.61
18	19.2	0.46

▶ **PROBLEMS**

2.1   Determine the resistivity of intrinsic silicon assuming that the electron mobility is $1300 \text{ cm}^2 \text{ V}^{-1} \text{s}^{-1}$ and the hole mobility is $500 \text{ cm}^2 \text{ V}^{-1} \text{s}^{-1}$.

[239 kohm cm]

2.2   Determine the resistivity of intrinsic germanium if $n_i = 2.5 \times 10^{13}$ atoms cm^{-3}, the electron mobility is $3800 \text{ cm}^2 \text{ V}^{-1} \text{s}^{-1}$, and the hole mobility is $1800 \text{ cm}^2 \text{ V}^{-1} \text{s}^{-1}$.

[45 ohm cm]

2.3   A bar of germanium 2 cm long has a voltage of 1 V placed across it. If the electron velocity is 19 m s^{-1}, determine the electron mobility.

[$3800 \text{ cm}^2 \text{ V}^{-1} \text{s}^{-1}$]

2.4   A sample of silicon is doped with 3×10^{17} atoms cm^{-3} of boron. Determine the number of holes and electrons and the resistivity at room temperature assuming that the electron mobility is $500 \text{ cm}^2 \text{ V}^{-1} \text{s}^{-1}$.

[3×10^{17}, 700, 0.04 ohm cm]

2.5 Determine the number of impurity atoms required to produce 10 ohm cm n-type silicon, assuming that $\mu_n = 1300$ cm^2 V^{-1} s^{-1} and $\mu_p = 500$ cm^2 V^{-1} s^{-1} and also determine the number of minority carriers.

$$[4.8 \times 10^{14} \text{ cm}^{-3}, 4.4 \times 10^5 \text{ cm}^{-3}]$$

2.6 Sketch the position of the depletion layer with respect to the metallurgical junction for a p$^+$n diode with no external voltage applied, where the p$^+$ represents a high impurity concentration ($>10^{19}$ cm^{-3}) and the n represents a low concentration ($<10^{15}$ cm^{-3}).

2.7 When a reverse bias is applied to a p$^+$n junction into which region does the depletion layer expand?

2.8 The n$^+$-region of an n$^+$p junction diode has an impurity concentration of 5×10^{19} atoms cm^{-3} and the p region has 1×10^{16} atoms cm^{-3}. Determine the number of minority carriers in each region. Which region contributes the most carriers to the saturation current under reverse bias conditions?

$$[4, 2 \times 10^4]$$

2.9 The saturation current for a diode at 20°C is 0.2 pA. Determine the forward current for an applied bias of 0.55 V at 20°C and 100°C.

$$[0.57 \text{ mA}, 1.36 \text{ mA}]$$

2.10 The turn-on voltage for a silicon rectifier diode is 0.6 V and the saturation current is 1 pA at 300 K. Determine the saturation current for a germanium rectifier diode for which the turn-on voltage is 0.2 V for the same forward current as for that of the silicon diode.

$$[4.8 \text{ }\mu\text{A}]$$

2.11 Select an appropriate value of saturation current for the generic diode (**dbreak** from the PSpice **breakout.slb**) to produce diode characteristics which have a turn-on voltage of 0.6 V and 0.2 V at 1 mA at 300 K and verify using PSpice. (Set the series resistance to 10 ohms when verifying with PSpice.)

$$[\sim 1 \times 10^{-13} \text{ A}, 4.6 \times 10^{-7} \text{ A}]$$

2.12 Use PSpice to investigate the effect of the series resistance on the forward characteristic curves of a junction diode. Use values of 10 Ω and 100 Ω. (Use **DC Sweep...** with **rs** as a parameter and plot **logI$_F$ versus V$_F$**.)

2.13 Select a Zener diode (D1N750) from the **eval.slb** library and modify the breakdown voltage value to 5.6 V. Use the **DC Sweep...** from -5.8 V to $+1$ V to verify the reverse I–V characteristics.

2.14 Examine the SCR (2N1595) circuit with PSpice. Set the dc gate voltage (V_2) to 1 V and the sine wave source (V_1) to 50 V amplitude and 50 Hz frequency. From **Analysis** select **Transient...** and set the **Print Step** to 100 μs and the **Final Time** to 40 ms. Run the simulation and examine the output voltage across the 1 kΩ resistor.

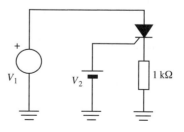

CHAPTER 3

Diode Applications

The diode is one of the simplest semiconductor devices and its ability to allow current to flow in one direction but not in the other has a number of applications in electronics. The most widespread application is in the use of pn junction diodes for the conversion of ac to dc in power supplies. The rectifier circuit takes advantage of the switch-like behaviour of the diode for voltages of different polarity.

Another important application is in the use of the reverse voltage breakdown of Zener diodes to establish a voltage reference for voltage regulation purposes.

Diode limiters and diode clamps are useful for restricting the amplitude of a signal or for wave shaping, and for changing signal levels.

The non-linear variation of the current with voltage when forward biased can be used in telecommunication circuits for mixing or modulation and also the junction capacitance is used extensively as a tuning element in frequency selective circuits.

The four-layer pnpn structures are used in power electronics to control resistive heating in furnace applications and as part of the drive circuitry for motors, but these applications form the basis of a subject in its own right and will not be covered in this text.

3.1 The Diode as a Rectifier

The process of converting ac to dc is known as rectification, and the diodes used for this purpose are known as rectifiers. The rectifier diode is simply an n^+p or p^+n junction diode for which the most important properties are low series resistance and an adequate reverse breakdown voltage to withstand the applied ac voltage when the junction is reverse biased. These two requirements are often conflicting, because low resistance requires a high impurity concentration in order to reduce the resistivity, while high breakdown voltage requires a low impurity concentration. Obviously there must be a compromise and the methods of manufacture and the area of the junction can be varied to control the electrical properties to match the requirements.

Half-Wave Rectifier

In its simplest form the *half-wave rectifier* comprises a diode, a load resistance and the source of ac voltage as shown in Figure 3.1.

A sine wave input with a peak amplitude of V_P is applied to the diode. When the input voltage is increasing from 0 V in a positive direction the diode conducts when

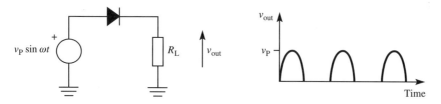

FIGURE 3.1   Half-wave rectifier and output waveform.

the voltage exceeds the turn-on voltage (~0.6 V to 0.7 V). The diode continues to conduct while the input exceeds this voltage. For negative excursions of the input voltage the diode is reverse biased and does not conduct, and it is effectively an open circuit. As a result the voltage across the load is a series of half sine waves in the positive-going direction, with a peak amplitude of approximately V_P. The actual voltage is $V_P - 0.6$ V (or 0.7 V). The output voltage of this *half-wave rectifier* is unidirectional, but is pulsating and for conversion to a steady dc voltage requires *smoothing* and *regulation*.

Transformer Coupled

For a power supply the source of the input is likely to be a sine wave from a transformer as shown in Figure 3.2.

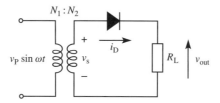

FIGURE 3.2   Transformer-coupled half-wave rectifier.

The transformer has a number of advantages. First it allows a voltage to be applied to the diode which is different from the input by a suitable choice of the turns ratio (N_2/N_1). Typically the dc voltage for many computer-based circuits is 5 V and for analog circuits, which use operational amplifiers, it is ±15 V. The voltage at the secondary of the transformer is $v_s = (N_2/N_1)V_P \sin \omega t$, where V_P is the peak amplitude of the incoming alternating line voltage. The turns ratio is adjusted to provide an appropriate secondary voltage: about 12 V for a 5 V supply and about 24 V for a 15 V supply.

A further advantage is that the transformer provides electrical isolation between the primary and secondary circuits. This is important for the safety of equipment operating from line voltages.

Finally the transformer can provide voltages which are 180° out of phase with each other, and this feature is used for the *full-wave rectifier*.

Full-Wave Rectifier

The addition of a second diode and a centre-tapped transformer allows the output to be generated for both the positive and negative halves of the sine wave as shown in Figure 3.3.

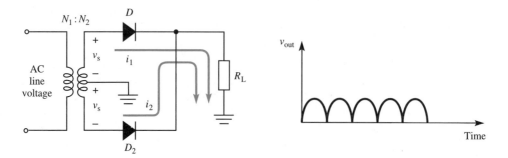

FIGURE 3.3   Full-wave rectifier circuit.

The diode D_1 conducts when the upper terminal of the transformer is positive and the current i_1 flows through the load resistor. The diode D_2 conducts on the next half cycle and the current i_2 flows in the same direction as i_1 through the load. Thus the voltage across the load is a unidirectional pulsating voltage for both negative and positive halves of the input sine wave.

The peak value of the secondary voltage for each half of the secondary is half the total; that is, for a 1:1 turns ratio the peak value of half the secondary is $V_{primary}/2$.

Bridge Rectifier

The *full-wave rectifier* in Figure 3.3 requires a transformer with a centre-tap on the secondary winding. While this is not difficult, it does add to the cost of the transformer. In the circuit shown in Figure 3.4 an additional two diodes are used to achieve full-wave rectification without the need for a centre-tapped transformer. The configuration of four diodes is known as a *bridge rectifier* (cf. the configuration of the Wheatstone bridge).

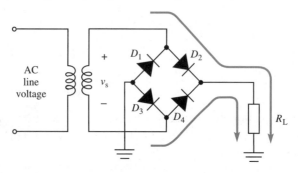

FIGURE 3.4   Full-wave rectifier with a diode bridge.

During the positive half cycle of v_S the current flows through D_2, the load resistor R_L, through the ground connection and back to the transformer via D_3. During the negative half cycle current flows through D_4, R_L and D_1. Thus the current through the load is unidirectional for both negative and positive cycles of the ac voltage. The four diodes are available in a single package with four external terminals.

Peak Inverse Voltage

For the half-wave rectifier the diode is subjected to a reverse bias equal to the peak of the secondary voltage during the half cycle when it is not conducting. It is important that the voltage does not exceed the breakdown voltage of the diode, and the diode should be able to withstand the *peak inverse voltage* (PIV). For the half-wave diode rectifier the PIV is equal to the peak voltage of the secondary:

$$PIV = V_{Psec} \quad \text{half wave} \tag{3.1}$$

For the full-wave rectifier in Figure 3.3 the PIV is equal to twice the peak voltage of the secondary, that is:

$$PIV = 2V_{Psec} \quad \text{full wave (centre-tap transformer)} \tag{3.2}$$

For the bridge rectifier:

$$PIV = V_{Psec} \quad \text{full wave (bridge rectifier)} \tag{3.3}$$

For most low-voltage transistor-based circuitry the PIV is not a serious problem, because the minimum breakdown voltage is likely to exceed 50 V to 60 V. However, there is increasing interest in operating electronic circuits at ac line voltage levels of 120 V (USA) to 230 V (Europe), and for these levels more care needs to be exercised in the selection of the diodes. Similarly for television and oscilloscopes the voltages required to power the cathode ray tube is likely to be several tens of thousands of volts. The diodes required for these applications require special manufacturing steps and special packaging to withstand the high voltages.

▷ **EXAMPLE 3.1** ——————————————————————————

Determine the peak value for the output voltage and the PIV for the rectifier circuit in Figure 3.5.

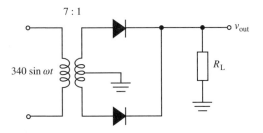

FIGURE 3.5   Circuit for Example 3.1.

▷ *Solution*

The turns ratio is:

$$\frac{N_2}{N_1} = \frac{1}{7} = 0.143$$

The total voltage applied to each diode is:

$$V_{Psec} = \frac{1}{2}\frac{N_2}{N_1}V_P$$

$$V_{Psec} = 0.0714(340\ V) \approx 24.3\ V$$

The peak output voltage is:

$$V_{Pout} = 24.3V - 0.7\ V = 23.6\ V$$

The PIV is:

$$PIV = 2V_{Psec} = 48.6\ V$$

| 3.2 | ▷ **Capacitive Filter** |

The voltage across the load resistor is unidirectional, that is it has been converted to dc, but it is pulsating and has very limited application, for example it could be used in a battery charger without any further processing. However, the voltage supply for electronic circuits must be constant and free from any variations in value. For both the half-wave and the full-wave rectifier some means must be found to maintain the current flow during the periods when the supply voltage drops to zero. A means for storing charge is required which can be released when the voltage drops below some given level. The simplest storage element is a capacitor and if a capacitor is placed in parallel with the load then it will store charge which it releases when the voltage drops below the level to which the capacitor is charged. An alternative explanation is that the capacitor acts as a filter to remove the alternating component from the output voltage. The larger the value of capacitance the greater its ability to filter out the varying or the alternating component of voltage. A half-wave rectifier with a capacitive filter is shown in Figure 3.6(a) and the simulated output from PSpice is shown in Figure 3.6(b)

In Figure 3.6(b) the output waveform with no capacitive filter (0 μF) is included for reference. With the capacitor present the output voltage rises at time 0 s as the capacitor charges to reach a value of ~20 V. At 5 ms (50 Hz supply) when the input sine wave starts to decrease the current to the load resistor is now being provided by the capacitor. In the case of the 20 μF capacitor the voltage drops to ~10 V before the diode starts to conduct again to recharge the capacitor to 20 V. With the 50 μF capacitor the output voltage only drops to ~15 V.

From this simulation it can be seen that the output is converted from a pulsating waveform to a dc value with a superimposed triangular-shaped *ripple*. The amplitude of

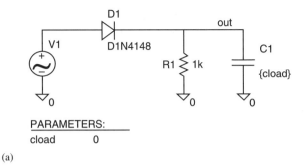

(a)

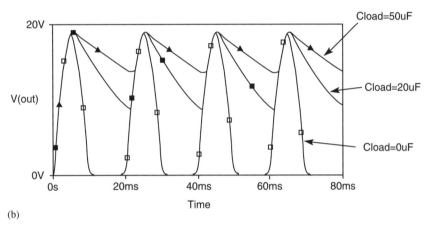

(b)

FIGURE 3.6 Half-wave rectifier with capacitive filter, (a) schematic and (b) the output waveforms for different values of load capacitor.

the ripple is dependent on the value of the filter capacitor, with a large value reducing the amount of ripple. For a given value of capacitance the amount of ripple is dependent on the current which the capacitor has to provide during the period when the diode is not conducting. Thus if the load current is increased, that is the load resistance is reduced, then the ripple increases. It should also be apparent that the amount of ripple will be less for a full-wave rectifier because the diode conducts every half cycle rather than every cycle as is the case for the half-wave rectifier.

Ripple

An analytical expression can be obtained for the dc output voltage by considering the behaviour of the voltage across the capacitor. When the diode conducts, the capacitor charges and the voltage rises to the peak value. When the voltage applied to the diode drops below that across the capacitor, the diode ceases to conduct, and the current is supplied by the capacitor. The capacitor now discharges through the resistive load. (Note that the resistive load simulates the piece of equipment which may be attached to the output of the rectifier circuit.) Referring to the shape of the output waveform in Figure 3.6 it can be seen that the waveform is approximately triangular. If this approximation is assumed then the analysis is greatly simplified. The waveform to be analyzed is shown in Figure 3.7.

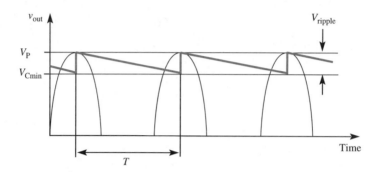

FIGURE 3.7   Simplified output waveform from the half-wave rectifier.

In Figure 3.7 the charging of the capacitor is assumed to be instantaneous, and the discharge is assumed to be linear. Provided that the ripple voltage is small then the error caused by ignoring the time required to charge the capacitor is small. When the capacitor discharges through a resistor then the normal exponential relationship between voltage and time applies, that is:

$$C_C = V_P \exp\left(\frac{-1}{R_L C}\right) \tag{3.4}$$

where V_P is the peak value after taking account of the voltage drop across the diode. After a time T the voltage on the capacitor is:

$$V_{Cmin} = V_P \exp\left(\frac{-T}{R_L C}\right) \tag{3.5}$$

The expansion of a general exponential term as a series gives:

$$\exp(-x) = 1 - x + \frac{x^2}{2!} - \frac{x^3}{3!} + \cdots$$

and provided that x is small, then:

$$\exp(-x) \approx 1 - x \tag{3.6}$$

that is, the exponential term is assumed to be linear. Substituting $T/R_L C$ for x in equation 3.6 gives:

$$V_{Cmin} \approx V_P\left(1 - \frac{T}{R_L C}\right) \tag{3.7}$$

In this approximation of the series expansion the voltage is directly proportional to time. This approximation applies if $R_L C \gg T$, that is if x is small.

The peak-to-peak ripple voltage is:

$$V_{\text{ripple}} = V_P - V_{\text{Cmin}}$$

$$V_{\text{ripple}} = V_P - V_P\left(1 - \frac{T}{R_L C}\right)$$

$$V_{\text{ripple}} = V_P \frac{T}{R_L C}$$

and

$$V_{\text{ripple}} = V_P \frac{T}{f R_L C} \tag{3.8}$$

where the frequency $f = 1/T$.

The simplified output from the rectifier with a capacitive filter is shown in Figure 3.8.

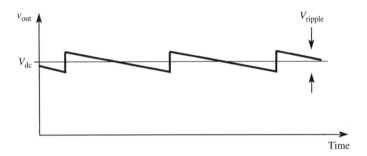

FIGURE 3.8   Rectifier output with a capacitive filter.

The dc voltage is equal to the peak voltage minus half the ripple voltage, that is:

$$V_{\text{dc}} = V_P - \frac{V_{\text{ripple}}}{2}$$

$$V_{\text{dc}} = V_P\left(1 - \frac{1}{2 f R_L C}\right) \quad \text{half wave} \tag{3.9}$$

This equation applies for the half-wave rectifier. The same method of analysis may be used for the full-wave rectifier, but now the capacitor is only allowed to discharge for half the period of the ac line voltage, that is use $T/2$ in place of T. Then the dc voltage is:

$$V_{\text{dc}} = V_P\left(1 - \frac{1}{4 f R_L C}\right) \quad \text{full wave} \tag{3.10}$$

Ripple Factor

The amount of ripple in the dc output voltage is a measure of the effectiveness of the capacitive filter in removing the ac component. A figure of merit for the rectifier is the *ripple factor* defined as:

$$r = \frac{V_{r(rms)}}{V_{dc}} \tag{3.11}$$

The rms value of a saw-tooth wave is (see Appendix 1):

$$V_{r(rms)} = \frac{V_{ripple}}{2\sqrt{3}} \tag{3.12}$$

and

$$V_{r(rms)} = \frac{V_P}{fR_LC} \frac{1}{2\sqrt{3}} \quad \text{half wave} \tag{3.13a}$$

and

$$V_{r(rms)} = \frac{V_P}{2fR_LC} \frac{1}{2\sqrt{3}} \quad \text{full wave} \tag{3.13b}$$

▷ **EXAMPLE 3.2** —————————————————————————

Determine the ripple factor for the bridge rectifier shown in Figure 3.9 if the input is a 240 V$_{rms}$. 50 Hz sine wave.

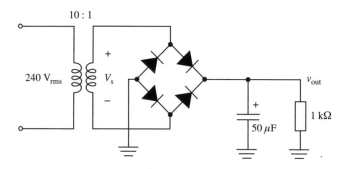

FIGURE 3.9   Bridge rectifier for Example 3.2.

▷ **Solution**

The peak primary voltage is:

$$V_{Pprimary} = \sqrt{2}\,(240) = 339 \text{ V}$$

The peak secondary voltage is:

$$V_{Psec} = \left(\frac{1}{10}\right)(339 \text{ V}) = 33.9 \text{ V}$$

The peak rectified output voltage is:

$$V_{Pout} = V_{Psec} - 2V_D$$
$$V_{Pout} = 33.9 - 1.4 = 32.5 \text{ V}$$

The filtered dc is:

$$V_{dc} = V_{Pout}\left(1 - \frac{1}{4fR_LC}\right)$$

$$V_{dc} = 32.5\left(1 - \frac{0.005}{(1 \text{ k}\Omega)(50 \,\mu\text{F})}\right) = 29.25 \text{ V}$$

The rms ripple voltage is:

$$V_{r(rms)} = 0.0029 \frac{V_{Pout}}{R_LC}$$

$$V_{r(rms)} = 0.0029 \frac{32.5}{(1 \text{ k}\Omega)(50 \,\mu\text{F})} = 1.88 \text{ V}$$

The ripple factor is:

$$r = \frac{V_{r(rms)}}{V_{dc}} = \frac{1.88}{29.25} = 0.06$$

or expressed as a percentage the ripple factor is 6%.

Capacitor Selection

The ripple factor provides a means for calculating the smoothing capacitor, because there is a direct relationship between the amount of ripple and the value of the smoothing capacitor. Thus the value of the capacitor may be obtained from the equation for the ripple factor as follows:

$$r = \frac{V_{r(rms)}}{V_{dc}}$$

$$= \frac{V_{Pout}}{2\sqrt{3}fR_LC} \frac{1}{V_{dc}} \quad \text{half wave} \tag{3.14}$$

For 'good' conversion of the ac to dc the dc output voltage is approximately equal to the peak value of the pulsating voltage from the rectifier, that is:

$$V_{Pout} \approx V_{dc}$$

and thus:

$$r \approx \frac{1}{2\sqrt{3}fR_LC} \quad \text{half wave}$$

$$\approx \frac{1}{4\sqrt{3}fR_LC} \quad \text{full wave}$$

or:

$$C = \frac{1}{2\sqrt{3}fR_Lr} \quad \text{half wave}$$

$$= \frac{1}{4\sqrt{3}fR_Lr} \quad \text{full wave} \tag{3.15}$$

▶ **EXAMPLE 3.3** ────────────────────────

Determine the value of the smoothing capacitor to give a ripple factor of not greater than 5% for a full-wave rectifier when supplying a load at 5 V with a current of 10 mA from a 60 Hz ac line supply.

▶ *Solution*

The load current of 10 mA at 5 V represents a load resistance of:

$$R_L = \frac{5 \text{ V}}{10 \text{ mA}} = 500 \ \Omega$$

For a ripple factor of 5% or 0.05 the capacitor is:

$$C = \frac{1}{4\sqrt{3}fR_LC} = \frac{1}{4\sqrt{3}(60)(500)(0.05)} = 96 \ \mu F$$

Surge Current

The current in a rectifier diode which has a capacitive filter takes the form of a series of pulses, as shown in Figure 3.10.

Current only flows through the diode when the voltage at the anode is greater than the voltage at its cathode, that is the voltage applied to the anode exceeds the voltage on the

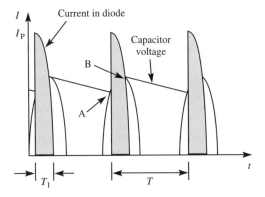

FIGURE 3.10   Repetitive surge current in rectifier diodes.

capacitor. As the capacitor discharges and the output voltage falls, a point is reached at A in the diagram when the rising voltage on the anode (from the secondary of the transformer) is equal to the voltage on the capacitor. At this point the diode becomes forward biased and current starts to flow. This current replaces charge lost by the capacitor when it was supplying the load. It stops flowing at point B when the voltage applied to the anode drops below that on the capacitor.

The current replaces the charge lost when the diode is not conducting and if T is the period of discharge and T_1 the period of charge, then:

$$Q_{\text{discharge}} = Q_{\text{charge}}$$

and to a first approximation the charge can be expressed as a product of current and time as:

$$I_{\text{dc}}T = I_{\text{P}}T_1$$

This relationship assumes that the current pulse through the diode – I_{P} for a duration T_1 – is rectangular. From Figure 13.2 the diode current is more nearly triangular and a more accurate expression for Q_{charge} is $I_{\text{P}}T_1/2$. Thus:

$$I_{\text{P}} = I_{\text{dc}}\,\frac{2T}{T_1} \tag{3.16}$$

where I_{dc} is the average current flowing into the load and I_{P} is the peak current flowing in the diode in order to replace the charge in the capacitor. Equation 3.16 is only approximate, but can be used to ascertain the trend in the value of the peak current as the size of the capacitor is increased. With a larger-value capacitor there is less ripple voltage, and point A moves closer to point B, that is T_1 decreases. From the above relationship for I_{P} this would suggest that as T_1 decreases then I_{P} increases. Usually the resistance in the circuit, for example in the transformer and the series resistance of the diode, will limit the current, but it may be necessary to include an additional small-value surge resistor to limit the current to a value which the diode can handle.

Equation 3.16 may be written in terms of the conduction angle θ as:

$$I_P = I_{dc}2\,\frac{2\pi}{\theta} = I_{dc}\,\frac{720°}{\theta°} \qquad \text{half wave} \tag{3.17a}$$

$$I_P \cong I_{dc}2\,\frac{\pi}{\theta} = I_{dc}\,\frac{360°}{\theta°} \qquad \text{full wave} \tag{3.17b}$$

where 2π is the angular representation for the period T and θ is the conduction angle for the period T_1. The conduction angle may be expressed as (see Appendix 1):

$$\theta = \cos^{-1}\frac{V_P - V_r}{V_P} \tag{3.18}$$

▷ **EXAMPLE 3.4** ────────────────────────────

A load is supplied with 24 V at 100 mA with a ripple factor of 2% from a 50 Hz supply. Determine the peak current and voltage ratings of the diode.

▷ *Solution*

With the aid of equations 3.11 and 3.12:

$$V_T = 2\pi3rV_{dc} = 2\sqrt{3}(0.02)(24\text{ V}) = 1.66\text{ V}$$

$$V_P = V_{dc} + V_r = 24\text{ V} + \frac{1.66\text{ V}}{2} = 24.83\text{ V}$$

From equation 3.18:

$$\theta = \cos^{-1}\frac{24.83\text{ V} - 1.66\text{ V}}{24.83\text{ V}} = 21°$$

From equation 3.17:

$$I_P = I_{dc}\,\frac{720°}{\theta°} = 100\text{ mA}\,\frac{720°}{21°} \approx 3.4\text{ A}$$

The peak inverse voltage is reached when the supply reaches its maximum negative value of 24 V plus the voltage across the load of 24 V, that is 48 V.

The average current rating is 100 mA with a peak surge rating of 3.4 A.

──

The above example provides a means for estimating the surge current which occurs during each cycle of the 50 Hz or 60 Hz supply. It does not take account of the surge which may occur at switch-on. At switch-on when the capacitor is completely discharged it acts as a short circuit to the voltage supplied by the diode. The current is particularly large if switch-on coincides with the peak of the ac line voltage. The maximum current is able to flow to charge the capacitor. As the voltage across the capacitor increases then the current decreases, until the current drops to zero when the capacitor is fully charged. The switch-on surge is largely determined by the resistance in the transformer secondary, and it can damage the diode if the resistance is insufficient to limit the current to a safe value. A small-value resistor can be placed in series with the diode to limit this current if the ratio of peak secondary voltage and the resistance of the transformer secondary exceeds the diode surge current, but this will result in a voltage drop and some power dissipation across the resistor.

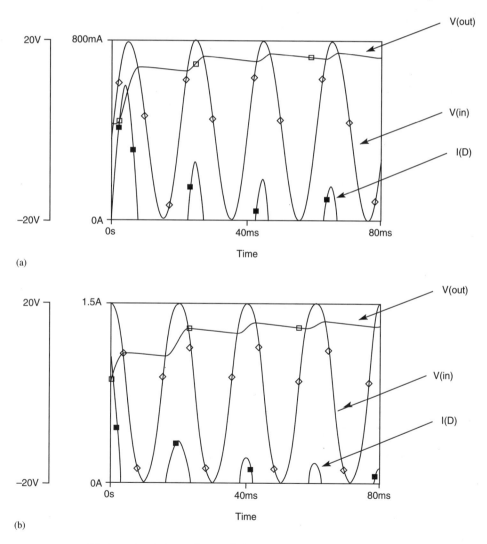

FIGURE 3.11   PSpice simulations for a half-wave rectifier with (a) $C_{load} = 200\,\mu F$, (b) $C_{load} = 200\,\mu F$ and a phase shift of 90°.

Some examples of the surge current as obtained with PSpice for different values of capacitor, and for two different switch-on times, are shown in Figure 3.11 based on the circuit shown in Figure 3.6.

For the simulation a 1 Ω resistor is placed between the source and the diode to act as a surge resistor and to limit the current to reasonable levels. Also an initial condition flag is attached to the output which is set to 0 V to ensure that the capacitor is fully discharged at the instant the ac voltage is applied.

Comparing (a) and (b) it can be seen that there is a considerable difference in the initial surge of current at switch-on, with approximately 600 mA in (a) where switch-on coincides with the zero crossing of the input ac voltage, and 1.25 A in (b) where switch-on coincides with the peak of the input voltage.

Diode Ratings

For rectifier diodes there are two ratings of importance to the design engineer – the maximum reverse voltage and the maximum forward current. The maximum reverse voltage is determined by the peak inverse voltage (PIV) associated with the particular rectifier circuit. The maximum reverse voltage of the diode should exceed the PIV. For most low-voltage supplies this is not a problem since the maximum reverse voltage is likely to exceed 50 V, and is typically 100–200 V. It becomes more of a problem for high-voltage supplies for power electronics and for extra high-voltage supplies for cathode ray tubes. For these applications the PIV may be in excess of 1000 V.

The maximum current rating is of greater interest, because even for low-voltage supplies, say 5 V, the current required may be several amps. Commercial rectifier diodes have two current ratings, one related to the average, or dc, current required by the load, and the other related to the peak surge current which can be safely provided during one cycle of the incoming mains supply. The two values are widely different with a small rectifier diode having a typical average rating of 1 A and a peak surge rating of 30 A, or a medium-size rectifier having an average rating of 40 A and a surge rating of 800 A.

Improved Filtering

The simple diode rectifier with a capacitive filter is not suitable for providing the dc for amplifiers or computer circuits because the ripple factor is too large, even with very large capacitors. Additional filtering can be achieved with the inclusion of an inductor, but suitable inductors for low-frequency use are large and expensive, and much more efficient smoothing can be achieved with electronic methods, rather than *LC* filters. These voltage supplies will be considered in more detail in Chapter 18.

3.3   Limiting and Clamping

The non-linear behaviour of diodes can be used to shape a signal by limiting the voltage excursion in either a positive or negative direction, or both, and by changing the dc level associated with an alternating signal. These operations are referred to as *limiting and clamping*.

Diode Limiter

The diode limiter (or *clipper*) is so named because it prevents a signal exceeding a certain value. Some typical limiter circuits are shown in Figure 3.12 and limit the output voltage to $V_1 + 0.6$ V (or $-V_2 - 0.6$ V), where the 0.6 V is the diode turn-on voltage.

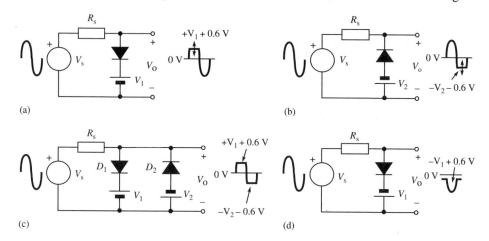

(a)   (b)   (c)   (d)

FIGURE 3.12   Diode limiter circuits.

In diagram 3.12(a) the diode provides clipping of the positive-going signal at $V_1 + 0.6$ V. As the input signal (V_s) increases the diode is prevented from conducting by the presence of the potential from V_1 which is in such a direction as to reverse bias the diode. When the source voltage exceeds V_1 by an amount equal to the turn-on voltage (0.6 V) then the diode conducts, current flows from the source and there is a voltage drop across R_s. The output voltage (V_o) is held at $V_1 + 0.6$ V. Changing the polarity of both the diode and the biasing voltage as in diagram 3.12(b) results in the negative-going waveform being clipped. In both cases the level at which clipping takes place rises or falls with the level of V_1 or V_2. In 3.12(c) clipping occurs for both positive- and negative-going directions and again the level at which clipping takes place can be varied by changing V_1 and/or V_2.

In 3.12(d) the diode is forward biased by the presence of V_1 and the input signal is largely dropped across R_s as the input rises from zero during the positive half cycle. It is not until the input passes into the negative half cycle that the negative voltage applied to the anode eventually exceeds $-V_1 + 0.6$ V and causes the diode to become reverse biased.

The diode and resistor can be interchanged to produce a series rather than a shunt clipping circuit as shown in Figure 3.13.

The voltage V_1 reverse biases the diode so that as the input rises from zero during the positive half cycle the diode does not conduct, and the output remains at a constant

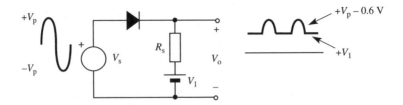

FIGURE 3.13   Series diode limiter.

value of $+V_1$. At some value of the input the diode becomes forward biased, current flows through the diode and R_s and the output voltage rises. The positive peak value is $+V_P - 0.6$ V, where V_P is the peak value of the input.

Limiting and clamping are not very sophisticated operations, and it is important to realize that in the examples above, the operations are rather idealized. In practice the source and source resistor may represent a complex amplifier or signal processing circuit. The diode is added to this circuit to shape the output waveform in some way.

A common requirement is to shape a waveform produced by an operational amplifier which has an output which swings above and below 0 V by ±13 V to one which is entirely positive and varies from 0 V to +5 V. Such a waveform could then be used to operate digital circuitry.

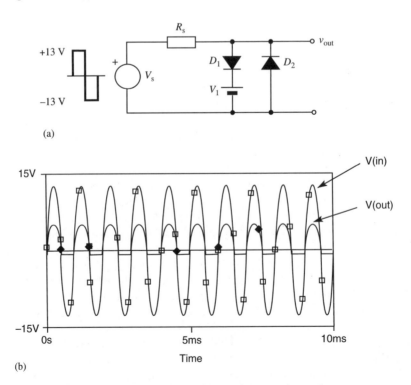

FIGURE 3.14   Alternating waveform converted to unidirectional waveform.

▶ **EXAMPLE 3.5** ――――――――――――――――――――――――――

Design a limiter to convert a signal which alternates between ±13 V to one which is suitable for a digital circuit for which the signal must not exceed −0.7 V and +5.5 V and use PSpice to confirm the design.

▶ *Solution*

The circuit is shown in Figure 3.14(a) with the resultant waveform predicted by PSpice shown in 3.14(b). The bias voltage V_1 is set to 4.3 V. The diode D_2 conducts when the input is negative and the excess voltage is dropped across R_s. The current when $V_s = -13$ V is $(13\text{ V} - 0.7\text{ V})/R_s$. If $R_s = 1$ kΩ, then the current is 12.3 mA. The diode D_1 together with the dc voltage V_1 limit the output during the positive excursion of V_s to $V_1 + 0.7$ V = 4.3 V + 0.7 V = 5.0 V.

―――――――――――――――――――――――――――――――――――

Zener Diode Limiter

The Zener diode can be used to limit a signal to a desired level without the need for offset voltages. The basic circuit is shown in Figure 3.15.

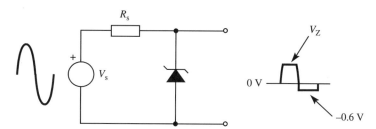

FIGURE 3.15   Zener diode limiter.

The amount of limiting is determined by the value of the Zener voltage. Two diodes can be used in series, but in opposite directions, to provide limiting for both positive and negative half cycles of the source voltage.

Diode Clamp

The diode *clamp* is used to add a dc level to a signal. It is sometimes referred to as a *dc restorer*. The most common example of the use of a dc restorer is in a television receiver where the dc level of the composite video signal must be re-established after it has passed through a number of capacitively coupled stages. This is illustrated in Figure 3.16.

The composite video signal consists of the high-frequency information for each line of the television picture, plus blanking and horizontal synchronization pulses. After amplification through capacitively coupled stages this signal varies symmetrically above and below the zero-voltage reference. In order to extract the blanking and synchronization pulses it is necessary to process the signal so that all parts are positively referenced with respect to ground. This is the purpose of the dc restoration circuit illustrated in Figure 3.16.

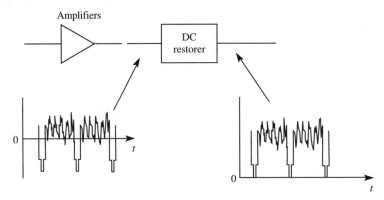

FIGURE 3.16   DC restoration of the composite video signal in a television receiver.

An example of a clamping circuit is shown in Figure 3.17(a) with a much simpler input signal, and the resultant output from a PSpice simulation is shown in 3.17(b). The input frequency is 10 kHz sine wave with an amplitude of 5 V.

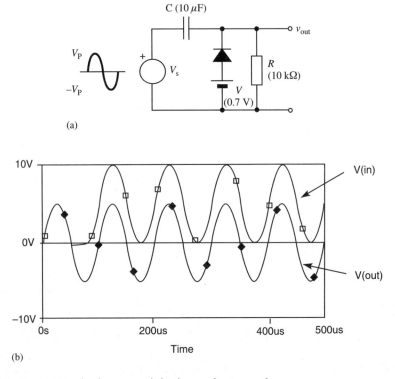

FIGURE 3.17   (a) Diode clamper and (b) the resultant waveforms.

The bipolar input signal of ±5 V is converted to a unipolar output signal which varies from 0 V to +10 V – the output is clamped at 0 V. The dc source in 3.17(a) is set to 0.7 V to correct for the diode turn-on voltage. The output signal can be clamped at levels above 0 V by increasing the value of the dc source.

The operation of the clamping circuit is explained by reference to Figure 3.18.

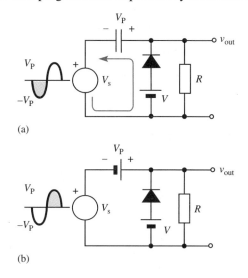

(a)

(b)

FIGURE 3.18   Clamping operation for (a) negative-going cycle and (b) positive-going cycle.

During the negative-going cycle, shown in Figure 3.18(a), the diode conducts and current flows, which charges the capacitor to the peak voltage V_P with the polarity shown. Provided that the value of the load resistance is large (large RC time constant) then the capacitor will not discharge to any great extent (notice that the polarity of the voltage on the capacitor provides a reverse bias for the diode so that it does not discharge through the diode). When the input polarity becomes positive the diode is non-conducting and the output voltage is $V_s + V_P$, where V_P is the voltage across the capacitor. The capacitor may be regarded as a battery with a value V_P, as shown in Figure 3.18(b). The dc supply V in this example simply cancels the diode turn-on voltage. If V is increased then it contributes to the voltage which appears across the capacitor, and raises v_{out} above 0 V.

▶ **EXAMPLE 3.6** ——————————————————————————————

Design a clamping circuit to provide a negative-going signal clamped at 0 V for an input signal with ±5 V peak values. Assume a capacitor value of 10 µF and a load resistance of 10 kΩ. Verify with PSpice.

▶ *Solution*

The circuit is shown in Figure 3.19.

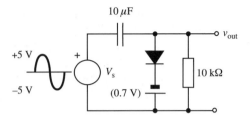

FIGURE 3.19   Diode clamp for negative-going signal.

3.4   Voltage Multipliers

The clamping action of a diode circuit can be used to increase or multiply the peak value of the rectified voltage without having to increase the peak voltage from the secondary of the transformer. The most commonly used voltage multipliers are voltage doublers and triplers.

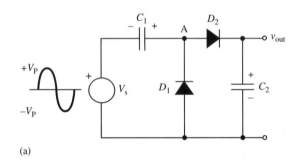

(a)

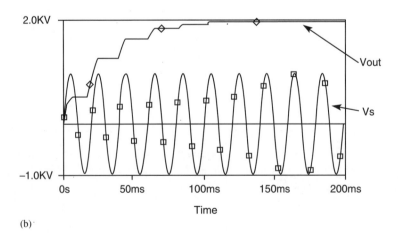

(b)

FIGURE 3.20   (a) Voltage doubler circuit and (b) the waveforms obtained from PSpice.

Voltage Doubler

A voltage doubler is shown in Figure 3.20. The diode D_1 and capacitor C_1 act as a clamp so that the voltage at node A is positive going and is clamped at 0 V with a peak value of $2V_P$. The diode D_2 conducts and charges C_2 to $2V_P$ when the voltage at node A is at its peak value. As the voltage at A falls below $2V_P$ then D_2 stops conducting as the voltage on C_2 is now greater than the voltage on its anode. The capacitor C_2 provides the current until the next peak value of voltage occurs at node A. Thus the voltage at the output is twice the peak value of the input, that is 2 kV, as seen in Figure 3.20(b).

Notice that it takes a few cycles of the input for the output to reach the maximum value.

Voltage Tripler

A voltage tripler is shown in Figure 3.21(a) with the output waveform in 3.21(b).

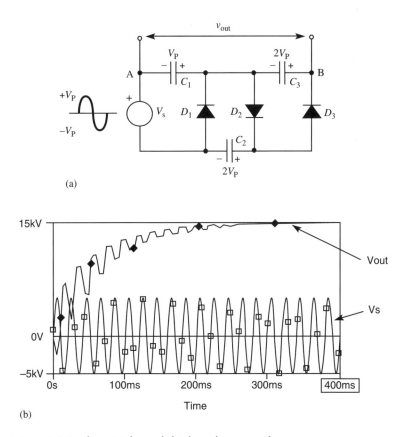

(a)

(b)

FIGURE 3.21   (a) A voltage tripler and (b) the voltage waveforms.

The operation is as follows. On the negative half cycle C_1 charges through D_1 to V_P. On the positive half cycle D_2 conducts and the peak voltage is V_P from the source (V_s) and V_P from C_1, that is the capacitor C_2 is charged to $2V_P$. During the next negative half cycle D_3 conducts and voltage which is applied to C_3 is the sum of V_P from V_s and $2V_P$ from C_2 less the voltage on C_1, that is C_3 is charged to $2V_P$. The output voltage is obtained between nodes A and B and has a value of $3V_P$.

The waveforms obtained from PSpice are shown in Figure 3.21(b). For an input of ±5 kV the output after about 150 ms is 15 kV. The same principle can be extended to higher-order multipliers.

Voltage multipliers are used extensively in large-screen television where voltages of 25 kV and above are required to create a bright television image. Voltage multipliers are also used in low-voltage ICs when a non-standard voltage may be required for some of the IC circuitry. An example is in battery-powered computer circuits where the voltage may be limited to 6 V. However, for some EPROM chips voltages in excess of 20 V may be required to program the chips. The higher voltage is obtained by means of an on-board oscillator and voltage multiplier.

3.5 Voltage Regulator

The reverse breakdown voltage of junction diodes provides a useful voltage reference for voltage regulators. Reverse voltage breakdown for a junction diode can vary from <10 V to >100 V, but it is the <10 V diodes which are most suitable for voltage regulation. For this range of voltages the breakdown is predominantly caused by the Zener effect.

Zener Diode Regulator

The Zener diode has a well-defined breakdown voltage, which can be varied by controlling the number of impurity atoms which are used to form the p- and n-type regions of the diode. Unlike the rectifier diode where the operating point is the dividing line between forward and reverse bias, the operating point for the Zener diode is the voltage at which breakdown occurs. The current–voltage characteristics for the diode are shown in Figure 3.22.

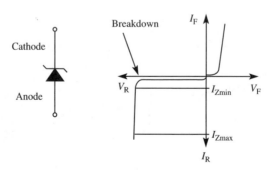

FIGURE 3.22   Zener diode symbol and I–V characteristic curves.

When used as a regulator, or as a voltage reference, the diode is maintained in a state of breakdown. In this condition, provided a minimum current (I_{Zmin}) is allowed to flow, and provided that the maximum (I_{Zmax}) is not exceeded, then the voltage across the diode (V_Z) remains almost constant. There is a small variation because of the internal resistance in the diode. Thus the equivalent circuit of the diode when reverse biased is of a battery equal to V_Z in series with a resistor r_Z. A simple voltage regulator is shown in Figure 3.23.

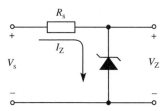

FIGURE 3.23   Zener diode voltage regulator.

The conditions which are necessary for operation are that the input voltage (V_s) must be greater than V_Z and that the current I_Z must be greater than I_{Zmin} and less than I_{Zmax}. Applying Kirchhoff's voltage laws (KVL) to the circuit gives:

$$V_s = I_Z R_s + V_Z \tag{3.19}$$

The minimum current is given by:

$$I_{Zmin} = \frac{V_s - V_Z}{R_s}$$

In practice the voltage regulator is used to provide a stable voltage from a variable supply, that is if V_s is assumed to vary from V_{smin} to V_{smax}, then the regulator provides a stable value equal to V_Z. For this to happen the minimum current must flow when V_s is at its minimum value, that is:

$$I_{Zmin} = \frac{V_{Smin} - V_Z}{R_s} \tag{3.20}$$

and similarly the maximum current occurs when the supply is at its maximum value of V_{smax}. Thus:

$$I_{Zmax} = \frac{V_{Smax} - V_Z}{R_s} \tag{3.21}$$

The value of the series resistor must be sufficiently small to allow the minimum Zener diode current to flow when the input voltage is at its minimum value to ensure that equation 3.18 is satisfied. However, it must be sufficiently large to ensure that when the input voltage is at its maximum value the Zener current does not exceed its maximum permitted value. This value of R_s is predicted by equation 3.21.

Zener diodes are categorized by the voltage breakdown and their power dissipation. The power is determined as the product of the Zener voltage and the maximum current which the diode can handle without being destroyed:

$$P_{diss} = I_{Zmax} V_Z \qquad\qquad (3.22)$$

▶ **EXAMPLE 3.7** ————————————————————

Determine the series resistor (R_s) required for a Zener diode regulator with an output voltage of 5.6 V, if the supply voltage (V_s) varies from 10 V to 50 V. The minimum Zener current is 3 mA. Determine also the maximum Zener current and the power dissipation.

▶ *Solution*

The series resistance is:

$$R_s = \frac{V_{smin} - V_Z}{I_{Zmin}}$$

$$R_s = \frac{10\ V - 5.6\ V}{3\ mA} = 1.46\ k\Omega$$

Notice that the series resistor must be determined first, and that it is determined by the need to ensure that the minimum current flows when the supply is also at a minimum value.

When the supply is at its maximum value, then for the given value of R_s the maximum current flows:

$$I_{Zmax} = \frac{V_{smax} - V_Z}{R_s}$$

$$I_{Zmax} = \frac{50\ V - 5.6\ V}{1.46\ mA} = 30\ mA$$

The power dissipation is:

$$P_{Zdiss} = (30\ mA)(5.6\ V) = 168\ mW$$

▶ **EXAMPLE 3.8** ————————————————————

Repeat Example 3.7 if the maximum load current is 20 mA, and determine the minimum and maximum values for R_s.

▶ *Solution*

The minimum diode current (3 mA) must flow when the input voltage is at the minimum value of 10 V, that is:

$$R_s \leqslant \frac{V_{in(min)} - V_Z}{I_{Z(min)}}$$

$$R_s \leqslant 1.46\ k\Omega$$

Notice that if R_s is less than 1.46 kΩ then the current for the minimum input voltage will be greater than 3 mA, but this does not matter. All that matters is that the current is greater than 3 mA.

If the maximum load current is 20 mA then R_s must satisfy the maximum load current when the input is at its maximum value, but must also ensure that the maximum Zener current is not exceeded, that is:

$$R_s \geq \frac{V_{in(max)} - V_Z}{I_{Z(max)} + I_{L(max)}}$$

$$R_s \geq \frac{50 \text{ V} - 5.6 \text{ V}}{30 \text{ mA} + 20 \text{ mA}} \geq 888 \text{ }\Omega$$

Voltage Regulation

As the input voltage varies the output voltage (V_Z) remains constant as the variations in V_s are absorbed across the series resistor R_s. While the current through the diode increases and decreases it is assumed that the voltage across the diode remains constant.

In a real Zener diode the voltage across the terminals of the diode varies as the current varies. This is caused by the voltage drop across the internal series resistance associated with the semiconductor material. Thus for the above example the series resistance (r_Z) could be 10–20 Ω and this would result in a voltage drop which would be added to the Zener voltage. The *voltage regulation* is a means of specifying the amount of change which takes place in the output when the input changes. It is affected by the value of the external series resistor (R_s) and the internal Zener diode resistance (r_Z). The relationship may be obtained as follows.

Applying KVL to the input gives:

$$V_s = I_Z R_s + I_Z r_Z + V_Z$$

and to the output:

$$v_{out} = I_Z r_Z + V_Z$$

and by substitution:

$$v_{out} = \frac{r_Z(V_s - V_Z)}{R_s + r_Z} + V_Z$$

$$v_{out} = \frac{r_Z V_s}{R_s + r_Z} - \frac{r_Z V_Z}{R_s + r_Z} + V_Z$$

Now the voltage regulation can be defined as $\Delta v_{out} / \Delta V_s$, that is:

$$\frac{\Delta v_{out}}{\Delta V_s} = \frac{\Delta}{\Delta V_s}\left(\frac{r_Z V_s}{R_s + r_Z} - \frac{r_Z V_Z}{R_s + r_Z} + V_Z\right) = \frac{r_Z}{R_s + r_Z}$$

and since $R_s \gg r_Z$ then:

$$\frac{\Delta v_{\text{out}}}{\Delta V_s} = \frac{r_Z}{R_s} \tag{3.23}$$

Equation 3.23 provides a means for calculating the regulation in the output voltage for a change to the input voltage.

▷ **EXAMPLE 3.9** ───────────────────────────────

For the regulator in Example 3.7 assume that the series resistance of the Zener diode is 20 Ω. Determine the change in the output voltage when V_s varies from 10 V to 50 V.

▷ *Solution*

The equivalent circuit of the regulator is shown in Figure 3.24.

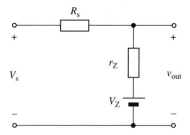

FIGURE 3.24   Regulator with the Zener diode replaced by an equivalent circuit.

When V_s is 10 V the current through the Zener diode is 3 mA and v_{out} is:

$$v_{\text{out}} = V_Z + I_Z r_Z$$

$$v_{\text{out}} = 5.6 \text{ V} + (3 \text{ mA})(20 \text{ }\Omega) = 5.66 \text{ V}$$

and when V_s is 50 V the current is 30 mA and v_{out} is:

$$v_{\text{out}} = 5.6 \text{ V} + (30 \text{ mA})(20 \text{ }\Omega) = 6.2 \text{ V}$$

For the above example the output voltage varies from 5.66 V to 6.2 V when the input varies from 10 V to 50 V, that is:

$$\frac{\Delta v_{\text{out}}}{\Delta V_s} = \frac{6.2 \text{ V} - 5.66 \text{ V}}{50 \text{ V} - 10 \text{ V}} = 0.0135$$

This compares with the predicted value based on the ratio r_Z/R_s from equation 3.23 of:

$$\frac{r_Z}{R_s} = \frac{20 \text{ }\Omega}{1460 \text{ }\Omega} = 0.0137$$

Regulator With Load Resistance

The Zener diode regulator is often used to regulate the voltage to an electronic circuit which can be represented by a load resistance (R_L) as shown in Figure 3.25.

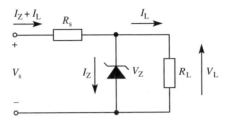

FIGURE 3.25   Regulator with load.

Applying KVL to the input gives:

$$V_s = (I_Z + I_L)R_s + V_Z \tag{3.24}$$

For a particular value of load resistance the value of R_s is determined by V_{smin} and I_{Zmin} as follows:

$$R_s = \frac{V_{smin} - V_Z}{I_L + I_{Zmin}} \tag{3.25}$$

▶ **EXAMPLE 3.10** ─────────────────────────────

Determine the value of the series resistor and the maximum Zener diode current for a regulator which is required to provide a load current of 10 mA stabilized at 12 V from a dc supply which varies from 24 V to 30 V. Assume that the minimum Zener diode current is 2 mA and that $r_Z = 0\ \Omega$.

▶ *Solution*

The series resistor is:

$$R_s = \frac{24\ V - 12\ V}{10\ mA + 2\ mA} = 1\ k\Omega$$

The maximum current is:

$$I_{Zmax} = \frac{V_{smax} - V_Z - I_L R_s}{R_s}$$

$$I_{Zmax} = \frac{30\ V - 12\ V - (10\ mA)(1\ k\Omega)}{1\ k\Omega} = 8\ mA$$

Regulator With Variable Load

If the load resistor is variable then for a given value of series resistance (R_s), as the load increases in value (load current decreases), then the excess current must be absorbed by the Zener diode. As the load resistance decreases (load current increases) then the current is diverted from the Zener diode to the load. The regulator will only maintain regulation while the Zener diode current remains greater than I_{Zmin} and less than I_{Zmax}.

▷ **EXAMPLE 3.11** ——————————————————————————

Determine the minimum and maximum values of the load resistance for the circuit in Figure 3.26. The following values apply: $V_s = 24$ V, $V_Z = 10$ V, $I_{Zmin} = 3$ mA, $I_{Zmax} = 50$ mA and $r_Z = 0$ Ω.

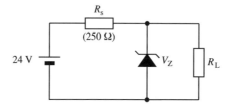

FIGURE 3.26   Voltage regulator for Example 3.10.

▷ **Solution**

The maximum current through the Zener diode occurs when $I_L = 0$, that is:

$$I_{Zmax} = \frac{24 \text{ V} - 10 \text{ V}}{250 \text{ }\Omega} = 56 \text{ mA}$$

This exceeds the maximum specified for the diode; therefore, the load resistance must not be allowed to become infinite (open circuit).

With $I_{Zmax} = 50$ mA, but with the regulator capable of supplying 56 mA, then the load must absorb the additional 6 mA. That is, the maximum value of the load resistance is:

$$R_{Lmax} = \frac{10 \text{ V}}{6 \text{ mA}} = 1.66 \text{ k}\Omega$$

As R_L decreases and the load current increases the current through the Zener diode decreases. The diode current must not be allowed to go below I_{Zmin} (3 mA). Thus:

$$24 = (3 \text{ mA})(250) + (I_{Lmax})(250) + 10$$

and:

$$I_{Lmax} = \frac{24 \text{ V} - 10 \text{ V} - 0.75 \text{ V}}{250 \text{ }\Omega} = 53 \text{ mA}$$

The minimum value of R_L is 10 V/53 mA = 189 Ω.

The Zener diode provides the means for producing a simple regulated power supply, but it is not adequate for high-gain amplifier circuits or complex digital circuits. For these circuits a more complex regulator is required, but the Zener diode can still form the basis for a stable reference voltage for a negative feedback circuit. This type of regulator will be considered in the chapter on power supplies (Chapter 18).

▶ PROBLEMS

3.1   The ac line voltage on the primary of the transformer attached to a half-wave rectifier is 240 V_{rms}. If the forward voltage drop across the diode is 0.7 V determine the peak voltage across the load and the PIV for the diode

[41.7 V, 42.4 V]

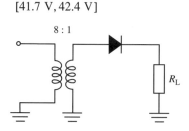

3.2   If the primary ac line voltage is 100 V_{rms}, determine the peak voltage across the load and the PIV for the diodes assuming that the turns ratio applies to each half of the secondary.

[~1413 V, 2828 V]

3.3   The ac line voltage at the primary is 110 V_{rms} at 60 Hz and the forward voltage drop across the diodes is 0.6 V. Determine the peak voltage across the load resistor and the PIV for the diodes.

[37.7 V, 38.9 V]

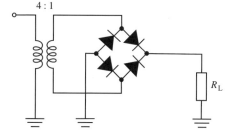

3.4   If the forward voltage drop across the diode is 0.7 V determine the peak voltage across the load, the peak-to-peak ripple voltage and the dc load voltage.

[55.9 V, 2.24 V, 54.8 V]

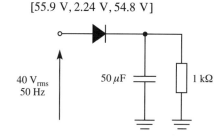

3.5   A full-wave bridge rectifier operates from 110 V_{rms}, 60 Hz line voltage through a 5:1 step-down transformer. It is required to supply 24 V dc at 100 mA. Determine the filter capacitor required and the percentage ripple.

[91 μF, 13.3%]

NB Calculating the capacitor from the percentage ripple will give a value of 71 μF. The error results from the assumption that $V_P = V_{dc}$.

3.6 Determine the maximum load current for a half-wave rectifier supplied with 50 Hz line voltage, which has a dc voltage of 12 V and a smoothing capacitor of 1000 μF, if the percentage ripple is not to exceed 2%.

[41.5 mA]

3.7 A half-wave rectifier with a 500 μF capacitor and a 200 Ω load operates from a 50 V_{rms}, 50 Hz supply. If the capacitor recharges during each cycle in 5 ms estimate the peak current rating of the diode.

[1.27 A]

3.8 Repeat Problem 3.7 for a full-wave rectifier.

3.9 Use PSpice to investigate the half-wave rectifier shown. Set the source to 24 V peak with a frequency of 50 Hz and examine the output voltage across the load resistance and the current through the diode. Increase the value of the load resistance to 2 kΩ.

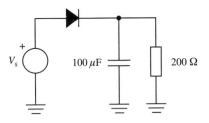

3.10 Sketch the output waveforms for each of the clipping circuits shown below and verify with the aid of PSpice.

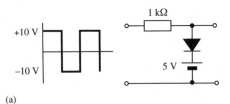

(a)

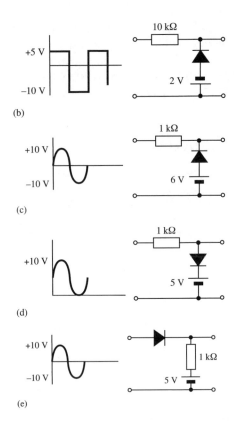

(b)

(c)

(d)

(e)

3.11 Sketch the output waveforms for each of the clamping circuits shown and verify with the aid of PSpice.

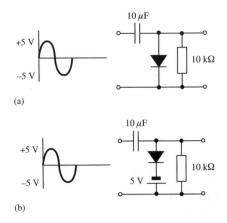

(a)

(b)

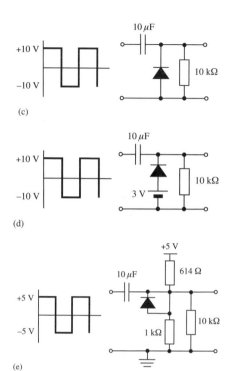

(c)

(d)

(e)

3.12 Use PSpice to examine the output voltage of the voltage tripler shown, if the input is a sine wave with an amplitude of 50 V and frequency of 1 kHz. Determine the polarity of the dc output voltage with respect to the two output terminals A and B, and the value of the output voltage. (Simulate for at least 10 cycles of the 1 kHz waveform.)

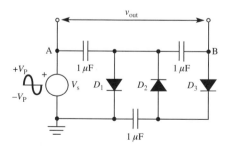

3.13 Add another diode and capacitor to the circuit for Problem 3.12 to produce a voltage quadrupler and use PSpice to verify the output.

3.14 For the simple Zener diode regulator as illustrated in Figure 3.23 the minimum Zener current is 5 mA, the maximum value is 20 mA, $R_s = 270\ \Omega$, $V_Z = 6.8$ V and $R_Z = 0\ \Omega$. Determine the minimum and maximum values of the dc input voltage which can be regulated.

[8.2 V, 12.2 V]

3.15 Repeat Problem 3.14 if the Zener diode internal resistance (r_Z) is 10 Ω, and determine the variation of v_{out} and compare the value obtained with that predicted by equation 3.21.

[6.85 V–7 V, $\Delta v_{out} = 0.155$ V

3.16 For the regulator shown in Figure 3.25 which includes a load resistor, $I_{Zmin} = 2$ mA, $I_{Zmax} = 40$ mA, $V_{DCin} = 30$ V. Assuming that $r_Z = 0\ \Omega$, determine the minimum and maximum values of the load resistance.

[146 Ω, 381 Ω]

3.17 For the simple Zener diode regulator as illustrated in Figure 3.25 the dc input voltage is 30 V, the Zener voltage is 12 V, $R_s = 100\ \Omega$, $R_Z = 0\ \Omega$ and the load resistance varies from 100 Ω to 250 Ω. Determine

(a) the minimum and maximum Zener diode currents
(b) the power rating for the diode
(c) the minimum power rating for the series resistor.

[(a) 60 mA, 132 mA,
(b) 1.58 W, (c) 3.2 W]

3.18   For Problem 3.17 determine the Zener diode current if the load is accidently disconnected.

[180 mA]

3.19   Repeat Problem 3.17 if in addition the dc input voltage varies from 25 V to 35 V.

[(a) 10 mA, 182 mA, (b) 2.2 W, (c) 5.3 W]

3.20   A 5.6 V Zener diode has a maximum power rating of 500 mW. The minimum sustaining diode current is 3 mA. Determine a suitable range of values for R_s if the dc input voltage varies from 10 V to 15 V and the load resistance varies from 400 Ω to 800 Ω. Assume that $R_Z = 0$ Ω.

[98 Ω–259 Ω]

3.21   Use PSpice to simulate the Zener diode regulator shown in Figure 3.25 for which the dc input is a nominal 12 V, but which varies from 10 V to 14 V. Use the Zener diode D1N750 from the **eval.slb** for which $V_Z = 4.7$ V. The series resistance is 200 Ω and the load resistance is 300 Ω. For the input use the sine wave source (VSIN) with voff = 12 V, vampl = 2 V and freq = 50 (or 60) and use transient analysis for at least 2 cycles. Use Probe to compare the input and output waveforms.

CHAPTER 4

Bipolar Junction Transistor

4.1 Introduction

The bipolar junction transistor (BJT) is a three-terminal current-operated semiconductor device. It has a low input impedance and a high output impedance and when the same current flows in both input and output it can provide power gain through virtue of the different impedance levels. A small selection of transistors is shown in Figure 1.2 in Chapter 1.

Transistors come in all shapes and sizes to satisfy a wide variety of needs. The most obvious difference is the size. Transistors which are required to dissipate several watts are large in order to dissipate heat. In the majority of cases the actual piece of semiconductor material within the metal package is considerably smaller than the package. The large package is required to increase the surface area, and thus improve the flow of heat away from the active semiconductor chip. In many applications the surface area is further increased by attaching the transistor package to a metal heat sink.

There are many hundreds, if not thousands, of different transistor types, many of which have very similar characteristics. There is the obvious difference of power rating, where the transistor properties are optimized to provide the maximum current gain at the required current and voltage levels. The currents can range from a few milliamps to many tens of amps. Voltage breakdown levels also need to be satisfied. Typically a pn junction will break down at a few tens of volts, but for high-power applications the device may be required to operate at hundreds of volts, for example motor control circuits for consumer products may operate directly from the ac line voltage. Transistors can be optimized for low-noise applications, for use in high-quality pre-amplifiers which are required to amplify very small signals. Another parameter of interest is the frequency response. Most low-power, small-signal transistors are able to amplify signals at frequencies of hundreds of MHz, but with careful design transistors can be used to amplify microwave signals at thousands of MHz. For these purposes the packaging is often more important than the silicon chip, and special subminiature packages are available to reduce stray capacitance and lead inductance.

The important factors in selecting a transistor are gain, maximum collector current, maximum collector voltage, maximum power rating and maximum frequency, or switching speed for switching applications. There are many other parameters, but for most applications these are probably the most important.

For many analog applications an integrated circuit operational amplifier may be a preferred choice rather than a discrete transistor circuit. Provided that an operational amplifier is available for the frequency range required and is of adequate power rating, then it may be easier to incorporate into the design, and because it is likely to require fewer components, the final product will be cheaper to manufacture. Operational amplifiers are now available for a very wide range of frequencies and power ratings and are used extensively in electronic circuits. However, transistors are still generally required for very high frequencies (>100 MHz), and for medium to large power ratings (>10 W), but new analog integrated circuits are continually being developed to replace circuits which may have originally used discrete components.

However, even for operational amplifiers it is necessary to understand how discrete component transistor circuits operate in order to design and develop new operational amplifiers, and of course to understand how operational amplifiers operate.

This chapter deals with the operation of the transistor in terms of the materials used to construct it, and the formation of pn junctions. The relationship between the current in the three terminals is explained in terms of current gains. Finally there is an important section on the electrical *equivalent circuit* which is required for circuit analysis and computer simulation. It is not strictly necessary to know how a transistor is manufactured or how it operates in order to design it into circuits, so that the section on transistor operation may be regarded as optional. Also it is not necessary to know about the more complex equivalent circuits which are required for computer simulation, and so these sections are also optional.

4.2 Transistor Operation

This section deals with the physical operation of the transistor in terms of current flow in p- and n-type semiconductors, and although greatly simplified, is not strictly necessary for an understanding of the electrical operation and circuit applications of the transistor. Apart from Figure 4.1, which identifies the transistor symbols, the remainder of this section is optional.

Physical Structure (optional)

The bipolar transistor consists of a sandwich of n-, p- and n-type semiconductor material (or p-, n- and p-type), as shown in Figure 4.1.

The starting material for the npn transistor is a wafer of n-type silicon into which a p-type impurity (boron) is introduced at high temperature by the process of diffusion. This results in the impurity profile shown in Figure 4.2(a).

The graph shows the variation of the impurity concentration with distance from the silicon surface. The background impurity concentration of the wafer (N_c) may be 1×10^{15} atoms cm^{-3} while that of the boron at the surface (N_b) may be 1×10^{17} atoms cm^{-3}. The point at which the boron profile intersects the background concentration is the position of the base–collector junction. From the surface to this point the concentration of the boron exceeds that of the background and, therefore, this region is p-type. Next an n-type impurity (phosphorus or arsenic) is added by the same high-

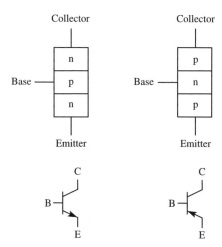

FIGURE 4.1   Schematic of structure and symbols for the npn and pnp bipolar transistors.

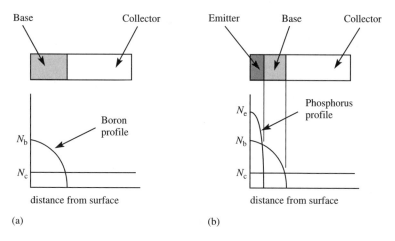

(a)   (b)

FIGURE 4.2   Simplified diagram showing the stages in the fabrication of an npn transistor.

temperature process so that the concentration at the surface becomes n-type (N_e) with a value of approximately 1×10^{20} atoms cm^{-3}. The point at which the phosphorus profile in Figure 4.2(b) crosses the boron profile is the position of the emitter–base junction.

The junction depth for the base may typically be 1–2 micrometres (1 micrometre is 1×10^{-6} m), while that of the emitter–base junction may be 0.5–1 micrometre. Thus the width of the base region may be 0.5-1 micrometre. These dimensions will obviously depend on the end use of the transistor. For high-current, high-voltage devices the dimensions will be considerable larger, but the base width will not change too much. For high-frequency devices the dimensions, particularly the base width, will be much smaller. The distribution of impurities in the collector region may also be more complex than the uniform distribution shown in Figure 4.2. In order to avoid too large a resistance in series with the collector, which affects high-frequency performance and power dissipation in high-power devices, the n-type impurity concentration 5–10 micrometres from the surface is increased to lower the resistivity from this point to the collector contact on the back surface of the wafer (see Figure 4.5 below).

Operation (optional)

In normal operation the emitter–base junction is forward biased, while the base–collector junction is reverse biased. The polarities of the bias voltages for an npn transistor are shown in Figure 4.3.

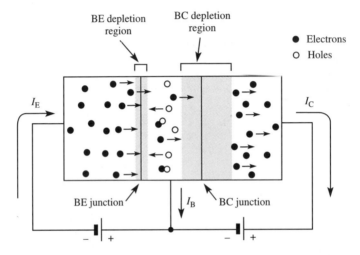

FIGURE 4.3   An NPN transistor with forward-biased emitter and reverse-biased collector.

When a forward bias is applied to the base–emitter junction, the negative voltage applied to the emitter drives the free electrons in the emitter towards the base–emitter junction, where they cross into the base region. Some holes in the base are attracted to the negative voltage on the emitter and cross the junction in the opposite direction. These electrons and holes form the emitter current. However, because there are many more electrons in the emitter than there are holes in the base, the current is carried predominantly by the electrons. When the electrons reach the base they diffuse across towards the collector junction, and on reaching the edge of the base–collector depletion layer they are accelerated by the electric field in the depletion layer to the collector, where they form the collector current. Thus electrons enter at the emitter contact, are injected into the base, diffuse across the base, are accelerated across the collector depletion layer and return to the external circuit as collector current. The current flow shown in Figure 4.3 is electron flow.

NB By convention the 'positive' current flow is in the opposite direction.

The small number of holes which cross into the emitter from the base forms an unwanted base current, since they do not contribute to the current in the collector circuit. In addition some of the electrons which enter the base from the emitter collide with holes which exist in the base, and they too contribute to the base current. These two components of base current have a controlling influence on the current gain of the transistor. The former can be minimized by ensuring that the number of electrons in the emitter is much greater than the number of holes in the base. This is achieved by making the resistivity of the emitter much smaller than that of the base, by controlling the impurity concentrations in the two regions as illustrated in Figure 4.2(b). The second effect, recombination in the base, is minimized by making the base-width small to reduce the possibility of collision and to enable the electrons to travel through this region as quickly as possible.

In addition to the currents described above there is also the leakage current across the reverse-biased collector junction (I_{CBO}). The small number of holes which are always present in n-type material and the small number of electrons in the p-type base material are naturally attracted across the base–collector junction by the electric field which exists there. These carriers contribute to the collector current but are independent of the base–emitter voltage. The currents are illustrated in Figure 4.4 where the width of the shaded areas corresponds to the relative magnitudes of the different components.

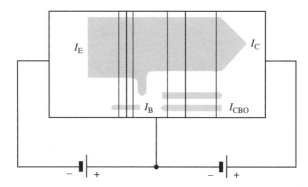

FIGURE 4.4   Current flow in an npn transistor.

The base current consists of the hole current flowing from the base into the emitter and the small fraction of electrons which recombine with holes when crossing the base. The leakage current I_{CBO} is measured with the emitter open circuit so that the only carriers flowing are the holes and electrons flowing across the base–collector junction which form the leakage for the reverse-biased base-collector junction.

Practical Structure (optional)

In Figures 4.3 and 4.4 it would appear that the transistor is formed in a small cube of silicon, whereas in practice many hundreds of transistors are fabricated on silicon wafers with the emitter and base regions being formed by a high-temperature diffusion process, which results in the impurity profiles shown in Figure 4.2. The regions of silicon which form the emitter and base are defined by photolithography and a plan view and resultant cross-section of typical small-signal npn transistor device is shown in Figure 4.5.

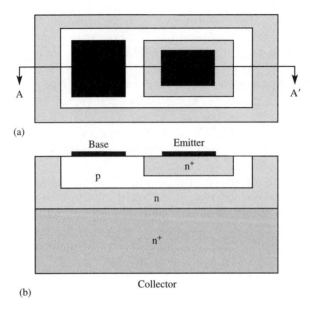

(a)

(b)

FIGURE 4.5   (a) Plan view of an npn bipolar transistor and (b) the cross-section.

The geometrical shapes are obtained by photographic methods and the simple rectangular shape shown would be typical for a low-power small-signal transistor. Other geometries are possible and complex interdigitated structures are used for power transistors. An important parameter is the base width, and as previously shown in Figure 4.2 it is typically less than $0.5\,\mu m$ $(1\,\mu m = 10^{-6}\,m)$. The wafer may be $200-250\,\mu m$ thick and in order to minimize the resistance of the silicon between the active region immediately beneath the emitter and the collector contact on the rear of the wafer, the bulk of the wafer is very low-resistance material, with only a thin layer on the surface of high-resistance material required for the operation of the transistor.

4.3 Transistor Parameters

The two most common circuit configurations are the common-base and the common-emitter as illustrated in Figure 4.6.

The relationship between the emitter, base and collector currents is:

$$I_E = I_B + I_C \tag{4.1}$$

For good transistor design the base current is small so that $I_E \approx I_C$.

Common-Base Current Gain

For the common-base circuit, Figure 4.6(a), $I_E \approx I_C$ or more precisely:

$I_C = \alpha I_E$ where α is the *common-base current gain*

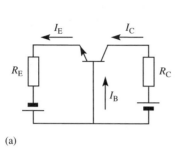
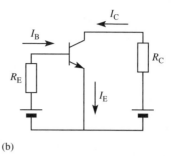

(a)   (b)

FIGURE 4.6   (a) Common-base and (b) common-emitter circuits with 'conventional' current flow.

and

$$\alpha = \frac{I_C}{I_E} \tag{4.2}$$

If the collector leakage current is included then the collector current is:

$$I_C = \alpha I_E + I_{CBO} \tag{4.3}$$

In practice the leakage current for a silicon transistor is very small, but like the leakage current of a reverse-biased diode it increases with increasing temperature, approximately *doubling for every 10°C rise in temperature*, and may become important for a power-output stage where the junction temperature may by between 100 and 150°C.

For a good transistor α is very close to unity, and values of between 0.95 to 0.995 are typical.

Common-Emitter Current Gain

Using the relationship:

$$I_C - I_{CBO} = \alpha I_E \text{ from equation 4.3}$$

and dividing through by α gives:

$$\frac{I_C}{\alpha} - \frac{I_{CBO}}{\alpha} = I_E$$

Substituting $I_B + I_C = I_E$ and collecting together the terms in I_C gives:

$$I_C\left(\frac{1}{\alpha} - 1\right) = I_B + \frac{I_{CBO}}{\alpha}$$

or

$$I_C = \frac{\alpha I_B}{(1-\alpha)} + \frac{I_{CBO}}{(1-\alpha)} \tag{4.4}$$

If a value of 0.995 is assumed for α, then the collector current is:

$$I_C = 199I_B + 200I_{CBO}$$

The collector current is proportional to the base current and the constant of proportionality is known as the *common-emitter gain* β, where β is given by:

$$\beta = \frac{\alpha}{(1-\alpha)} \tag{4.5}$$

The other term is the collector to emitter leakage current (I_{CEO}) with the base open circuit, that is:

$$I_{CEO} = \frac{I_{CBO}}{(1-\alpha)} \tag{4.6}$$

and

$$I_C = \beta I_B + I_{CEO} \tag{4.7}$$

The *common-emitter leakage current* is considerably larger than the *common-base leakage current*, but is still much smaller than the collector current and is only of significance for power devices.

▶ **EXAMPLE 4.1** _____

Determine α, β and I_E if the collector current is 2.5 mA for a base current of 22.5 μA.

▶ *Solution*

$$\beta = \frac{I_C}{I_B} = \frac{2.5 \text{ mA}}{22.5 \text{ μA}} = 111$$

$$\alpha = \frac{\beta}{(1+\beta)} = \frac{111}{112} = 0.991$$

$$I_E = \frac{I_C}{\alpha} = \frac{2.5 \text{ mA}}{0.991} = 2.52 \text{ mA}$$

Notice that $I_C \approx I_E$. For dc analysis this is generally a good approximation, and if one or other current is determined then it is usually sufficient to assume that the other is equal in value.

DC Analysis

With the aid of equation 4.7 it is now possible to perform a simple dc analysis of a transistor circuit. Consider the common-emitter circuit shown in Figure 4.7.

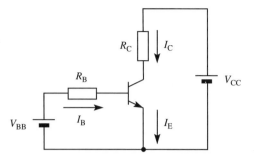

FIGURE 4.7   Common-emitter stage with dc voltages and currents.

The voltage across the *forward-biased base–emitter junction is approximately 0.7 V* so that applying KVL to the base circuit results in the relationship:

$$V_{BB} = I_B R_B + V_{BE}$$

and

$$I_B = \frac{V_{BB} - V_{BE}}{R_B}$$

Now $I_C \approx \beta I_B$ from equation 4.7 if $I_{CEO} \ll \beta I_B$ and

$$I_C = \frac{\beta(V_{BB} - V_{BE})}{R_B}$$

or

$$I_C = \frac{V_{BB} - V_{BE}}{R_B/\beta} \tag{4.8}$$

This equation provides the means for calculating the collector current based on the conditions in the base circuit. It is this type of analysis which forms the basis for determining the dc conditions in a transistor circuit.

Apart from the collector current it is also useful to determine the voltage across the collector and emitter of the transistor (V_{CE}). For the circuit shown, application of KVL to the collector–emitter circuit gives:

$$V_{CC} = I_C R_C + V_{CE}$$

or

$$V_{CE} = V_{CC} - I_C R_C \tag{4.9}$$

▶ **EXAMPLE 4.2** ——————————————————————————

For the circuit shown in Figure 4.7 $R_B = 100 \text{ k}\Omega$, $R_C = 2 \text{ k}\Omega$, $V_{BB} = 3 \text{ V}$, $V_{CC} = 9 \text{ V}$ and $\beta = 120$. Determine the collector current and V_{CE}.

▶ *Solution*

$$I_B = \frac{V_{BB} - V_{BE}}{R_B} = \frac{3 - 0.7}{100k} = 23 \ \mu A$$

$$I_C = \beta I_B = (120)(23 \ \mu A) = 2.76 \ mA$$

$$V_{CE} = V_{CC} - I_C R_C = 9 - (2.76 \ mA)(2k) = 3.48 \ V$$

Common-Emitter Output Characteristics

The relationship between the collector current and the collector–emitter voltage is important because it provides a visual picture of the way in which the dc conditions are likely to affect the output signal when the transistor is used as an amplifier. This variation can be visualized with the aid of a *load-line*. The load-line can be obtained with the aid of PSpice, or by hand if a set of output characteristics is available for a particular transistor. The circuit schematic for PSpice is shown in Figure 4.8. The transistor is the generic type from the breakout library and the gain is set to 120.

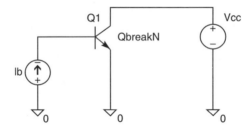

FIGURE 4.8   PSpice circuit schematic for obtaining the output characteristics.

The current–voltage characteristics are first obtained with a nested **DC Sweep...** with the first level of nesting being the variation of V_{CC} from 0 to 9 V in 0.05 V steps, and the second level being the current source I_B from 5 to 40 µA in 5 µA steps. The characteristics are shown in Figure 4.9. The x axis represents the collector–emitter voltage V_{CE}.

When a collector resistance (R_C) is included in the circuit as in Figure 4.7, then a load-line can be superimposed on the output characteristics as shown in Figure 4.10. This is obtained from the Probe data for the circuit shown in Figure 4.8, by plotting the following function (from equation 4.9):

$$4.5 \ mA = V_{CE}/2k$$

where the collector resistance is 2 kΩ, and 4.5 mA is the maximum collector current when $V_{CE} = 0$ V, that is $I_{Cmax} = 9 \ V/2 \ k\Omega$.

Note that if a set of collector output characteristics is available, then the load-line is constructed by drawing a straight line from $I_C = 0$, $V_{CE} = V_{CC}$ having a slope of $-1/R_C$.

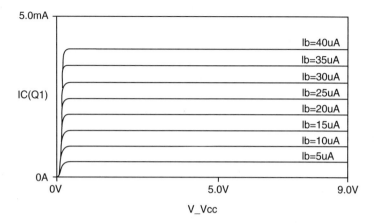

FIGURE 4.9   Collector output characteristics.

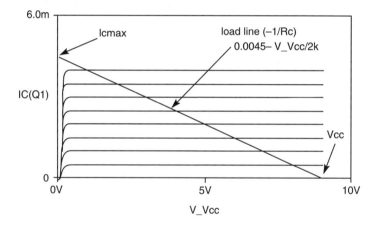

FIGURE 4.10   Output characteristics with a superimposed load-line.

Saturation and Cut-off

The load-line shown in Figure 4.10 assumes that a collector resistance R_C is present in the circuit. The collector–emitter voltage is given by equation 4.9 as:

$$V_{CE} = V_{CC} - I_C R_C$$

For a given value of resistance the collector–emitter voltage will depend on the value of the collector current, which in turn is determined by the base current. If the base current is large, then I_C is large and V_{CE} is very small. In Figure 4.10 this corresponds to I_C approaching $I_{Cmax} = 4.5$ mA as V_{CE} approaches 0 V. In this condition with I_C large and V_{CE} very small the transistor is said to be in *saturation*.

If the current in the base is zero, then the collector current is reduced to the leakage current I_{CEO} from equation 4.3. Under these conditions the collector current is very small and there is very little voltage drop across R_C; the collector–emitter voltage is approximately equal to V_{CC} and the transistor is said to be *cutoff*.

The load-line describes graphically the way in which the collector–emitter voltage varies with the collector current when a collector resistor is present. The two extremes are represented by the intercepts with the x and y axis, as *cut-off* and *saturation* respectively.

Early Voltage

In Figure 4.9 the variation of the collector current for values of V_{CE} greater than about 1 V remains constant; with further increase of V_{CE}, the characteristic curves are horizontal. These characteristics were obtained for a generic, or idealized, transistor. For a real transistor this is not the case, and the current increases slightly with increasing V_{CE}. The effect can be described by means of the *Early voltage*, after J. M. Early, who first investigated the effect. The Early voltage is used in computer simulation programs and the effect can be observed by setting the value of VAF in the **model...** term under **Edit** for the transistor in PSpice and rerunning the simulation to obtain the $I–V$ characteristics with the Early voltage present. The results are shown in Figure 4.11 for a value of VAF of 50 V, with some added construction lines.

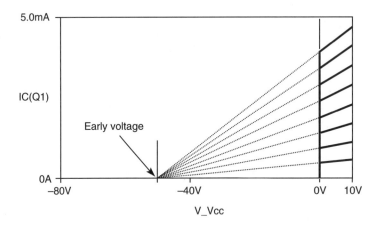

FIGURE 4.11   DC characteristics with the Early voltage set to 50 V.

In Figures 4.9 and 4.10 the variation of collector current with V_{CE} for $V_{CE} > 1$ V is shown as a set of horizontal straight lines, that is the collector current remains at a constant value for each value of base current.

In Figure 4.11 the $I–V$ curves are present in the positive voltage portion of the graph, from 0 V to 10 V, on the right-hand side of the diagram. Note that the collector current is no longer constant with V_{CE}, but increases with increasing V_{CE}. Additional construction lines (dotted) have been added to enable the slope of each curve to be projected back to a point on the negative voltage axis, where they intersect the axis at −50 V. It can be seen that the slope of the $I_C–V_{CE}$ curves can be constructed by projecting a series of lines from the Early voltage across the current axis. In a real device the intersection with the Early voltage is not so precise, but is sufficiently accurate to allow the effect to be described by a mathematical expression for use in a computer simulation program. This is discussed in more detail later in this chapter. The intersection corresponds to the value of the Early voltage used in the **model...** statement. In practical devices the Early voltage is >100 V. The output resistance of the transistor can be expressed as the slope of the I_C versus V_{CE} curve and it can be obtained from a knowledge of the Early voltage. The higher the voltage, the higher the output resistance.

▷ **EXAMPLE 4.3** ————————————————————————

Estimate the output resistance of a transistor for which the collector current is 100 µA at $V_{CE} = 5$ V and the Early voltage is 110 V.

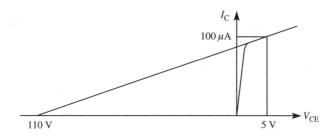

▷ *Solution*

The line which extends from 110 V through the operating point of 100 µA and 5 V describes the output resistance as:

$$slope = \frac{(110 + 5)}{100 \ \mu A} = 1.15 \ M\Omega$$

or approximately:

$$slope = \frac{VAF}{I_C} = \frac{110}{100 \ \mu A} = 1.1 \ M\Omega$$

Knowledge of the Early voltage provides a means for estimating the output resistance as:

$$r_o = \frac{VAF}{I_C} \qquad\qquad (4.10)$$

4.4 ▷ Maximum Ratings

The reverse breakdown voltages of the emitter–base and collector–base junctions are specified in the manufacturer's data sheets. Typically the emitter–base breakdown voltage is between 7 and 10 V. This is determined by the high concentration of impurities in the base next to the emitter, as shown in Figure 4.2. The breakdown voltage of the collector–base junction is much higher and can range from several tens of volts to hundreds of volts. The external voltages which are applied to the transistor must not exceed these voltages.

In addition to the breakdown voltage another important parameter is the amount of power which can be dissipated. The power dissipation is defined as the product of V_{CE} and I_C as:

$$P_{diss} = V_{CE}I_C \text{ watts} \tag{4.11}$$

For a given power dissipation the product of V_{CE} and I_C must not exceed this value.

▷ **EXAMPLE 4.4** ──────────────────────────

A transistor has a power rating of 250 mW and is required to operate with $V_{CE} = 5$ V. Determine the maximum collector current which it can handle and the minimum collector load resistance.

▷ *Solution*

$$I_{Cmax} = \frac{P_{diss}}{V_{CE}} = \frac{250 \times 10^{-3}}{5} = 50 \text{ mA}$$

The collector load is given by:

$$R_{load} = \frac{V_{CE}}{I_C} = \frac{5}{50 \text{ mA}} = 100 \text{ } \Omega$$

Transistors which are intended for use as power amplifiers usually have a derating factor specified in the data sheet for temperatures above 25°C. Thus if the device is operating in an enclosure in which the temperature can rise above 25°C, the room-temperature power rating must be reduced by the derating factor.

▷ **EXAMPLE 4.5** ──────────────────────────

A 10 W medium-power transistor has a derating factor of 57 mW °C^{-1}. Determine the power dissipation and maximum collector current if $V_{CE} = 15$ V at an operating temperature of 70°C.

▷ *Solution*

The change in temperature is:

$$\Delta T = 70 - 25 = 45°C$$

Above 25°C the power rating must be derated at the rate of 57 mW °C^{-1}. Thus the change in power is:

$$\Delta P = (57 \text{ mW } °C^{-1})(45) = 2.56 \text{ W}$$

and the power dissipation at 70°C is:

$$P_{\text{diss}}(70°C) = 10 \text{ W} - 2.56 \text{ W} = 7.44 \text{ W}$$

and the maximum collector current is:

$$I_{\text{Cmax}} = \frac{P_{\text{diss}}(70°C)}{V_{\text{CE}}} = \frac{7.44}{15} = 0.496 \text{ A}$$

The power rating is considered further in Chapter 14.

4.5   Equivalent Circuits

For dc analysis it is sufficient to assume that there is a voltage drop of approximately 0.7 V across the base–emitter junction and that the current in the collector circuit is βI_B, as illustrated in Example 4.2. However, for small-signal analysis it is necessary to include the resistive and capacitive effects associated with the bulk semiconductor material and the pn junctions. The transistor needs to be replaced by an *equivalent electrical circuit* so that mesh and nodal analysis can be used to determine the small-signal currents and voltages. A simple equivalent circuit is required for hand calculations to allow a circuit designer to assess quickly the properties of a particular circuit, and a more complex circuit is required for use in computer simulation software. For computer simulation the circuit model should be as accurate as possible to ensure that the simulated results are a good representation of the real transistor in an actual circuit. Because the calculations are performed by computer, the complexity of the resultant equivalent circuit is not a serious problem, although if it is too complex it can increase the time required to simulate a circuit containing many transistors.

For bipolar transistors the computer simulation models are based on the *Ebers–Moll* and *Gummel–Poon* models. A brief description of these models is given below, but this material may be regarded as optional, because these models are only of value for computer simulation, and as such are made available to the user as a library of models for a wide variety of transistors. Knowledge of these models is of value when it becomes necessary to modify or create an equivalent circuit for a transistor which may not be available in a device library.

Simple Equivalent Circuit

There are a large number of possible equivalent circuits for the bipolar transistor ranging from *y parameters*, *h parameters*, *r parameters*, *hybrid π*, *equivalent Tee* to the Ebers–Moll and Gummel–Poon models which are described in more detail below. Many of the early equivalent circuits were developed from the 'black box' concept with input and output terminals as shown in Figure 4.12.

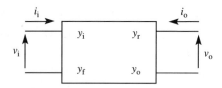

FIGURE 4.12   Network representation of a transistor using y parameters.

The operation ·of the network contained within the box is then described in terms of the terminal currents and voltages. Thus for the y parameters the equations for the network are:

$$i_i = y_i v_i + y_r v_o$$
$$i_o = y_f v_i + y_o v_o \qquad (4.12)$$

while for a set of h parameters the equations for the network are:

$$v_i = h_i i_i + h_r v_o$$
$$i_o = h_f i_i + h_o v_o \qquad (4.13)$$

where the subscripts i, r, f, o refer to input, reverse, forward and output, respectively, and are used to identify the current, voltage, admittance, impedance or gain factor in the network which is used to represent the transistor. The network, or as in this case, the transistor, is described in terms of either a set of y parameters or a mixed set of admittance, impedance and dimensionless h parameters. For the admittance parameters the subscripts refer to the input admittance (y_i), the reverse-transfer admittance (y_r), the forward-transfer admittance (y_f) and the output admittance (y_o). For the h parameters they refer to the input impedance (h_i), the reverse voltage feedback (h_r), the forward current gain (h_f) and the output admittance (h_o).

Values for the y or h parameters may be obtained from measurements of the small-signal currents and voltages performed on the transistors with the appropriate dc bias applied. They are often quoted in manufacturers' data sheets. The h parameters are probably the most common and the current–voltage relationships in equations 4.13 can be represented as an electrical circuit, as shown in Figure 4.13.

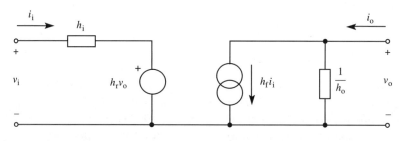

FIGURE 4.13   Equivalent circuit based on the h parameters.

It can be seen that the input consists of an impedance (h_i) in series with a vol. controlled voltage source ($h_r v_o$). Thus by applying KVL to the input circuit in Fig 4.13 the input voltage can be written as:

$$v_i = h_i i_j + h_f v_o$$

For the output there is a current generator ($h_f i_i$) and a shunt impedance, and the output current can be written as:

$$i_o = h_h i_j + h_o v_o$$

The circuit shown in Figure 4.13 is an accurate representation of equations 4.13.

In practical devices the reverse transfer voltage parameter h_r is very small so that the voltage which is generated in the input by this parameter can be ignored. In terms of the physical transistor this means that variations of the collector voltage have little influence on the flow of current across the base–emitter junction. The output admittance can also be ignored for most practical transistor circuits, because the resulting resistance is usually greater than 50 kΩ and as will be seen in a later chapter on small-signal analysis the presence of this resistor in the equivalent circuit does not greatly affect the analysis. This output resistance has already been explained in terms of the Early voltage. Thus the equivalent circuit reduces to two components, as shown in Figure 4.14.

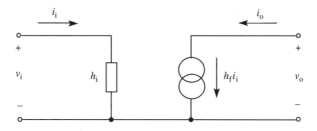

FIGURE 4.14   Simplified h parameter equivalent circuit.

r Parameters

The h parameters are derived from a network representation of the transistor and are not directly related to the physical structure of the transistor in terms of forward-biased junctions and current flow from the emitter to the collector. The r parameter circuit is an attempt to associate the physical processes in the transistor to electrical components. The main components are resistors to represent both bulk resistance within the transistor and the resistance of a forward-biased base–emitter junction. The current gain is represented by a current generator. The r parameter equivalent circuit for the common-emitter configuration is shown in Figure 4.15 together with the ideal transistor symbol to show how the parameters relate to the structure of the transistor.

The base resistance r_b represents the resistance of the semiconductor material between the active region immediately beneath the emitter, and the base contact. The forward-biased base–emitter junction is represented by the resistor r_e, which is

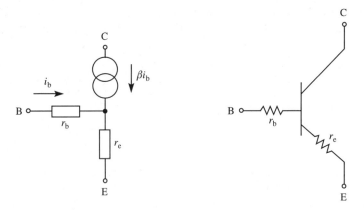

FIGURE 4.15   Relationship between the *r* parameters and the transistor symbol.

obtained from the slope of the I_E–V_{BE} graph. The value is the same as that obtained for the forward-biased diode at 25°C (equation 2.13) as:

$$r_e = \frac{26 \text{ mV}}{I_E} \tag{4.14}$$

The base resistance r_b is typically 10–$100 \ \Omega$ for small-signal transistors and with base currents of a few tens of microamps the dc voltage drop across r_b is very small. For simple hand calculations it can be ignored. Thus the *r* parameter equivalent circuit for simple hand calculations reduces to two components, the resistance of the forward-biased base–emitter junction and a current generator βi_b to represent the collector current which is generated when base current flows, as shown in Figure 4.16.

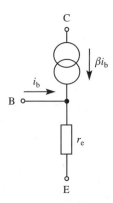

FIGURE 4.16   Simplified *r* parameter equivalent circuit for the common-emitter configuration.

This circuit is adequate for simple hand calculations to determine voltage gain or input impedance.

The simplified h parameter circuit of Figure 4.14 could also have been used, but for many applications the r parameter circuit is slightly easier to use, and for the remainder of this text it will be used for hand calculations.

▷ **EXAMPLE 4.6** ────────────────────────────

Repeat the exercise described in Example 4.2 where $V_{BB} = 3$ V, $R_B = 100$ kΩ, $R_C = 2$ kΩ, $V_{CC} = 9$ V, $\beta = 120$, and in addition an ac input signal of 1 sin ωt (this represents a small-signal) is inserted in series with V_{BB} as shown in Figure 4.17. Determine the circuit amplification.

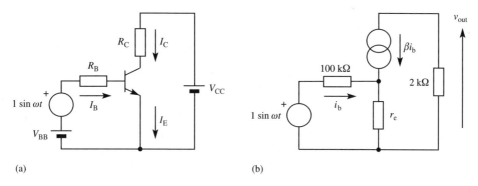

(a)   (b)

FIGURE 4.17   Common-emitter amplifier with (a) the dc bias circuit with an ac signal source in the base circuit and (b) the small-signal circuit.

▷ *Solution*

From Example 4.2 the collector current is 2.76 mA.

In Figure 4.17(b) the dc sources are replaced by short circuits and the transistor is replaced by its equivalent circuit. The small-signal base–emitter resistance is:

$$r_e = \frac{26 \text{ mV}}{I_E} = \frac{26 \text{ mV}}{2.76 \text{ mA}} = 9.42 \ \Omega$$

Applying KVL to the base circuit gives:

$$1 \sin \omega t = i_b (100 \text{ k}\Omega + 9.42 \ \Omega) + \beta i_b 9.42$$

$$i_b = \frac{1 \sin \omega t}{100 \text{ k}\Omega + (1 + \beta)9.42 \ \Omega} \approx 10 \times 10^{-6} \sin \omega t$$

For the output circuit:

$$v_{out} = -\beta i_b 2 \text{ k}\Omega = -(120)(10 \times 10^{-6} \sin \omega t)(2 \text{ k}\Omega) = -2.4 \sin \omega t$$

and thus the voltage gain is:

$$\frac{v_{out}}{v_{in}} = \frac{-2.4 \sin \omega t}{1 \sin \omega t} = -2.4$$

The current gain is:

$$\frac{i_c}{i_b} = \frac{\beta i_b}{i_b} = 120$$

Notice the negative sign for the output voltage which results from the sign convention adopted in Figure 4.17. It indicates that there is a phase change of 180° between the input and the output. This will be considered again in Chapter 6 when small-signal analysis is considered in much more detail.

Hybrid π Model (optional)

The simple h parameter model (Figure 4.14) or the r parameter model (Figure 4.16) are perfectly adequate for hand-based calculations for low-frequency analysis. However, capacitive effects are associated with the pn junctions, and there are the equivalent of capacitive delays incurred in the transport of charge carriers from the emitter to the collector through the base. These capacitive effects have a considerable influence on the high-frequency performance of a transistor amplifier. An equivalent circuit which is suitable for use with hand-based calculations for high frequencies is the *hybrid π*, as shown in Figure 4.18. The circuit is based on a development of the simplified h parameter model, where $r_{b'e}$ is approximately equal to h_{ie}, where h_{ie} is the input impedance for the common-emitter. In this circuit there is an internal base contact, B′, for the ideal transistor, and an external base contact, B, which represents the actual terminal to the base. Between these two there is a resistance $r_{bb'}$ which represents the ohmic resistance between the ideal transistor and the actual base connection. It is typically 10 Ω to 100 Ω. The resistance r_o is equivalent to $1/h_o$ in the h parameter equivalent circuit, and can often be neglected, because it is generally much larger than the external load. High-frequency effects are catered for with the addition of two capacitors which represent the junction capacitance of the base–emitter and the base–collector.

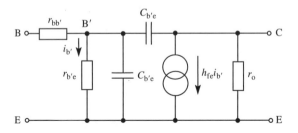

FIGURE 4.18   Hybrid π model of the bipolar transistor.

Further information on the hybrid π may be found in Appendix 4, and there is further reference to it in Chapter 10, when the high-frequency performance of bipolar transistors is considered.

Ebers–Moll (optional)

For hand calculations it is adequate to assume that the dc voltage drop across the forward-biased base–emitter junction is 0.7 V and that the small-signal model consists of a current generator (βi_b) and a small-signal resistance (r_e). However, such a model is far too simple for the accurate simulation of complex circuits, such as an operational amplifier, where the overall gain may be 100000 or more, and where a small error in the simulated dc voltage or current in the input stages will be amplified to such an extent that the output transistors will probably be cut off or in saturation. For accurate computer simulation a much more detailed equivalent circuit is necessary.

The Ebers–Moll model starts from the basis that the transistor consists of two diodes which are back to back, as shown in Figure 4.19.

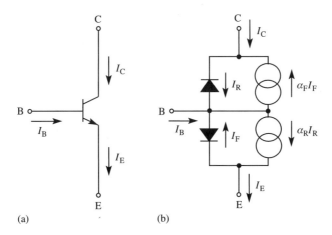

FIGURE 4.19 Basic Ebers–Moll equivalent circuit of an npn transistor with the transistor symbol in (a) and the equivalent circuit in (b).

The current generator in the collector ($\alpha_F I_F$) represents the electron current which is injected by the emitter. There is also a current across the reverse-biased base–collector junction represented by I_R and, therefore, the collector current is:

$$I_C = -\alpha_F I_F + I_R \tag{4.15}$$

A positive current convention is adopted where the collector current flows into the collector, whereas in reality electrons flow out of the collector. This apparent contradiction does not matter provided that a consistent sign convention is used throughout. Some of the holes which form part of the current I_R will diffuse across the base to the collector and form the current in the emitter $\alpha_R I_R$, where α_R is the reverse common-base current gain factor (the current gain which would be measured if the transistor terminals were reversed). The manufacturing process is such that $\alpha_F \gg \alpha_R$, and also because the area of the collector is larger than that of the emitter $I_{CO} \gg I_{EO}$. (Note that these currents are I_{CO} and I_{EO} and not I_{CBO} or I_{CEO}.) It can be shown that:

$$\alpha_F I_{EO} = \alpha_R I_{CO} = I_S \tag{4.16}$$

where I_S is the saturation current for a diode and is typically 1×10^{-12} to 1×10^{-15} A. The emitter current is given by:

$$I_E = \alpha_R I_R - I_F \tag{4.17}$$

and

$$I_B = I_R + I_F - \alpha_F I_F - \alpha_R I_R$$

or

$$I_B = (1 - \alpha_F)I_F + (1 - \alpha_R)I_R \tag{4.18}$$

The relationship between diode current and junction voltage described in Chapter 2 can be used to give:

$$I_F = I_{EO}\left[\exp\left(\frac{V_{BE}}{V_T}\right) - 1\right] \tag{4.19}$$

$$I_R = I_{CO}\left[\exp\left(\frac{V_{BC}}{V_T}\right) - 1\right] \tag{4.20}$$

Substituting these values into the equations for I_C, I_E and I_B yields the following:

$$I_C = -\alpha_F I_{EO}\left[\exp\left(\frac{V_{BE}}{V_T}\right) - 1\right] + I_{CO}\left[\exp\left(\frac{V_{BC}}{V_T}\right) - 1\right]$$

or

$$I_C = -I_S\left[\exp\left(\frac{V_{BE}}{V_T}\right) - 1\right] + \frac{I_S}{\alpha_R}\left[\exp\left(\frac{V_{BC}}{V_T}\right) - 1\right] \tag{4.21}$$

$$I_E = \alpha_R I_{CO}\left[\exp\left(\frac{V_{BC}}{V_T}\right) - 1\right] - I_{EO}\left[\exp\left(\frac{V_{BE}}{V_T}\right) - 1\right]$$

or

$$I_E = -\frac{I_S}{\alpha_F}\left[\exp\left(\frac{V_{BE}}{V_T}\right) - 1\right] + I_S\left[\exp\left(\frac{V_{BC}}{V_T}\right) - 1\right] \tag{4.22}$$

$$I_B = (1 - \alpha_F)I_{EO}\left[\exp\left(\frac{V_{BE}}{V_T}\right) - 1\right] + (1 - \alpha_R)I_{CO}\left[\exp\left(\frac{V_{BC}}{V_T}\right) - 1\right]$$

$$I_B = \frac{(1 - \alpha_F)}{\alpha_F}I_S\left[\exp\left(\frac{V_{BE}}{V_T}\right) - 1\right] + \frac{(1 - \alpha_R)}{\alpha_R}I_S\left[\exp\left(\frac{V_{BC}}{V_T}\right) - 1\right]$$

or

$$I_{\mathrm{B}} = \frac{I_{\mathrm{S}}}{\beta_{\mathrm{F}}}\left[\exp\left(\frac{V_{\mathrm{BE}}}{V_{\mathrm{T}}}\right) - 1\right] + \frac{I_{\mathrm{S}}}{\beta_{\mathrm{R}}}\left[\exp\left(\frac{V_{\mathrm{BC}}}{V_{\mathrm{T}}}\right) - 1\right] \tag{4.23}$$

where $\beta_{\mathrm{F}} = \alpha_{\mathrm{F}}/(1 - \alpha_{\mathrm{F}})$ and $\beta_{\mathrm{R}} = \alpha_{\mathrm{R}}/(1 - \alpha_{\mathrm{R}})$ are the common-emitter current gains for forward and reverse operation.

Equations 4.21, 4.22 and 4.23 are the equations which are used in circuit simulators to obtain the dc currents and voltages in a bipolar transistor. They are presented here to illustrate how the dc model for the transistor is developed and are not intended to be memorized or used in calculations. The I–V curves shown in Figures 4.9 and 4.10 can be obtained with these equations, and they are the equations which are used in the SPICE model for the bipolar transistor.

Operating Regions (optional)

When a transistor is used in a circuit the external components determine the voltages on the two junctions according to the above equations. There are three modes of operation:

1 Base–emitter junction forward biased and the base–collector junction reverse biased. This is the normal condition for small-signal amplification.
2 Both junctions forward biased. Under these conditions both the emitter and the collector inject carriers into the base region. The transistor is said to be in saturation. It is a condition which can exist when the transistor is used as a switch, either as a discrete component or in an integrated circuit. It can also occur if the input signal is too large.
3 Both junctions reverse biased (or a more likely situation is for the base–emitter junction to be at 0 V, with the collector junction reverse biased). The transistor does not conduct and is said to be cut off. This condition occurs when the transistor is used as a switch, or if the input signal is too large.

The Ebers–Moll equations can cope with each of these conditions, and models based on the equivalent circuit shown in Figure 4.19 can accurately model the dc conditions which exist when different voltages are applied to the three terminals.

▷ **EXAMPLE 4.7** ——————————————————————————

Create the Ebers–Moll model with PSpice components using the breakout diode and the current-controlled current source (F-source). Connect a current source to the base and a voltage source to the collector and use the **DC Sweep...** to sweep the collector voltage from 0 to 5 V and a nested base current sweep from 5 to 40 µA. Let the gains of the current sources be $\alpha_{\mathrm{F}} = 0.995$ and $\alpha_{\mathrm{R}} = 0.1$.

▷ *Solution*

The PSpice circuit schematic is shown in Figure 4.20.

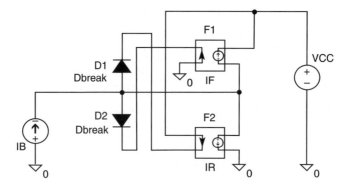

FIGURE 4.20   Circuit schematic from PSpice for simulating the Ebers–Moll model.

The cross-connections of the controlling current of the current sources are necessary to allow the correct current to be in series with the correct diode, as illustrated in Figure 4.19. The gain of the F1 source is set to 0.995 and the gain of the F2 source is set to 0.1.

The current-voltage characteristics are shown in Figure 4.21.

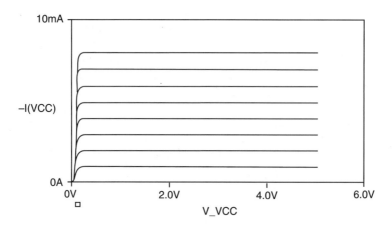

FIGURE 4.21   I–V characteristics produced by Probe.

It can be seen from Figure 4.21 that the model does accurately reproduce the current–voltage characteristics of the bipolar transistor. In the PSpice model of a transistor the equations 4.21–4.23 are used directly rather than diodes and controlled-current sources shown in Figure 4.20.

Second-Order Effects (optional)

The basic Ebers–Moll model shown in Figure 4.19 satisfies the dc conditions for an ideal transistor, but it neglects parasitic resistances, the Early effect, high-current effects and charge-storage effects. It is not possible in this text to consider these effects in detail, but it is worth while making a qualitative assessment of some of the more important factors to ensure that the correct parameters are used in equivalent circuit models required by circuit simulators such as PSpice.

Ohmic Resistance (optional)

The ideal model with parasitic resistance is shown in Figure 4.22.

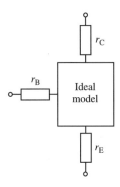

FIGURE 4.22   Ideal model with parasitic resistance.

The 'ideal model' is described by equations 4.21–4.23, and the stray resistors are added to this model. The resistor r_C is associated with the collector region from immediately beneath the emitter to the collector contact. However, in a real device it is not constant and is affected by collector current and collector voltage. It causes the collector saturation voltage V_{CEsat} to increase and is of particular concern when modelling transistors in switching circuits. Typical values may be a few tens of ohms.

The resistor r_E (not to be confused with the small-signal resistor r_e) is usually very small and constant, and is almost entirely due to the emitter contact resistance and is typically less than 10 Ω.

The base resistance r_B is an important parameter because it affects the high-frequency performance. It is associated with the base region from immediately beneath the emitter to the base contact and can have values ranging from 10 to 100 Ω.

Early Effect (optional)

The Early effect is associated with the voltage across the base–collector junction and the depletion layer which is produced by this voltage. The depletion layer extends into the collector region and also a smaller distance into the base region (see Figure 4.3). The active width of the base region extends from the edge of the emitter depletion layer to the edge of the collector depletion layer. The emitter depletion layer is narrow and does not vary greatly, but the collector depletion layer is much wider and varies with the collector voltage. Therefore, the width of the base region also varies. It is this variation which produces the change in the output I–V characteristics, as seen in Figure 4.11.

The change in the depletion-layer width affects the saturation current I_S, while the change in the base width affects the current gain factor β_F. For the analytical model described by equations 4.21, 4.22, 4.23 the Early effect is modelled by making the saturation current and the current gain dependent on the base–collector voltage, as follows:

$$I_{S(V_{BC})} \approx \frac{I_S}{1 + V_{BC}/V_A} \tag{4.24}$$

where V_A is the Early voltage. The current gain is:

$$\beta_{F(V_{BC})} \approx \frac{\beta_F}{1 + V_{BC}/V_A} \tag{4.25}$$

These values are used in place of I_S and β_F in the equations 4.21–4.23 for the Ebers–Moll model. For a given value of Early voltage this modification results in the saturation current and the gain being dependent on the base–collector voltage, and the current–voltage curves shown in Figure 4.21 would have a finite slope instead of the horizontal lines shown.

Charge-Storage Effects (optional)

The capacitance associated with the depletion layers of the emitter–base and base–collector junctions (the same as for the pn junction diode described in Section 2.3) together with the time taken for carriers to cross the base region result in charge-storage effects which affect the high frequency and switching speed of the transistor. The equivalent circuit can be modified to model these effects with the addition of capacitors across each of the junctions, as shown in Figure 4.23.

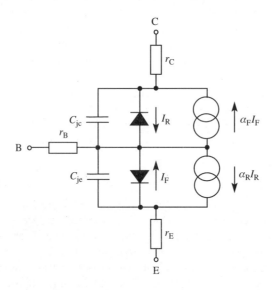

FIGURE 4.23   Equivalent circuit with junction capacitance.

The capacitors are voltage dependent and decrease in value with increasing voltage. For the emitter–base junction the capacitor (C_{je}) is in parallel with the forward-biased junction and consequently its impedance is swamped by the low resistance of the forward-biased junction. However, the collector–base junction has a very large impedance because it is reverse biased and at high frequencies the impedance of the junction capacitance (C_{jc}) provides a low-impedance path which effectively bypasses the junction and consequently affects the performance of the transistor at high frequencies.

Gummel–Poon (optional)

The Gummel–Poon[1] model takes account of additional second-order effects, notably the reduction of the common-emitter current gain (β_F) at low currents, a more accurate description of the base-width modulation and also effects which occur at high currents which cause the current gain to decrease. The model is basically the same as the Ebers–Moll, but the analytical equations governing the diode currents are further developed with the addition of extra terms.

Some of the more important terms which form part of the final Gummel–Poon model in SPICE and its various derivatives, are as follows:

Term	Description	Default
BF	Forward common-emitter gain	##
IKF	High-current roll-off knee for β_F (A)	∞
IS	Saturation current (A)	1×10^{-16}
VAF	Forward Early voltage (V)	∞
CJE	Zero-bias base–emitter capacitance (F)	0
CJC	Zero-bias collector–base capacitance (F)	0
RB	Zero-bias base resistance (Ω)	0

Notice that many of the default values are either zero or infinity, that is they are assumed not to exist. In PSpice there is the choice of named transistor types from the library or a default generic type. For the named devices the library models contain appropriate values for the Gummel–Poon parameters, which have been obtained from a complex series of measurements. If the required device is not in the library then a model can usually be generated for the device by reference to the manufacturer's data sheets, and the inclusion of appropriate parameters in the model for the generic model.

▷ **EXAMPLE 4.8** ─────────────────────────────────

Identify the above parameters in the model statement for the npn transistor 2N2222 and the pnp transistor 2N2907A from the PSpice **eval.slb** library.

▷ *Solution*

From the PSpice circuit simulator open a schematic window and place the two transistors in the schematic. Save the schematic under a temporary name. Select each transistor in turn (point and click) and from the **Edit** command select **Model...** and then **Edit Instance Model...** and view the parameters which are displayed. The following values should be observed:

Parameter	2N2222	2N2907A
Ikf	0.2847	1.079
Is	14.34f	0.65f
Bf	255.9	231.7
Vaf	74.03	115.7
Cjc	7.3p	14.7p
Cje	22p	19.8p
Rb	10	10

Some of the values appear to be remarkably precise with, for example, the knee current I_{kf} recorded as 0.2847 A. This is because the results are obtained with the aid of a computer-based measurement system. Some of the results are measured directly from dc and ac data, while others, including I_{kf}, are obtained by extrapolation and optimization. The final results recorded are the computer averages of a large number of measurements and in the above examples the computer-generated averages have obviously been used without any rounding of the values. The same applies for the values of the Early voltage: 74 V and 115 V would have been quite adequate. Notice that the saturation currents are very small with 1 fA being 10^{-15} A. The junction capacitors and base resistance are important for modelling transistors in circuits at high frequencies.

▷ PROBLEMS

4.1 Use equation 4.3 to show that
$I_E = (1 + \beta)I_B + I_{CEO}$.

4.2 A certain transistor exhibits an α of 0.965. Determine I_C and I_B when $I_E = 6.5$ mA.

4.3 Determine β if $I_E = 5.2$ mA and $I_B = 23$ μA.

[225]

4.4 A transistor has a collector–base leakage current of 20 nA and β of 150. Determine:

(a) the common-base current gain
(b) the collector–emitter leakage current

(c) I_C when $I_B = 10$ μA.

[(a) 0.9934, (b) 3.03 μA, (c) 1.508 mA]

4.5 Determine I_E, I_C, I_B and V_{CE} for the circuit shown if $V_{BE} = 0.7$ V and $\beta = 180$.

[2.775 mA, 2.76 mA, 15.3 μA, 3.24 V]

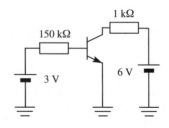

4.6 If $\alpha = 0.995$ and $V_{BE} = 0.7$ V, determine V_{CB} for the circuit shown.

[6.4 V]

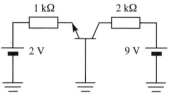

4.7 Determine V_{CE}, V_{CB} if $V_{BE} = 0.7$ V and $\beta = 120$ for the circuit shown.

[4.8 V, 4.1 V]

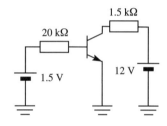

4.8 Repeat Problem 4.7 for the pnp configuration shown.

[−7.3 V, −6.6 V]

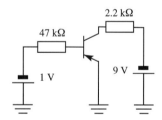

4.9 Determine β if $V_{BE} = 0.7$ V and $V_{CE} = 5$ V for the circuit shown.

[~51]

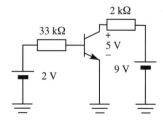

4.10 Determine which of the circuits in Problems 4.7 and 4.8 will saturate when β increases to 320.

4.11 A particular point on the I–V output curve of a transistor corresponds to I_C and V_{CE} of 2 mA and 6 V respectively. If the Early voltage is 50 V determine the collector current at another point on the same output curve where $V_{CE} = 10$ V.

[2.14 mA]

4.12 A transistor for the differential input stage of an operational amplifier has a dc collector current of 10 μA and the Early voltage is 120 V. Determine the output resistance at the collector.

[12 MΩ]

4.13 A transistor is rated at 10 W at 25°C. If the collector current is 1.5 A, determine the maximum value of V_{CE} and the minimum load resistance to avoid exceeding the maximum power dissipation.

[6.6 V, 4.4 Ω]

4.14 A 1 W transistor is derated linearly from 1 W at 25°C to 0 W at 150°C. Determine the power dissipation if it operates at 100°C.

[0.4 W]

4.15 Determine the voltage gain if $V_{BE} = 0.7$ V and $\beta = 150$ in the circuit shown.

[−2.9]

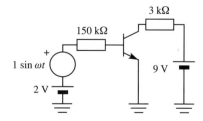

4.16 Draw the Ebers–Moll equivalent circuit for a pnp transistor.

4.17 Use PSpice to simulate the Ebers–Moll model for a pnp transistor and obtain the $I–V$ output curve for a voltage range of $0–5$ V for V_{CE} with a base current of 20 μA. Assume that $\alpha_F = 0.998$ and $\alpha_R = 0.1$.

4.18 Use PSpice to simulate the circuit shown using the breakout transistor with $\beta = 220$ and $I_S = 1 \times 10^{-12}$ A. Obtain V_{BE} and V_{CE} for temperatures of 27°C (default), 70°C and 150°C.

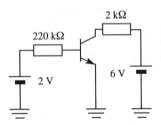

4.19 In Problem 4.18 the temperature change is 123°C. From the recorded values of V_{BE} verify that it decreases at approximately 2 mV per °C.

4.20 Simulate the circuit shown in Problem 4.15 with PSpice with a sine wave generator with a frequency of 1 kHz and amplitude of 1 V. Observe the output at the collector and verify that the gain is -2.9.

Reference

(1) P. Antognetti, G. Massobrio, *Semiconductor Device Modelling with Spice*, McGraw-Hill, Singapore (1988).

CHAPTER 5

Bipolar Junction Transistor: DC Biasing

The normal operating condition for the bipolar transistor for analog purposes is for the emitter–base junction to be forward biased and for the collector–base junction to be reverse biased. An additional requirement is that the current and voltage in the collector circuit must be chosen to ensure that an input signal is amplified without distortion. These conditions are achieved by the application of dc voltages to the junctions, or dc biasing. In Chapter 4 the characteristic I–V curves were obtained by means of two battery supplies, one for the base and one for the collector. For most applications this is not practical and the biasing needs to be achieved with a single dc supply.

An important requirement is that the dc currents and voltages are independent of the transistor parameters. This is particularly important for a circuit which is to be mass produced. The parameters of transistors of a given type are not all identical. An example is the value of β, which may vary by a factor of 3:1. In the manufacturer's data sheets these variations are identified by the minimum and maximum values quoted for the different parameters. To understand the effect of these changes the circuit analysis of dc bias circuits must include an examination of the likely effects of variation in current gain, leakage current and base–emitter voltage. All of these are likely to vary from one transistor to another, and also with temperature.

The circuit simulator can be useful in examining the worst conditions which may occur in a particular design, but some knowledge of how changes in different parameters will affect the dc conditions is necessary in order to identify the critical components to examine in a worst-case analysis with a circuit simulator.

For the hand-based calculations to be introduced in this and following chapters it is assumed that the reader is familiar with Kirchhoff's voltage and current laws.

5.1 Base Bias

The most commonly used transistor configuration is the common-emitter. For this configuration the polarities of the voltages required for an npn transistor are positive for both the base and the collector. The base-bias circuit in Figure 5.1 satisfies these conditions.

The relationship between the base and collector currents, the resistors and the supply voltage is obtained by applying KVL to the base and collector circuits. For the base circuit:

$$V_{CC} = I_B R_B + V_{BE}$$

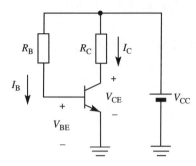

FIGURE 5.1   Schematic of a single-supply base-bias circuit.

and

$$I_B = \frac{V_{CC} - V_{BE}}{R_B}$$

From the definition for the common-emitter current gain the collector current can be obtained as:

$$I_C = \beta I_B = \frac{\beta(V_{CC} - V_{BE})}{R_B} \tag{5.1a}$$

or

$$I_C = \frac{V_{CC} - V_{BE}}{R_B/\beta} \tag{5.1b}$$

The collector-emitter voltage is:

$$V_{CE} = V_{CC} = I_C R_C \tag{5.2}$$

The circuit and the resultant equations are very simple, but there is a serious problem. It can be seen from equation 5.1 that the collector current is directly proportional to the common-emitter current gain (β). The β of a transistor can vary by a factor of 2 or 3 within a batch, and consequently I_C and V_{CE} also vary. While the base-bias circuit can be designed to provide the correct bias for a particular value of β, it is not ideal for mass production when the value of β may vary from transistor to transistor.

▶ **EXAMPLE 5.1** ──────────────────────────────

For the circuit shown in Figure 5.1 $R_B = 470$ kΩ, $R_C = 2.2$ kΩ and $V_{CC} = 9$ V. Determine the spread of I_C and V_{CE} for a base-bias circuit if β varies from 80 to 240.

▷ *Solution*

From equation 5.1

$$I_C = \frac{\beta(9 - 0.7)}{470k}$$

and

$$I_C\big|_{\beta=80} = 1.4 \text{ mA}$$

$$I_C\big|_{\beta=240} = 4.24 \text{ mA}$$

and

$$V_{CE}\big|_{\beta=80} = 5.92 \text{ V}$$

$$V_{CE}\big|_{\beta=240} = -0.328 \text{ V}$$

It can be seen that there is a very large change in the current, but more importantly V_{CE} also changes. For good biasing V_{CE} should be approximately half the supply voltage to avoid distortion of the signal applied to the input. This is indeed the case for $\beta = 80$; however, the $\beta = 240$, V_{CE} appears to be negative. What this implies is that the transistor is driven into saturation for this value of β and the transistor is effectively acting as a short circuit, and V_{CE} is in fact 0 V. The predicted current of 4.24 mA is not in fact obtained, because this is greater than is possible with a 9 V supply and a 2.2 kΩ resistor. When $\beta = 240$ the transistor is driven into saturation and no longer acts as an amplifier.

Load-Line

An understanding of the operation of the circuit can be obtained from consideration of the load-line diagram for the transistor. The load-line is the reciprocal of the collector resistance drawn on the $I_C - V_{CE}$ characteristic curves for the transistor as shown in Figure 5.2.

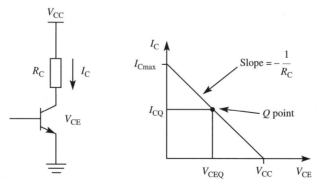

FIGURE 5.2   Collector circuit and the load-line superimposed on the *I–V* axis.

The equation for the load-line is obtained from equation 5.2 by rearranging the terms so that I_C is a function of V_{CE}, as shown below:

$$I_C = -\frac{1}{R_C} V_{CE} + \frac{V_{CC}}{R_C} \tag{5.3}$$

This equation describes an equation of a straight line ($y = mx + c$), where m is the slope and c the intercept on the y axis. For equation 5.3 when the collector current is zero the collector–emitter voltage is V_{CC}. When $V_{CE} = 0$ V, that is the transistor is saturated, the collector current reaches a maximum that is limited by the collector resistance and is $I_{Cmax} = V_{CC}/R_C$. The slope of the line joining I_{Cmax} and V_{CC} is:

$$\text{slope} = \frac{I_{Cmax} - 0}{0 - V_{CC}} = \frac{V_{CC}/R_C}{-V_{CC}} = -\frac{1}{R_C}$$

The operating point lies on this line, for example at the Q point, as shown in Figure 5.2 where the collector current is I_{CQ} and the collector–emitter voltage is V_{CEQ}. In Example 5.1 this point is defined by $I_C = 1.4$ mA and $V_{CE} = 5.92$ V when $\beta = 80$. When $\beta = 240$ the Q point moves to the end of the load line where $I_C = I_{Cmax}$ and $V_{CE} = 0$ V. In practice the Q point would be a little below this with $I_C < I_{Cmax}$ and $V_{CE} \approx 0.2$ V.

▷ **EXAMPLE 5.2** ————————————————————————

The base resistor for a base-bias circuit is 560 kΩ, the collector resistor is 1.8 kΩ and the supply voltage is 12 V. As a result of a change in temperature from 25°C to 75°C β changes from 120 to 240. Determine the Q point for each temperature and sketch the load-line diagram.

▷ *Solution*

At 25°C the collector current is:

$$I_C = \frac{12 - 0.7}{560k/120} = 2.42 \text{ mA}$$

and the collector–emitter voltage is:

$$V_{CE} = 12 - (2.42 \text{ mA})(1.8k) = 7.64 \text{ V}$$

At 75°C the collector current is:

$$I_C = \frac{12 - 0.7}{560k/240} = 4.84 \text{ mA}$$

and the collector–emitter voltage is:

$$V_{CE} = 12 - (4.28 \text{ mA})(1.8k) = 4.29 \text{ V}$$

The load-line is constructed by noting that the maximum collector current is:

$$I_{Cmax} = 12 \text{ V}/1.8 \text{ k}\Omega = 6.7 \text{ mA}$$

and that V_{CEmax} is 12 V. Thus the load-line is drawn between these two points on the I_C-V_{CE} characteristics as shown below.

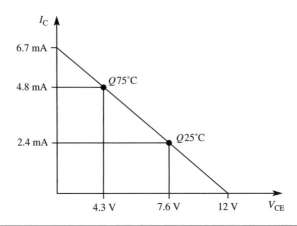

Ideally the bias circuit should maintain the position of the Q point when β, or any other parameters, vary. The base-bias circuit is very simple but it is not very good at stabilizing the Q point.

5.2 ⟩ Emitter Bias

In the emitter bias, or self-bias, or voltage divider bias, the base voltage rather than the base current is determined by the bias circuit. The circuit is shown in Figure 5.3.

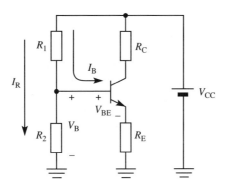

FIGURE 5.3 Schematic of emitter-bias circuit.

The required positive voltage for the base circuit is obtained from the potential divider circuit across the main supply (V_{CC}). The circuit is more complicated than the base-bias circuit because of the resistor in the emitter lead (R_E). The emitter current flows through this resistor generating a voltage V_{RE}. If the current flowing in the potential divider resistors (I_R) is much greater than the base current (I_B) then the expression for the voltage across R_2 is simply:

$$V_B = \frac{V_{CC}R_2}{R_1 + R_2} \tag{5.4}$$

and applying KVL to the base circuit gives:

$$V_B = V_{BE} + I_E R_E \tag{5.5}$$

Negative Feedback

Equation 5.5 can be rewritten in terms of V_{BE} as follows:

$$V_{BE} = V_B - I_E R_E \tag{5.6}$$

In this equation V_B is assumed to be constant and the current through the transistor is governed by the value of V_{BE}. Thus the equation is balanced when the LHS (V_{BE}) is equal to the difference between a constant voltage and a transistor-dependent voltage on the RHS ($V_B - I_E R_E$). The emitter current is determined by V_{BE} through the diode equation:

$$I_E = I_{EO} \exp\left(\frac{V_{BE}}{V_T}\right)$$

where I_{EO} is the saturation current for the emitter–base junction. Thus in equation 5.6 if there is any tendency for I_E to increase, for example as a result of an increase in temperature, then the voltage $I_E R_E$ increases which produces a reduction in V_{BE}. This reduction in V_{BE} has the effect of reducing I_E because of the exponential relationship between I_E and V_{BE}. Thus the external factor which is trying to bring about an increase in I_E is counteracted by a balancing action resulting from equation 5.6. This balancing action is an example of negative feedback which acts to stabilize the emitter current.

Feedback is an important effect for many applications, from mechanical speed governors for constant speed motor drives to the driver of a car who governs the speed of the engine by means of the car accelerator based on visual feedback of what is on the road ahead. In electronic circuits the feedback is a little more subtle, but never the less is still present, and is a necessary feature of good design.

Thévenin Equivalent Analysis

The approximation used to derive equation 5.4, that is $I_R \gg I_B$, is important in the arguments used above about negative feedback and stability. From equation 5.5 the emitter current is:

$$I_E = \frac{V_B - V_{BE}}{R_E} \tag{5.7}$$

which suggests that the emitter current, and hence the collector current, are independent of β. This relationship depends on the presence of the inequality between I_R and I_B. In practice it is difficult to ignore totally the existence of the base current flowing in the potential divider and, therefore, the analysis should include the presence of I_B. One approach is to replace the potential divider by its Thévenin equivalent circuit as shown in Figure 5.4.

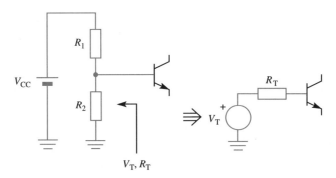

FIGURE 5.4   Transformation of the potential divider into the Thévenin equivalent.

The Thévenin equivalent is obtained by viewing the potential divider from the base terminal of the transistor, as shown in Figure 5.4. The Thévenin voltage is given by:

$$V_T = \frac{V_{CC}R_2}{R_1 + R_2} \tag{5.8}$$

and the Thévenin resistance by:

$$R_T = \frac{R_1 R_2}{R_1 + R_2} \tag{5.9}$$

The potential divider is now replaced by the Thévenin equivalent circuit of a resistor and voltage source so that the circuit is as shown in Figure 5.5. Applying KVL to the base circuit gives:

$$V_T = I_B R_T + V_{BE} + I_E R_E$$

but

$$I_E = (1 + \beta)I_B$$

and on substituting for I_E and rearranging, the base current is:

$$I_B = \frac{V_T - V_{BE}}{R_T + (1 + \beta)R_E}$$

and since $I_C = \beta I_B$ then with some rearrangement:

$$I_C = \frac{V_T - V_{BE}}{R_E + R_T/(1 + \beta)} \tag{5.10}$$

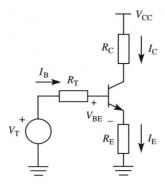

FIGURE 5.5   Emitter bias with the Thévenin equivalent of the potential divider.

Notice that equation 5.10 reduces to 5.7 if $R_E \gg R_T/(1 + \beta)$. Equation 5.7 represents an ideal situation, while equation 5.10 represents the practical position which takes account of the presence of base current which flows through the potential divider. The presence of the potential divider is represented by R_T.

▶ **EXAMPLE 5.3** ──────────────────────────────────

For the emitter-bias circuit shown in Figure 5.3 $R_1 = 22$ kΩ, $R_2 = 11$ kΩ, $R_E = 1$ kΩ, $R_C = 1.2$ kΩ, $V_{CC} = 9$ V and $\beta = 120$. Determine I_C and V_{CE}. Assume that $V_{BE} = 0.7$ V.

▶ *Solution*

From equations 5.8 and 5.9:

$V_T = 3$ V and $R_T = 7.3$ kΩ

Then from equation 5.10:

$$I_C \approx \frac{3\ \text{V} - 0.7\ \text{V}}{1\ \text{k}\Omega + (7.3\ \text{k}\Omega/120)} = 2.16\ \text{mA} \qquad \text{since } \beta \gg 1$$

and:

$$V_{CE} = V_{CC} - I_C R_C - I_E R_E$$
$$= 9\ \text{V} - (2.16\ \text{mA})(1.2\ \text{k}\Omega) - (2.16\ \text{mA})(1\ \text{k}\Omega) \text{ since } I_{Cr} \approx I_E$$
$$= 4.2\ \text{V}$$

In the calculation of V_{CE} it is perfectly justified to assume that $I_C \approx I_E$ since $\beta \gg 1$. This assumption simplifies the calculation.

▶ EXAMPLE 5.4

For the emitter-bias circuit described in Example 5.3 determine I_C and V_{CE} if $\beta = 240$.

▷ Solution

Repeating the calculations gives:

$$I_C \approx \frac{3\ \text{V} - 0.7\ \text{V}}{1\ \text{k}\Omega + (7.3\ \text{k}\Omega/240)} = 2.23\ \text{mA}$$

and

$$V_{CE} = 9\ \text{V} - (2.23\ \text{mA})(1.2\ \text{k}\Omega + 1\ \text{k}\Omega) = 4.1\ \text{V}$$

Q-Point Stability

It can be seen from the results of Examples 5.3 and 5.4 that a doubling of the transistor current gain has only produced a small change in I_C (about 3%). Thus it would appear that the emitter-bias circuit is very successful in stabilizing the current and voltage against changes in gain. This is true provided $R_T/(1 + \beta) \ll R_E$ which is the case in the above examples. Consider what happens if the potential divider resistors are much larger, and R_T is much larger The approximation which reduces equation 5.10 to that of equation 5.7 no longer applies and the effect of changes in the gain become much more significant.

▶ EXAMPLE 5.5

Repeat the calculations in Examples 5.3 and 5.4 if $R_1 = 220\ \text{k}\Omega$ and $R_2 = 110\ \text{k}\Omega$.

▷ Solution

The Thévenin resistance and voltage are:

$$R_T = 73.3\ \text{k}\Omega \text{ and } V_T = 3\ \text{V}$$

and the collector current is:

$$I_C|_{\beta=120} \approx \frac{3\ \text{V} - 0.7\ \text{V}}{1\ \text{k}\Omega + (73.3\ \text{k}\Omega/120)} = 1.42\ \text{mA}$$

$$V_{CE}|_{\beta=120} = 9\ \text{V} - (1.42\ \text{mA})(1.2\ \text{k}\Omega + 1\ \text{k}\Omega) = 5.9\ \text{V}$$

and for $\beta = 240$

$$I_C|_{\beta=240} \approx \frac{3\ \text{V} - 0.7\ \text{V}}{1\ \text{k}\Omega + (73.3\ \text{k}\Omega/240)} = 1.76\ \text{mA}$$

$$V_{CE}|_{\beta=240} = 9\ \text{V} - (1.76\ \text{mA})(2.2\ \text{k}\Omega) = 5.13\ \text{V}$$

The change in collector current is now greater, from 1.42 mA to 1.76 mA or 24%. Thus good stability is closely related to the value of the potential divider resistors with respect to the value of the emitter resistance.

The stability of the bias circuit is determined by the inequality:

$$R_E \gg R_T/\beta$$

How much larger than R_T/β does R_E have to be to satisfy this inequality? For engineering purposes a factor of 10 is often adequate and a factor of 100 is usually more than adequate. Obviously a particular design specification may place more stringent conditions on the stability of the operating point, in which case greater care will need to be taken, for example the common-emitter gain of the transistor may have to be preselected, or the circuit may have to be temperature stabilized, or there may be an input impedance restriction.

▶ **EXAMPLE 5.6** ─────────────────────────────

Use PSpice to verify the above bias circuits. Select a transistor from the breakout library (qbreak). Use the schematic editor to produce the circuit required for Example 5.3. Include a sine wave source at the input.

▶ *Solution*

The dc operating point information is obtained from the output text file created by PSpice and also by enabling the **Bias Point Detail** in the **Setup**. A sine wave source is included with a frequency of 1 kHz and amplitude of 10 mV together with two capacitors to allow the effect of change in the Q point to be noted on an output waveform. The analysis performed is **Transient** with a step size of 10 µs and duration of 2 ms.

From **Edit** select **Model...** and set β at 120, that is bf = 120. The circuit schematic for the simulation is shown below.

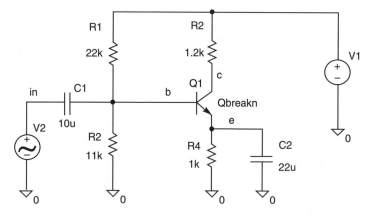

Repeat the analysis for $\beta = 240$ and observe the dc values at the base, emitter and collector for both values. Also note the output sine wave to see whether any distortion is introduced when β is changed.

Emitter-Bias Load-Line

The equation for the load-line is slightly modified with the addition of the emitter resistor. Applying KVL to the collector–emitter branch of the circuit gives:

$$V_{CC} = I_C R_C + V_{CE} + I_E R_E$$

or

$$I_C = \frac{1}{R_C + R_E} V_{CE} + \frac{V_{CC}}{R_C + R_E} \tag{5.11}$$

and

$$I_{Cmax} = \frac{V_{CC}}{R_C + R_E} \tag{5.12}$$

and

$$\text{slope} = -\frac{1}{R_C + R_E} \tag{5.13}$$

The load-line is constructed by drawing a line through V_{CC} on the x axis and I_{Cmax} on the y axis. The slope of the line now includes the emitter resistor.

Emitter-Bias Design

The procedures described above involving the calculation of the Thévenin voltage and Thévenin resistance are used to **analyze** a bias circuit. However, in addition to being able to analyze a circuit it is also necessary to **design** a bias circuit to satisfy a particular design specification. The main requirement of a bias circuit is that the Q point should not change when β or the temperature change. In the above examples the effect of changing β has been examined. While the effect of temperature has not been examined, it is implicit in the changes to β, since β approximately doubles for a 50°C rise in temperature. The emitter-bias circuit does not provide complete protection from changes in temperature, but it is more than adequate for the majority of applications.

The starting point for establishing a set of design rules for the emitter-bias circuit is the inequality $R_E \gg R_T/\beta$. For most applications a factor of 10 is usually sufficient to satisfy this condition, that is:

$$R_E = 10 R_T / \beta \tag{5.14}$$

Minimum and maximum values of β may be obtained from the manufacturer's data sheets. Some examples for small-signal npn transistors are shown in the table below:

Type	β_{min}	β_{max}
BC108	110	800
BC184	250	800
2N918	20	200
2N2222	100	300

It can be seen that there are considerable variations in the value of β. It is necessary, therefore, to decide on the correct value to be used for the relationship between R_E and R_T in equation 5.14.

The conditions to be satisfied for the emitter current to be independent of β were identified for equation 5.7, where it was assumed that the current in the potential divider was much greater than the base current. Consider now an emitter-bias circuit in which the gain varies. Since $I_B = I_C/\beta$ the base current is smaller for larger values of gain. Thus if the circuit were designed for the *minimum* value of gain, that is:

$$I_R = 10I_B \,|_{\beta=\min} \quad \text{using the engineering approximation for 10 for '} \gg \text{'}$$

then I_B is reduced still further if the *maximum* value of β is used. However, if instead the *maximum* value of gain had been used to determine I_B for the design value then for a smaller value of gain, I_B would increase and the inequality ($I_B \ll I_R$) would no longer hold.

Thus the value of β for establishing the relationship between R_E and R_T is the *minimum* value.

A relationship exists between the voltage at the base (V_B) and the voltage across the emitter resistor as given in equation 5.6 and reproduced here for convenience:

$$V_B = V_{BE} + I_E R_E$$

The emitter current is determined by the required operating point conditions, V_{BE} is typically 0.6 V to 0.8 V (0.6 V if the collector current is less than 1 mA and 0.8 V for values in excess of about 3 mA) and what remains is the choice of R_E. It is the voltage across R_E which is of interest and it should be equal to or greater than V_{BE}. It should typically be $1-2$ V. Alternatively it can be set to *one-tenth* of the supply voltage where the supply may typically be between 9 and 20 V. With a knowledge of V_{RE} and I_E it is possible to determine R_E and then R_T. The potential divider resistors R_1 and R_2 are then obtained from the relationships for R_T and V_T.

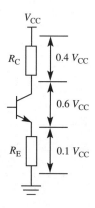

FIGURE 5.6   Distribution of the dc voltages.

The final component for which a value is required is the collector resistance (R_C). The value of R_C determines the voltage gain, but it also establishes the collector–emitter voltage and thus the operating point. The dc voltage across the transistor is important because it determines the maximum variation of the collector–emitter voltage when an alternating signal is applied to the input. The slope of the load-line is determined by the value of R_C as shown in Figure 5.2. For a given value of current the load-line determines the value of V_{CE}. As indicated in Figure 5.2 the Q point needs to be positioned at approximately the mid-point of the load-line. With 1/10th of the supply voltage appearing across R_E it is convenient to divide the remaining voltages, as shown in Figure 5.6.

The distribution of voltages is chosen for convenience rather than perfect symmetry, and assuming equal collector and emitter currents, then:

$$R_C = 4R_E \tag{5.15}$$

The design rules for the emitter bias circuit may be listed as follows:

1 Identify the collector current (I_C)
2 Identify the supply voltage (V_{CC})
3 Identify the minimum value of gain (β_{min})
4 Let $V_{RE} = V_{CC}/10$
5 Calculate $R_E = V_{RE}/I_C$ (assume $I_E = I_C$)
6 Calculate $V_T = V_{BE} + V_{RE}$
7 Calculate $R_T = 0.1\beta_{min}R_E$ (from equation 5.14)
8 Calculate $R_1 = V_{CC}R_T/V_T$
9 Calculate $R_2 = V_{CC}R_T/(V_{CC} - V_T)$
10 Calculate $R_C = 4R_E$.

The above design steps result in an emitter-bias circuit which ensures good stability of the operating point for variations of β. The stability of the circuit could be further improved if in step 7 a factor of 0.01 were used instead of 0.1. This would reduce the values of the potential divider resistors which would increase current consumption through them and reduce the input impedance of the circuit. These changes would obviously have to be taken into account in deciding whether the improved stability justifies the increase in current consumption and the reduction in the input impedance.

▷ **EXAMPLE 5.7** ────────────────────────────────────

Design an emitter-bias circuit to operate from 12 V with a collector current of 2 mA if the common-emitter gain varies from 80 to 240.

▷ *Solution*

$$V_{RE} = \frac{V_{CC}}{10} = 1.2 \text{ V}$$

The base voltage is:

$$V_T = V_{BE} + V_{RE} = 0.7 \text{ V} + 1.2 \text{ V} = 1.9 \text{ V}$$

Then

$$R_E = \frac{V_{RE}}{I_C} = \frac{1.2\ \text{V}}{2\ \text{mA}} = 600\ \Omega$$

$$R_T = 0.1\beta_{min}R_E = (0.1)(80)(600) = 4.8\ \text{k}\Omega$$

and

$$R_1 = \frac{V_{CC}}{V_T}R_T = \frac{12\ \text{V}}{1.9\ \text{V}}\ 4.8\ \text{k}\Omega = 30.3\ \text{k}\Omega$$

and

$$R_2 = \frac{V_{CC}}{(V_{CC} - V_T)}R_T = \frac{12\ \text{V}}{(12\ \text{V} - 1.9\ \text{V})}\ 4.8\ \text{k}\Omega = 5.7\ \text{k}\Omega$$

The collector resistance is:

$$R_C = 4R_E = 2.4\ \text{k}\Omega$$

This completes the design.

► EXAMPLE 5.8

Use PSpice to determine the effectiveness of the above design and to determine the values of I_C for each value of β.

► Solution

The circuit schematic is the same as that shown in Example 5.6. The following values were obtained:

Parameter	Design value	Simulated values	
		$\beta = 80$	$\beta = 240$
V_C (V)	8.4	7.95	7.69
V_B (V)	1.9	1.81	1.87
V_E (V)	1.2	1.02	1.08
I_C (mA)	2.0	1.69	1.79

The differences noted between the design values and simulated values arise from the assumption that for the design $R_E \gg R_T/\beta$, and as a consequence the effect of R_T is ignored. In the simulation the effect of the base resistors (and hence R_T) is obviously included and as a result the collector current is slightly lower than the design value.

For a real design the values obtained for the resistors may not match the preferred range of values available from a manufacturer. Consequently it will be necessary to perform an analysis or simulation with the nearest preferred values in place, rather than the exact calculated values, to ensure that the Q point is sufficiently close to the design value to satisfy the design specification.

5.3 Collector Bias

For many transistor amplifier applications a high input impedance is desirable but this is not always achievable with the emitter bias because of the need to satisfy the inequality $R_T \ll \beta R_E$. The collector bias offers an alternative biasing arrangement. The circuit is shown in Figure 5.7.

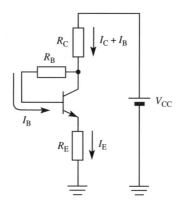

FIGURE 5.7   Schematic of a collector-bias circuit.

The difference between the collector bias and base bias (Figure 5.1), apart from the inclusion of the emitter resistor, is the connection of the base resistor to the collector, rather than directly to the power supply. This arrangement provides negative feedback, because any tendency for the current to change produces a change in the collector voltage, which forces a change to the base current. Thus if there is an increase in the temperature, which would have the effect of increasing the collector current, the collector voltage will reduce as the voltage drop across R_C increases, and cause a reduction in the base current, which then modifies I_C.

To obtain an expression for the collector current apply KVL to the collector and base circuit as follows:

$$V_{CC} = (I_C + I_B)R_C + I_B R_B + V_{BE} + I_E R_E$$

$$\approx I_C R_C + I_B R_B + V_{BE} + I_E R_E \quad \text{since } I_C \gg I_B$$

$$\approx I_C(R_C + R_E) + \frac{I_C}{\beta} R_B + V_{BE} \quad \text{since } I_E \approx I_C$$

and

$$I_C = \frac{V_{CC} - V_{BE}}{R_C + R_E + (R_B/\beta)} \tag{5.16}$$

This equation is similar to equation 5.10 with V_{CC} replacing V_T and with the extra resistance R_C in the denominator. A similar inequality exists in the denominator, and if I_C is to be independent of β then $R_C + R_E \gg R_B/\beta$.

A useful relationship for the design of the collector-bias circuit is:

$$R_C = R_B/\beta \tag{5.17}$$

Substituting into equation 5.16 gives:

$$I_C = \frac{V_{CC} - V_{BE}}{2R_C + R_E}$$

This relationship establishes a Q point which is approximately mid-way along the load-line. It does not, however, follow that because β does not appear in equation 5.17 the collector current is independent of β. An example will illustrate the effectiveness of this circuit to stabilize against changes in β.

▶ **EXAMPLE 5.9** ─────────────────────────────

For the collector-bias circuit shown in Figure 5.7 $R_C = 2.4$ kΩ, $R_B = 190$ kΩ, $R_E = 600$ Ω, $V_{CC} = 9$ V and β varies from 80 to 240. Determine the collector current for each value of β.

▶ *Solution*

Using equation 5.12 the collector current is:

$$I_C = \frac{V_{CC} - V_{BE}}{R_C + R_E + (R_B/\beta)}$$

$$I_C = \frac{9 \text{ V} - 0.7 \text{ V}}{2.4\text{k} + 600 + (190\text{k}/80)}$$

$$I_C|_{\beta=80} = 1.54 \text{ mA}$$

and

$$I_C|_{\beta=240} = \frac{9 - 0.7}{2.4\text{k} + 600 + (190\text{k}/240)}$$

$$I_C|_{\beta=240} = 2.18 \text{ mA}$$

From a comparison of the results from Example 5.8 it can be seen that the variation of I_C is much greater with collector bias and the differences are seen more clearly in the table below:

Bias circuit	$\beta = 80$	$\beta = 240$	% change
Emitter	1.69 mA	1.79 mA	5.5
Collector	1.54 mA	2.18 mA	41.5

While the change in I_C with collector bias is large it need not be unacceptable and the circuit does have the advantage of a much higher input impedance than the emitter-bias circuit.

5.4 PNP Transistor Biasing

An emitter-bias circuit for a pnp transistor is shown in Figure 5.8.

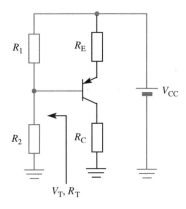

FIGURE 5.8   Schematic of an emitter-bias circuit for a pnp transistor.

The approach adopted for analyzing the circuit is the same as that used for the emitter-bias circuit for the npn transistor, that is to transform the potential divider to its Thévenin equivalent. This is done by viewing the potential divider from the base of the transistor as shown in the diagram. The values of R_T and V_T are the same as for the npn circuit, that is:

$$V_T = \frac{V_{CC} R_2}{R_1 + R_2}$$

$$R_T = \frac{R_1 R_2}{R_1 + R_2}$$

The circuit is then redrawn with the potential divider replaced by the Thévenin equivalent as shown in Figure 5.9.

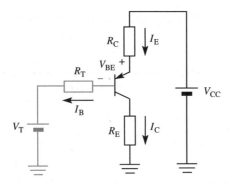

FIGURE 5.9   Thévenin equivalent of emitter bias for a pnp transistor.

The relationship between the base current and the supply voltage is now obtained by application of KVL to the base circuit which gives the following:

$$V_{CC} = V_T + I_B R_T + V_{BE} + (1 + \beta)I_B R_E \quad \text{since } I_E = (1 + \beta)I_B$$

and

$$I_B = \frac{V_{CC} - V_{BE} - V_T}{R_T + \beta R_E} \quad \text{since } \beta \gg 1$$

and

$$I_C = \frac{V_{CC} - V_{BE} - V_T}{R_E + R_T/\beta} \tag{5.18}$$

The equation is of the same format as that for equations 5.10 and 5.16. The main variable in all of the circuits is the gain, and the same inequality condition applies for all of the circuits, that is R_T/β must be small compared with the other resistive term in the denominator if the collector current is to be independent of the gain. The engineering approximation of using a factor of 10 for the inequality is usually adequate, but a factor of 100 is better, although external impedance levels may influence the choice, rather than the dc stability.

▶ **EXAMPLE 5.10** ──────────────────────────────────

For the circuit shown in Figure 5.8 $R_1 = 6.8$ kΩ, $R_2 = 15$ kΩ, $R_E = 680$ Ω, $R_C = 1.8$ kΩ, $V_{CC} = 10$ V and $\beta = 150$. Determine V_{CE}.

▶ *Solution*

The Thévenin voltage and resistance are:

$$V_T = \frac{(10)(15 \text{ k}\Omega)}{(6.8 \text{ k}\Omega + 15 \text{ k}\Omega)} = 6.88 \text{ V}$$

$$R_T = \frac{(6.8\ k\Omega)(15\ k\Omega)}{(6.8\ k\Omega + 15\ k\Omega)} = 4.68\ k\Omega$$

and

$$I_C = \frac{10\ V - 0.7\ V - 6.88\ V}{680 + (4.68\ k\Omega/150)} = 3.4\ mA \quad \text{using equation 5.16}$$

The collector–emitter voltage is obtained by applying KVL to the collector–emitter branch of the circuit. Then:

$$V_{CE} = 10\ V - (3.4\ mA)(680\ \Omega + 1.8\ k\Omega) = 1.56\ V$$

PROBLEMS

5.1 For the base-bias circuit shown assume that $\beta = 150$ and $V_{BE} = 0.7$ V and determine I_C and V_{CE}.

[3.6 mA, 5.5 V]

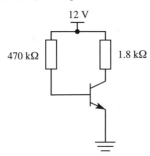

5.2 Repeat Problem 5.1 for $\beta = 350$ and comment on the result.

5.3 Determine I_C and V_{CE} for the circuit shown if $\beta = 120$ and $V_{BE} = 0.7$ V.

[2.2 mA, 3.47 V]

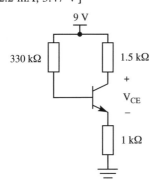

5.4 Determine a suitable value for R_B to provide a V_{CE} of 5 V for the circuit shown if $\beta = 180$ and $V_{BE} = 0.7$ V.

[560 kΩ]

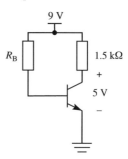

5.5 Determine I_C and V_{CE} for the circuit shown if $\beta = 200$ and $V_{BE} = 0.7$ V.

[6.4 mA, 3.4 V]

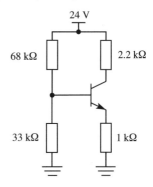

5.6 Determine the value of R_C to ensure that $V_{CE} \geq 3$ V if β varies from 80 to 240. Assume that $V_{BE} = 0.7$ V.

[1.65 kΩ]

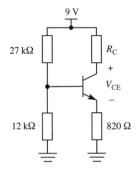

5.7 Determine the change in V_{CE} if β varies from 120 to 420. Assume that $V_{BE} = 0.7$ V.

[4.72 V to 3.76 V]

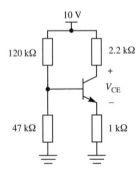

5.8 Select a suitable value for R_E to provide a collector current of 2 mA if $\beta = 150$ and $V_{BE} = 0.7$ V.

[1018 Ω]

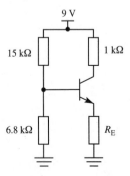

5.9 For the circuit shown $\beta = 120$ and $V_{BE} = 0.7$ V. Verify that the collector current is 1.1 mA and determine the value of R_E which is required to provide a collector current of 2 mA.

[500 Ω]

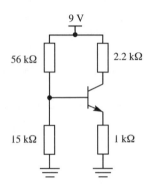

5.10 Design a base potential divider bias circuit to provide a collector current of 3 mA from a supply of 12 V for an npn transistor where the minimum and maximum values of β are specified as 120 and 320 respectively. Determine values for R_1, R_2, R_E and R_C.

[30.3 kΩ, 5.7 kΩ, 400 Ω, 1.6 kΩ]

5.11 Determine I_C and V_{CE} if $\beta = 200$ and $V_{BE} = 0.7$ V for the collector-bias circuit shown.

[2.64 mA, 6.4 V]

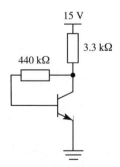

5.12 For the circuit shown determine V_{CE} for $\beta = 120$ and 350 if $V_{BE} = 0.7$ V.

[6.35 V, 3.6 V]

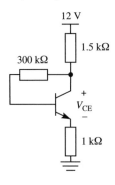

5.13 If $V_{CE} = 5$ V, $\beta = 250$ and $V_{BE} = 0.7$ V, determine R_B.

[537 kΩ]

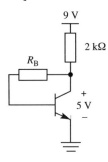

5.14 Determine the collector current for the pnp emitter-biased circuit if $\beta = 120$ and $V_{BE} = 0.7$ V.

[1.94 mA]

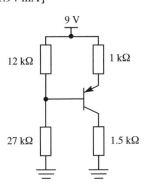

5.15 Determine V_C if $\beta = 150$ and $V_{BE} = 0.65$ V.

[4.4 V]

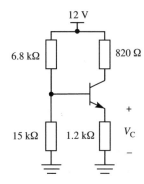

5.16 For the circuit shown $V_C = 4$ V. Determine the value of R_E if $\beta = 200$ and $V_{BE} = 0.7$ V.

[790 Ω]

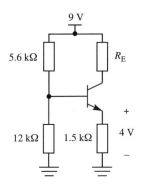

5.17 For Problem 5.10 the nearest preferred values for the resistors are:

$R_1 = 33$ kΩ, $R_2 = 5.6$ kΩ,
$R_E = 390$ Ω and $R_C = 1.5$ kΩ

Use PSpice to observe the value of the collector current for the two extremes of β, that is 120 and 320, with $V_{CC} = 12$ V.

[2.19 mA, 2.33 mA]

5.18 Use PSpice to verify the parameters in Problems 5.3, 5.7, 5.12 and 5.15.

[P5.3 $I_C = 2.18$ mA, $V_{CE} = 3.52$ V]

[P5.7 $V_{CE} = 4.95$ V and 4.01 V]

[P5.12 $V_{CE} = 6.37$ V and 3.66 V]

[P5.15 $V_C = 4.11$ V]

CHAPTER 6

Bipolar Junction Transistor: Small-Signal Analysis

The amplification of analog signals is an important application for bipolar transistors. The dc bias circuits described in Chapter 5 are used to establish the dc operating point so that the Q point is mid-way along the load-line. When a signal is applied to the input the current passing through the transistor varies in response to the variation of the input signal. The variation in the current produces a corresponding variation in the output voltage and provided that the correct operating point has been selected then the output voltage will be an accurate representation of the input voltage. It is assumed that these variations of voltage and current do not affect the dc conditions. The variations are assumed to be small in comparison with the dc currents and voltages. For simplicity the input signals are assumed to be sinusoidal, and the variations are sufficiently small so that the non-linear parameters associated with the transistor, such as r_e, or β, are assumed to be constant. It is this assumption which forms the basis of *small-signal analysis*.

There are many factors which influence the small-signal properties of transistor amplifiers and this chapter is intended to provide a basic understanding of amplifiers with one or two transistors. It is not practical to cover every possible circuit. In many instances the analytical methods developed for simple circuits can be extended to the more complex circuits. However, there are still many specialist applications where the simple theory is difficult to apply, and such circuits will have to be left to a more advanced course. For example, analog signals can have a very large dynamic range, that is the signal to be processed may range from microvolts up to several volts. Special purpose logarithmic amplifiers can be designed for such applications which compress the signals into a more suitable range. Similarly there are anti-logarithmic amplifiers to decompress the signals. A signal may vary in frequency from a few hertz to many megahertz, for example television signals, and very wide-band amplifiers are required for such signals, which may include special frequency compensation circuitry. For very low-level signals account must be taken of electrical noise created in resistors and pn junctions. Special low-noise transistors are available for these applications. Usually only the first stage of amplification needs to employ such transistors, because once the signal has been raised above the noise threshold then standard devices may be used for further processing. These and many other specialist circuits can be found in more specialist texts.

The bipolar transistor has three terminals, and this immediately leads to the possibility of three different circuit configurations. Each configuration results in a unique set of small-signal properties. For two-transistor circuits the different configurations can be combined in different ways, to produce amplifiers with many different small-signal properties.

Throughout this text simple hand-based calculations, using simple equivalent circuits, are used to estimate gain and impedance levels, which can then be examined in more detail with a circuit simulator

6.1   Small-Signal Circuits

The variations in current and voltage in a transistor amplifier can best be illustrated by reference to the load-line diagram, as shown in Figure 6.1.

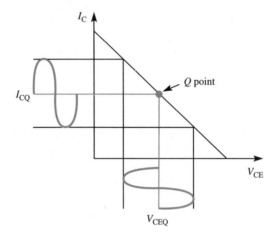

FIGURE 6.1   Load-line and the variation of the collector current and the collector–emitter voltage.

When the collector current varies there is a corresponding variation in the collector–emitter voltage, as shown in Figure 6.1. It is important that the dc bias point is positioned near the middle of the load-line to ensure maximum swing of the current and voltage. An example of a common-emitter amplifier for which this load-line diagram may apply is shown in Figure 6.2.

The input v_s is amplified to produce an output v_{out} which is 180° out of phase with the input (note the polarity of the sine waves at input and output). The capacitor C_B is included to provide dc isolation between the connection to the potential divider and the signal source, which may have a dc path to ground. The capacitor is assumed to have a very low value of reactance at the operating frequency of the source (at 1 kHz a 10 µF capacitor has a reactance of 15.9 Ω) so that it may be regarded as a short circuit in comparison with the input impedance of the transistor. Similarly the capacitor C_E also acts as a short circuit at the frequency of the source so that there is a low-impedance path to ground in parallel with the emitter resistance R_E. The emitter resistor is said to be *decoupled*.

The dc voltage supply V_{CC} is assumed to have a very small internal resistance, typically 0.01 Ω. Thus to all intents the dc supply looks like a short circuit as far as the

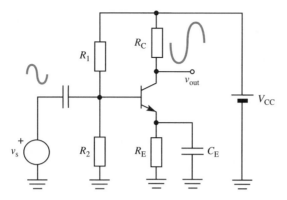

FIGURE 6.2   Schematic of a common-emitter amplifier.

alternating signal is concerned. The circuit of Figure 6.2 can be redrawn with C_B, C_E and the dc supply replaced by short circuits. The components which remain are those which affect the progress of the signal from the input to the output. The resultant *ac circuit* is shown in Figure 6.3.

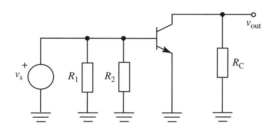

FIGURE 6.3   AC circuit for the common-emitter amplifier.

   Notice that the two potential divider resistors now appear in parallel across the input, that the emitter is connected directly to ground, and that the collector resistor is connected to ground with the output voltage appearing across it. Substituting short circuits for the capacitors and the dc power supply greatly simplifies the circuit for the purposes of small-signal analysis. Notice that this circuit only applies at frequencies for which the capacitors and the power supply behave as short circuits. For very low frequencies the reactance of the capacitors increases and they would then have to be included. At very high frequencies the reactive impedance associated with the inductance of the leads to the capacitors and to the dc power supply, and the effect of stray capacitance, may become significant. However, for the purposes of this chapter it is assumed that the circuit shown in Figure 6.3 applies, and forms the basis of simple hand calculations. A more accurate representation which includes coupling and decoupling capacitors, stray capacitance and lead inductance can be analyzed with a circuit simulator (but even with a circuit simulator it would be necessary to add stray capacitance and stray inductance as discrete components).

It is very important to be able to make an estimate of the circuit performance before using a simulator.

A simulator will always provide an answer, but if the input data is wrong, then the output data is also wrong.

▷ **EXAMPLE 6.1** ———————————————————————

Draw the small-signal ac circuits for the amplifiers shown in Figure 6.4.

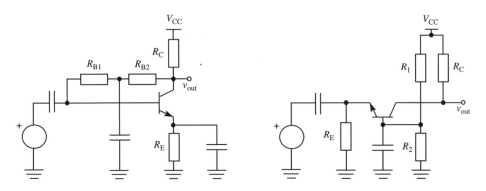

FIGURE 6.4   Schematic of common-emitter and common-base amplifiers.

▷ *Solution*

Hint: Short-circuit all capacitors and replace V_{CC} with a ground connection.

———————————————————————————————————

6.2 ▷ Common-Emitter

For the *common-emitter amplifier* (CE amplifier) shown in Figure 6.2 with its small-signal circuit in Figure 6.3 the emitter terminal is common to both the input and the output. In order to analyze the circuit in Figure 6.3 it is first necessary to replace the transistor by an equivalent circuit.

Equivalent Circuit

The circuit based on the r parameters as described in Chapter 4 and shown in Figure 4.16 is easy to use and adequate for simple hand calculations. The ac circuit is redrawn with the transistor replaced by the r-parameter equivalent circuit, as shown in Figure 6.5.

It is a simple matter to apply KVL to the circuit in order to determine small-signal voltage gain or impedance levels.

An important parameter for any amplifier is the voltage gain. This is defined as the ratio of the output voltage to the input voltage. From Figure 6.5 the input voltage v_s

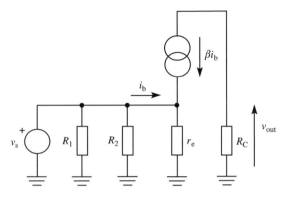

FIGURE 6.5   Small-signal equivalent circuit of common-emitter amplifier.

must be balanced by the voltage developed across r_e as a result of the currents i_b and βi_b flowing through it, and can be expressed as:

$$v_s = i_b r_e + \beta i_b r_e$$

and for the output:

$$v_{out} = -\beta i_b R_C$$

where the negative sign follows from the direction adopted for the current generator and the direction adopted for the output voltage. It describes the 180° phase change which takes place in the common-emitter amplifier which is illustrated by the sine waves shown in Figure 6.2.

The gain is given by:

$$A_v = \frac{v_{out}}{v_s} = \frac{-\beta i_b R_C}{i_b(1+\beta)r_e}$$

and

$$A_v \approx \frac{R_C}{r_e} \quad \text{since } \beta \gg 1 \tag{6.1}$$

The gain is the ratio of the collector resistance and the internal resistance of the forward-biased base-emitter junction where $r_e = 26 \text{ mV}/I_E$. Therefore, the gain is dependent on the dc current, with larger values of gain being obtained for larger values of I_E.

▶ **EXAMPLE 6.2** ─────────────────────────────────

For the common-emitter amplifier circuit shown in Figure 6.2 $R_1 = 27$ kΩ, $R_2 = 11$ kΩ, $R_C = 1.5$ kΩ, $R_E = 820$ Ω and $V_{CC} = 9$ V. Determine the voltage gain. Assume that $\beta = 100$.

▶ *Solution*

DC Solution

The emitter current is obtained as follows:

$$R_T = \frac{(27k)(11k)}{(27k + 11k)} = 7.8 \text{ k}\Omega$$

and

$$V_T = \frac{(9)(11k)}{(27k + 11k)} = 2.6 \text{ V}$$

and

$$I_C = \frac{(2.6 - 0.7)}{820 + (7.8k/100)} = 2.1 \text{ mA}$$

AC Solution

The small-signal emitter resistance is:

$$r_e = \frac{25 \text{ mV}}{2.1 \text{ mA}} \approx 12 \ \Omega$$

and the voltage gain is:

$$A_v = -\frac{R_C}{r_e} = -\frac{1.5k}{12} = -125$$

AC Load-Line

In Chapter 5 the dc load-line was obtained for the common-emitter configuration with and without an emitter resistor. With an emitter resistor the slope from equation 5.13 is $-1/(R_C + R_E)$, and the dc load-line is drawn between $V_{CC}/(R_C + R_E)$ and V_{CC}. However, from the ac circuit shown in Figure 6.3 the emitter resistor (R_E) does not appear and the small-signal voltage across the collector and emitter is:

$$v_{ce} = i_c R_C$$

This equation can be represented by an *ac load-line* with a slope of $-1/R_C$. This value of slope is greater than the slope of the dc load-line. Because it also represents the variation of v_{out} for all values of the input, including 0 V, then the ac load-line must intersect the dc load-line at the Q point because this corresponds to the condition for zero small-signal input. The combined dc and ac load-lines are shown in Figure 6.6.

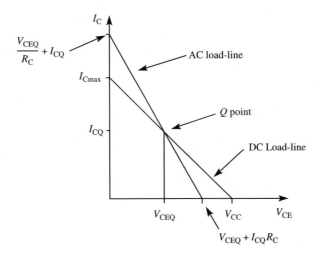

FIGURE 6.6   AC and dc load-lines.

The ac load-line intersects the dc load-line at the Q point. The intersection of the ac load-line with the current axis is:

$$i_{Cmax} = \frac{V_{CEO}}{R_C} + I_{CQ} \tag{6.2}$$

and with the voltage axis:

$$v_{CEmax} = V_{CEQ} + I_{CQ}R_C \tag{6.3}$$

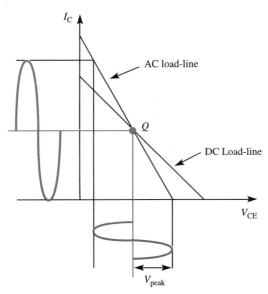

FIGURE 6.7   Variation of the small-signal voltage and current.

The small-signal variation of the collector current and the collector–emitter voltage across the ac load are different to those which may be obtained from the dc load-line. The ac variations need to be superimposed onto the ac load-line rather than the dc load-line, as illustrated in Figure 6.7.

Notice that the ac signal current may exceed the maximum dc current, and that the maximum variation of the collector–emitter voltage (v_{peak}) is less than would be predicted by the dc load-line.

▷ **EXAMPLE 6.3** ——————————————————————————————

For the circuit described in Example 6.2 determine the peak small-signal output voltage.

▷ *Solution*

DC Conditions

The collector current is 2.1 mA. The collector–emitter voltage is:

$$V_{CEQ} = 9 \text{ V} - (2.1 \text{ mA})(1.5 \text{ k}\Omega + 820) \approx 4.1 \text{ V}$$

AC Load-Line

From equation 6.2 the maximum small-signal current through the load is:

$$i_{Cmax} = \frac{4.1 \text{ V}}{1.5 \text{ k}\Omega} + 2.1 \text{ mA} \approx 4.8 \text{ mA}$$

and the maximum voltage is:

$$v_{CEmax} = 4.1 \text{ V} + (2.1 \text{ mA})(1.5 \text{ k}\Omega) \approx 7.2 \text{ V}$$

To determine the peak voltage across the load refer to Figure 6.7 and note that the peak variation could extend from 0 V to 4.1 V and from 4.1 V to 7.2 V. This would provide peaks of:

negative: 4.1 V − 0 V = 4.1 V

or

positive: 7.2 V − 4.1 V = 3.1 V

For a symmetrical sinusoidal wave the maximum is limited by the swing in the positive direction, that is:

$$v_{peak} = 3.1 \text{ V}$$

With reference to Figure 6.6 the maximum swing may be obtained in both positive and negative directions if $i_{Cmax} = 2I_{CQ}$. Substituting this value into equation 6.2 gives:

$$2I_{CQ} = \frac{V_{CEO}}{R_C} + I_{CQ}$$

and

$$I_{CQ} = \frac{V_{CEQ}}{R_C} \tag{6.4}$$

Substituting this value into the dc load-line equation gives:

$$V_{CC} = V_{CEQ} + \left(\frac{V_{CEQ}}{R_C}\right)(R_C + R_E)$$

and:

$$V_{CEQ} = \frac{V_{CC}}{2 + R_E/R_C} \tag{6.5}$$

$$I_{CEQ} = \frac{V_{CC}}{2R_C + R_E} \tag{6.6}$$

These relationships are rather idealized because they take no account of the non-linearity of the transistor characteristics as saturation or cut-off are approached. In practice it is not possible to achieve these maximum values predicted by this simple use of the load-line because the small-signal conditions assumed are being exceeded. However, the ac load-line provides a useful indicator of the maximum current and voltage swings, even though in practice it would not be possible to achieve these levels without some distortion.

▷ **EXAMPLE 6.4** —————————————————————————————————

Use PSpice to examine the circuit described in Example 6.3. First apply a 5 mV, 1 kHz sine wave to the input and use transient analysis to examine the output and to determine the gain, then increase the amplitude to 20 mV.

▷ *Solution*

The schematic for PSpice is shown in Figure 6.8. The default QbreakN transistor is used for which $\beta = 100$. The base coupling capacitor has a value of 10 µF and the emitter decoupling capacitor is 100 µF. Use **Transient Analysis...** to observe the output waveform with input signal levels of 5 mV and 20 mV and with a frequency of 1 kHz. Set the **Print time:** to 10 µs and the **Final time:** to 2 ms.

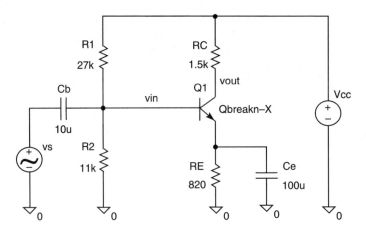

FIGURE 6.8 Schematic for Example 6.4 from PSpice.

From the output waveform shown in Figure 6.9 obtained from Probe with an input of 5 mV, the gain is measured as 112. When 20 mV is applied to the input the output is as shown in Figure 6.10 opposite.

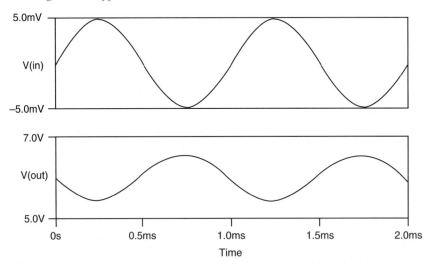

FIGURE 6.9   Output obtained from Probe with an input of 5 mV.

It is apparent that the output waveform shown in Figure 6.10 is severely distorted even though the peak output with a gain of 112 and a 20 mV input is only 2.2 V, which is less than the 3.1 V predicted from the load-line in Example 6.3. This is a result of the non-linear behaviour of the transistor and represents a departure from the small-signal conditions which are assumed when calculating the gain. The non-linearity of the

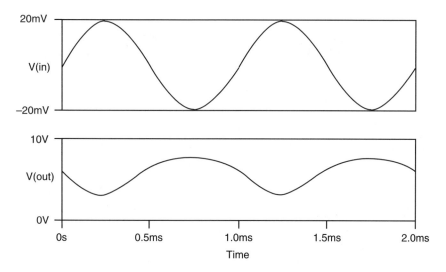

FIGURE 6.10   Output from Probe with an input of 20 mV.

relationship between the base–emitter voltage and the collector current is difficult to handle with the simple models used for hand-based calculations. The more sophisticated model used in the simulator is able to handle such non-linear conditions resulting from large input signal conditions.

Input Resistance

The input resistance for the common-emitter amplifier is obtained from the equivalent circuit by viewing the circuit from the position of the input voltage source as shown in Figure 6.11.

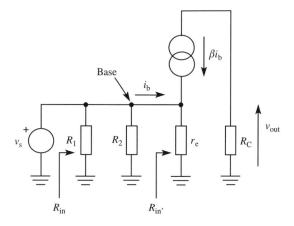

FIGURE 6.11   Equivalent circuit used to determine the input resistance.

First consider the input ($R_{in'}$) as seen from the base of the transistor as shown in Figure 6.11. It would appear that it should simply be r_e, but this does not take account of the fact that there are two currents flowing through r_e, i_b and βi_b. The impedance is given by:

$$R_{in'} = \frac{v_s}{i_b} \tag{6.7}$$

From KVL the source voltage is:

$$v_s = i_b r_e + \beta i_b r_e \tag{6.8}$$

Substituting into equation 6.7 and rearranging the terms gives:

$$R_{in'} = (1 + \beta)r_e \tag{6.9}$$

Notice that the resistance is not simply r_e, but $(1 + \beta)r_e$. With a typical value of β of 100 and for $I_E = 1$ mA, this results in $r_e = 26\ \Omega$, and the input resistance is 2.6 kΩ.

The equivalent circuit, as seen by the source, can now be simplified to that shown in Figure 6.12.

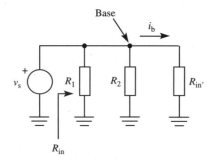

FIGURE 6.12   Modified equivalent circuit for the calculation of the input resistance.

By inspection the input resistance as seen by the source is:

$$R_{in} = R_1 \| R_2 \| R_{in'} \tag{6.10}$$

▶ **EXAMPLE 6.5** ────────────────────────────────

For the common-emitter circuit $R_1 = 68$ kΩ, $R_2 = 33$ kΩ, $R_E = 1$ kΩ, $R_C = 1.8$ kΩ, $V_{CC} = 12$ V and $\beta = 120$. Determine the input resistance and the voltage gain.

▶ *Solution*

DC Condition

The Thévenin voltage and resistance are:

$$V_T = \frac{(12\ \text{V})(33\text{k})}{(33\text{k} + 68\text{k})} = 3.9\ \text{V}$$

and

$$R_T = \frac{(33k)(68k)}{(33k + 68k)} = 22.2 \text{ k}\Omega$$

and the collector current is:

$$I_C = \frac{(3.9 \text{ V} - 0.7 \text{ V})}{1k + (22.2k/120)} = 2.7 \text{ mA}$$

The small-signal resistance is:

$$r_e = \frac{26 \text{ mV}}{2.7 \text{ mA}} = 9.6 \text{ }\Omega$$

The voltage gain is:

$$A_v = -\frac{1.8 \text{ k}\Omega}{9.6 \text{ }\Omega} = -187$$

The input resistance is:

$$R_{in'} \approx \beta r_e = (120)(9.6 \text{ }\Omega) = 1.2 \text{ k}\Omega$$

and the input resistance as seen by the source is:

$$R_{in} = 68 \text{ k}\Omega \,\|\, 33 \text{ k}\Omega \,\|\, 1.2 \text{ k}\Omega = 1138 \text{ }\Omega$$

▶ EXAMPLE 6.6 ───────────────────────────────

Use PSpice to verify the results of Example 6.5.

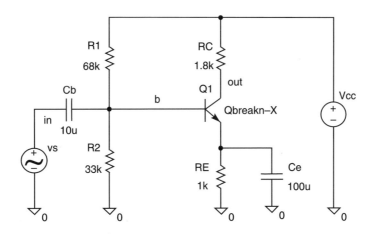

FIGURE 6.13   Schematic for Example 6.6 from PSpice.

▶ *Solution*

The circuit schematic for PSpice is shown in Figure 6.13. For the generic transistor model use **Edit** and select **Model...** to set β to 120 (bf = 120). Use **Transient Analysis...** to observe the output waveform with an input signal of 1 mV amplitude and a frequency of 1 kHz. The dc supply (V_{CC}) is 12 V. For the transient **setup...** let the **Print time:** be 10 μs and the **Final time:** be 2 ms.

The gain is obtained from Probe by plotting v_{out} and v_{in} to obtain the output and input sine waves as shown in Figure 6.14. A gain of 180 is measured with the **cursor** within Probe.

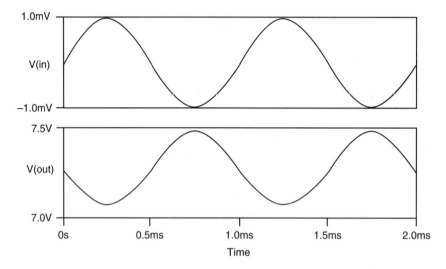

FIGURE 6.14   Input and output voltage waveforms as obtained with Probe.

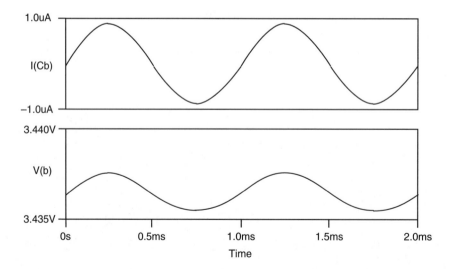

FIGURE 6.15   Input voltage and input current as obtained with Probe.

The input resistance is measured by observing the input voltage (v_b) and the input current, for example the current through C_b listed as I(Cb), as shown in Figure 6.15.

The input resistance is then obtained as the ratio of the peak values of the voltage and the current. A value of 1.13 kΩ should be obtained. (Note that the input resistance may be plotted directly by replacing the default parameter of 'Time' for the X-axis with V(b), and then plotting I(Cb) versus V(b).)

A small amount of phase difference may be observed between the input voltage and the input small-signal current in Figure 6.15. This is caused by the presence of a reactive component from the capacitor C_b. At 1 kHz the reactance of the 10 µF capacitor is 15.9 Ω. Try repeating the exercise with the input frequency set to 10 kHz. The reactive component will be smaller and the phase difference will decrease, and a more accurate value of the input resistance will be obtained.

Output Resistance

The output resistance is obtained with reference to Figure 6.16.

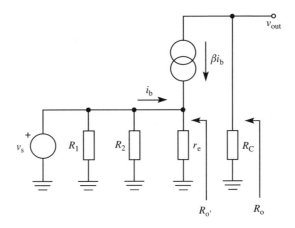

FIGURE 6.16   Equivalent circuit used to determine the output resistance.

The output resistance ($R_{o'}$) as seen by the collector resistance R_C includes the current generator (βi_b). The small-signal resistance of an ideal current generator is ∞ Ω. Thus as far as R_C is concerned it is connected to an open circuit (this follows the normal practice for the analysis of electrical networks of replacing current generators by open circuits when calculating impedance). The presence of r_e in the emitter part of the circuit has no effect on the output resistance, because the current generator is between the collector terminal and the emitter. Therefore, the output resistance as seen from the output (v_{out}) is simply:

$$R_{out} = R_C \tag{6.11}$$

Effect of a Load Resistor

The single-stage common-emitter amplifier considered above and shown in Figure 6.2 is used to increase the amplitude of a signal in order that it may be processed further. The output from the amplifier would be connected to additional circuitry, which could be additional amplification or some other form of signal processing, but most importantly the circuit which follows presents the amplifier with an output load. This is shown in Figure 6.17 by means of a single load resistor R_L, which in practice could be the input resistance of another stage of amplification.

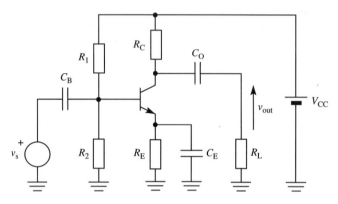

FIGURE 6.17   Common-emitter amplifier with external load.

The load is assumed to be a simple resistor, and for hand calculations this is an adequate representation. Notice the need for another coupling capacitor (C_O) to isolate the dc currents and voltages associated with the transistor from the load. The equivalent circuit is now modified to include the load resistor in parallel with the collector resistor, as shown in Figure 6.18.

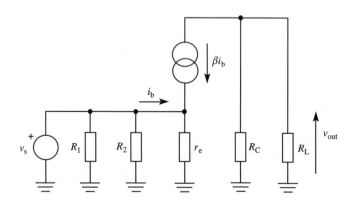

FIGURE 6.18   Equivalent circuit of common-emitter amplifier with an external load.

The voltage gain is now given by:

$$A_v = -\frac{R_C \| R_L}{r_e} \tag{6.12}$$

where $R_C \| R_L$ is the parallel combination of the two resistors.

EXAMPLE 6.7

For the circuit in Example 6.6 determine the voltage gain if a resistive load of 1 kΩ is attached to the collector by means of a coupling capacitor.

Solution

With a load of 1 kΩ the gain is:

$$A_v = -\frac{R_C \| R_L}{r_e} = -\frac{1.8\text{k} \| 1\text{k}}{9.6} = -67$$

Notice that the gain has been reduced with the addition of an external load resistance.

Effect of the Source Resistance

In the above calculations the source is assumed to have zero internal resistance. In practice this may not be the case. The modified circuit with a source resistance is shown in Figure 6.19.

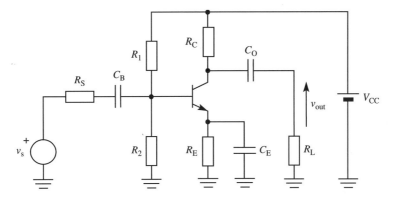

FIGURE 6.19 Common-emitter amplifier with source resistance and external load.

The equivalent circuit is shown in Figure 6.20.

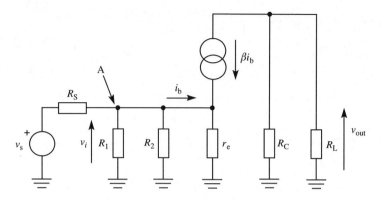

FIGURE 6.20 Equivalent circuit for common-emitter amplifier with source resistance and external load.

From the equivalent circuit in Figure 6.20 the voltage gain from point A at the input to the output is simply:

$$A_v|_A = \frac{v_{out}}{v_i} = - \frac{R_C \| R_L}{r_e} \tag{6.13}$$

but the voltage v_i is generated by v_s. The input can be replaced by the input resistance R_{in} as shown in Figure 6.21.

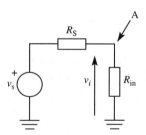

FIGURE 6.21 Modified input to show the effect of the source resistance.

By inspection the relationship between v_i and v_s is:

$$v_i = \frac{v_s R_{in}}{(R_S + R_{in})} \tag{6.14}$$

where R_S and R_{in} act as a potential divider across the source. The overall gain is given by:

$$A_v = \frac{v_{out}}{v_s} = \frac{v_{out}}{v_i} \frac{v_i}{v_s}$$

$$= A_v|_A \frac{v_i}{v_s}$$

and by substitution from equations 6.13 and 6.14 the gain is:

$$A_v = -\frac{R_C \| R_L}{r_e} \frac{R_{in}}{(R_S + R_{in})} \tag{6.15}$$

Equation 6.15 provides a complete expression for the gain with an external load and a source resistance. Notice that the effect of R_S is minimized if $R_{in} \gg R_S$ and for this reason, it is usually desirable for the input resistance of an amplifier to be large.

EXAMPLE 6.8

For the circuit in Example 6.7 the source resistance is 1 kΩ. Determine the voltage gain.

Solution

From Example 6.5 the input resistance is 1.14 kΩ and thus the gain is:

$$A_v = -\frac{R_C \| R_L}{r_e} \frac{R_{in}}{(R_S + R_{in})} = -(67) \frac{1.05k}{(1k + 1.14k)} = -36$$

Notice that the gain is reduced by the presence of the load resistor and the presence of a finite source resistance.

Effect of Emitter Resistance

In the circuits above the emitter resistance is assumed to be fully decoupled and does not appear in the equivalent circuit. However, the presence of a resistor in the emitter circuit changes the properties of the amplifier very significantly and offers a number of attractive features. These features arise from the effect of negative feedback.

The emitter resistor is part of the dc bias circuit for the transistor, but it can also be used as part of the small-signal circuit. The dc requirements may demand a certain value of resistance, but the small-signal requirements may require a different value. These requirements often conflict and result in different values of resistance. However, the difficulty is easily resolved by having two resistors and decoupling only one of them. The circuit with a partially decoupled emitter is shown in Figure 6.22.

The sum of the two resistors $(R_{E1} + R_{E2})$ is used for dc analysis for calculating the collector current, but for ac analysis only R_{E1} need be considered because the decoupling capacitor acts as a short circuit for R_{E2}. The equivalent circuit is shown in Figure 6.23.

The output voltage is:

$$v_{out} = -\beta i_b R_C$$

Notice that this is the same expression which is obtained for a fully decoupled emitter resistor. The presence of R_{E1} in the emitter does not affect the expression for the output voltage.

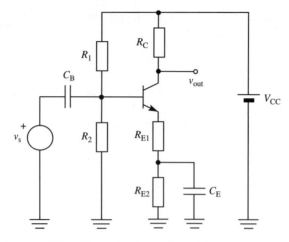

FIGURE 6.22   Emitter amplifier with partial decoupling of the emitter resistor.

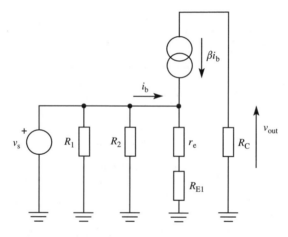

FIGURE 6.23   Equivalent circuit for the common-emitter amplifier with a partially decoupled emitter resistance.

The input voltage can be expressed as:

$$v_s = i_b(r_e + R_{E1}) + \beta i_b(r_e + R_{E1}) \tag{6.16}$$

and the voltage gain is:

$$A_v = \frac{v_{out}}{v_s} = -\frac{\beta R_C}{(1+\beta)(r_e + R_{E1})}$$

and since $\beta \gg 1$ then

$$A_v \approx -\frac{R_C}{(r_e + R_{E1})} \tag{6.17}$$

From a comparison with equation 6.1 it is seen that the emitter resistor R_{E1} has simply been added to the internal resistance of the transistor (r_e). This observation can be applied to the other equations for gain with and without a load or source resistance (equations 6.12 and 6.15).

An important point to note is that the gain is reduced by the presence of the emitter resistor. The internal resistance of the transistor is usually a few tens of ohms, and if $R_{E1} \gg r_e$ then the gain approximates to:

$$A_v \approx -\frac{R_C}{R_{E1}} \tag{6.18}$$

The importance of this equation is that the influence of the transistor has been removed from the expression for the gain, that is the gain is dependent on the ratio of two external resistors, rather than the internal resistance of the transistor. In practice the approximation may not apply because the inequality may be difficult to achieve, but the presence of R_{E1} does improve the stability of the gain and reduces the dependency of the gain on the properties of the transistor.

The input resistance as observed at the base of the transistor with a fully decoupled emitter from equation 6.9 is $(1 + \beta)r_e$. From equation 6.16 above the input resistance from the same point in the equivalent circuit shown in Figure 6.23 is:

$$R_{in'} = \frac{v_s}{i_b} = (1 + \beta)(r_e + R_{E1})$$

$$\approx \beta(r_e + R_{E1}) \tag{6.19}$$

The equation indicates that the input resistance is increased by the presence of R_{E1} as a result of it being added to r_e and then multiplied by β. Thus the input resistance as seen from the source, and which includes the bias resistors, is:

$$R_{in} = R_1 \| R_2 \| R_{in'} \tag{6.20}$$

The output resistance remains unchanged as R_C.

The increase of the input resistance and the reduction in the gain have been examined by the application of KVL to the equivalent circuit. These effects could also have been explained by the application of negative feedback theory which predicts certain improvements for amplifiers which have negative feedback. Some of these improvements are:

■ stabilize the gain
■ change input and output impedance levels
■ increase the frequency bandwidth
■ reduce the level of electrical noise
■ reduce distortion.

Without going into the theory of feedback, and accepting that the presence of an emitter resistor does in fact result in negative feedback, then the gain has been stabilized by making it independent of the transistor, provided that $R_{E1} \gg r_e$. Also the input resistance has been changed and in this instance it has been increased. Different forms of negative feedback affect the input and output impedances differently. The feedback, resulting from the presence of a series resistance like R_{E1}, increases the input resistance. Negative feedback also reduces distortion resulting from non-linearities in the transistor which produced the distortion shown in Figure 6.10. The effects of negative feedback and other forms of feedback will be considered in a later chapter.

▶ **EXAMPLE 6.9** ────────────────────────────────

For the circuit shown in Figure 6.22 $R_1 = 68$ kΩ, $R_2 = 47$ kΩ, $R_C = 2.2$ kΩ, $R_{E1} = 220$ Ω, $R_{E2} = 1.8$ kΩ, $C_B = 4.7$ μF, $C_E = 22$ μF, $V_{CC} = 9$ V and $\beta = 220$. Determine the collector current, the voltage gain, and the input resistance.

▶ *Solution*

DC Conditions

The Thévenin resistance for the base potential divider is:

$$R_T = 68 \text{ k}\Omega \parallel 47 \text{ k}\Omega \approx 27.8 \text{ k}\Omega$$

and the Thévenin voltage is:

$$V_T = \frac{(9 \text{ V})(47 \text{ k}\Omega)}{(68 \text{ k}\Omega + 47 \text{ k}\Omega)} \approx 3.7 \text{ V}$$

The collector current is:

$$I_C = \frac{(3.7 \text{ V} - 0.7 \text{ V})}{2.02 \text{ k}\Omega + (27.8 \text{ k}\Omega/220)} \approx 1.4 \text{ mA}$$

AC Conditions

The small-signal emitter resistance is:

$$r_e = \frac{0.026 \text{ V}}{1.4 \text{ mA}} \approx 18.6 \text{ }\Omega$$

From equation 6.14 the voltage gain is:

$$A_v = -\frac{2.2 \text{ k}\Omega}{18.6 \text{ }\Omega + 220 \text{ }\Omega} = -9.2$$

Note that an approximate expression for the gain is given by the ratio of the collector and the emitter resistors, as given by equation 6.18, that is:

$$A_v = -\frac{2.2 \text{ k}\Omega}{220 \text{ }\Omega} = -10$$

The input resistance is obtained from equation 6.20 as the resistance in the emitter multiplied by β and in parallel with the base potential divider, that is:

$$R_{in} = 27.8 \text{ k}\Omega \, \| \, [(220)(18.6 \ \Omega + 220 \ \Omega)] \approx 18 \text{ k}\Omega$$

Notice that the gain is small and the resistance considerably larger than those in Example 6.5 where the emitter is fully decoupled.

EXAMPLE 6.10

Use PSpice to simulate the circuit described in Example 6.9 and measure the gain and the input resistance for a sine wave input of 10 mV and 1 kHz.

Solution

The circuit schematic is shown in Figure 6.24.

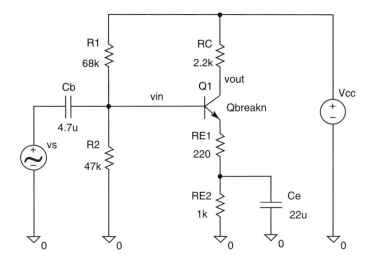

FIGURE 6.24   Schematic circuit for a common-emitter amplifier with an emitter resistor.

The sinusoidal input and output voltage waveforms are shown in Figure 6.25.

With the aid of the cursor it is possible to measure the gain as −9.4, which is in close agreement with the calculated value of −9.2.

The input resistance is obtained by observing the small-signal input voltage and input current, as shown in Figure 6.26. The input current is taken as being the current flowing through the base coupling capacitor.

With the aid of the cursor it is possible to measure the peak-to-peak input voltage and input current as 19.89 mV and 1.1 µA, which gives an input resistance of 18 kΩ.

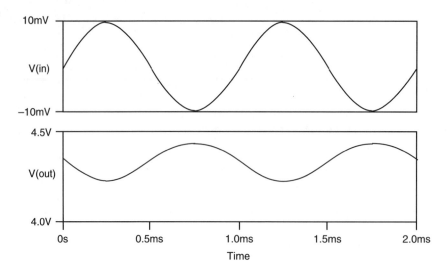

FIGURE 6.25   Input and output waveforms from Probe.

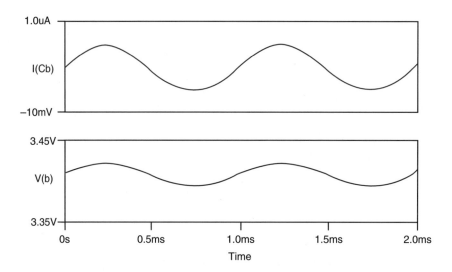

FIGURE 6.26   Probe waveforms used to determine the input resistance.

NB The output waveform is less distorted as a result of the negative feedback introduced by the emitter resistor.

6.3   Common-Base

For the *common-base* (CB) configuration the base terminal is common to both the small-signal input and output as shown in Figure 6.27.

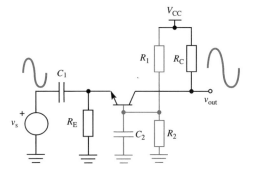

FIGURE 6.27 Schematic of common-base amplifier.

Notice that the input and output sine waves are in phase. Increasing the positive bias to the emitter reduces the emitter current which causes the collector voltage to rise, and thus there is no phase inversion.

The dc biasing is achieved with R_1, R_2 and R_E. The dc analysis or design of the dc biasing is the same as for the common-emitter circuit, because R_1 and R_2 act as a base potential divider in exactly the same manner as for the common-emitter stage, with R_E providing the dc negative feedback. This can be seen more readily by repositioning the bias components, as shown in Figure 6.28.

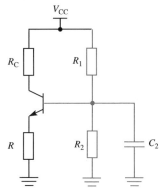

FIGURE 6.28 Common-base circuit redrawn to show the similarity of the biasing arrangements with the common-emitter circuit.

Notice that the capacitor C_2 grounds the base for small-signal currents and acts as a short circuit for R_2. Notice also that with the base effectively shorted to ground and with the power supply also acting as a small-signal short circuit, R_1 is also shorted to ground by the power supply. Thus neither R_1 nor R_2 appears in the ac small-signal circuit.

Equivalent Circuit

The ac small-signal circuit contains R_E, the transistor with the base connected to ground and R_C. Replacing the transistor with the common-emitter r parameter equivalent circuit (as shown in Figure 4.16) results in the circuit shown in Figure 6.29.

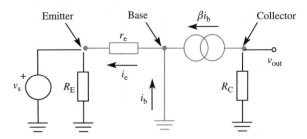

FIGURE 6.29   Common-base equivalent circuit with the common-emitter r parameter circuit.

The circuit shown in Figure 6.29 is an accurate representation of the common-base circuit, but it is not ideal for circuit analysis because the input current is i_e and the output current is βi_b. By convention the signal source (v_s) is shown with its positive terminal attached to the input of the network, which results in an input current in the opposite direction to the small-signal emitter current (i_e) shown in Figure 6.29. Ideally i_e should be flowing in the opposite direction if it is to be the input signal current. It would also be more convenient for the output current generator to be proportional to the input current i_e rather than i_b. These changes can be made as follows.

The base current can be expressed as a function of the emitter current:

$$i_b = \frac{i_e}{1 + \beta}$$

so that the current generator becomes:

$$\beta i_b = \frac{\beta}{1 + \beta} i_e = \alpha i_e \quad \text{since } \alpha = \frac{\beta}{1 + \beta} \tag{6.21}$$

The circuit can be redrawn with the input current flowing from the source into the network, the base connected directly to ground, the current generator described in terms of common-base current gain (α) and also the direction changed to take account of the change of direction of the emitter current. The resultant circuit is shown in Figure 6.30.

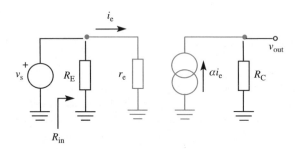

FIGURE 6.30   Common-base equivalent circuit.

This circuit does not use 'conventional current flow', but it is easier to understand and much simpler to analyze since the input current flows in the direction expected, and the input and output networks are completely separate. In addition the output current is a function of the input current.

Voltage Gain

The input and output voltages are obtained by inspection as:

$$v_{out} = \alpha i_e R_C$$
$$v_s = i_e r_e$$

and the gain is:

$$A_v = \frac{v_{out}}{v_s} = \frac{\alpha R_C}{r_e} \approx \frac{R_C}{r_e} \quad \text{since } \alpha \approx 1 \tag{6.22}$$

The expression for voltage gain is the same for the common-base and the common-emitter, with the exception of the negative sign for the common-emitter. The presence of an external load resistor is taken into account by replacing R_C in equation 6.22 with $R_C \| R_L$ where R_L is the load resistance.

Input Resistance

By inspection of Figure 6.30 the input resistance is:

$$R_{in} = \frac{R_E r_e}{R_E + r_e} \approx r_e \quad \text{since } r_e \ll R_E \tag{6.23}$$

Remembering that r_e is inversely proportional to the emitter current and that for a current of 1 mA the resistance is 26 Ω, it is apparent that the input resistance of the common-base amplifier is very low.

The output resistance is the same as for the common-emitter and is simply R_C.

A possible application of the common-base amplifier is to amplify radio frequency signals which are carried on a coaxial cable. The radio frequency impedance of coaxial cable is very low, typically 50 Ω. To avoid reflections occurring in the cable it is necessary to terminate the cable in an impedance which closely matches that of its characteristic impedance, that is 50 Ω. This can be achieved with the common-base amplifier with its low input impedance. The CB amplifier is also useful as an active load in a circuit configuration known as a cascode. This will be considered in a later chapter on field effect transistors.

▷ **EXAMPLE 6.11** _____

For the common-base amplifier shown in Figure 6.31 $V_{CC} = 12$ V, $R_C = 1.8$ kΩ, $R_1 = 18$ kΩ, $R_2 = 3.3$ kΩ, $R_E = 560$ Ω, $R_L = 4.7$ kΩ and $\beta = 150$. Determine the output voltage and the input resistance if the input is 1 mV$_{rms}$ sine wave.

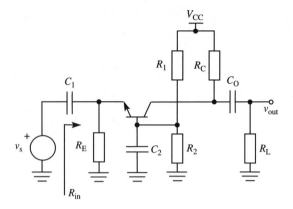

FIGURE 6.31   Common-base amplifier.

▷ Solution

DC Condition

The Thévenin voltage across R_2 is:

$$V_T = \frac{(12 \text{ V})(3.3\text{k})}{(3.3\text{k} + 18\text{k})} = 1.86 \text{ V}$$

and the Thévenin resistance is:

$$R_T = 3.3\text{k} \parallel 18\text{k} = 2.8 \text{ k}\Omega$$

The collector current is:

$$I_C = \frac{(1.86 - 0.7)}{560 + (2.8\text{k}/150)} = 2.0 \text{ mA}$$

and

$$r_e = \frac{0.026}{2 \text{ mA}} = 13 \text{ }\Omega$$

AC Conditions

The voltage gain is:

$$A_v = \frac{R_C \parallel R_L}{r_e} = \frac{1.8\text{k} \parallel 4.7\text{k}}{13} \approx 100$$

and the output voltage is:

$$v_{out} = A_v v_s = 100 \text{ mV}$$

The input resistance is:

$$R_{in} = R_E \| r_e = 560 \| 13 \approx 13 \ \Omega$$

Emitter Follower

For the *emitter follower* (EF) or *common-collector* (CC) circuit the collector terminal is common to the input and the output. The circuit for the emitter follower is shown in Figure 6.32.

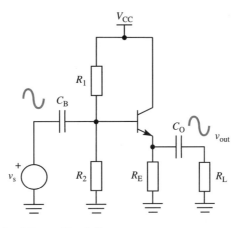

FIGURE 6.32 Schematic of the emitter follower.

At first glance the collector does not appear to be common with the input, but notice that the output is taken from the emitter and that the collector is shorted to ground through the low ac impedance of the power supply. Notice that there is no phase inversion shown for the output waveform. This is because as the input signal increases the current through the transistor increases which causes the output voltage across R_E and R_L to increase; the emitter *follows* the base. A load resistance is included because an important feature of this circuit is the magnitude of the load. For the common-emitter and the common-base the load resistance is usually large and can often be ignored in comparison with the collector resistance, but this is often not the case for the emitter follower.

The dc biasing is the same as for the common-emitter, although the operating point may be very different with the transistor operating at much higher current levels.

Voltage Gain

The equivalent circuit is shown in Figure 6.33.

With reference to the equivalent circuit the output voltage is:

$$v_{out} = i_b(R_E \| R_L) + \beta i_b(R_E \| R_L) \tag{6.24}$$

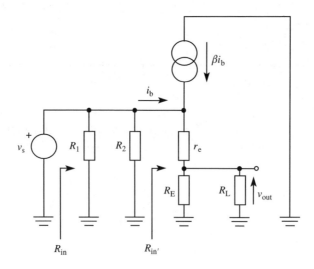

FIGURE 6.33   Equivalent circuit for the emitter follower.

and the input voltage is:

$$v_s = i_b(r_e + R_E \| R_L) + \beta i_b(r_e + R_E \| R_L) \tag{6.25}$$

The voltage gain is:

$$A_v = \frac{v_{out}}{v_s} = \frac{R_E \| R_L}{r_e + R_E \| R_L} \approx 1 \quad \text{since } R_E \| R_L \gg r_e \tag{6.26}$$

The emitter follower usually operates at 1 mA or greater so that $r_e \ll 26\ \Omega$ and the approximation in equation 6.26 is generally valid. Thus the voltage gain of the emitter follower is unity.

Input Resistance

The input resistance is the same as that obtained for the common-emitter stage with the partially decoupled emitter resistance, that is:

$$R_{in'} = (1 + \beta)(r_e + R_E \| R_L) \tag{6.27a}$$

and the input resistance as seen by the source is:

$$R_{in} = R_{in'} \| R_1 \| R_2 \tag{6.27b}$$

Depending on the value of R_L the input resistance $(R_{in'})$ can be large, and provided that the bias resistors do not shunt this resistance the overall value of input resistance is also large.

Output Resistance

At first glance it may appear that the output resistance is simply R_E, in the same way that the output resistance for the CE and CB amplifiers is R_C. However, this would be incorrect because the signal path now contains r_e and the external base circuit, and also the current generator. The approach used here is the repeated use of Thévenin's theorem to reduce the circuit to a simple Thévenin equivalent circuit.

From the Thévenin network in Figure 6.34 the output resistance is defined as:

$$R_o = \frac{v_{oc}}{i_{sc}} \tag{6.28}$$

where i_{sc} is the short circuit current, and where R_o is the output resistance which would be measured at the output of a circuit which comprises a resistor and a voltage generator. The object of the transformation of the CB circuit is to simplify it so that v_{oc} and i_{sc} may be obtained.

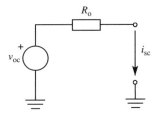

FIGURE 6.34   Thévenin circuit.

The first step is to remove the load and the emitter resistors and then to determine the Thévenin voltage and resistance as seen from the emitter terminal as shown in Figure 6.35(a).

The bias and source resistors as seen from the base at AA′ in Figure 6.35(a) are replaced by the Thévenin circuit and redrawn as in Figure 6.35(b). The Thévenin resistance R_T for the base circuit is:

$$R_T = R_S \,\|\, R_1 \,\|\, R_2 \tag{6.29}$$

Now from the definition for the resistance of the Thévenin circuit the output resistance is the ratio of the open-circuit voltage and the short-circuit current. The open-circuit voltage in Figure 6.35(b) is simply:

$$v_{oc} = V_T \tag{6.30}$$

since there is no current flow in either R_T or r_e.

The short circuit current from Figure 6.35(b) is:

$$i_{sc} = i_b + \beta i_b \tag{6.31}$$

The current which flows with the short circuit present is generated by V_T, that is:

$$V_T = i_b R_T + i_b r_e + \beta i_b r_e$$

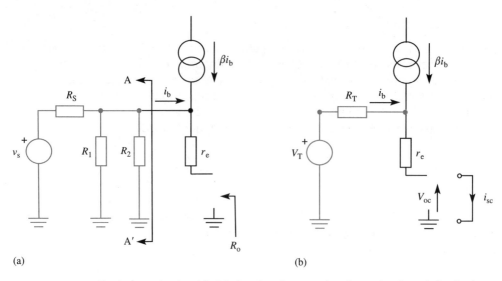

(a)   (b)

FIGURE 6.35   Equivalent circuit with (a) showing the complete base circuit and (b) the base circuit replaced by the Thévenin equivalent.

so that:

$$i_b = \frac{V_T}{R_T + (1+\beta)r_e} \qquad (6.32)$$

Substituting for i_b in equation 6.31 gives:

$$i_{sc} = \frac{(1+\beta)V_T}{R_T + (1+\beta)r_e}$$

and from equations 6.28 and 6.30 the output resistance is:

$$R_o = \frac{v_{oc}}{i_{sc}} = \frac{R_T}{(1+\beta)} + r_e \qquad (6.33)$$

The effective output resistance is:

$$R_{oeff} = R_E \| R_o \qquad (6.34)$$

where R_E is the emitter resistor.

From equation 6.29 R_T is typically a few thousand ohms and with a value of at least 100 for β then from equation 6.33 the output resistance looking back into the emitter is usually only a few tens of ohms.

The important features of the emitter follower are high input resistance, low output resistance and a voltage gain of approximately unity. It acts as an electronic transformer and provides a means for matching the high output resistance of a CE or CB amplifier to the low impedance of a transducer or coaxial cable. It is often know as a *buffer amplifier*.

6.5 ▷ Impedance Reflection

With reference to the emitter follower or the common-emitter amplifier it is seen that the input impedance as seen from the base includes the resistance which is in the emitter, either r_e or $r_e + R_E$, multiplied by $(1 + \beta)$. For the emitter follower the output resistance when viewed from the emitter contains the resistance in the base circuit $(R_S \| R_1 \| R_2)$ divided by $(1 + \beta)$.

This impedance transformation is known as *impedance reflection*. The impedance in the emitter is reflected into the base circuit by multiplying it by $(1 + \beta)$, while the impedance in the base is reflected into the emitter circuit by dividing it by $(1 + \beta)$.

▷ **EXAMPLE 6.12** ————————————————————————

For the circuit in Figure 6.36 determine the voltage gain, the input resistance and the output resistance when $V_{CC} = 9$ V, $R_B = 150$ kΩ, $R_E = 1$ kΩ, $R_S = 600$ Ω and $\beta = 100$.

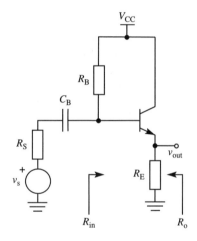

FIGURE 6.36   Schematic of an emitter follower.

▷ *Solution*

DC Conditions

The collector current is:

$$I_C = \frac{9 - 0.7}{1k + (150k/100)} = 3.3 \text{ mA}$$

and

$$r_e = \frac{0.026}{3.3 \text{ mA}} = 7.9 \text{ Ω}$$

AC Conditions

The voltage gain is:

$$A_v = \frac{R_E}{r_e + R_E} = \frac{1000}{7.9 + 1000} = 0.992 \approx 1$$

The input resistance is:

$$R_{in} = 160k \,\|\, [\beta(r_e + R_E)] = 60 \text{ k}\Omega$$

The output resistance is:

$$R_o = \left(\frac{R_T}{\beta} + r_e\right) \Big\| R_E$$

$$R_o = \left(\frac{600 \,\|\, 150k}{100} + 7.9\right) \Big\| 1k = 13.7 \ \Omega$$

Notice the voltage gain of approximately unity, the high input resistance and the low output resistance. The value of an external load can have a very significant effect on these values, as illustrated in the next example.

► EXAMPLE 6.13 ——————————————————————

Determine the voltage gain, input resistance and output resistance for the circuit described in Example 6.12, but with a load resistor (R_L) of 50 Ω capacitively coupled to the emitter, as shown in Figure 6.37.

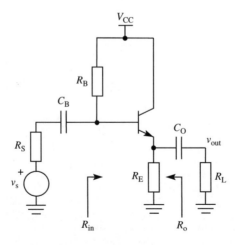

FIGURE 6.37   Emitter follower with an external load.

Solution

The dc conditions remain unchanged and R_e remains equal to 7.9 Ω. The voltage gain, however, is now given as:

$$A_v = \frac{R_E \| R_L}{r_e + R_E \| R_L} = \frac{1000 \| 50}{7.9 + 1000 \| 50} = 0.858$$

The input resistance is

$$R_{in} = 150\text{k} \| [\beta(r_e + R_E \| R_L)] = 5.3 \text{ k}\Omega$$

The output resistance remains unchanged at 13.7 Ω.

While the voltage gain has been reduced it is still close to unity, but there has been a large change in the input resistance. As will be seen in a later chapter this can have a significant effect on the frequency response when calculating the value of coupling capacitors.

6.6 Multistage Amplifier

Multistage amplifiers are used to increase the gain or to modify the characteristics of the amplifier to satisfy a particular specification, such as increasing or lowering the impedance levels at the input and output. A common requirement is that an amplifier should have a high input resistance to avoid loading the source, and a low output resistance to avoid reducing the gain when a low-resistance load is attached. This can be achieved with a common-emitter input stage and an emitter follower output stage. A common-base stage followed by an emitter follower has low input impedance and a low output impedance.

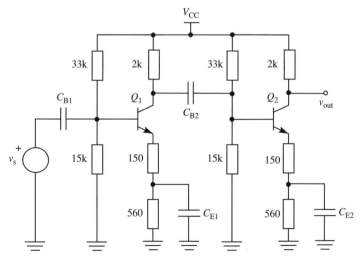

FIGURE 6.38   Two-stage common-emitter amplifier.

Two-Stage CE–CE

An example of a two-stage CE amplifier is shown in Figure 6.38.

Each stage is identical, although this is not a requirement for a two-stage amplifier, and is only used here for simplicity in the calculations. The partially decoupled emitter resistor stabilizes the gain and increases the input resistance. The small-signal circuit is shown in Figure 6.39.

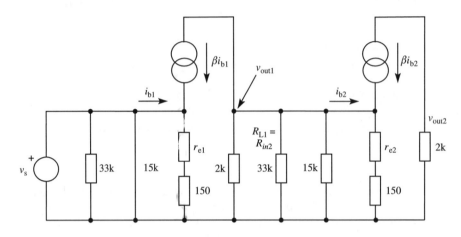

FIGURE 6.39   Small-signal equivalent circuit of two-stage amplifier.

If in the above example the current gain (β) is 150 and the emitter current is 3 mA, then the following calculations can be made.

The small-signal emitter resistance is:

$$r_e = \frac{0.026}{3 \text{ mA}} = 8.7 \ \Omega$$

The output voltage for the first stage is:

$$v_{out1} = -\beta i_{b1} R_{c1} \| R_{L1}$$

where R_{L1} is the external load resistance for the first stage which in this case represents the input resistance of the second stage as shown in Figure 6.39.

The input voltage is:

$$v_{in1} = v_s = (i_{b1} + \beta i_{b1})(r_{e1} + R_{E1})$$
$$\approx \beta i_{b1}(r_{e1} + R_{E1})$$

and the voltage gain for the first stage is:

$$A_{v1} = \frac{v_{out1}}{v_{in1}} = -\frac{2k \| R_{L1}}{8.7 + 150}$$

The input resistance of the second stage is:

$$R_{in2} = 33k \| 15k \| [(1 + \beta)(r_{e2} + R_{E2})]$$
$$= 33k \| 15k \| [(151)(8.7 + 150)] \approx 7.2 \text{ k}\Omega$$

and

$$A_{v1} = -\frac{2k \| 7.2k}{158.7} \approx -9.86$$

For the second stage the gain is:

$$A_{v2} = -\frac{2k}{8.7 + 150} = -12.6$$

The total gain is the product of A_{v1} and A_{v2}, that is:

$$A_{vtotal} = (-9.86)(-12.6)$$
$$= 124$$

Notice that the negative sign associated with the single stage disappears with two stages and the output voltage is now in phase with the input.

The input resistance is:

$$R_{in1} = 33k \| 15k \| [(1 + \beta)(r_{e1} + R_{E1})]$$
$$= 33k \| 15k \| 24k$$
$$= 7.2 \text{ k}\Omega$$

and the output resistance is:

$$R_o = 2 \text{ k}\Omega$$

Two-Stage CE–EF

In Figure 6.40 a common-emitter stage provides the gain while the emitter follower provides a low-output resistance. Let the emitter currents be 2 mA for Q_1 and 1.5 mA for Q_2 and let $\beta = 100$ for both transistors.

In the diagram Q_1 provides the gain as a CE stage while Q_2 acts as an emitter follower to provide a low-output impedance. The overall gain is that of the first stage since the gain of the emitter follower is approximately unity. The small-signal equivalent circuit is shown in Figure 6.41.

The output voltage for Q_1 is:

$$v_{out1} = -\beta i_{b1} R_{C1} \| R_{L1}$$

where R_{L1} is the load resistance for the first stage, which is equal to the input resistance of the second stage.

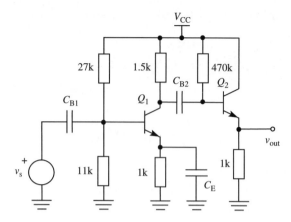

FIGURE 6.40   Two-stage amplifier with common-emitter gain and emitter follower output.

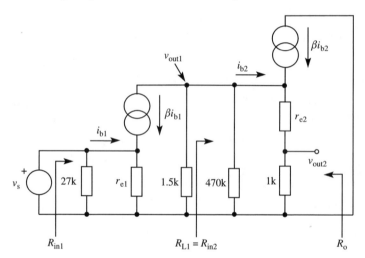

FIGURE 6.41   Small-signal equivalent circuit of two-stage amplifier with emitter follower output.

The input voltage for the first stage is:

$$v_{in1} = v_s = (i_{b1} + \beta i_{b1})r_{e1}$$
$$\approx i_{b1}\beta r_{e1}$$

Then the voltage gain of the first stage is:

$$A_{v1} = -\frac{v_{out1}}{v_{in1}} = -\frac{R_{C1} \| R_{L1}}{r_{e1}}$$

The small-signal emitter resistances are:

$$r_{e1} = \frac{0.026}{2 \text{ mA}} = 13 \ \Omega \quad \text{and} \quad r_{e2} = \frac{0.026}{1.5 \text{ mA}} = 17.3 \ \Omega$$

The load resistance (R_{L1}) is the input resistance (R_{in2}) of the emitter follower and is:

$$R_{in2} = 470k \, \| \, [(1 + \beta)(r_{e2} + R_{E2})]$$
$$R_{in2} = 470k \, \| \, [(101)(17.3 + 1000)] \approx 84 \; k\Omega$$

and the gain of the first stage is:

$$A_{v1} = -\frac{1.5k \, \| \, 84k}{13}$$

$$= -113$$

The gain of the second stage is unity so that the total gain is:

$$A_{vtotal} = (-113)(1) = -113$$

Notice that even though there are two stages the output is out of phase with the input, as indicated by the negative sign, because the emitter follower does not introduce a phase change.

The output resistance is:

$$R_o = (1k) \, \| \left(\frac{R_T}{\beta} + r_{e2}\right)$$

where the Thévenin resistance is:

$$R_T = 1.5k \, \| \, 470k \approx 1.5 \; k\Omega$$

and

$$R_o = (1k) \, \| \left(\frac{1.5k}{100} + 17.3\right) \approx 31 \; \Omega$$

The input resistance is:

$$R_{in1} = 27k \, \| \, 11k \, \| \, [(1 + \beta)r_{e1}]$$
$$= 27k \, \| \, 11k \, \| \, 1.3k = 1.1 \; k\Omega$$

Two-Stage CB–CE

The circuit in Figure 6.42 has a low input impedance from the common-base stage followed by a common-emitter with a partially decoupled emitter resistor. The emitter current for Q_1 is 2.4 mA and for Q_2 it is 2 mA, and $\beta = 120$ for both transistors.

The small-signal equivalent circuit is shown in Figure 6.43.

The output voltage of the first stage is:

$$v_{out1} = \alpha i_{e1}(1.8k \, \| \, R_{L1})$$
$$\approx i_{e1}(1.8k \, \| \, R_{in2}) \quad \text{since } \alpha \approx 1 \text{ and } R_{L1} = R_{in2}$$

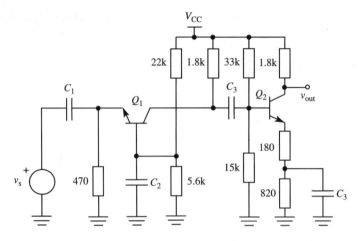

FIGURE 6.42   Two-stage amplifier with common-base input and common-emitter output.

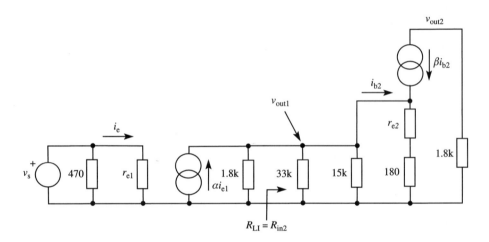

FIGURE 6.43   Small-signal equivalent circuit with common-base input.

The input voltage for the first stage is:

$$v_{in1} = i_{e1}r_{e1}$$

The voltage gain of the first stage is:

$$A_{v1} = \frac{v_{out1}}{v_{in1}} = \frac{1.8k \| R_{in2}}{r_{e1}}$$

The input resistance of the second stage is:

$$R_{in2} = 33k \| 15k \| [(1 + 120)(r_{e2} + 180)]$$

The small-signal emitter resistances are:

$$r_{e1} = \frac{0.026}{2.4 \text{ mA}} \approx 11 \ \Omega \quad \text{and} \quad r_{e2} = \frac{0.026}{2 \text{ mA}} = 13 \ \Omega$$

and the gain of the first stage is:

$$A_{v1} = \frac{1.8k \,\|\, 23.3k}{11} = 130$$

and for the second stage:

$$A_{v2} = -\frac{1.8k}{(r_{e2} + 180)}$$

$$= -\frac{1.8k}{193}$$

$$= -9.3$$

The total gain is:

$$A_{vtotal} = (130)(-9.3) = -1209$$

The input resistance is:

$$R_{in1} = 470 \,\|\, r_{e1} \approx 11 \ \Omega$$

▶ **EXAMPLE 6.14** ————————————————————

For the circuit shown in Figure 6.44 $V_{CC} = 9$ V and the common-emitter current gain for both transistors is 120. Determine the voltage gain by hand calculations

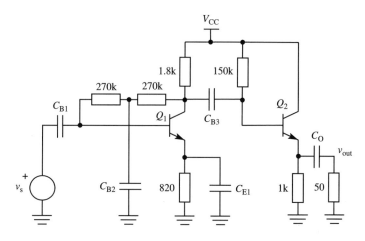

FIGURE 6.44   Two-stage amplifier with collector bias.

and verify using PSpice. For the purposes of PSpice set all the capacitors to 100 µF and use a sine wave input of 1 kHz.

▷ *Solution*

DC Conditions

For Q_1 the collector current is:

$$I_{C1} = \frac{(9-0.7)}{[820 + 1.8k + (540k/120)]} = 1.2 \text{ mA}$$

and for Q_2 the collector current is:

$$I_{C2} = \frac{(9-0.7)}{[1k + (150k/120)]} = 3.7 \text{ mA}$$

and the small-signal emitter resistances are:

$$r_{e1} = \frac{0.026}{1.2 \text{ mA}} = 21.7 \ \Omega \quad \text{and} \quad r_{e2} = \frac{0.026}{3.7 \text{ mA}} = 7 \ \Omega$$

AC Conditions

The small-signal circuit is shown in Figure 6.45.

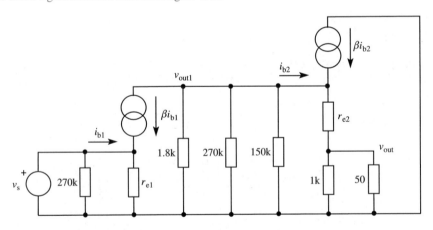

FIGURE 6.45   Small-signal equivalent circuit.

From Figure 6.45 note the position of the collector-bias resistors for the first stage (2×270 kΩ) and also the presence of the 50 Ω load at the output of the second stage.

The voltage gain for the first stage is:

$$A_{v1} = -\frac{1.8k \parallel 270k \parallel R_{in2}}{21.7}$$

and the input impedance for the second stage is:

$$R_{in2} = 150k \,\|\, (1 + \beta)(7 + 1k \,\|\, 50) = -6.3 \text{ k}\Omega$$

and the gain is:

$$A_{v1} = -\frac{1.8k \,\|\, 270k \,\|\, 6.3k}{21.7} \approx -64$$

The gain of the second stage is approximately unity but because of the 50 Ω load will be reduced. Check the gain as:

$$A_{v2} = \frac{1k \,\|\, 50}{7 + 1k \,\|\, 50} = 0.872$$

The presence of the 50 Ω load has a significant effect on the gain. The total gain is:

$$A_{vtotal} = A_{v1} A_{v2} = (-64)(0.872) \approx -56$$

PSpice

The circuit schematic for PSpice is shown in Figure 6.46 and the output waveform in Figure 6.47.

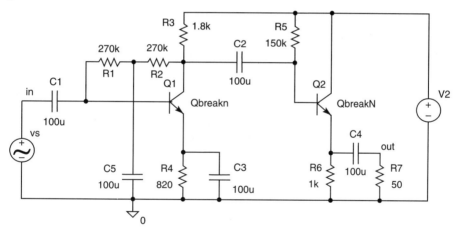

FIGURE 6.46 Schematic for PSpice.

The values of the collector currents are obtained from the output text file and the voltage gain is obtained from the ratio of the amplitudes of the input and output sine waves. The values obtained are as follows:

Parameter	Calculated	PSpice
I_{C1}	1.2 mA	1.2 mA
I_{C2}	3.7 mA	3.8 mA
A_{vtotal}	−56	−53

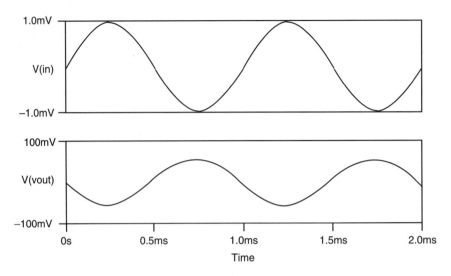

FIGURE 6.47   Input and output sine waves from PSpice.

Direct Coupled Multistage Amplifier

A reduction in the number of components may be achieved by connecting the input of the second stage directly to the output of the first stage. This removes the need for a coupling capacitor and the bias resistors for the second stage. However, it may present a problem for some configurations since the dc voltage at the collector of the first stage is typically $V_{CC}/2$ and this is generally much too large to apply directly to the input, for example the base, of the second stage. The problem is resolved by using a pnp transistor for the second stage, as shown in Figure 6.48, for a CE–CE amplifier.

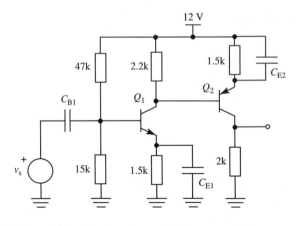

FIGURE 6.48   CE–CE amplifier with directly coupled pnp second stage.

For the npn transistor the dc voltage levels increase in a positive direction, that is towards the dc supply of 12 V, from the emitter to the base and then to the collector. For the pnp transistor the dc voltage decreases from the emitter to the base and then to the collector, and thus the collector of Q_2 is at a lower dc level than the collector of Q_1. If an npn transistor were used for Q_2 then its collector would be at a higher level than that of Q_1 and consequently the output voltage swing would be restricted.

For Q_1 the collector current, assuming that $\beta = 150$, is:

$$I_{C1} = \frac{2.9\ V - 0.7\ V}{1.5k + (8.9\ k\Omega/150)} = 1.4\ mA$$

The voltage at the collector of Q_1 is:

$$V_{C1} = 12\ V - (1.4\ mA)(2.2\ k\Omega) \approx 8.9\ V$$

Ignoring the base current of Q_2 simplifies the calculation so that the voltage at the emitter of Q_2 is 8.9 V + 0.7 V = 9.6 V. The collector current of Q_2, if $\beta = 150$, is:

$$I_{C2} \approx \frac{12\ V - 9.6\ V}{1.5\ k\Omega} \approx 1.6\ mA$$

and

$$V_{C2} = (1.6\ mA)(2\ k\Omega) = 3.2\ V$$

The equivalent circuit is shown in Figure 6.49.

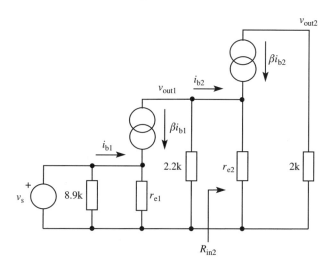

FIGURE 6.49 Equivalent circuit for CE–CE amplifier.

Notice that the relative positions of the current generator and the small-signal resistance (r_e) of each transistor are the same for both the npn and pnp devices.

The small-signal resistance values are:

$$r_{e1} = \frac{0.026}{1.4 \text{ mA}} \approx 18.6 \ \Omega$$

and

$$r_{e2} = \frac{0.026}{1.6 \text{ mA}} \approx 16 \ \Omega$$

The gain of the first stage is:

$$A_{v1} = -\frac{R_{C1} \| R_{in2}}{r_{e1}} = -\frac{2.2\text{k} \| (1+\beta)r_{e2}}{r_{e1}} \approx -62$$

The gain for the second stage is:

$$A_{v2} = -\frac{R_{C2}}{r_{e2}} = -\frac{2\text{k}}{16} = -125$$

and the overall gain is

$$A_{vT} = (-62)(-125) = 7750$$

There are many other examples of the use of direct coupling between transistor stages which reduce the number of components, for example the base of an emitter follower stage may be connected directly to the collector of a gain stage, as shown in Figure 6.50, provided that a correct choice of emitter resistor is made.

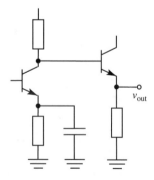

FIGURE 6.50   Direct coupling of an emitter follower to a gain stage.

One of the problems of the direct coupling shown above is that of dc stability. If the circuit is likely to be subjected to large temperature changes, then great care must be taken, because a small change in collector current, and hence voltage, in the first stage is amplified by the second stage. More complex forms of direct coupling are possible which include dc feedback which compensates for such changes, but with the availability of a wide range of operational amplifiers, which are the ultimate design examples of direct coupling, it is generally not necessary to design discrete transistor stages for this purpose.

▷ PROBLEMS

6.1 For the circuit shown determine the voltage gain if $\beta = 120$. Assume that the capacitors act as short circuits.

[−150]

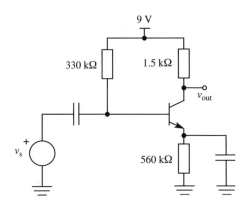

6.2 For the circuit in Problem 6.1 draw the dc and ac load-lines and estimate the peak small-signal output voltage.

[3.75 V]

6.3 Use PSpice to verify the performance of the circuit in Problem 6.1 and obtain values for I_C, V_{CE} and A_v for a sine wave input of 1 mV at 1 kHz. Let the capacitors be 100 μF. Increase the input to 10 mV and 15 mV and observe the output.

[$I_C = 2.47$ mA, $V_{CE} = 3.89$ V, $A_v = -142$]

[sine wave with $v_{peak} = 1.38$ V, distorted sine wave]

6.4 Determine the voltage gain and the input resistance for the circuit shown if $\beta = 150$.

[$A_v = -189$, $R_{in} = 1.04$ kΩ]

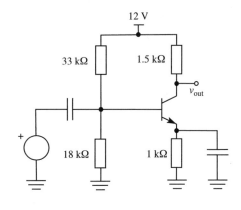

6.5 Use PSpice to verify the values for Problem 6.4 with a sine wave input of 1 mV at 1 kHz. Measure the input resistance as $v_{in}/i_{base\ capacitance}$.

[$I_C = 3.16$ mA, $A_v = -181$, $R_{in} = 1.15$ kΩ]

6.6 For the circuit shown determine R_{in} and A_v if $\beta = 200$.

[$I_C = 3.5$ mA, $A_v = -131$, $R_{in} = 1.47$ kΩ]

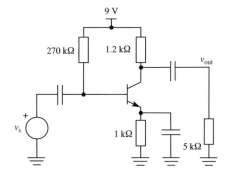

6.7 For the circuit shown determine the voltage gain from the source to the output if $\beta = 180$.

$$[I_C = 2.7 \text{ mA}, A_v = -97]$$

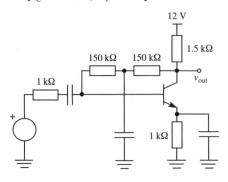

6.8 Use PSpice to verify the values obtained in Problem 6.7.

$$[I_C = 2.68 \text{ mA}, A_v = -96.5]$$

6.9 For the circuit shown determine the voltage gain if $\beta = 150$.

$$[I_C = 2.7 \text{ mA}, A_v = -89]$$

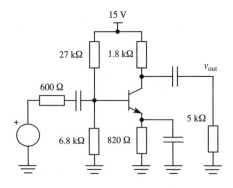

6.10 Use PSpice to verify the values obtained in Problem 6.9.

$$[A_v = -86.5]$$

6.11 For the circuit shown determine the input resistance and the voltage gain if $\beta = 100$.

$$[I_C = 1.3 \text{ mA}, A_v = -22.3, R_{in} = 6.6 \text{ k}\Omega]$$

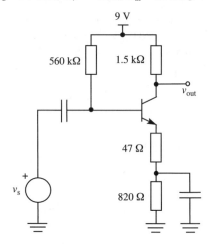

6.12 Use PSpice to verify the values obtained in Problem 6.11.

$$[A_v = -22, R_{in} = 6.7 \text{ k}\Omega]$$

6.13 Determine the voltage gain and the input resistance if $\beta = 100$ for the circuit shown.

$$[I_C = 1.8 \text{ mA}, A_v = -7.8,$$
$$R_{in} = 17.8 \text{ k}\Omega]$$

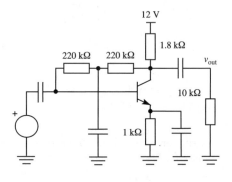

6.14 Use PSpice to verify the values obtained in Problem 6.13.

$$[A_v = -7.6, R_{in} = 18 \text{ k}\Omega]$$

6.15 Determine the gain and input resistance for the circuit shown if $\beta = 100$.

$$[I_C = 2 \text{ mA}, A_v = -64, R_{in} = 1 \text{ k}\Omega]$$

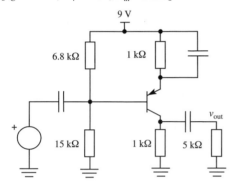

6.16 For the common-base circuit shown determine the voltage gain if $\beta = 100$.

$$[I_C = 1.35 \text{ mA}, A_v = 114]$$

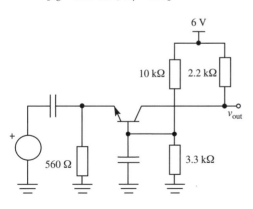

6.17 For the circuit shown determine the voltage gain if $\beta = 100$.

$$[I_C = 4.2 \text{ mA}, A_v = 32]$$

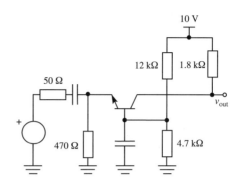

6.18 Use PSpice to verify the values obtained in Problem 6.17.

$$[A_v = 31]$$

6.19 For the emitter follower determine the voltage gain, the input resistance and the output resistance if $\beta = 150$.

$$[I_C = 4.5 \text{ mA}, A_v = 0.891,$$
$$R_{in} = 7.8 \text{ k}\Omega, R_{out} = 9.7 \text{ }\Omega]$$

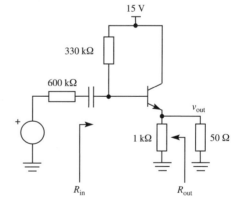

6.20 Use PSpice to verify the values of voltage gain and input resistance obtained in Problem 6.19.

$$[A_v = 0.82, R_{in} = 7.9 \text{ k}\Omega]$$

CHAPTER 7

Field Effect Transistor: DC Biasing

The bipolar transistor is a low-impedance current-operated device. The field effect transistor, on the other hand, is a high-impedance voltage-operated device. For the bipolar transistor current flow results from the movement of both holes and electrons – it is bipolar. Current flow in the field effect transistor only involves one carrier type, either electrons for n-channel devices, or holes for p-channel devices – it is sometimes referred to as a unipolar device.

There are two basic types of field effect transistor, one involving a pn junction (JFET) and the other involving a metal–oxide–semiconductor structure (MOSFET or MOST). Both classifications have very similar current–voltage characteristics, but the equations describing these characteristics are different for each device. The junction device is more widely used as a discrete component, while the metal–oxide device is used extensively in integrated circuits. There are a small number of special applications where a discrete MOSFET is used to achieve very high input impedance. One such application is the vertical channel MOS (or VMOS) device which is used extensively for power amplifiers.

The description of the physical structure and physical operation is not necessary for an understanding of how the devices operate in a circuit and so is optional, apart from a description of the symbols in Sections 7.1 for the JFET and 7.4 for the MOSFET below. Similarly the description of the complex equivalent circuit for CAD programs is not necessary and may be disregarded.

7.1 Physical Structure of the JFET

A very much simplified drawing of the junction field effect transistor is shown in Figure 7.1. There are two types of device. An n-channel device in which current flow takes place in n-type material, and a p-channel device in which current flow takes place in p-type material. The symbols for each device are shown alongside the physical representations.

The physical representations are rather idealized and do not represent practical devices, but they are adequate to explain how the devices operate. The diagrams represent small rectangular bars of semiconductor (silicon), n-type or p-type depending on the polarity required for the final device. For the n-channel device a p-type region is formed around the middle of the bar. The terminals are named drain, source and gate as shown, and the appropriate symbol for each transistor is also shown. For normal active operation the polarity of V_{GG} is such that the pn junction is reverse biased, and the polarity of V_{DD} is such that the majority carriers in the channel (electrons for n-channel, holes for p-channel) are attracted from the source to the drain.

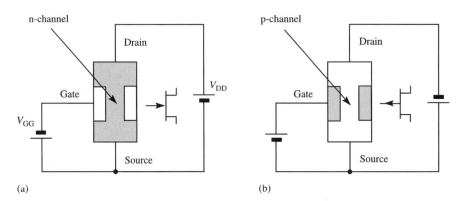

FIGURE 7.1   Drawing of (a) an n-channel JFET and its symbol and (b) a p-channel JFET.

Device Operation (optional)

In Figure 7.2 the drain is shorted to ground so that when the pn junction between the gate and channel is reverse biased a uniform field exists across the pn junction along the length of the channel.

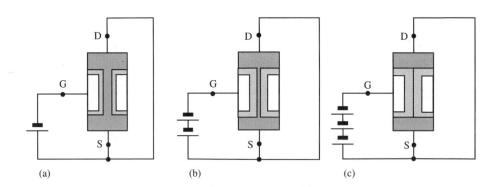

FIGURE 7.2   JFET with increasing values of reverse bias from (a) to (c) applied between the gate and the source.

The impurity concentration in the gate region is much greater than the concentration in the channel region, and the depletion layer around the junction extends further into the channel region than into the gate region. Diagrams (a), (b) and (c) in Figure 7.2 illustrate the effect of increasing the voltage applied between the gate and source. The width of the depletion layers increase until in (c) they meet and the channel is completely *pinched off*. There is a relationship between the width of the depletion layer, the impurity concentration in the channel and the applied voltage. The voltage required to pinch off the channel is referred to as the *pinch-off voltage*, V_P or V_{GSoff}. This voltage typically ranges from 1 to 4 V for small-signal transistors.

Now consider what happens if a voltage is applied to the drain, while the gate is connected to ground. With no voltage applied to the gate the channel is unobstructed by any depletion layers and current flows between the source and drain. The finite resistivity of the channel causes it to act as a resistor with a voltage drop along its length. Thus at the drain end of the channel a voltage exists between the channel and the grounded gate; the gate is at 0 V and the drain end of the channel is at a $+V_{DD}$. A positively biased channel has the same effect as the negatively biased gate of Figure 7.2, that is the gate–channel pn junction is reverse biased. The effect of changing the drain voltage is shown in Figure 7.3(a), (b) and (c).

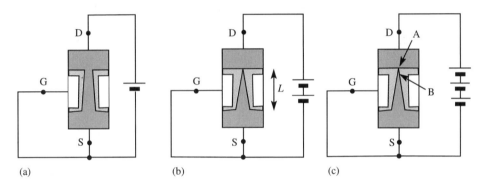

FIGURE 7.3   Variation of the depletion layer with increasing drain voltage from (a), (b) to (c).

Assuming that the resistance of the channel is uniform then the voltage drop along the channel is linear which results in a tapering of the depletion layers either side of the channel (for simplicity it is shown as a linear taper). At some value of the drain voltage the depletion layers meet at the drain end of the channel. This occurs when the voltage across the pn junction is equal to the pinch-off voltage. In Figure 7.3(b) the depletion layers are just touching at the drain end of the channel, while in Figure 7.3(c), with a further increase in V_{DD} the depletion layers merge between points A and B.

Practical Structures (optional)

The structure shown above suggests that the transistor is made from individual pieces of semiconductor, whereas in practice they are mass produced on silicon wafers using the processes which are used to produce integrated circuits. The simplified plan view and cross-section of an integrated circuit device are shown in Figure 7.4.

The plan view is of a simple rectangular device but in practice it could be circular, star shaped or interdigitated. Two important geometrical parameters are the length (L) and width (W) of the channel. The transconductance (g_m), and therefore the gain, is proportional to the width and inversely proportional to the length. Typically the length may be 2–5 µm (1 µm = 10^{-6} m), while the width may be between 100 and 1000 times greater. If the rectangular shape shown in Figure 7.4 were retained with this kind of

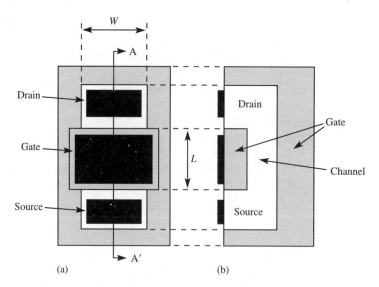

FIGURE 7.4   (a) Plan view and (b) cross-section of JFET fabricated using integrated circuit technology.

aspect ratio it would result in an unwieldy piece of silicon which would be difficult to package. Circular or star-shaped geometries are used to improve the packaging for large aspect ratio structures.

7.2   JFET Electrical Characteristics

I–V Curves

When the drain voltage is applied, electrons flow in the n-type channel from the source towards the drain. This corresponds to the conventional positive flow of current from the positive terminal of the drain–source battery (V_{DD}) into the drain of the device. Below pinch-off (as depicted in Figure 7.3(a)) the current flows unrestricted through the channel. However, the depletion layer which extends into the channel reduces the cross-sectional area of the channel and increases its resistance, cf.

$$R_{channel} = \frac{\rho L}{A} \ \text{ohms} \tag{7.1}$$

where ρ is the resistivity of the channel material, L is the channel length and A is the cross-sectional area. The depletion layer decreases the cross-sectional area.

As the drain–source voltage increases from zero the current also increases as it would in a simple resistor. However, the depletion layer increases in width and reduces the cross-sectional area of the channel. Therefore, the resistance increases and the rate of increase of the current diminishes, until a point is reached when the channel is pinched off at the drain end as illustrated in Figure 7.3(b). The current does not cease to flow when pinch-off occurs. The electrons are able to cross the depletion layer and travel on to the drain contact in much the same way that the electrons cross the collector–base depletion layer in the bipolar transistor. For further increase of the drain voltage the pinched-off region at the drain end of the channel increases, as shown in Figure 7.3(c), and extends towards the source end of the channel. Because there are no free electrons in the depletion layer it is capable of supporting a large electric field. Consequently further increase in the drain voltage results in a larger voltage drop across the depletion layer between points A and B, rather than the region of the channel between point B and the source. Because the drain current is controlled by the voltage across the region between the source and point B, the current remains constant for further increase of V_{DS}. The variation of the drain current (I_D) with drain to source voltage (V_{DS}) is shown in Figure 7.5.

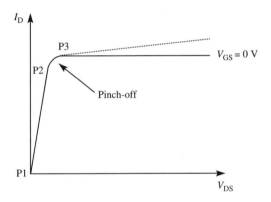

FIGURE 7.5   Variation of drain current with drain to source voltage.

The current increases linearly from P1 to P2, as it would in a simple resistor, and then between P2 and P3, as pinch-off is reached, the rate of increase reduces until beyond P3 the current remains constant, as represented by the solid line for $V_{GS} = 0$ V. In practical transistors the current continues to increase slightly beyond P3, as shown by the dotted line. For the solid line, however, it is assumed that beyond P3 the channel is pinched off, and the voltage between the source and the point in the channel at which it is pinched off (point B in Figure 7.3(c)) does not change. Further increase in V_{DS} beyond P3 is assumed to be dropped across the depletion layer between A and B. The current is said to be *saturated* and its value is determined by the voltage between the source and point B in the channel. In practice the continued increase in drain voltage causes the width of the depletion layer at the drain end of the channel to increase and for the point B in Figure 7.3(c) to continue to move towards the source end of the channel. Thus the length (L) of the active channel is reduced with increasing drain voltage, and based on equation 7.1, the resistance of the channel decreases, which results in an increase in the current, shown by the dotted line in Figure 7.5.

If instead of grounding the gate, a voltage is applied between the gate and the source, then a family of curves can be generated as shown in Figure 7.6.

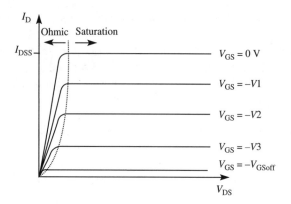

FIGURE 7.6   Family of I_D–V_{DS} curves for an n-channel JFET.

With increasing gate voltage the drain current in the saturation region decreases until for $V_{GS} = -V_{GSoff}$ (or $-V_P$) the channel is completely pinched off and the current is reduced to a very small value. This corresponds to the situation illustrated in Figure 7.2(c). At the other extreme the maximum current (I_{DSS}) flows when the gate voltage is 0 V. Note that for practical transistors all of the curves in Figure 7.6 would show an increasing current in the saturation regions.

The dividing line between the ohmic and saturation regions is a parabola (shown as a dotted line in Figure 7.6) which is described by:

$$I_D = I_{DSS}\left(\frac{V_{DS(sat)}}{V_P}\right)^2 \tag{7.2}$$

where $V_{DS(sat)}$ is the value of the source to drain voltage which is required just to produce a pinch-off of the channel at any particular value of V_{GS}. It is the difference between V_{GS} and V_P, that is:

$$V_{DS(sat)} = V_{GS} - V_P \tag{7.3}$$

The curves in Figure 7.6 can be presented in an alternative fashion by plotting the drain current against gate voltage. The resultant graph is shown in Figure 7.7.

It can be shown that the curve in Figure 7.7 is described by:

$$I_D = I_{DSS}\left(1 - \frac{V_{GS}}{V_P}\right)^2 \tag{7.4}$$

where I_{DSS} is the saturation current and V_P is the pinch-off voltage. These two parameters appear in manufacturers' data sheets.

Equation 7.4 only applies in the saturation region of Figure 7.6 which is defined as:

$$V_{DS} \geq (V_{GS} - V_P) \tag{7.5}$$

A different expression applies for the ohmic region shown in Figure 7.6, but since the transistor does not normally operate in this region it is not necessary to develop such an expression for normal circuit applications.

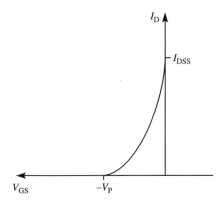

FIGURE 7.7   Variation of drain current with gate to source voltage for a constant value of drain voltage.

▶ **EXAMPLE 7.1** ——————————————————————————————

If $I_{DSS} = 8$ mA and $V_P = -4$ V, determine the drain current when $V_{GS} = -1.5$ V.

▶ *Solution*

Substituting the values into equation 7.4 gives:

$$I_D = 8 \text{ mA}\left(1 - \frac{-1.5 \text{ V}}{-4 \text{ V}}\right)^2 = 3.1 \text{ mA}$$

Transconductance

For the bipolar transistor the important transfer parameter is the current gain defined as:

$$\beta = \frac{di_C}{di_B} \approx \frac{I_C}{I_B}$$

which is dimensionless since it is the ratio of two currents. For the field effect transistor the important parameter is the transconductance which has the units of siemens (1/ohm) and is defined as:

$$g_m = \frac{di_D}{dv_{GS}} \text{ siemens} \tag{7.6}$$

The current gain of the bipolar transistor remains relatively constant with variation of the collector current; however, the g_m of the FET varies with the *dc* value of the drain current as shown in Figure 7.8.

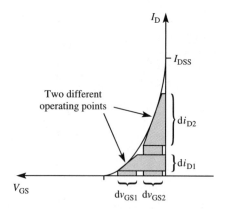

FIGURE 7.8   Variation of g_m with dc operating point for the JFET.

The variation of drain current with gate voltage is given by equation 7.4 and because it is not linear the slope of the graph dv_{GS}/di_D varies with the operating point. As the current increases so the transconductance increases, until it reaches a maximum at I_{DSS}, as shown in Figure 7.8. An analytical expression for the transconductance is obtained by differentiating equation 7.4 with respect to V_{GS} as follows:

$$g_m = \frac{dI_D}{dV_{GS}} = \frac{d}{dV_{GS}}\left[I_{DSS}\left(1 - \frac{V_{GS}}{V_P}\right)^2\right]$$

$$g_m = \frac{2I_{DSS}}{V_P}\left(1 - \frac{V_{GS}}{V_P}\right) \tag{7.7}$$

or

$$g_m = g_{mo}\left(1 - \frac{V_{GS}}{V_P}\right) \tag{7.8}$$

where g_{mo} is the maximum value of the transconductance given by:

$$g_{mo} = \frac{2I_{DSS}}{V_P} \tag{7.9}$$

▶ **EXAMPLE 7.2** ────────────────────────────

From the manufacturer's data $V_P = -3$ V to -5 V and I_{DSS} varies from 5 mA to 30 mA. Determine the variation in g_m if $V_{GS} = -1$ V.

▶ *Solution*

From equation 7.7:

$$g_m\big|_{\text{IDSS}=5\,\text{mA}} = \frac{(2)(5\text{ mA})}{3}\left(1 - \frac{-1}{-3}\right) = 2.22 \times 10^{-3}\text{ siemens}$$

or

$$g_m = 2220\ \mu\text{S}$$

and

$$g_m\big|_{\text{IDSS}=30\,\text{mA}} = \frac{(2)(30\text{ mA})}{5}\left(1 - \frac{-1}{-5}\right) = 9600\ \mu\text{S}$$

The variation in the transconductance may be caused by changes in the dc operating point, but in addition there will be changes which are caused by the variation in the manufacturing process as illustrated above. Manufacturers' data provide information on the minimum and maximum values of g_m.

Equivalent Circuit (optional)

A pn junction diode exists between the gate and the channel which under normal operating conditions is reverse biased. Current flow in the channel is controlled by the gate voltage and the drain to source voltage. It has been seen that there are two distinct operating regions, a linear region for small values of drain voltage and a saturation region for larger values. These two modes of operation are described by different current–voltage equations which allow the drain current to be determined over the full range of V_{DS}. One of these equations is described by equation 7.4.

The reverse-biased pn junctions exhibit very large values of resistance which can usually be ignored, but the capacitance associated with the junctions can have a considerable effect on the performance of the transistor at high frequencies.

For dc and transient simulation the transistor is represented by the circuit shown in Figure 7.9. In the real JFET there is only one gate diode, but its effect is distributed along the length of the channel. To simplify the presentation of a diode which is distributed along the length of the channel it is divided into two separate diodes which are placed either end of the channel. Similarly the capacitance associated with the diode is also divided into two. Under dc or transient operating conditions the current through the diodes is the leakage current associated with reverse-biased diodes, while the capacitance is represented by a voltage-dependent equation which results in a reduction in the value of the capacitance for increasing gate voltage. The resistance of the bulk silicon between the end of the channel and the drain or source contact is represented by the resistors r_d and r_s.

The circuit in Figure 7.9 is used to establish the current flowing through the JFET for dc and large-signal or transient conditions. For small-signal analysis the drain current is

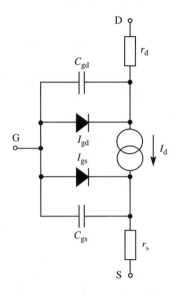

FIGURE 7.9   DC and transient equivalent circuit for the JFET.

assumed to remain constant, and the circuit is simplified by removing the diodes and replacing them with large-value resistors. In addition it is convenient to introduce the transconductance in order to determine the small-signal current from the small-signal input gate voltage. To account for the fact that practical transistors have a finite output resistance rather than the infinite resistance predicted by simple theory for saturation, the circuit also includes a resistance (r_{ds}) in parallel with the current generator. The small-signal circuit is shown in Figure 7.10.

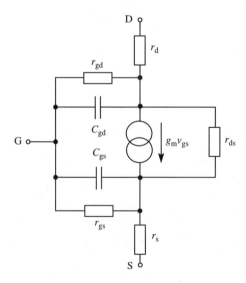

FIGURE 7.10   Small-signal equivalent JFET circuit.

Simplified Equivalent Circuit

While the circuits in Figures 7.9 and 7.10 are perfectly acceptable for a CAD program, they are too complex for simple hand calculations. The resistors r_{gs} and r_{gd} are very large (>10 MΩ) and the resistors r_s and r_d are very small (<100 Ω) and they can all be ignored for approximate hand calculations. The junction capacitors C_{gs} and C_{gd} are important for high-frequency analysis (>1 MHz), but for low frequencies they too can be ignored. The source to drain resistance r_{ds} may or may not be important, and depends on the value of the external drain resistor. Thus the approximate equivalent circuit for low frequencies reduces to the circuit shown in Figure 7.11.

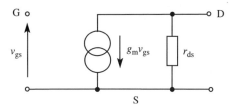

FIGURE 7.11   Simplified equivalent circuit for low-frequency use.

The absence of a component between the gate and source may look rather strange, but remember that the JFET is a voltage-operated device and it is the voltage which appears at the gate which is important.

SPICE Parameters for JFETs (optional)

Equation 7.4 is greatly simplified and the complete equation which is derived from consideration of the physical structure and the conduction processes within the channel region is much more complex. The SPICE version of equation 7.4 is:

$$I_D = \text{BETA}(1 + \text{LAMBDA} \cdot V_{DS})(V_{TO} - V_{GS})^2 \tag{7.10a}$$

Rewriting equation 7.4 in the form of equation 7.10a gives:

$$I_D = \frac{I_{DSS}}{V_P^2}(V_P - V_{GS})^2 \tag{7.10b}$$

and by comparing equation 7.10b with equation 7.10a, it is seen that:

$$\text{BETA} = I_{DSS}/V_P^2 \tag{7.11}$$

and that $V_{TO} = V_P$. The term LAMBDA in equation 7.10a describes the modulation of the channel length with V_{DS}, as illustrated in Figure 7.3(c) where the length of region AB, and hence the length of the channel, changes as the drain voltage changes. It is similar in concept to the Early voltage associated with the bipolar transistor. The default value of LAMBDA is zero, which reduces equation 7.10a to the simplified form in equation 7.10b. Notice that SPICE does not directly model the channel resistance r_{ds}, but rather the channel length modulation term, LAMBDA. There is no simple relationship between LAMBDA and r_{ds}, as there is between the Early voltage and the output resistance. Some of the important SPICE parameters which can be used to modify the generic devices from the PSpice breakout library are as follows:

Parameter	Description	Default
VTO	Threshold voltage (pinch-off)	-2.0 V
BETA	Transconductance parameter	1×10^{-4}
LAMBDA	Channel length modulation	0
CGS	Gate–source capacitance	0
CGD	Gate–rain capacitance	0

Note that the manufacturer' data usually specify I_{DSS} and V_P (or V_{GSoff}) rather than BETA and V_{TO}.

▷ **EXAMPLE 7.3** ⎯⎯⎯⎯⎯⎯⎯⎯⎯⎯⎯⎯⎯⎯⎯⎯⎯⎯⎯⎯⎯

Use PSpice to produce the I_D–V_{DS} characteristic curves for a JFET and examine the effect of changing the value of LAMBDA from 0 V^{-1} to 1×10^{-2} V^{-1} for a transistor with $I_{DSS} = 10$ mA and $V_P = -4$ V.

▷ *Solution*

Produce the schematic shown in Figure 7.12 with PSpice using the generic JFET from the breakout library (JbreakN)

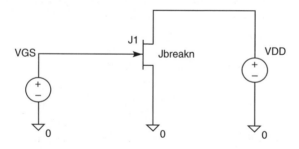

FIGURE 7.12 Circuit schematic for PSpice.

From equation 7.11:

$$BETA = \frac{10 \text{ mA}}{(-4 \text{ V})^2} = 6.25 \times 10^{-4} \text{ AV}^{-2}$$

Select **Model...** from **Edit** and enter vto $= -4$, BETA $= 6.25$e-4 and LAMBDA $= 0$ in the **Edit model** menu box.

From **Setup…** select **DC Sweep…** and select **Voltage source**. Enter the name VDD for the drain voltage supply with a start value of 0 V, end value of 12 V and increment of 0.2 V; select the nested sweep with VGS as the other source with a start value of 0 V, end value of −4 V and increment of −1 V.

The $I_D - V_{DD}$ graph with LAMBDA = 0 is shown Figure 7.13.

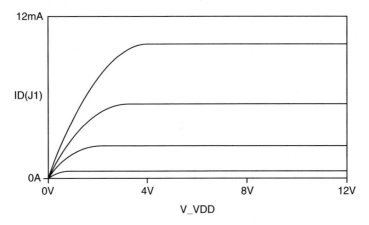

FIGURE 7.13   Output from Probe for LAMBDA = 0.

Change the value of LAMBDA to 1e-2 in the **Model…** menu box and rerun PSpice to obtain the output shown in Figure 7.14.

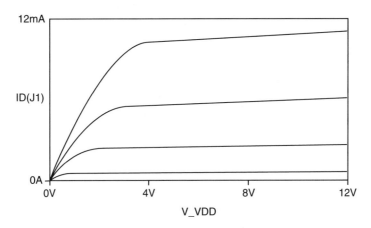

FIGURE 7.14   Output from Probe for LAMBDA = 1×10^{-2}.

Notice that there is now a finite slope to the graph in the saturation region. In practice the value of LAMBDA is $<1 \times 10^{-3}$ so that the slope shown in the above graph is much greater than that observed with the majority of practical devices.

Biasing the JFET

Correct dc biasing is necessary to ensure that the operating point is centrally positioned on the load-line to provide maximum voltage swing at the drain terminal. Biasing of the JFET is made easy by the fact that the gate must be negatively biased with respect to the source (cf. the positive bias required for the base of the common-emitter bipolar transistor). Simply placing a resistor in series with the source produces a voltage drop across the resistor with the polarity as shown in Figure 7.15.

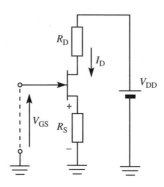

FIGURE 7.15   Self-bias circuit for JFET.

If the gate were to be connected to ground as suggested by the dotted line in Figure 7.15, then the voltage drop across R_S is of the correct polarity to reverse-bias the gate–channel junction. However, connecting the gate to ground directly is not practical because this short-circuits the input and there is no means of applying an input signal.

The gate to source pn junction is reverse biased and consequently the dc current flowing in the gate circuit is very small (a few picoamps associated with the leakage current). A dc connection between the gate and ground can be provided with a large-value of the reverse biased pn junction resistor as shown in Figure 7.16.

The gate resistor is typically 1 MΩ and even with a gate leakage current of 1 nA the

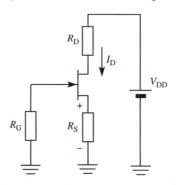

FIGURE 7.16   Self-bias circuit with gate resistor (R_G).

dc voltage drop across R_G is only 1 mV. This gate resistor provides the required dc connection, and also allows a signal to be applied to the gate.

DC Operating Point

For the circuit shown in Figure 7.16, provided that $V_\mathrm{DS} > 1$ V or 2 V the transistor operates in the saturation region, and the drain current is given by equation 7.4. This equation describes the transfer curve $I_\mathrm{D} - V_\mathrm{GS}$ as shown in Figure 7.7. In addition the voltage across R_S establishes the gate to source voltage:

$$V_\mathrm{GS} = -I_\mathrm{D} R_\mathrm{S} \tag{7.12}$$

Substituting the value of V_GS from equation 7.12 into equation 7.4 results in a quadratic equation in terms of the drain current with two possible roots as follows:

$$I_\mathrm{D} = I_\mathrm{DSS}\left(1 + \frac{I_\mathrm{D} R_\mathrm{S}}{V_\mathrm{P}}\right)^2 \tag{7.13}$$

which yields

$$I_\mathrm{D}^2 R_\mathrm{S}^2 + I_\mathrm{D}\left(2 V_\mathrm{P} R_\mathrm{S} - \frac{V_\mathrm{P}^2}{I_\mathrm{DSS}}\right) + V_\mathrm{P}^2 = 0 \tag{7.14}$$

cf. $ax^2 + bx + c = 0$

which gives:

$$x = \frac{-b \pm \sqrt{b^2 - 4ac}}{2a}$$

from which two values of x (or I_D) are obtained. The 'sensible' value is chosen as the current at the operating point based on the conditions that $I_\mathrm{DQ} < I_\mathrm{DSS}$ or $V_\mathrm{GSQ} < V_\mathrm{P}$.

As an alternative to the analytical approach, and the possibility of introducing an error in solving the quadratic equation, it is a relatively simple matter to use a graphical

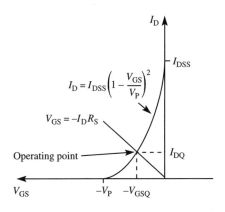

FIGURE 7.17 Graphical approach for obtaining the operating point.

means to solve for I_D and V_{GS}. The equation for the drain current results in the transfer curve shown in Figure 7.7, while equation 7.12 for V_{GS} is a straight line passing through the origin. The two curves can be combined as shown in Figure 7.17.

The values of I_D and V_{GS} for the two graphs are equal at the point of intersection. This represents the operating point for the transistor. This graphical approach can be used to find the operating point as shown in the following example.

▷ EXAMPLE 7.4 ───────────────────────────

For the circuit shown in Figure 7.16 $R_D = 4.7$ kΩ, $R_S = 680$ Ω, $I_{DSS} = 8$ mA, $V_P = -4$ V and $V_{DD} = 18$ V. Determine graphically the dc operating values of I_D, V_{GS} and V_{DS}.

▷ *Solution*

The drain current given by equation 7.4 is:

$$I_D = 8 \text{ mA}\left(1 - \frac{V_{GS}}{-4 \text{ V}}\right)^2$$

The values of I_D are calculated for a number of values of V_{GS} as follows:

V_{GS}	I_D
0	8 mA
−1	4.5 mA
−2	2 mA
−3	0.5 mA
−4	0 mA

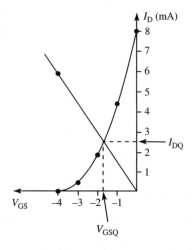

FIGURE 7.18   Transfer curve and load-line.

A graph of this data is shown in Figure 7.18.

The straight line for $V_{GS} = -I_D\ 680\ \Omega$ is also shown, which passes through the origin, and a construction point at $V_{GS} = -4$ V and $I_D = 5.88$ mA. The intersection of this line and the curve provide values for the operating current and voltage. From the graph these are:

$$V_{GSQ} \approx -1.7\ \text{V}$$
$$I_{DQ} \approx 2{,}6\ \text{mA}$$

The value of the source-drain voltage is:

$$V_{DSQ} = 10 - [(2.6\ \text{mA})(4.7\ \text{k}\Omega) + (2{,}6\ \text{mA})(680\ \Omega)] = 4\ \text{V}$$

▶ EXAMPLE 7.5 ───────────────────────

For the transistor in Example 7.4 determine the transconductance at the operating point.

▶ *Solution*

The transconductance is given by equation 7.7 as:

$$g_m = \frac{(2)(8\ \text{mA})}{4}\left(1 - \frac{-1.7}{-4}\right) = 2.3\ \text{mS}$$

Mid-Point Biasing

When designing a JFET circuit it is necessary to select an appropriate value for the drain current and source to drain voltage. The maximum drain current is I_{DSS} and a convenient value for the operating current is $I_{DSS}/2$. The drain to source voltage should be approximately half of the supply voltage to ensure a large symmetrical voltage swing. Thus for *mid-point bias* the current is:

$$I_D = \frac{I_{DSS}}{2} \tag{7.15}$$

Substituting into equation 7.4 gives:

$$\frac{I_{DSS}}{2} = I_{DSS}\left(1 - \frac{V_{GS}}{V_P}\right)^2$$

and

$$\frac{V_{GS}}{V_P} = 1 - \frac{1}{\sqrt{2}}$$

or

$$V_{GS} = 0.293 V_P$$

or for simplicity, for design purposes, V_{GS} is:

$$V_{GS} \approx \frac{V_P}{4} = 0.25 V_P \qquad\qquad (7.16)$$

The drain resistor is selected to give:

$$I_D R_D = \frac{V_{DD}}{2}$$

or

$$R_D = \frac{V_{DD}}{2 I_D} \qquad\qquad (7.17)$$

Equations 7.15, 7.16 and 7.17 provide a set of design criteria which can be used to optimize the dc conditions for the JFET amplifier.

▷ **EXAMPLE 7.6** —————————————————————————

From the manufacturer's data sheets the following parameters are obtained: $I_{DSS} = 10$ mA and $V_P = -4$ V. Design a single-stage JFET amplifier to operate from an 18 V supply.

▷ *Solution*

A suitable value for the drain current is:

$$I_D = \frac{I_{DSS}}{2} = 5 \text{ mA}$$

From equation 7.16 the gate to source voltage is:

$$V_{GS} = \frac{V_P}{4} = \frac{-4 \text{ V}}{4} = -1 \text{ V}$$

Therefore, the source resistance is:

$$R_S = \frac{V_{GS}}{I_D} = \frac{1 \text{ V}}{5 \text{ mA}} = 200 \ \Omega$$

The drain resistor is selected to give a voltage drop equal to half the supply voltage, that is:

$$R_D = \frac{V_{DD}}{2 I_D} = \frac{18}{(2)(5 \text{ mA})} = 1.8 \text{ k}\Omega$$

The transconductance is:

$$g_m = \frac{2 I_{DSS}}{V_P} \left(1 - \frac{V_{GS}}{V_P} \right)$$

$$g_m = \frac{(2)(10 \text{ mA})}{4}\left(1 - \frac{1}{4}\right) = 3.75 \text{ mS}$$

Operating Point Stability

Ideally the biasing of the transistor should offer some degree of stability to changes in the transistor parameters, particularly the values of I_{DSS} and V_P. These values can vary as a result of variations in the resistivity of the channel and the thickness of the channel. For example, consider a JFET for which the minimum and maximum values are as follows:

$V_{Pmin} = -3 \text{ V} \quad I_{DSSmin} = 6 \text{ mA}$
$V_{Pmax} = -5 \text{ V} \quad I_{DSSmax} = 10 \text{ mA}$

Assume that the source resistance is 600 Ω. The two transfer curves and the source resistor load-line are shown in Figure 7.19

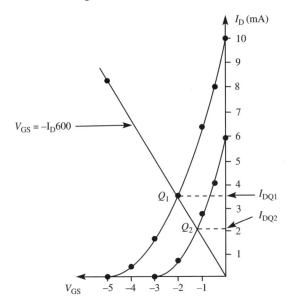

FIGURE 7.19   Transfer curves and load-line for two extremes of the transistor parameters.

For the single load-line, which describes the relationship between V_{GS} and I_D, there are two possible operating points, Q_1 and Q_2, which result in two different values of drain current. These range from a minimum value of approximately 2.1 mA to a maximum value of approximately 3.4 mA, that is a change of 62%. Provided that the amplitude of the signal being processed is not too large then this change may be acceptable, but where the maximum amplitude is required then changes of this magnitude could result in distortion for one or other extremes of the transistor parameters.

The stability of the operating point can be improved by use of a voltage divider to provide the gate bias.

Potential Divider Bias

The circuit with potential divider biasing is shown in Figure 7.20.

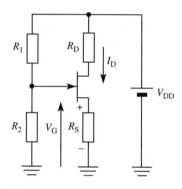

FIGURE 7.20 Potential divider JFET biasing.

There is no gate current and the voltage across R_2 is simply:

$$V_G = \frac{V_{DD} R_2}{R_1 + R_2}$$

and the source to gate voltage is:

$$V_{GS} = V_G - I_D R_S \tag{7.18}$$

Equation 7.18 is used in place of equation 7.12 for the graphical solution of the operating point. For the previous example illustrated in Figure 7.19, assume that a potential divider results in a gate voltage $V_G = 3$ V; then the origin of the load-line is on the x axis at the point corresponding to 3 V as shown in Figure 7.21. An additional construction point for the load-line is obtained on the y axis by setting $V_{GS} = 0$ V and calculating the drain current. A straight line is drawn through these two construction points to represent the load-line.

Notice that it is necessary to adjust the value of R_S in order to obtain values of current which are similar to the previous example, and a value of 2 kΩ produces values of I_D between 2 mA and 3 mA. From Figure 7.21 it is seen that the current varies from approximately 2.1 mA to 2.75 mA, or a change of 31%. The improvement results from the fact that the slope of the load-line is reduced. Increasing the value of V_G reduces the slope still further; however, it also increases the voltage drop across R_S and as a result limits the maximum voltage available across the transistor and the drain resistor.

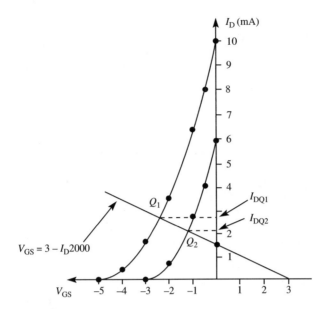

FIGURE 7.21   Graphical solution of the operating point with a potential divider.

▷ **EXAMPLE 7.7** _____

For the circuit shown in Figure 7.20 $R_1 = 560$ kΩ, $R_2 = 150$ kΩ, $R_S = 1$ kΩ, $R_D = 1.8$ kΩ, $I_{DSS} = 12$ mA, $V_P = -3.5$ V and $V_{DD} = 18$ V. Determine the dc operating values of I_D, V_{GS} and V_{DS}.

▷ *Solution*

The drain current is given by:

$$I_D = 12 \text{ mA}\left(1 - \frac{V_{GS}}{-3.5 \text{ V}}\right)^2$$

and values of V_{GS} and I_D are obtained as follows:

V_{GS}	I_D
0	12 mA
-0.5	8.8 mA
-1	6.1 mA
-2	2.2 mA
-3	0.24 mA
-3.5	0 mA

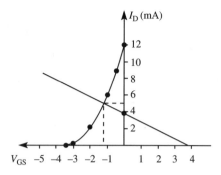

FIGURE 7.22   Transfer curve and load-line for potential divider bias.

The transconductance curve resulting from this data is shown in Figure 7.22 together with the load-line described by:

$$V_{GS} = 3.8 \text{ V} - (I_D)(1 \text{ k}\Omega)$$

where 3.8 V is the voltage across R_2 in the potential divider.

From Figure 7.22 the intersection of the two curves results in the quiescent values for the current and voltage of:

$$I_D \approx 5 \text{ mA}$$
$$V_{GS} \approx 1.3 \text{ V}$$

The drain to source voltage is:

$$V_{DS} = 18 \text{ V} - (5 \text{ mA})(1.8 \text{ k}\Omega + 1 \text{ k}\Omega) = 4 \text{ V}$$

7.4 ▷ MOS Transistor

When a voltage is applied to a metal electrode which is positioned very close to the surface of a semiconductor then the properties of the surface layer are changed. The electrons or holes in the semiconductor are attracted or repelled depending on the polarity of the electric field. This results in the creation or extinction of a channel at the surface of the semiconductor. If the channel is formed between two diffused regions (source and drain) then current will flow between the source and drain when the channel is present. The magnitude of the current is a function of the resistance of the channel region, and this is controlled by the voltage applied to the external electrode. The external electrode is a layer of metal which is deposited on a layer of dielectric which is formed on the surface of the semiconductor. The dielectric is a layer of silicon oxide and the electrode is usually polycrystalline silicon.

For the purposes of circuit analysis knowledge of the structure and operation of the MOSFET is not necessary and they are, therefore, listed as optional.

Physical Structure (optional)

A cross-section of the basic MOSFET is shown in Figure 7.23.

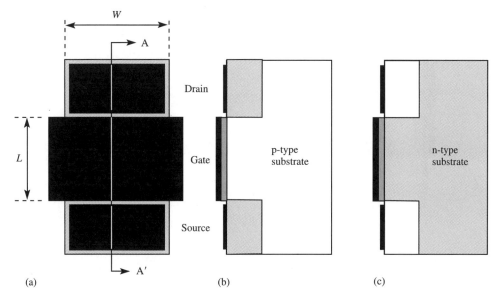

FIGURE 7.23   Basic structure of a MOSFET in (a) plan view, (b) cross-section of an n-channel device and (c) cross-section of a p-channel device.

The different regions have the same identities as those in the JFET, that is source, drain and gate. The source and drain regions are of opposite conductivity type to that of the substrate. Thus with a p-type substrate the source and drain regions are n-type, and the channel, which is formed beneath the gate, is n-type, and the device is an n-channel device. For an n-type substrate the source and drain are p-type to give a p-channel device.

Depletion Mode (optional)

Apart from the difference in the conductivity type of the channel there is another difference resulting from the presence or absence of a channel for zero gate bias. If a channel is present when the gate voltage is zero then the device is known as a *depletion-mode MOSFET*. The channel is formed during the manufacturing process and the application of a suitable gate voltage *depletes* the channel. The basic structure is the same as for the device shown in Figure 7.23, and is shown in more detail in Figure 7.24.

When drain current flows in the channel a voltage drop occurs along the length of the channel which has the same effect as the voltage drop in the channel of the JFET, and as for the JFET, a point is reached when the channel becomes pinched off as shown in Figure 7.25.

With an n-channel device a negative voltage is required to repel the electrons which form the n-type channel. For the p-channel device a positive voltage is required.

Unlike the JFET the voltage applied to the gate can have both polarities, because the gate oxide is a perfect insulator. For the JFET the pn junction would become forward

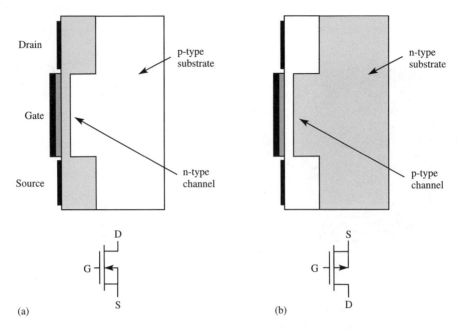

FIGURE 7.24   Physical structure and symbols for the (a) n-channel depletion-mode device and (b) p-channel depletion-mode device.

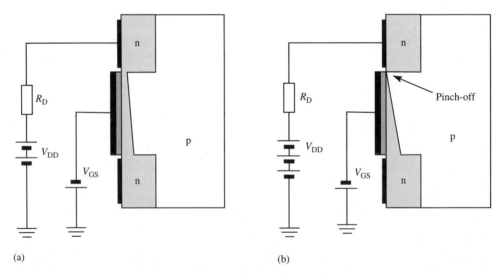

FIGURE 7.25   Reduction in the thickness of the channel (a) as drain current flows and (b) eventual pinch-off.

biased if the polarity reversed. For an n-channel device a positive gate voltage *enhances* the channel while for negative voltages the channel is depleted (the polarities are reversed for p-channel devices).

The absence of a pn junction between the gate and the channel allows the gate voltage to have both negative and positive polarities, as shown in Figure 7.26. Notice the position of the $V_{GS} = 0$ V curve. For a JFET this would represent the maximum drain current. For the MOSFET the current can be increased by applying a positive gate voltage.

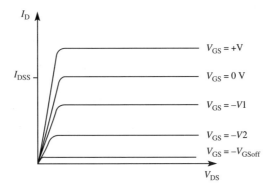

FIGURE 7.26   Drain characteristics for an n-channel depletion-mode MOSFET.

The saturation current I_{DSS} is defined as the current which flows when $V_{GS} = 0$ V. When V_{GS} is positive for an n-channel device more electrons are attracted to the surface of the semiconductor to increase, or enhance, the thickness of the channel region and to increase the amount of current which can flow.

The transfer characteristic for the depletion device is shown in Figure 7.27.

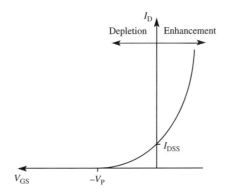

FIGURE 7.27   Transfer characteristic for the n-channel depletion-mode MOSFET.

I–V Characteristics

The current flow in the depletion MOSFET is described by the same equation as that used for the JFET, that is:

$$I_D = I_{DSS}\left(1 - \frac{V_{GS}}{V_P}\right)^2 \tag{7.19}$$

The same square-law dependency exists for both devices and the equation accurately predicts the current for both negative and positive values of V_{GS}. Notice that $(1 - V_{GS}/V_P) > 1$ for positive values of V_{GS} (since V_P is negative) and therefore I_D is greater than I_{DSS}.

► **EXAMPLE 7.8** ────────────────────────────────────

For an n-channel depletion-mode MOSFET $I_{DSS} = 15$ mA, $V_P = -5$ V. Assuming that it is operated in the saturation region, determine the drain current when $V_{GS} = -3.5$ V and when $V_{GS} = +1.5$ V.

► *Solution*

From equation 7.19 with $V_{GS} = -3.5$ V:

$$I_D = 15 \text{ mA}\left(1 - \frac{-3.5 \text{ V}}{-5 \text{ V}}\right)^2 = 1.35 \text{ mA}$$

and when $V_{GS} = +1.5$ V:

$$I_D = 15 \text{ mA}\left(1 - \frac{1.5 \text{ V}}{-5 \text{ V}}\right)^2 = 25.3 \text{ mA}$$

► **EXAMPLE 7.9** ────────────────────────────────────

For the same values repeat the example assuming a p-channel MOSFET.

► *Solution*

For the p-channel MOSFET the pinch-off voltage is positive, that is $V_P = +5$ V. Then the current for $V_{GS} = -3.5$ V is:

$$I_D = 15 \text{ mA}\left(1 - \frac{-3.5 \text{ V}}{5 \text{ V}}\right)^2 = 43.3 \text{ mA}$$

and when $V_{GS} = +1.5$ V:

$$I_D = 15 \text{ mA}\left(1 - \frac{1.5 \text{ V}}{5 \text{ V}}\right)^2 = 7.35 \text{ mA}$$

Depletion-mode Biasing

The biasing methods used for the JFET may also be used for the depletion-mode MOSFET, but in addition it is possible to use zero-voltage bias. This arises because the gate voltage on a depletion-mode MOSFET can be positive, negative and zero for current to flow as shown in Figure 7.27. Remember that the gate voltage for the n-channel JFET cannot be positive (negative for the p-channel JFET) or else the pn junction becomes forward biased. For the MOSFET symmetrical operation is achieved for zero-voltage bias.

A bias circuit with zero-voltage bias is shown in Figure 7.28.

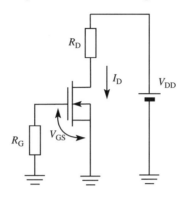

FIGURE 7.28   Zero-biased depletion-mode MOSFET.

With zero gate bias ($V_{GS} = 0$ V) the drain current is I_{DSS} and the drain to source voltage is:

$$V_{DS} = V_{DD} - I_{DSS} R_D \qquad (7.20)$$

The gate resistor R_G provides the dc path for the gate to source dc voltage and in practice is simply a large value (~10 MΩ).

EXAMPLE 7.10

A depletion-mode n-channel MOSFET has a pinch-off voltage of -4 V and the saturation current is 10 mA. Determine the drain to source voltage for a zero-bias circuit if $V_{DD} = 12$ V, $R_G = 10$ MΩ and $R_D = 470$ Ω.

Solution

With $V_{GS} = 0$ V the drain current is I_{DSS} and V_{DS} is:

$$V_{DS} = V_{DD} - I_{DSS} R_D$$
$$V_{DS} = 12 \text{ V} - (10 \text{ mA})(470 \ \Omega) = 7.3 \text{ V}$$

Enhancement Mode (optional)

For the enhancement-mode MOSFET no channel exists between the source and drain for zero gate voltage. The construction is the same as for the depletion-mode devices and is illustrated in Figure 7.29, together with the symbols for n-channel and p-channel devices.

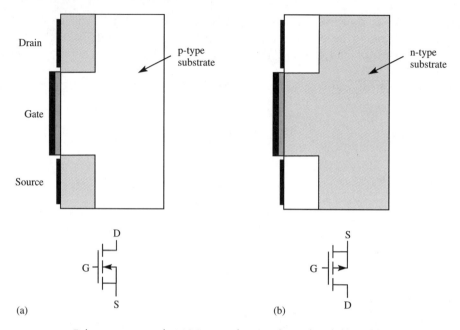

FIGURE 7.29   Enhancement-mode MOSFET with (a) n-channel and (b) p-channel.

The symbol is shown with a discontinuous line between source and drain to represent the absence of a continuous current path for zero gate voltage. The channel is formed for n-channel devices when a positive voltage is applied to the gate, and for the p-channel when a negative voltage is applied. The transfer characteristics are shown in Figure 7.30.

Because the channel does not exist for zero gate voltage there is no I_{DSS} for the enhancement MOSFET as there is for the depletion device or the JFET. Notice also that

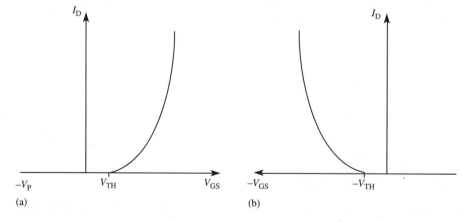

FIGURE 7.30   Transfer curves for (a) the n-channel enhancement MOSFET and (b) the p-channel enhancement MOSFET.

the current does not begin to flow until the gate voltage reaches a threshold value V_{TH}, which is positive for the n-channel device and negative for the p-channel device.

I–V Characteristics

The equation which describes the current in the enhancement MOSFET is:

$$I_D = \frac{K_P}{2} \frac{W}{L} (V_{GS} - V_{TH})^2 \tag{7.21}$$

where K_P is a transconductance parameter which is related to the physical properties of the semiconductor, W is the width of the channel, L is the channel length and V_{TH} is the threshold voltage. The same square-law relationship is observed for the current for the enhancement MOSFET as for the JFET. An alternative and simpler form of the equation is:

$$I_D = K(V_{GS} - V_{TH})^2 \tag{7.22}$$

where K is a transconductance factor which can be obtained from the manufacturer's data if the drain current is specified for a particular value of gate to source voltage.

▶ **EXAMPLE 7.11** ──────────────────────────────────

The data sheet for an enhancement-mode MOSFET specifies $I_D = 9$ mA for $V_{GS} = 8$ V, and $V_{TH} = 1$ V. Determine the value of the drain current when $V_{GS} = 3$ V.

▶ *Solution*

From equation 7.22:

$$K = \frac{I_D}{(V_{GS} - V_{TH})^2} = \frac{9 \text{ mA}}{(8 \text{ V} - 1 \text{ V})^2} = 0.18 \text{ mA V}^{-2}$$

The value of I_D at a gate to source voltage of 3 V is:

$$I_D = 0.18 \times 10^{-3}(3 \text{ V} - 1 \text{ V})^2 = 0.72 \text{ mA}$$

Enhancement Mode Biasing

The enhancement-mode MOSFET does not conduct unless the input gate voltage exceeds the threshold voltage, and therefore a bias circuit is required to supply a voltage which exceeds V_{TH}. For the n-channel device this voltage is positive and can be provided with a potential divider as shown in Figure 7.31.

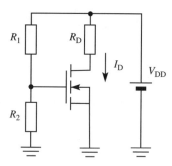

FIGURE 7.31   Enhancement-mode MOSFET with potential divider biasing.

The gate to source voltage is:

$$V_{GS} = V_{DD}\left(\frac{R_2}{R_1 + R_2}\right)$$

$$V_{DS} = V_{DD} - I_D R_D$$

where

$$I_D = K(V_{GS} - V_{TH})^2 \quad \text{from equation 7.22}$$

Enhancement-mode transistors are used extensively for power applications, both as a switch and as a linear gain device in power amplifiers. Another important application is as a switch in digital integrated circuits where the gate voltage is either $<V_{TH}$ in which case no current flows and the device is OFF, or the voltage is $>V_{TH}$ in which case the device is ON and drain current flows. The majority of very large-scale integrated circuits are based on the enhancement-mode MOSFET.

Equivalent Circuit

For small-signal analysis the equivalent circuit consists of the capacitance between the gate electrode and the channel and a voltage-controlled current generator, as shown in Figure 7.32.

The finite output resistance is represented by r_{ds}, while the resistors r_d and r_s represent the bulk resistance of the material between the end of the channel and the drain and source contacts. For simple hand calculations the circuit can be reduced to the current generator and the output resistance as shown in Figure 7.33.

If the high-frequency response is important then the two capacitors are added, one (C_{gd}) between the gate and drain and the other (C_{gs}) between the gate and source.

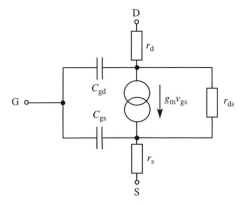

FIGURE 7.32   Small-signal equivalent circuit for the MOSFET.

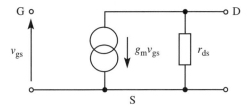

FIGURE 7.33   Simplified equivalent circuit for the MOSFET.

SPICE Parameters for MOSFETs (optional)

The original model for MOSFETs was based on an enhancement-mode device because these devices were to be used in MOS integrated circuits. The mathematical equations used in SPICE are based on equation 7.21 (which is an extreme simplification of the actual equation used in SPICE) rather than equation 7.22 and therefore include the dimensions for the channel. There are three models – LEVEL I, LEVEL II and LEVEL III. The LEVEL I model is based on a long-channel approximation of the MOSFET, which applies for channel lengths in excess of 1 μm. The LEVEL II model includes second-order effects which are ignored in the LEVEL I model to improve accuracy and as a result is mathematically much more complex. Consequently the computation time when using the LEVEL II model is increased, particularly when the simulation is performed for a circuit which contains a large number of devices. The LEVEL III model is an attempt to achieve the accuracy of the LEVEL II model without sacrificing computational speed. It is based on empirical relationships derived from experimental data, and consequently has no direct physical relationship to the device. A number of other proprietary models also exist which have been developed by individual manufacturers to 'improve' the model for their particular devices and to provide better agreement between theory and experiment.

When using SPICE the same model is used for both depletion and enhancement transistors. For the depletion-mode MOSFET it is possible to specify I_{DSS} and V_P (or

V_{TH}) as for a JFET where I_{DSS} corresponds to the value of I_D for zero gate voltage. From equation 7.21 the value of I_{DSS} is:

$$I_{DSS} = \frac{K_P}{2} \frac{W}{L} V_{TH}^2 \tag{7.23}$$

The transconductance is given by:

$$g_m = \frac{dI_D}{dV_{GS}} = \frac{d}{dV_{GS}} \left(\frac{K_P}{2} \frac{W}{L} (V_{GS} - V_{TH})^2 \right)$$

and

$$g_m = K_P \frac{W}{L} (V_{GS} - V_{TH}) \tag{7.24}$$

or substituting for K_P from equation 7.21 gives:

$$g_m = \frac{2I_D(V_{GS} - V_{TH})}{(V_{GS} - V_{TH})^2}$$

and

$$g_m = \frac{2I_D}{(V_{GS} - V_{TH})} \tag{7.25}$$

From equation 7.25 it can be seen that the transconductance for the MOSFET varies with the drain current in a similar manner to the transconductance of the JFET.

From equation 7.24 it is seen that the transconductance is proportional to the width and inversely proportional to the length. Thus to obtain large values of gain for analog amplifiers it is necessary to make the channel length as small as possible, while the width is made as large as possible.

The SPICE model has been developed for transient analysis and does not accurately model small-signal effects, particularly for depletion-mode devices. The best results are likely to be obtained with the default LEVEL I model. There is no advantage to be gained by using the other models for small-signal analysis. The most important SPICE parameters are as follows:

Parameter	Description	Default
VTO	Threshold voltage (pinch-off)	1.0 V
KP	Transconductance parameter	2×10^{-5}
LAMBDA	Channel length modulation	0
CGDO	Gate–source capacitance	0
CGSO	Gate–drain capacitance	0
W	Channel width	1
L	Channel length	1

Notice that the default values for the channel dimensions are both unity.

▷ **EXAMPLE 7.12** ―――――――――――――――――――――――――――

Use PSpice to plot the transfer curve for a depletion-mode MOSFET for which $I_{DSS} = 12$ mA and $V_{TH} = -3.5$ V. Assume that $W = L = 1$.

▷ *Solution*

From the breakout library select MbreakN. Notice that the transistor is described as an enhancement transistor. This does not prevent it being used as a depletion device, since it is the information contained in the **Model...** statement from **Edit** which determines how the device operates during the simulation. The information required for the model is K_P and V_{TH}.

From equation 7.23:

$$K_P = \frac{2I_{DSS}}{V_{TH}^2} = \frac{(2)(12 \text{ mA})}{3.5^2} = 1.96 \text{ mA V}^{-2}$$

Use the **Model Editor** to add kp = 1.96e-3 and vto = -3.5 to the **Model...** description. The schematic for simulation is shown in Figure 7.34.

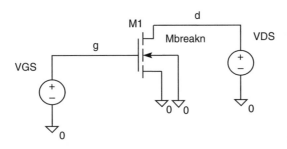

FIGURE 7.34   Circuit schematic for PSpice.

Use the **DC Sweep...** to sweep V_{GS} from -3.5 V to $+3$ V in 0.1 V steps. The resulting transfer curve of $I_D - V_{GS}$ is shown in Figure 7.35.

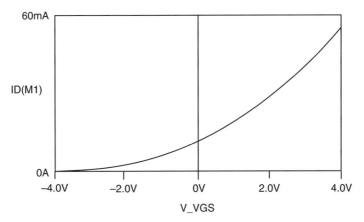

FIGURE 7.35   Output from Probe for a depletion-mode MOSFET.

▶ **EXAMPLE 7.13**

Use PSpice to plot the transfer curve for an enhancement-mode MOSFET for which $K_P = 0.37$ mA V^{-2} and $V_{TH} = 1$ V. Verify that $I_D = 9$ mA when $V_{GS} = 8$ V (cf. Example 7.11). Assume that $W = L = 1$.

▶ *Solution*

Select MbreakN from the breakout library and select **Model...** from **Edit** and add kp = 0.37e-3 and vto = 1.

The output from Probe is shown in Figure 7.36.

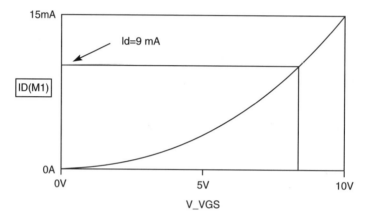

FIGURE 7.36   Output from Probe for an enhancement-mode MOSFET.

With the aid of the cursor the current is measured as 9 mA as shown.

Handling Precautions

Static electricity can destroy a MOSFET even before electrical power is applied to the circuit. The gate oxide is an almost perfect insulator and if charge is applied to the gate in the form of an electrostatic discharge (ESD), then because of the small value of the gate capacitance the voltage between the gate and the channel can reach many thousands of volts. The voltage breakdown of the gate oxide is typically <100 V and the consequence of inadvertent ESD to the gate electrode is the permanent breakdown of the gate oxide. The most common source of ESD is from the human operator where electrostatic potentials of many thousands of volts can be generated in dry atmospheres.

Some of the precautions to be taken when handling MOSFETs are:

1   Before handling a MOSFET discharge static electricity by touching ground. For technicians who are regularly involved in handling MOSFETs it is desirable to use a conductive wrist strap with high-resistance conductive lead to attach to the laboratory earth while working on circuits which contain MOS devices.

2   MOSFETs are normally packaged with their leads placed in conductive foam and this should be retained until they are to be placed in the circuit board.

3   Always switch off the power to the circuit before inserting or removing a MOSFET (or any other device).

▷ PROBLEMS ─────────────

7.1   The manufacturer's data for a JFET specify V_P as -2 V and I_{DSS} as 6 mA. Determine the standard value of a suitable source resistance to provide a drain current of 3 mA for a self-bias common-source amplifier.

[200 Ω]

7.2   For the JFET shown $V_P = -6$ V and $I_{DSS} = 10$ mA. Determine the drain current and the gate to source voltage.

[2.8 mA, -2.8 V]

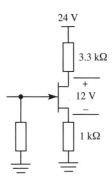

24 V

3.3 kΩ

+
12 V
−

1 kΩ

7.3   For an n-channel JFET $I_{DSS} = 1$ mA and $V_P = -7$ V. Determine I_D and $V_{DS(sat)}$ when $V_{GS} = -4$ V.

[2.2 mA, 3 V]

7.4   An n-channel JFET has a pinch-off voltage of -6 V and I_{DSS} of 8 mA. Determine the value of V_{GS} and I_D if $V_{DS(sat)}$ is 4 V.

[-2 V, 3.5 mA]

7.5   A p-channel JFET has a pinch-off voltage of 4 V and the saturation current I_{DSS} is 6 mA. Determine the drain current for gate to source voltages of 0 V, 2 V and 3 V.

[6 mA, 1.5 mA, 0.375 mA]

7.6   For an n-channel JFET $V_P = -5$ V and I_{DSS} ranges from a minimum value of 6 mA to a maximum of 12 mA. Determine the range of drain current for $V_{GS} = -2$ V.

[2.16 mA, 4.32 mA]

7.7   For the circuit shown $V_P = -8$ V and $I_{DSS} = 10$ mA. Determine V_{DS} and check whether the transistor is in the ohmic or saturation region.

[5.7 V]

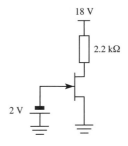

18 V

2.2 kΩ

2 V

7.8 For the circuit shown $V_P = -4$ V and $I_{DSS} = 12$ mA. Determine the value of V_{GS} which is required to produce a drain to source voltage of 8 V.

[−1.29 V]

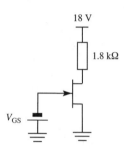

7.9 For the circuit shown the minimum value of I_{DSS} is 12 mA and $V_P = -6$ V. Determine I_D and V_{DS} for V_{GS} of −1 V, −2 V and −3 V and determine whether the operating point is in the saturation or ohmic region in each case.

[8.3 mA, 2.55 V; 5.3 mA, 7 V; 3 mA, 10.5 V]

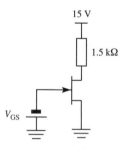

7.10 For the device in Problem 7.9 the maximum value of I_{DSS} is 24 mA. Determine how many of the operating points controlled by the three values of V_{GS} are in the saturation region.

[one at $V_{GS} = -3$ V]

7.11 For the JFET shown $V_P = -4$ V and $I_{DSS} = 12$ mA. Determine the value of V_{GS} so that $V_{DS} = 8$ V.

[−1.53 V]

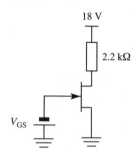

7.12 For the self-bias circuit shown $V_P = -4$ V and $I_{DSS} = 10$ mA. Determine I_D and V_{GS} algebraically.

[2.78 mA, −1.89 V]

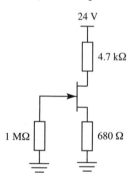

7.13 For the self-bias circuit shown in Problem 7.12 determine the drain current and the gate voltage by graphical means.

7.14 For the self-bias circuit shown the manufacturer's data for the JFET specify $V_P = -7$ V and I_{DSS} as varying from 8 mA to 14 mA. Use the algebraic method to determine the range of drain current.

[2.83 mA to 3.5 mA]

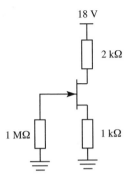

7.15 For the JFET in the circuit shown the manufacturer's data gives $V_P = -6$ V and I_{DSS} ranges from 6 mA to 15 mA. Use the graphical approach to determine the range of drain current and V_{DS}.

[4.8 mA, 12 V, 6.3 mA, 8.25 V]

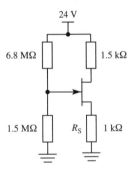

7.16 For the circuit shown in Problem 7.15 assume that the bias arrangement is changed to self-bias with the removal of the 6.8 MΩ resistor and changing R_S to 330 Ω. Determine the new values of I_D and V_{DS}.

[3.6 mA, 17.4 V, 6.3 mA, 12.5 V]

7.17 Use mid-point biasing to design a JFET amplifier stage for when $V_P = -8$ V, $I_{DSS} = 16$ mA and $V_{DD} = 24$ V. Obtain values for R_S, R_D, and g_m.

[250 Ω, 1.5 kΩ, 3 mS]

7.18 For a depletion-mode MOSFET $I_{DSS} = 10$ mA and $V_P = -4$ V. Determine the drain current and V_{DS} when $V_{GS} = -1$ V and $+3$ V.

[5.6 mA, 15.2 V, 30 mA, 3 V]

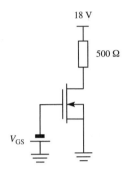

7.19 For the transistor in Problem 7.18 determine g_m when $I_D = 15$ mA.

[6.125 mS]

7.20 For an n-channel enhancement-mode MOSFET $K = 0.2$ mA V^{-2} and the threshold voltage is 1 V. Determine the drain current when $V_{GS} = 0.5$ V and 2 V.

[∗∗∗, 0.2 mA]

7.21 For a p-channel enhancement-mode MOSFET $K_P = 0.5 \times 10^{-5}$ A V^{-2}, the threshold voltage is -2 V, the channel width is 100 μm and the channel length is 1 μm. Determine the drain current and g_m when $V_{GS} = -5$ V.

[2.25 mA, 1.5 mS]

7.22 Based on equations 7.18 and 7.22 and values for $K = 0.5 \times 10^{-3}$ A V^{-2} and $V_{TH} = +2.5$ V, determine algebraically the values of I_D, V_{GS} and V_{DS}.

[3.45 mA, 5.1 V, 15.4 V]

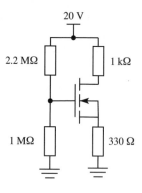

7.23 Use PSpice to determine the drain current, gate to source voltage and the drain to source voltage for the circuit shown in Problem 7.12.

7.24 With the aid of a J2N3819N from the PSpice eval library, determine the quiescent values for the drain current and the drain to source voltage for the circuit shown. Calculate the saturation current I_{DSS}.

[1.83 mA, 12.5 V, 11.7 mA]

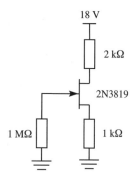

7.25 Use mid-point biasing for a JFET type 2N3819 to design a self-bias circuit for a 15 V

supply based on the manufacturer's data for typical values of $I_{DSS} = 11$ mA and $V_{GS(off)} = -5$ V. Select suitable values from the 10% preferred resistor range.

[220 Ω, 1.5 kΩ, 1 MΩ]

7.26 Use PSpice with a 2N3819 from the eval library to verify the design produced in Problem 7.25 and determine values for the drain current and drain to source voltage.

[4.9 mA, 6.6 V]

7.27 Use PSpice to obtain the quiescent values of I_D and V_{DS} for an n-channel enhancement-mode transistor from the breakout library. Use the model editor to set vto = 2 V, kp = 20×10^{-6}, $L = 2$ μm and $W = 2000$ μm. Run the simulator and verify that the following values are obtained: $I_D = 11.6$ mA, $V_{DS} = 12.4$ V.

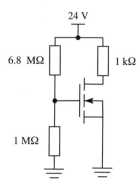

7.28 For the transistor in Problem 7.27 calculate the value of g_m at the quiescent operating point and verify from PSpice.

[21.6 mS]

CHAPTER 8

Field Effect Transistor: Small-Signal Analysis

The important feature to consider in deciding whether to use an FET or a bipolar transistor in a particular application is the input impedance. The high-impedance FET is preferred for many communication applications which may involve a high-Q tuned circuit. A parallel *RLC* circuit which is very frequency selective has a large effective parallel resistance at resonance. The signal from such a circuit can be coupled directly to the input of an FET, as illustrated in Figure 8.1(a) (the bias components are not shown). The Q of the tuned circuit is not reduced because of the very high input resistance of the JFET. The small input capacitance (C_{gs}) of the FET forms part of the main tuning capacitance of the *LC* circuit.

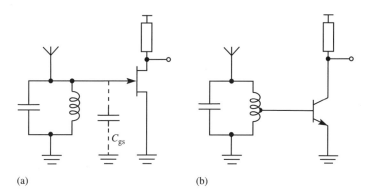

FIGURE 8.1 Comparative circuits showing the coupling of a high-impedance circuit to (a) a JFET and (b) a bipolar transistor.

By contrast, it is necessary to use transformer action to couple the same circuit to a bipolar transistor to transform the high impedance of the tuned circuit to the much lower input impedance of the common-emitter amplifier. Both circuits can provide the same degree of selectivity, but the FET is much simpler to use and does not require impedance matching.

A disadvantage of the FET is the small value of g_m in comparison with that for the bipolar transistor ($g_m = I_E/0.026$ for the bipolar transistor, see Appendix 4) and consequently the gain of individual FET amplifier stages is limited. This limitation can be overcome by combining the FET with a bipolar transistor, where the FET provides the high input resistance and the bipolar transistor provides the gain.

8.1
Small-Signal FET Amplifiers

The basic principles of small-signal calculations for the FET are the same as those for the bipolar circuits of Chapter 6; that is, that the transistor is correctly biased and that the amplitude of the input signal is not so great as to cause the transistor parameters to depart from their quiescent values. The effect is demonstrated graphically in Figure 8.2 where the transfer curves are shown for a JFET. The sine wave variation of the gate voltage represents the input, and the variation of the drain current represents the output. Because of the curvature of the V_G-I_D characteristic there is a non-linear relationship between the input and the output. This is very marked for large variations of V_G, as shown in Figure 8.2.

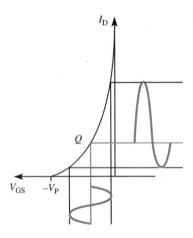

FIGURE 8.2   Transfer curves for JFET with sine wave input voltage.

Provided that the amplitude of the input voltage is not too great then the sine wave variation of the drain current and consequently the drain voltage is not too greatly distorted by the excursion of the input waveform along the parabolic transfer curve.

The most commonly used FET transistor configurations are the common-source and the common-drain, which are equivalent to the common-emitter and the emitter follower. There is no particular need for the common-gate which would be equivalent to the common-base configuration, and this will not be analyzed. (The common-gate is often used as an analog switch, with the switching signal being applied to the gate, and the signal being switched applied to the source and taken as output from the drain.)

8.2
Common-Source Amplifier

The most commonly used FET circuit configuration is the common-source amplifier which like the common-emitter provides voltage gain. It features a very high input impedance and a voltage gain of about -10.

JFET Common-Source Amplifier

The JFET common-source amplifier with self-bias is shown in Figure 8.3.

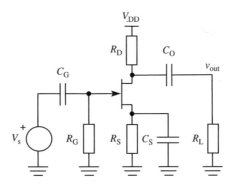

FIGURE 8.3   JFET common-source amplifier.

The source resistor is selected to obtain a suitable Q point on the $I_d–V_{GS}$ transfer curve. The load resistor R_L represents an external load, which could be the input resistance of another stage. The capacitors are assumed to be short circuits for the purposes of mid-frequency small-signal analysis and the ac circuit without the capacitors is shown in Figure 8.4.

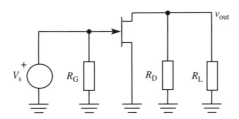

FIGURE 8.4   AC small-signal circuit.

For the purpose of simple hand calculations the equivalent circuit shown in Figure 7.11 is used to replace the transistor as shown in Figure 8.5.

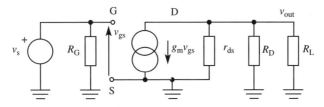

FIGURE 8.5   Equivalent circuit of the JFET common-source amplifier.

Notice the open circuit between the gate and the source. This does not mean that the signal does not propagate through the transistor. It is the voltage which appears between the gate and the source which is important and which drives the current generator $g_m v_{gs}$. The internal drain resistance r_{ds} of the transistor appears in parallel with the external drain and load resistors.

Voltage Gain

The output voltage is:

$$v_{out} = -g_m v_{gs}(r_{ds} \| R_D \| R_L) \qquad (8.1)$$

and the input voltage is:

$$v_s = v_{gs}$$

and the voltage gain is:

$$A_v = \frac{v_{out}}{v_s} = \frac{-g_m v_{gs}(r_{ds} \| R_D \| R_L)}{v_{gs}}$$

that is:

$$A_v = -g_m(r_{ds} \| R_D \| R_L) \qquad (8.2)$$

▷ **EXAMPLE 8.1** ——————————————————————————

For the circuit shown in Figure 8.3 $R_D = 2.2$ kΩ, $R_L = 10$ kΩ and for the transistor $g_m = 3.5$ mS and $r_{ds} = 12$ kΩ. Determine the voltage gain.

▷ *Solution*

$$A_v = -(3.5 \text{ mS})(12 \text{ k}\Omega \| 2{,}2 \text{ k}\Omega \| 10 \text{ k}\Omega)$$
$$= -5.5$$

Notice that the voltage gain is not very large. This is because of the limited value of the transconductance for the FET compared with what may be achieved with a bipolar transistor. (For the bipolar transistor the output current generator βi_b may be replaced by a voltage-controlled current generator $g_m v_{be}$ where $g_m = 1/r_e$, as described in Appendix 4.) At 1 mA $g_m = 39$ mS for a bipolar transistor compared with a typical value for the FET of 3.5 mS. For any given FET larger values of transconductance are obtained by operating with small values of gate voltage (refer to Figure 7.8), but the values are still below those obtained with bipolar transistors.

▷ **EXAMPLE 8.2** ——————————————————————————

Determine the voltage gain for the circuit shown in Figure 8.6. For the transistor $I_{DSS} = 12$ mA, $V_P = -4$ V and $r_{ds} = 20$ kΩ.

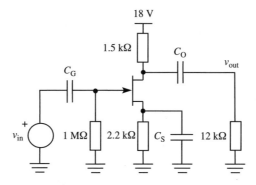

FIGURE 8.6   Common-source amplifier.

▷ *Solution*

DC Condition

Using the algebraic solution of the bias (equation 7.14) gives:

$$I_D = 12 \text{ mA}\left(1 - \frac{I_D 2200 \ \Omega}{4 \text{ V}}\right)^2$$

$$= 12 \text{ mA}(1 - I_D 550)^2$$
$$= 12 \text{ mA}(1 - I_D 1100 + I_D^2 3.025 \times 10^5)$$
$$= 12 \times 10^{-3} - I_D 13.2 + I_D^2 3.63 \times 10^3$$

Rearranging gives:

$$3.63 \times 10^3 I_D^2 - 14.21_D + 12 \times 10^{-3} = 0$$

which yields:

$$I_D = 2.67 \text{ mA or } 1.24 \text{ mA}$$

For the larger value $V_{GS} = (2.67 \text{ mA})(2.2 \text{ k}\Omega) = -5.87$ V which exceeds V_P. Therefore the drain current is 1.24 mA for which $V_{GS} = -2.7$ V.

AC Solution

The small-signal transconductance is given by:

$$g_m = \frac{2I_{DSS}}{V_P}\left(1 - \frac{V_{GS}}{V_P}\right)$$

$$g_m = \frac{24 \text{ mA}}{4 \text{ V}}\left(1 - \frac{2.7 \text{ V}}{4 \text{ V}}\right) = 1.95 \text{ mS}$$

The voltage gain is:

$$A_v = -g_m(r_{ds} \| R_D \| R_L)$$

$$A_v = -1.95 \text{ mS}(20\text{k} \| 1.5\text{k} \| 12\text{k}) = -2.4$$

▶ **EXAMPLE 8.3** ——————————————————————————————

Use PSpice to verify the value of I_D and the voltage gain for Example 8.2.

▶ *Solution*

For the PSpice JFET model a value is required for BETA, which is obtained from Equation 7.11 as:

$$\text{BETA} = \frac{I_{DSS}}{V_P^2} = \frac{12 \text{ mA}}{4^2} = 7.5 \times 10^{-4}$$

From the breakout library select JbreakN and select **Model...** from **Edit** to add the following:

vto = −4 beta = 7.5e-4

Apply a 1 mV, 1 kHz sine wave to the input and use transient analysis to examine the output. The circuit schematic is shown in Figure 8.7.

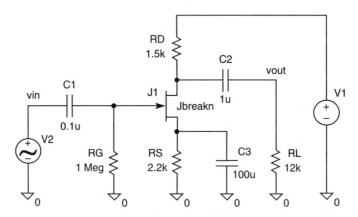

FIGURE 8.7   Circuit schematic for PSpice for JFET common-source amplifier.

The output from Probe for the input and output waveforms is shown in Figure 8.8. With the aid of the cursors the voltage gain is measured as:

$$A_v = \frac{v_{out}}{v_{in}} = -2.55$$

This is slightly higher than the calculated value but this is because the PSpice model statement does not include a value for the internal drain resistance r_{ds}. PSpice models channel length modulation by means of LAMBDA rather than with a discrete value of channel resistance, and there is no simple relationship between r_{ds} and LAMBDA.

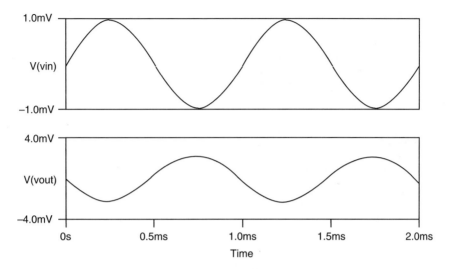

FIGURE 8.8   Input and output waveforms from common-source amplifier.

The dc drain current is obtained from the output file as:

$$I_D = \frac{V_{DD} - V_D}{R_D} = \frac{18\ V - 16.148\ V}{1.5\ k\Omega} = 1.23\ mA$$

MOSFET Common-Source Amplifier

A common-source amplifier with a depletion-mode MOSFET and zero bias is shown in Figure 8.9.

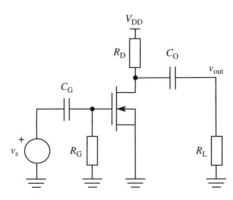

FIGURE 8.9   Depletion-mode MOSFET common-source amplifier.

The voltage gain expression is the same as that for the JFET, as given by equation 8.2.

▶ **EXAMPLE 8.4** _____

For the depletion-mode MOSFET used in Figure 8.9 $I_{DSS} = 10$ mA and $V_P = -3.5$ V. If $R_D = 1.5$ kΩ, $R_L = 5$ kΩ and $V_{DD} = 24$ V, determine the dc drain voltage and the small-signal voltage gain.

▶ *Solution*

DC Condition

With zero gate bias the drain current is I_{DSS} and the drain voltage is:

$V_D = V_{DD} - I_{DSS}R_D$

$V_D = 24 - (10 \text{ mA})(1.5 \text{ kΩ}) = 9$ V

AC Condition

The small-signal transconductance is:

$$g_m = \frac{2I_{DSS}}{V_P}\left(1 - \frac{V_{GS}}{V_P}\right)$$

$$g_m = \frac{20 \text{ mA}}{3.5 \text{ V}}\left(1 - \frac{0 \text{ V}}{3.5 \text{ V}}\right) = 5.7 \text{ mS}$$

The voltage gain is:

$A_v = -g_m(R_D \| R_L)$

$A_v = -5.7 \text{ mS}(1.5 \text{ kΩ} \| 5 \text{ kΩ}) = -6.6$

▶ **EXAMPLE 8.5** _____

Use PSpice to verify the value of the gain for the MOSFET described in Example 8.4.

▶ *Solution*

Select the MbreakN MOSFET transistor from the breakout library. Notice that the device is described as an n-channel enhancement transistor. The name given to the transistor does not affect the choice, because the simulation depends on the information contained in the **model** statement. For the MOSFET it is necessary to obtain a value of K_P, as given by equation 7.23, as:

$$K_P = \frac{2I_{DSS}}{V_P^2} = \frac{(2)(10 \text{ mA})}{3.5^2} = 1.63 \times 10^{-3}$$

From **Edit** select **Model...** and enter:

kp = 1.63e-3 vto = −3.5

The circuit schematic is shown in Figure 8.10 and the output waveform for a 1 mV peak input is shown in Figure 8.11.

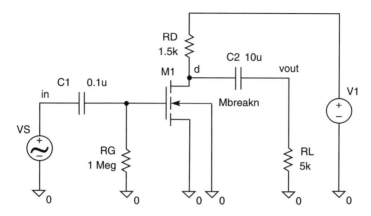

FIGURE 8.10   Circuit schematic from PSpice.

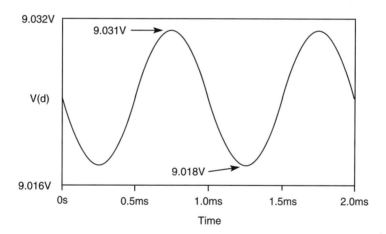

FIGURE 8.11   Output waveform at the drain.

From the output data file created by PSpice the following data is obtained :

$V_D = 9.02$ V

$I_D = 9.98$ mA

$g_m = 5.71$ mS

From the output waveform using Probe to measure the amplitude the gain is:

$A_v = -6.5$

NB A point to note is that the drain–source voltage (V_{DS}) must exceed the threshold voltage. This results from the condition required for pinch-off in the channel and it can be shown that for the transistor to be in the saturation region then $V_{DS} > (V_{GS} - V_{TH})$. For the transistor described in this example $V_{DS} > 3.5$ V.

▶ **EXAMPLE 8.6** ————————————————————————

Repeat Example 8.5 with $V_{DD} = 18$ V to demonstrate the effect of too small a value for V_{DS}.

▶ *Solution*

The new output waveform is shown in Figure 8.12.

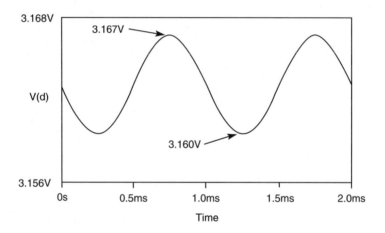

FIGURE 8.12   Output waveform with $V_{DD} = 18$ V.

From the output data created by PSpice the following values are obtained:

$$V_{DS} = 3.16 \text{ V}$$
$$I_D = 9.89 \text{ mA}$$
$$g_m = 5.16 \text{ mS}$$

and from the Probe waveform in Figure 8.12 the peak-to-peak variation is 7 mV. With an input peak-to-peak waveform of 2 mV the gain is:

$$A_v = -3.6$$

This is considerably less than the value obtained when $V_{DD} = 24$ V. This can be explained by noting the position of the dc operating point as shown in Figure 8.13 (obtained with the aid of Probe) using the dc sweep to obtain the output characteristics of the transistor and superimposing a load-line. The load-line is obtained by adding a trace of $(V_{DD} - V_{DS})/R_D$ or $(18 - V_{DS})/1.5$ kΩ.

It can be seen that the transistor is operating in a very non-linear part of the output characteristics. If the input voltage were to be increased beyond 1 mV the output would be severely distorted.

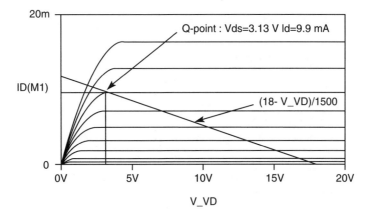

FIGURE 8.13 DC characteristics and load-line for the MOSFET circuit.

Effect of Source Resistance

In the common-source amplifier stage in Figure 8.14 the source resistor is not decoupled.

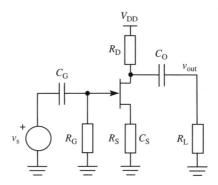

FIGURE 8.14 Common-source amplifier with source resistor.

The small-signal ac circuit is shown in Figure 8.15.

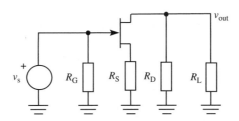

FIGURE 8.15 AC small-signal circuit for the common-source amplifier with source resistor.

Replacing the transistor with the equivalent circuit results in Figure 8.16.

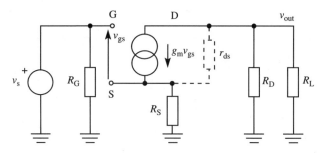

Equivalent circuit for common-source amplifier with source resistor.

The presence of the internal drain resistor r_{ds} complicates the analysis, because part of the current from the current generator flows through r_{ds} and part flows through R_S. This division of the current results in a more complex relationship for the output voltage which is not justified if r_{ds} is large in comparison with $R_D \parallel R_L$. For the majority of applications this is likely to be the situation and the analysis is greatly simplified if r_{ds} is ignored.

The output voltage assuming that r_{ds} is negligible is:

$$v_{out} = -g_m v_{gs}(R_D \parallel R_L) \tag{8.3}$$

and the input voltage is:

$$v_s = v_{gs} + g_m v_{gs} R_S \tag{8.4}$$

Notice the inclusion of the voltage (v_{gs}) which appears across the open-circuit gate–source terminals of the equivalent circuit for the transistor, to which is added the voltage generated across the source resistor by the current $g_m v_{gs}$.

The voltage gain is:

$$A_v = \frac{v_{out}}{v_s} = \frac{-g_m v_{gs}(R_D \parallel R_L)}{v_{gs} + g_m R_S}$$

and

$$A_v = -\frac{g_m(R_D \parallel R_L)}{1 + g_m R_S} \tag{8.5}$$

It can be seen from equation 8.5 that the gain is reduced by the presence of R_S in much the same way that the gain of the bipolar common-emitter amplifier with an undecoupled emitter resistor is reduced. For the bipolar circuit the attraction is the increased value of the input resistance and also a voltage gain which is less dependent on the transistor parameters. For the JFET the input resistance is already very large and although it is increased further, this is not so significant as for the bipolar circuit. Neither is the reduction in the gain particularly beneficial since the gain of JFET circuits is already low. One possible application of the circuit is as a phase-splitter.

JFET Phase-Splitter

The phase-splitter is shown in Figure 8.17.

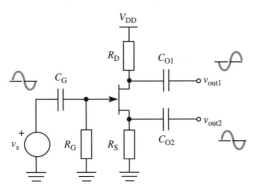

FIGURE 8.17   JFET phase-splitter.

There are two outputs, one from the drain and one from the source. Notice the polarities of the alternating waveforms at the input and the two outputs. The output at the drain shows the 180° phase shift, while the output at the source is in phase with the input. Thus the circuit provides two output waveforms which are 180° out of phase with each other and which are isolated from the input by the JFET. The values of R_D and R_S are adjusted to provide equal amplitudes of v_{out1} and v_{out2}.

The small-signal equivalent circuit is shown in Figure 8.18.

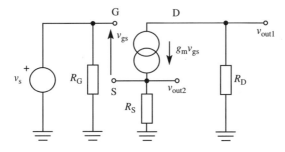

FIGURE 8.18   Small-signal equivalent circuit of the JFET phase-splitter.

The internal channel resistance r_{ds} is assumed to be negligible and external load resistors have also been ignored, although these could easily be included and would simply appear in parallel with R_D and R_S.

The output voltages are:

$$v_{out1} = -g_m v_{gs} R_D \tag{8.6}$$

and

$$v_{out2} = g_m v_{gs} R_S \tag{8.7}$$

The input voltage is:

$$v_s = v_{gs} + g_m v_{gs} R_S$$

and the voltage gains are:

$$A_{v1} = \frac{v_{out1}}{v_s} = -\frac{g_m R_D}{1 + g_m R_S} \tag{8.8}$$

and

$$A_{v2} = \frac{v_{out2}}{v_s} = -\frac{g_m R_S}{1 + g_m R_S} \tag{8.9}$$

The two values of gain are the same if $R_D = R_S$.

8.3 Common-Drain Amplifier

The common-drain amplifier or source follower is equivalent to the emitter follower bipolar transistor amplifier. Like the emitter follower it has a low output impedance and a voltage gain of approximately unity. The schematic of the circuit is shown in Figure 8.19.

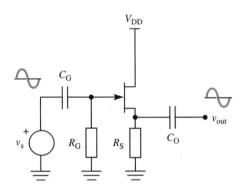

FIGURE 8.19 Schematic of JFET common-drain amplifier.

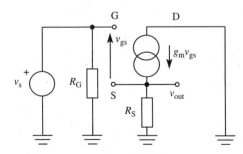

FIGURE 8.20 Small-signal equivalent circuit of the FET common-drain amplifier.

The output is taken from the source terminal of the FET and the signal is in phase with the input; the output follows the input. The small-signal circuit is shown in Figure 8.20.

Voltage Gain

The output voltage is:

$$v_{out} = g_m v_{gs} R_S \tag{8.10}$$

and the input voltage is:

$$v_s = v_{gs} + g_m v_{gs} R_S$$

The voltage gain is:

$$A_v = \frac{v_{out}}{v_s} = \frac{g_m R_S}{1 + g_m R_S} \tag{8.11}$$

and

$$A_v \approx 1 \quad \text{if } g_m R_S \gg 1 \tag{8.12}$$

Output Resistance

The output resistance may be obtained in a similar manner to that used for the bipolar emitter follower amplifier, that is to determine the open-circuit voltage and the short-circuit current at the source terminal with R_S removed. The equivalent circuit is shown in Figure 8.21.

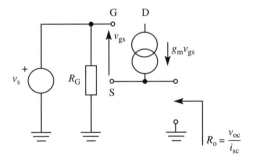

FIGURE 8.21   Modified equivalent circuit used to determine the output resistance.

The open-circuit voltage is:

$$v_{oc} = v_s$$

and the short-circuit current is:

$$i_{sc} = g_m v_{gs}$$

and under short-circuit condition $v_s = v_{gs}$. Thus the output resistance is:

$$R_o = \frac{v_{oc}}{i_{sc}} = \frac{v_s}{g_m v_s} = \frac{1}{g_m} \tag{8.13}$$

This represents the resistance 'seen' by the output 'looking into' the JFET from the source. The actual output resistance, as seen by an external load, is:

$$R_{oeff} = R_S \| (1/g_m) \tag{8.14}$$

Like the bipolar emitter follower the source follower is used as a buffer between a high-impedance source and a low-impedance load, although the output resistance is not as low as for an emitter follower.

▷ **EXAMPLE 8.7** ─────────────────────────────────

For the JFET in Figure 8.22 $g_m = 4 \times 10^{-3}$ S. Determine the input resistance, the voltage gain and the output resistance.

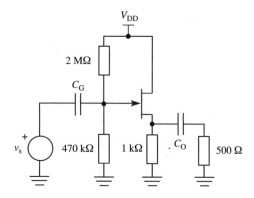

FIGURE 8.22   Schematic for Example 8.7.

▷ *Solution*

Input Resistance

By inspection the input resistance is simply the parallel combination of the two bias resistors which form the potential divider. That is:

$$R_{in} = 2 \text{ M}\Omega \| 470 \text{ k}\Omega = 380 \text{ k}\Omega$$

Voltage Gain

From equation 8.11 the voltage gain is:

$$A_v = \frac{(4 \times 10^{-3})(1 \text{ k}\Omega \| 500 \text{ }\Omega)}{1 + (4 \times 10^{-3})(1 \text{ k}\Omega \| 500 \text{ }\Omega)} = 0.57$$

where R_S is replaced by $R_S \| R_L$.

Notice that the gain for this particular example is $\ll 1$. This is because the inequality $g_m R_S \| R_L \ll 1$ does not hold for this example.

Output Resistance

From equation 8.14:

$$R_{oeff} = [1 \text{ k}\Omega] \| [1/(4 \times 10^{-3})] = 200 \ \Omega$$

Again notice that the output resistance is not particularly low. It would be much lower with an emitter follower.

8.4 ▶ Multistage Amplifier

The gain of a single FET stage amplifier is not very large. Larger values of gain could be obtained by using more than one FET stage, but a better solution is to combine an FET with a bipolar transistor. Some examples of multistage amplifiers, with FETs and with a combination of FET and bipolar transistors, are presented below as a series of examples.

▶ **EXAMPLE 8.8** ───────────────────────────────

For the two-stage common-source amplifier shown in Figure 8.23 $I_{DS} = 15$ mA and $V_P = -4$ V. Determine the drain current, the gate to source voltage and the overall gain.

▶ *Solution*

DC Analysis

Use the algebraic method to determine I_D and V_{GS}.

$$I_D = I_{DSS}\left(1 - \frac{I_D R_S}{V_P}\right)^2$$

$$I_D = 15 \text{ mA}\left(1 - \frac{I_D 560}{4}\right)^2$$

$$I_D = 15 \times 10^{-3} - 4.2 I_D + 294 I_D$$

and

$$294 I_D^2 - 5.2 I_D + 15 \times 10^{-3} = 0$$

so

$$I_D = \frac{5.2 \pm \sqrt{5.2^2 - (4)(294)(15 \times 10^{-3})}}{(2)(294)}$$

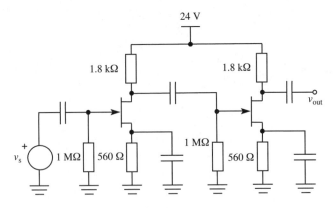

FIGURE 8.23   Two-stage JFET amplifier.

which gives $I_D = 14$ mA or 3.62 mA. Then:

$V_{GS} = -7.8$ V or -2 V

Since $V_P = -4$ V and since V_{GS} must be less than V_P then the correct values are:

$I_D = 3.62$ mA and $V_{GS} = -2$ V

AC Analysis

The small-signal transconductance is:

$$g_m = \frac{2I_{DSS}}{V_P}\left(1 - \frac{V_{GS}}{V_P}\right)$$

$$g_m = \frac{30\text{ mA}}{4}\left(1 - \frac{2}{4}\right) = 3.75\text{ mS}$$

The voltage gain of the first stage is:

$A_{v1} = -g_m R_D \| R_L$

where R_L is the input resistance of the second stage, that is the 1 MΩ gate resistor. Thus:

$A_{v1} = -(3.75 \times 10^{-3})(1.8\text{ kΩ} \| 1\text{ MΩ}) = -6.74$

and for the second stage:

$A_{v2} = -(3.75 \times 10^{-3})(1.8\text{ kΩ}) = -6.75$

The total gain is:

$A_{vT} = A_{v1}A_{v2} = 45.5$

▶ **EXAMPLE 8.9** ————————————————————————————————

The input voltage to the two-stage phase-splitter shown in Figure 8.24 is a 5 mV peak sine wave. Determine the output voltages.

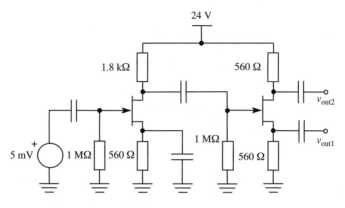

FIGURE 8.24   Two-stage phase-splitter.

▷ *Solution*

From Example 8.8 $I_D = 3.62$ mA, $V_{GS} = -2$ V and $g_m = 3.75$ mS. The voltage gain of the first stage is:

$$A_{v1} = -6.74$$

The gain at the drain of the second stage is:

$$A_{vD} = -\frac{g_m R_D}{1 + g_m R_S}$$

$$A_{vD} = -\frac{(3.75 \times 10^{-3})(560)}{1 + (3.75 \times 10^{-3})(560)} = -0.68$$

and the total gain at the drain is:

$$A_{vDTotal} = A_{v1} \times A_{vD} = 4.58$$

and at the source of the second stage:

$$A_{vS} = \frac{g_m R_S}{1 + g_m R_S}$$

$$A_{vS} = \frac{(3.75 \times 10^{-3})(560)}{1 + (3.75 \times 10^{-3})(560)} = 0.68$$

and the total gain at the source is:

$$A_{vSTotal} = A_{v1} \times A_{vS} = -4.58$$

The output voltages at the drain and source are out-of-phase sine waves with peak amplitudes of 23 mV.

▷ **EXAMPLE 8.10** ──────────────────────────────

Use PSpice to analyze the circuit shown in Figure 8.25 which uses a JFET for a high input impedance and a common-emitter bipolar stage for gain.

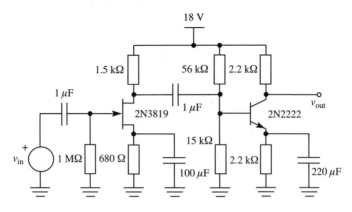

FIGURE 8.25 Common-source plus common-emitter amplifier.

Solution

The transistors are from the **eval.slb** library. The input is a 1 mV, 1 kHz sine wave and a transient analysis is performed with a step of 10 μs for 2 ms.

The following dc parameters are obtained from the output file:

JFET

BETA	1.3×10^{-3}
LAMBDA	2.25×10^{-3}
VTO	-3 V
gm	3.61 mS
ID	2.43 mA
VGS	-1.65 V

Bipolar

BF	256
IC	1.38 mA

The output waveform produced by Probe is a sine wave with an amplitude of 869 mV peak to peak which gives a gain of:

$$A_v = 434$$

EXAMPLE 8.11

Verify the values of I_D, I_C and A_v in Example 8.10.

▶ *Solution*

DC Conditions

Use the algebraic method to obtain the drain current for the FET.

$$I_D = I_{DSS}\left(1 - \frac{I_D 680}{3}\right)^2$$

where

$$I_{DSS} = (BETA)V_P^2 \quad \text{from equation 7.10a}$$
$$I_{DSS} = 1.3 \times 10^{-3} \times 3^2 = 11.7 \text{ mA}$$

and

$$I_D^2 601.4 - I_D 6.3 + 11.7 \times 10^{-3} = 0$$

so

$$I_D = \frac{6.3 \pm \sqrt{39.69 - 28.14}}{1202.8}$$

and

$$I_D = 8.06 \text{ mA} \quad \text{or } 2.4 \text{ mA}$$

The 2.4 mA would appear to be the correct value.
For the bipolar transistor the collector current is given by:

$$I_C = \frac{V_T - 0.7 \text{ V}}{2.2 \text{ k}\Omega + [(56 \text{ k}\Omega \| 15 \text{ k}\Omega)/\beta]}$$

$$I_C = \frac{3.8 \text{ V} - 0.7 \text{ V}}{2.2 \text{ k}\Omega + (11.8 \text{ k}\Omega/256)} = 1.38 \text{ mA}$$

which agrees with the value obtained with PSpice.

AC Conditions

The transconductance of the FET is given by:

$$g_m = \frac{2I_{DSS}}{V_P}\left(1 - \frac{I_D 680}{V_P}\right)$$

$$g_m = \frac{2 \times 11.7 \times 10^{-3}}{3}\left(1 - \frac{1.63}{3}\right) = 3.55 \text{ mS}$$

The gain of the first stage is:

$$A_{v1} = -g_m R_D \| R_{in2}$$

where R_{in2} is the input resistance of the second stage and is given by:

$$R_{in2} = 56 \text{ k}\Omega \,\|\, 15 \text{ k}\Omega \,\|\, \beta r_{e2}$$

where $r_{e2} = 0.026/1.38 \text{ mA} = 18.8 \text{ }\Omega$. Thus:

$$R_{in2} = 3.4 \text{ k}\Omega$$

and

$$A_{v1} = -3.7$$

The gain of the second stage is:

$$A_{v2} = -\frac{R_C}{r_e} = -\frac{2.2 \text{ k}\Omega}{18.8 \text{ }\Omega} = -117$$

and the overall gain is:

$$A_{vT} = A_{v1} \times A_{v2} = -3.7 \times -117 = 433$$

Cascoded Stages

In the above examples the two amplifier stages are *cascaded* together, with the output from one feeding into the input of the next stage. In the *cascode* the second stage is the load for the first stage. An example of an FET–bipolar cascode is shown in Figure 8.26.

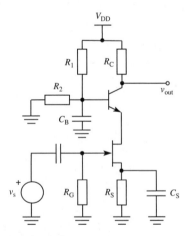

FIGURE 8.26   Cascode of FET and bipolar amplifier stages.

Notice that the base of the bipolar transistor is decoupled to ground by the capacitor C_B, so that the bipolar transistor is configured as a common-base stage. The resistors R_1 and R_2 provide the dc bias for the bipolar transistor. The load for the FET is the input resistance of a common-base transistor stage, which is low.

The equivalent circuit is developed as follows: starting from the signal source the coupling capacitor to the gate of the JFET is a short circuit and one end of R_G is connected to ground. The source of the JFET is short-circuited to ground, and the FET is simply a common-source stage. Notice that the potential divider resistors R_1 and R_2 are short-circuited by C_B and the low internal impedance of the power supply, and they do not appear in the small-signal circuit. The signal from the drain of the FET is applied to the emitter, and with the base grounded through C_B the bipolar transistor acts as a common-base stage with an input resistance r_e and a current generator $\alpha i_e \approx i_e$. The collector is connected to the load resistor R_C, the other end of which is connected to ground through the low impedance of the power supply. The complete small-signal equivalent circuit is shown in Figure 8.27.

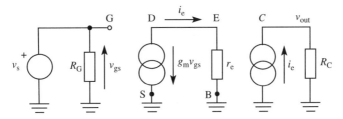

FIGURE 8.27 Small-signal equivalent circuit of the FET/bipolar cascode.

The voltage gain is obtained as follows:

$$v_{out} = i_e R_C$$

where

$$i_e = -g_m v_{gs}$$

and

$$v_s = v_{gs}$$

Then the gain is:

$$A_v = \frac{v_{out}}{v_s} = \frac{i_e R_C}{v_s} = \frac{-g_m v_{gs} R_C}{v_{gs}} = -g_m R_C \qquad (8.15)$$

This expression is of the same form as that for the single-stage common-source amplifier (equation 8.2). Notice that the bipolar transistor does not influence the gain expression. The importance of this configuration will become apparent in Chapter 9 on frequency response. Although the gain of the cascode is the same as that of the single stage, the high-frequency performance of the cascode is improved over that of the single stage.

▶ **EXAMPLE 8.12** ─────────────────────────────────────

Determine the gain of the cascode shown in Figure 8.28 with the aid of PSpice.

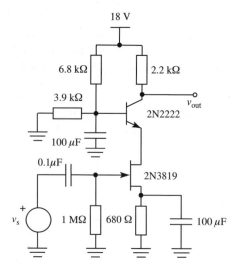

FIGURE 8.28   JFET–bipolar cascode.

▷ *Solution*

The dc current is determined by R_S (680 Ω) and is consequently the same as for Example 8.10, that is:

$$I_D = I_C = 2.4 \text{ mA}$$

The values of the potential divider resistors for the bipolar transistor are calculated by allowing a value of approximately 6 V to appear at the drain of the JFET. This results in a value of approximately 6.7 V at the base of the bipolar transistor and R_1 and R_2 are selected to provide this value. The collector resistance is chosen to provide a voltage drop of approximately 6 V so that $V_{CE} \approx 6$ V.

The transconductance is 3.55 mS and thus the voltage gain is:

$$A_v = -(3.55 \text{ mS})(2.2 \text{ k}\Omega) = -7.8$$

From PSpice the gain as determined by observing the output waveform for a 1 mV peak sine wave input is:

$$A_{vPSpice} = -7.75$$

From PSpice the voltages at the drain and collector are:

$$V_D = 5.8 \text{ V}$$

$$V_C = 12.7 \text{ V}$$

Therefore $V_{CE} = 6.9$ V

► PROBLEMS

8.1 (a) If the drain current of a JFET is given by:

$$I_D = I_{DSS}\left(1 - \frac{V_{GS}}{V_P}\right)^2$$

and g_m is defined as dI_D/dV_{GS} show that g_m is given by:

$$g_m = \frac{2I_{DSS}}{V_P}\sqrt{\frac{I_D}{I_{DSS}}}$$

(b) From the manufacturer's data it is seen that V_P varies from -4 V to -6 V and that I_{DSS} varies from 8 mA to 20 mA respectively. Determine the value of g_m for $V_{GS} = -1$ V.

[3000 μS, 5560 μS]

8.2 The curve shown of I_D–V_{GS} is for an n-channel JFET.

(a) Determine g_m graphically for $V_{GS} = -2$ V.
(b) Determine g_m algebraically for the same gate voltage.

[(b) 4800 μS]

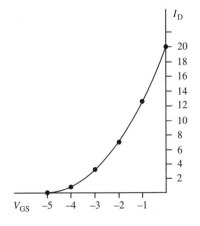

8.3 If the maximum value of transconductance for an n-channel JFET is 8000 μS and $I_{DSS} = 16$ mA, determine the pinch-off voltage and the transconductance when $V_{GS} = -1.5$ V.

[-4 V, 5000 μS]

8.4 For the n-channel JFET $I_{DS} = 16$ mA and $V_P = -4$ V. Determine

(a) the quiescent values of I_D, and V_{GS}
(b) the value of g_m
(c) the voltage gain.

[(a) 2.44 mA, 2.44 V, (b) 3120 μS, (c) -5.6]

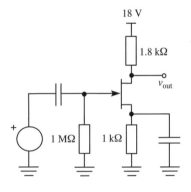

8.5 For the p-channel JFET $I_{DSS} = 12$ mA and $V_P = 3$ V. Determine

(a) the quiescent value of I_D
(b) the voltage gain.

[3 mA, -8.8]

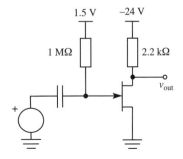

8.6 If for the JFET in Problem 8.5 $r_{ds} = 80$ kΩ determine the new value of the voltage gain.

[−8.56]

8.7 Identify the parameters in the SPICE model of the JFET which models I_{DSS} and r_{ds}.

8.8 For the circuit shown $I_{DSS} = 10$ mA and $V_P = -3$ V. Determine the quiescent values of I_D, V_{DS} and the voltage gain.

[2.9 mA, 1.4 V, −9.7]

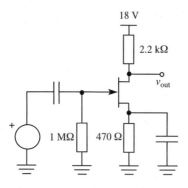

8.9 For the p-channel JFET $I_{DSS} = 15$ mA and $V_P = 5$ V. Determine by I_D graphical means, and then V_{DS} and the voltage gain.

[6.2 mA, 1.7 V, −5.9]

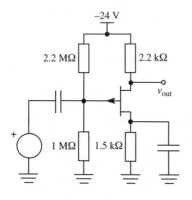

8.10 Determine the voltage gain for each of the circuits.

[0.91, 0.65]

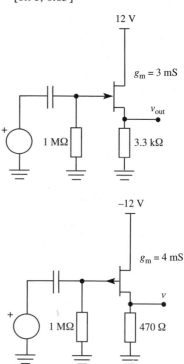

8.11 For the circuit shown $g_m = 3000$ μS and $r_{ds} = 70$ kΩ. Determine the output voltage v_{out} if $v_{in} = 10$ mV.

[92 mV]

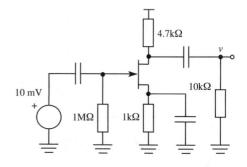

8.12 For the circuit shown $g_m = 4000$ μS and $r_{ds} = 60$ kΩ. Determine the voltage gain.

[−7.3]

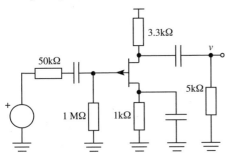

8.13 For the JFET phase-splitter $g_m = 3500$ μS. Determine the voltage gain for each output.

[−1.17, 0.78]

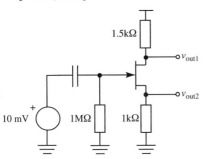

8.14 For the source follower shown $g_m = 4000$ μS. Determine the voltage gain and the output resistance R_{out}.

[0.705, 214 Ω]

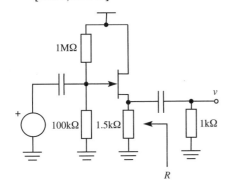

8.15 For the circuit in Problem 8.14 redraw the small-signal equivalent circuit to include r_{ds} and recalculate the voltage gain if $r_{ds} = 50$ kΩ.

[0.703]

8.16 For the p-channel source follower $g_m = 4500$ μS and $r_{ds} = 55$ kΩ. Determine the voltage gain.

[0.886]

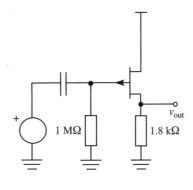

8.17 For the n-channel depletion-mode MOSFET $g_m = 3500$ μS and $r_{ds} = 80$ kΩ. Determine the voltage gain.

[−5.75]

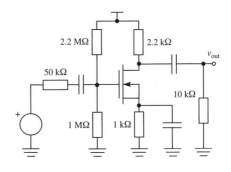

8.18   For the depletion-mode MOSFET
$I_{DSS} = 15$ mA and $V_P = -6$ V. Determine g_m
and the voltage gain.

[5000 µS, −23.5]

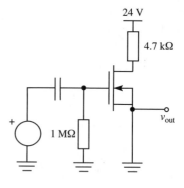

8.19   For the JFETS shown in the two-stage amplifier, $I_{DSS} = 14$ mA and $V_P = -5$ V, and the quiescent values of I_D are $I_{D1} = 4.5$ mA and $I_{D2} = 3$ mA. Determine the voltage gain.

[33]

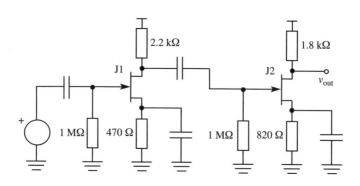

8.20   For the n-channel MOSFET $I_{DSS} = 12$ mA and $V_P = -4$ V, while for the npn transistor $\beta = 120$. Determine the voltage gain.

$[A_{v1} = -4, A_{v2} = -427, A_v = 1708]$

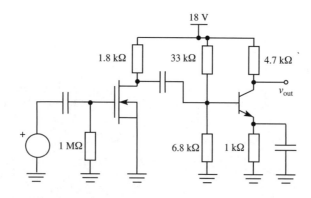

8.21 In the cascode shown the data for the MOSFET is $I_{DSS} = 10$ mA and $V_P = -2.5$ V, and for the bipolar transistor $\beta = 150$. Determine the voltage gain.

[−8]

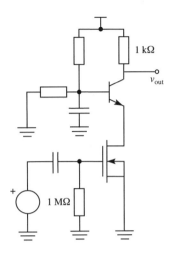

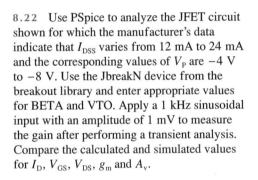

8.22 Use PSpice to analyze the JFET circuit shown for which the manufacturer's data indicate that I_{DSS} varies from 12 mA to 24 mA and the corresponding values of V_P are −4 V to −8 V. Use the JbreakN device from the breakout library and enter appropriate values for BETA and VTO. Apply a 1 kHz sinusoidal input with an amplitude of 1 mV to measure the gain after performing a transient analysis. Compare the calculated and simulated values for I_D, V_{GS}, V_{DS}, g_m and A_v.

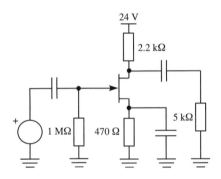

8.23 Explain the discrepancy which may exist between the calculated and simulated gains for the circuit in Problem 8.22 by noting the value of $V_{DS(sat)}$ for the two extremes of I_{DS}.

8.24 For the depletion-mode MOSFET $I_{DSS} = 4$ mA, $V_P = -2$ V and for the pnp transistor $\beta = 120$. Use PSpice to analyze the cascade circuit with devices from the breakout library, and compare the calculated and simulated values for V_{DS}, I_C, V_C and A_v.

[$A_v \approx 500$]

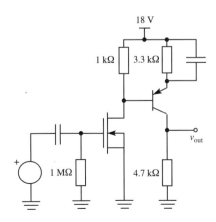

8.25 Use PSpice to analyze the cascode circuit shown, where $I_{DSS} = 8$ mA, $V_P = -3.5$ V and $\beta = 150$. Use devices from the breakout library and compare the calculated dc conditions and small-signal voltage gain with the simulated values.

[$A_v \approx -5.5$]

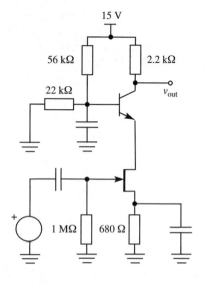

CHAPTER 9

Frequency Response of *RC* Networks

This chapter may be regarded as revision of ac analysis of resistor–capacitor networks. The material presented here may have been covered in an electrical principles or electrical theory course. It may be appropriate to try the problems at the end of this chapter to determine whether to continue with this chapter if the problems prove difficult to solve, or to move on to Chapter 10 if there are no difficulties.

A discrete component transistor amplifier usually contains capacitors to isolate the dc parts of the circuit and to decouple bias resistors. In the analysis developed so far it has been assumed that these capacitors act as short circuits and, therefore, their effect is not considered in the small-signal analysis of gain or impedance. However, in a practical amplifier the input signal may consist of many different frequency components: for example, an audio signal may range from 100 Hz up to 15 kHz, a video signal for television may range from 20 Hz up to 5 MHz and for these signal sources the reactance of the capacitors will vary very considerably over the complete range of frequencies. At low frequencies the reactance increases and any coupling capacitors will present an increasing impedance to the passage of a signal as the frequency decreases. A decoupling capacitor will no longer act as a complete short and thus the resistor which should be decoupled becomes part of the circuit and causes the gain to be reduced. At high frequencies the performance of the active devices begins to deteriorate and the gain decreases. In addition all components and connecting wires have stray capacitance associated with them. This stray capacitance provides a signal path to ground and as the frequency increases the impedance of this path decreases. Thus as the frequency increases the combined effect of deterioration in the performance of the active devices and the stray capacitors results in a loss of signal and a reduction in the gain.

The typical frequency response for a discrete component amplifier with capacitive or ac coupling is shown in Figure 9.1.

The gain decreases at both low frequency and high frequency, with a mid-band region where the gain is relatively constant. The two corner frequencies f_L and f_H define the mid-band region or *bandwidth*. The bandwidth is defined as:

$$BW = f_H - f_L \tag{9.1}$$

The bandwidth is chosen to suit a particular application. Thus for an audio amplifier it may extend from 20 Hz to 20 kHz, while for a video amplifier it could be 20 Hz to 5 MHz.

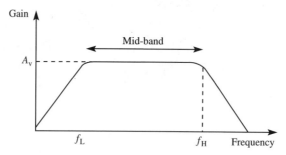

FIGURE 9.1   Frequency response of an ac-coupled amplifier.

Throughout the mid-band region the gain for the different frequency components of a signal is substantially the same for each component. Thus if a signal has frequency components which have rms values of 10 mV at 100 Hz, 200 mV at 1.5 kHz and 50 mV at 8 kHz, then after passing through an amplifier which has a gain of 10, the output components should have rms values of 100 mV at 100 Hz, 2000 mV at 1.5 kHz and 500 mV at 8 kHz. If the amplitude ratios are not maintained for each frequency component then the output will be distorted. In addition to the amplitude ratio being maintained, it is also important for the phase relationship between the different components to remain the same after amplification.

The relationship between the two corner frequencies and the capacitance in the circuit is important for understanding the operation of a circuit over a wide range of frequencies. The small-signal calculations performed so far apply only to the mid-band region where the capacitors are assumed to be either short circuits or open circuits, and as a result they have not entered into the expressions for gain or impedance. It is now necessary to consider how these capacitors affect the frequency performance of an amplifier, but it is also necessary to consider an alternative definition of gain.

9.1   The Decibel

The frequency response curve of gain versus frequency is an important means of characterizing an amplifier. Both the gain and the frequency can vary over wide ranges from less than 10 to values of greater than 100000 for the gain, and from less than 10 Hz to many tens of MHz for the frequency. A graph of gain versus frequency may have the appearance of that shown in Figure 9.2.

The graph is typical of what may be expected for an operational amplifier where the frequency response extends down to dc, as a result of there being no coupling capacitors between the different stages of an integrated circuit operational amplifier. Over the full frequency range the gain varies from 100000 at dc to 0 at several MHz. The large numerical values for the gain are often inconvenient to use on graphs, and also for mathematical reasons, which will become clear later, they are not ideal for conveying information on the nature of the frequency response of amplifiers.

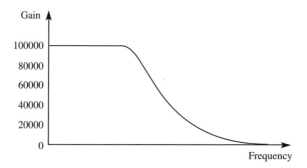

FIGURE 9.2   Linear representation of gain versus frequency.

A more convenient method for presenting the gain is to use the logarithm of the gain, that is log to the base 10 of the gain. Thus:

log(1)	0
log(100)	2
log(100000)	5

In addition to using a logarithmic scale for the y axis it is often desirable also to use a logarithmic scale for the frequency axis – the x axis. When the frequency response shown in Figure 9.2 is redrawn with logarithmic values it has the appearance shown in Figure 9.3.

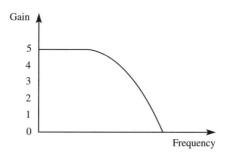

FIGURE 9.3   Frequency response with logarithmic x and y axis.

The use of power rather than voltage for measuring the gain provides the definition of the *bel* as:

$$\log(P_{gain}) = \log\left(\frac{P_{out}}{P_{in}}\right) \text{bel}$$

For most applications the bel is too large and a more practical value is the *decibel* (*dB*) which is:

$$10 \log(P_{\text{gain}}) = 10 \log\left(\frac{P_{\text{out}}}{P_{\text{in}}}\right) dB$$

Now

$$P_{\text{in}} = \frac{v_{\text{in}}^2}{R_{\text{in}}} \quad \text{and} \quad P_{\text{out}} = \frac{v_{\text{out}}^2}{R_{\text{out}}}$$

where R_{in} and R_{out} are the input and output resistances of an amplifier. Thus:

$$P_{\text{gain}} = \left(\frac{v_{\text{out}}}{v_{\text{in}}}\right)^2 \frac{R_{\text{out}}}{R_{\text{in}}}$$

and if $R_{\text{out}} = R_{\text{in}}$ then the power gain is the same as the voltage gain and:

$$P_{\text{gain}}(\text{dB}) = A_v(\text{dB}) = 10 \log\left(\frac{v_{\text{out}}}{v_{\text{in}}}\right)^2$$

or

$$A_v(\text{dB}) = 20 \log\left(\frac{v_{\text{out}}}{v_{\text{in}}}\right) \tag{9.2}$$

This definition of the voltage gain as 20 times the log of the voltage ratio is now an accepted method for defining the gain of an amplifier even though the input and output resistances may not in fact be equal. If $R_{\text{in}} \neq R_{\text{out}}$ then the power gain is not equal to the voltage gain, but this does not affect the definition of the voltage gain.

Some simple examples of voltage gain expressed in dB are as follows:

$v_{\text{out}}/v_{\text{in}}$	$20 \log(v_{\text{out}}/v_{\text{in}}) \text{dB}$
10	20
2	6
0.707	−3
0.1	−20

The significance of the −3 dB and −20 dB will be revealed below in the discussion of the frequency response of simple *RC* filters.

▶ **EXAMPLE 9.1** ─────────────────────────

The input voltage to an amplifier is 4 mV$_{\text{rms}}$ and the output voltage is 10 V$_{\text{rms}}$. Determine the voltage gain in dB.

▶ *Solution*

$$20 \log \frac{v_{out}}{v_{in}} = 20 \log \frac{10}{4 \times 10^{-3}} \approx 68 \text{ dB}$$

▶ **EXAMPLE 9.2**

An amplifier with a gain of 55 dB produces an output of 12 V_{rms}. Determine the input voltage.

▶ *Solution*

$$55 \text{ dB} = 20 \log \frac{v_{out}}{v_{in}} = 20 \log \frac{12}{v_{in}}$$

$$2.75 = \log \frac{12}{v_{in}}$$

That is:

$$10^{2.75} = \frac{12}{v_{in}}$$

and

$$v_{in} = \frac{12}{562.34} = 21.3 \text{ mV}$$

When amplifiers are cascaded and the coupling between the stages is ideal, that is input impedance matches output impedance, the overall gain may be obtained by simply adding the individual gains if they are expressed in decibels. Thus in Figure 9.4 the combined gain is 26 dB.

This may be confirmed since the total voltage gain is $10 \times 2 = 20$ and $20 \log 20 = 26.02$ dB.

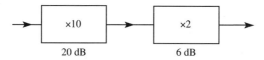

FIGURE 9.4   Cascade of two amplifiers with individual gains of 10 and 2.

▶ **EXAMPLE 9.3** ─────────────────────────────────

The input voltage to an amplifier is 1 mV$_{\text{rms}}$. The first stage of amplification produces a gain of 15 dB. This stage is then followed by a filter section with a gain of -6 dB. Finally there is an output stage with a gain of 45 dB. Determine

(a) the voltage either side of the filter
(b) the output voltage
(c) the overall gain.

▶ *Solution*

Let v_1 be the voltage at the input to the filter and v_2 at the output. Then:

(a) $20 \log \dfrac{v_1}{1 \text{ mV}} = 15$ dB

or

$\log \dfrac{v_1}{1 \times 10^{-3}} = 0.75$

and

$v_1 = 10^{0.75} \times 1 \times 10^{-3} = 5.62$ mV

(b) $20 \log \dfrac{v_2}{5.62 \times 10^{-3}} = -6$ dB

and

$v_2 = 10^{-0.3} \times 5.62 \times 10^{-3} = 2.82$ mV

(c) $20 \log \dfrac{v_{\text{out}}}{2.82 \text{ mV}} = 45$ dB

and

$v_{\text{out}} = 10^{2.25} \times 2.82 \times 10^{-3} \approx 500$ mV

A check can be made on the overall gain by simply summing the gains as expressed in dB. Thus:

$A_v = 15$ dB $- 6$ dB $+ 45$ dB $= 54$ dB

and the output voltage is:

54 dB $= 20 \log \dfrac{v_{\text{out}}}{1 \times 10^{-3}}$

and

$v_{\text{out}} = 10^{2.7} \times 1 \times 10^{-3} \approx 500$ mV

Log–Lin Frequency Plot

It is often convenient when plotting the gain versus frequency response of amplifiers to use *log–lin* graph paper. The linear *y* axis is used for the gain expressed in dB and the logarithmic *x* axis is used for frequency as shown in Figure 9.5. Most circuit simulators provide an option for plotting the results with either linear or logarithmic axes.

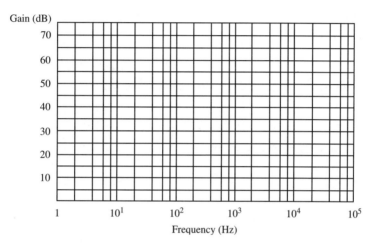

FIGURE 9.5   Example of log–lin graph paper used for gain versus frequency plots.

9.2 ▷ Basic *RC* Networks

The simplest *RC* network comprises a single resistor and capacitor. Two possible configurations are shown in Figure 9.6.

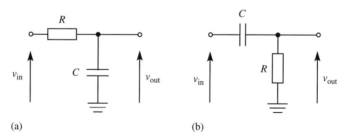

(a)   (b)

FIGURE 9.6   (a) Low-pass filter and (b) high-pass filter.

For the configuration shown in Figure 9.6(a) the capacitor may be considered an open circuit at low frequencies and thus $v_{out} = v_{in}$ at low frequencies. At high frequencies the capacitor acts as a short circuit and v_{out} is reduced to zero. The overall frequency response is shown in Figure 9.7(a). It is known as a *low-pass* filter because it passes low frequencies.

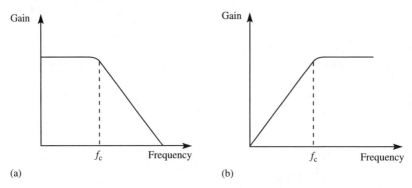

FIGURE 9.7 (a) Frequency response of the low-pass filter and (b) frequency response of the high-pass filter.

For the configuration shown in Figure 9.6(b) the capacitor blocks the low-frequency signal, while at high frequency it acts as a short circuit. The resulting frequency response is as shown in Figure 9.7(b). It is known as a *high-pass* filter.

9.3 High-Pass Filter

The high-pass filter of Figure 9.6(b) and its frequency response of Figure 9.7(b) results in the corner frequency f_L of Figure 9.1, and therefore has a similar effect to the coupling capacitors in a discrete component amplifier. To obtain a relationship between the values of the resistance and capacitance it is necessary to obtain an expression for the output voltage from the network in terms of the input voltage. From this expression the ratio of v_{out} to v_{in} can then be obtained.

From Figure 9.6(b) the output voltage is:

$$v_{out} = \frac{v_{in}}{R + 1/j\omega C} \cdot R \qquad (9.3)$$

where $\omega = 2\pi f$ radians.

From equation 9.3 the magnitude of v_{out} is (see Appendix 3):

$$|v_{out}| = \frac{R}{\sqrt{R^2 + (1/\omega C)^2}} |v_{in}| = \frac{\omega RC}{\sqrt{1 + (\omega RC)^2}} |v_{in}| \qquad (9.4)$$

Equation 9.4 shows that $v_{out} = 0$ when $\omega = 0$ (dc), and when ω is very large such that $\omega RC \gg 1$ then:

$$|v_{out}| \approx \frac{\omega RC}{\sqrt{(\omega RC)^2}} |v_{in}| = |v_{in}|$$

That is, v_{out} is zero at dc because the capacitor blocks the signal, and it is equal to v_{in}

at high frequency because the capacitor acts as a short circuit to a high-frequency signal. In between dc and some high-frequency value, the output voltage varies with the frequency. The value of the term ωRC changes with frequency: at low frequency it is $\ll 1$, because of the value of the capacitor, while at high frequencies it is $\gg 1$, because of the value of ω. At some intermediate value of frequency $\omega RC = 1$. Then:

$$|v_{out}| = \frac{1}{\sqrt{1+1}}\,|v_{in}| = \frac{1}{\sqrt{2}}\,|v_{in}| = 0.707\,|v_{in}| \tag{9.5}$$

At very high frequency the output voltage of the simple RC high-pass filter is equal to the input voltage, as shown in Figure 9.7(b) for frequencies $>f_c$. Equation 9.5 shows that the output voltage is reduced to 0.707 of its high-frequency value at a frequency for which $\omega RC = 1$, that is:

$$\omega RC = 1$$

or

$$f_c = \frac{1}{2\pi RC}\ \text{Hz} \tag{9.6}$$

and the ratio of v_{out} to v_{in} at f_c expressed in decibels is:

$$\left|\frac{v_{out}}{v_{in}}\right| \text{dB} = 20\,\log(0.707) = -3\ \text{dB} \tag{9.7}$$

The frequency f_c represents a unique frequency for the filter for which the voltage ratio is 3 dB lower than its maximum value of 0 dB which occurs when the frequency is sufficiently large for the capacitor to be assumed to be a short circuit. The *corner frequency* shown in Figure (9.7b) is defined by equation 9.6.

▷ **EXAMPLE 9.4** ──

In the high-pass filter of Figure 9.6(b) $R = 1.5$ kΩ and $C = 10$ nF. Determine

(a) the corner frequency
(b) the frequency at which $v_{out}/v_{in} = -1$ dB.

▷ *Solution*

(a) $f_c = \dfrac{1}{2\pi(1.5\ \text{k}\Omega)(10\ \text{nF})} = 10.6\ \text{kHz}$

(b) From equation 9.4

$$\left|\frac{v_{out}}{v_{in}}\right|(\text{dB}) = 20\,\log\!\left(\frac{\omega(1.5\ \text{k}\Omega)(10\ \text{nF})}{\sqrt{1+[\omega(1.5\ \text{k}\Omega)(10\ \text{nF})]^2}}\right) = -1$$

and

$$\frac{f(9.425 \times 10^{-5})}{\sqrt{1 + f^2(8.883 \times 10^{-9})}} = 10^{-0.05}$$

$$f^2(1.119 \times 10^{-8}) = 1 + f^2(8.883 \times 10^{-9})$$

$$f^2(2.307 \times 10^{-9}) = 1$$

or

$$f = 20.8 \text{ kHz}$$

▶ **EXAMPLE 9.5** _____

Use PSpice to verify the solutions of Example 9.4.

▶ *Solution*

The cursor is used within Probe (Figure 9.8) to measure the −3 dB frequency as 10.6 kHz and the frequency for which the output is reduced by 1 dB from the high-frequency value as 21 kHz.

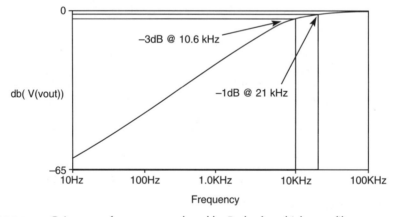

FIGURE 9.8   Gain versus frequency produced by Probe for a high-pass filter.

Equation 9.4 represents the mathematical relationship between v_{out} and v_{in} for any frequency for a high-pass filter, and it can be used to plot the complete frequency response for such an *RC* network. It can be used to predict the 'gain' at a particular frequency or to determine the frequency for a particular value of 'gain'.

NB The use of the word 'gain' for a passive circuit, such as an *RC* network, does not imply that the output voltage is greater than the input voltage. The term is simply being used to represent the ratio of v_{out} and v_{in}. For the case where $v_{out} = 0.707\ v_{in}$ the 'gain' is -3 dB, that is there is a reduction of 3 dB.

Roll-Off

It can be seen from the frequency response shown in Figure 9.8 that below the corner frequency the gain decreases linearly with decreasing frequency. This *roll-off* of the gain with frequency below the corner frequency is well defined. The rate at which the gain changes may be obtained with the aid of equation 9.4.

In Example 9.4 the corner frequency is 10.6 kHz. Consider what the gain is at 1.06 kHz, that is one decade lower. From equation 9.4 the gain is:

$$\left|\frac{v_{out}}{v_{in}}\right| = \frac{2\pi(1.06\text{ kHz})(1.5\text{ k}\Omega)(10\text{ nF})}{\sqrt{1 + [2\pi(1.06\text{ kHz})(1.5\text{ k}\Omega)(10\text{ nF})]^2}}$$

$$= 0.0994$$

$$= -20.04\text{ dB}$$

For a change of frequency of one decade the gain has changed by -20 dB. In reality the gain has changed by -20 dB $- (-3$ dB$) = -17$ dB because at the corner frequency of 10.6 kHz the gain is -3 dB, but the slope of the gain–frequency curve below the corner frequency is -20 dB per decade. This can be demonstrated by measuring the slope some distance from the corner frequency. Figure 9.9 shows the attenuation at 1 kHz and 100 Hz, and it can be seen that the slope is monotonic and changes by 20 dB for a change of one decade in the frequency.

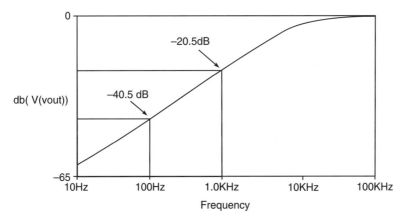

FIGURE 9.9   Diagram showing the 20 dB per decade roll-off of an *RC* network.

A single *RC* network results in a rate of change of 20 dB per decade beyond the corner frequency. Two *RC* networks produce a rate of change of 40 dB per decade, 20 dB from each network; three networks result in 60 dB per decade and so on. This rate of change has important consequences for filters which are required to separate different frequency components of a signal, as will be seen in a later chapter.

Effect of Source Resistance

If a source resistance R_s is present as shown in Figure 9.10 then the equation for v_{out} is modified.

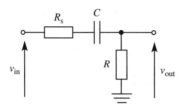

FIGURE 9.10   High-pass *RC* filter with a source resistance.

Equation 9.3 becomes:

$$v_{out} = \frac{v_{in}}{R_s + R + 1/j\omega C} \cdot R \tag{9.8}$$

and

$$|v_{out}| = \frac{\omega RC}{\sqrt{1 + [\omega(R_s + R)C]^2}} |v_{in}| \tag{9.9}$$

and the corner frequency is:

$$f_c = \frac{1}{2\pi(R_s + R)C} \text{ Hz} \tag{9.10}$$

Notice that the high-frequency 'gain' is:

$$|v_{out}| = \frac{\omega RC}{\sqrt{[\omega(R_s + R)C]^2}} |v_{in}| = \frac{R}{(R_s + R)} |v_{in}|$$

or

$$\left| \frac{v_{out}}{v_{in}} \right| = \frac{R}{(R_s + R)} \tag{9.11}$$

At high frequency the capacitor may be regarded as a short circuit and the source resistance together with the shunt resistor acts as a potential divider to reduce the value of v_{out} with respect to v_{in}.

▶ **EXAMPLE 9.6** ————————————————————

For the circuit shown in Figure 9.11 the overall mid-band voltage gain $|v_{out}/v_s|$ is 50. Determine

(a) the amplifier gain $|v_{out}/v_{in}|$
(b) the corner frequency.

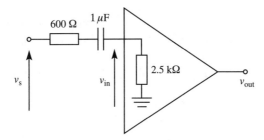

FIGURE 9.11   High-pass filter formed with the input resistance of an amplifier.

▶ *Solution*

(a) At mid-band the capacitor may be considered a short circuit and the overall voltage gain is:

$$\left|\frac{v_{out}}{v_s}\right| = \left|\frac{v_{in}}{v_s}\right|\left|\frac{v_{out}}{v_{in}}\right| = \left(\frac{R}{R_s + R}\right)\left|\frac{v_{out}}{v_{in}}\right|$$

Thus:

$$50 = \left(\frac{2500}{600 + 2500}\right)\left|\frac{v_{out}}{v_{in}}\right|$$

and

$$\left|\frac{v_{out}}{v_{in}}\right| = \frac{50}{0.806} = 62$$

(b) From equation 9.10 the corner frequency is:

$$f_c = \frac{1}{2\pi(600 + 2500)1 \times 10^{-6}} = 51 \text{ Hz}$$

9.4   Low-Pass Filter

The low-pass filter of Figure 9.6(a) and its corresponding frequency response curve in Figure 9.7(a) is responsible for the corner frequency f_H in Figure 9.1. The low-pass filter is often associated with stray capacitance from components and connecting wires, or pcb tracks, to the ground plane, and also the capacitance associated with pn junctions in the active devices. For certain applications, for example audio circuits, capacitance may be added to a circuit to restrict the high-frequency performance in order to reduce the effect of electrical noise. In either case, stray or deliberately added capacitance, the capacitance provides a path to ground for the signal, and as a result, the impedance to ground is reduced as the frequency increases, and therefore the gain decreases. A relationship between the frequency and the value of the resistance and capacitance may be obtained by considering the expression for the output voltage with respect to the input voltage, as for the high-pass filter.

From Figure 9.6(a) the output voltage is:

$$v_{out} = \frac{v_{in}}{R + 1/j\omega C} \cdot \frac{1}{j\omega C}$$

or

$$v_{out} = \frac{v_{in}}{j\omega RC + 1} \tag{9.12}$$

The magnitude of v_{out} is:

$$|v_{out}| = \frac{|v_{in}|}{\sqrt{1 + (\omega RC)^2}} \tag{9.13}$$

Equation 9.13 shows that $v_{out} = v_{in}$ when $\omega = 0$ (dc) and that when ω is very large such that $\omega RC \gg 1$ then v_{out} approaches zero. At some intermediate frequency $\omega RC = 1$ and then:

$$|v_{out}| = \frac{1}{\sqrt{1+1}}|v_{in}| = \frac{1}{\sqrt{2}}|v_{in}| = 0.707|v_{in}|$$

This equation is the same as equation 9.5 and represents the condition when the 'gain' is reduced by 3 dB. Notice that the expression for the corner frequency for the low-pass filter is the same as that for the high-pass filter, that is:

$$f_c = \frac{1}{2\pi RC} \text{ Hz} \tag{9.14}$$

This frequency corresponds to f_H in Figure 9.1.

▷ **EXAMPLE 9.7** ───────────────────────────────

In the low-pass filter of Figure 9.6(a) $R = 2$ kΩ and $C = 100$ pF. Determine

(a) the corner frequency
(b) the frequency at which $v_{out}/v_{in} = -2$ dB.

▷ *Solution*

(a) $f_c = \dfrac{1}{2\pi(2 \text{ k}\Omega)(100 \text{ pF})} \approx 796$ kHz

(b) From equation 9.13

$$\left|\frac{v_{out}}{v_{in}}\right| (dB) = 20 \log\left(\frac{1}{\sqrt{1 + [2\pi f(2 \text{ k}\Omega)(100 \text{ pF})]^2}}\right) = -2$$

$$\frac{1}{\sqrt{1 + f^2(1.58 \times 10^{-12})}} = 10^{-0.1}$$

and

$$0.369 = f^2 9.97 \times 10^{-13}$$

or

$$f = 608 \text{ kHz}$$

▷ **EXAMPLE 9.8** ───────────────────────────────

Use PSpice to verify the solution obtained in Example 9.7.

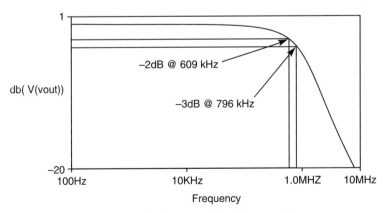

FIGURE 9.12   Variation of gain with frequency for a low-pass filter.

▶ *Solution*

In **Setup...** select **AC Analysis...** and set the **Total Pts** to 10, **Start Freq** to 100 Hz and the **Stop Freq** to 10 MHz. The output produced by Probe is shown in Figure 9.12.

Notice that the output voltage is equal to the input for low frequencies, which corresponds to 0 dB. The cursor is used to measure the frequency at which the output voltage is 3 dB lower than the input and also the frequency at which the 'gain' is −2 dB.

Effect of a Shunt Resistance

In a practical circuit the capacitance in the low-pass filter may represent the input capacitance of a bipolar transistor or FET and in parallel with this capacitor there is likely to be the input resistance of the device. For the FET it would be very large, but for the bipolar transistor it is a few thousand ohms. The circuit of the low-pass filter with shunt resistance is shown in Figure 9.13.

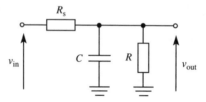

FIGURE 9.13 Low-pass *RC* filter with a shunt resistor.

From Figure 9.13 the output voltage is:

$$v_{out} = \frac{v_{in}}{R_s + R \| 1/j\omega C} \cdot R \| 1/j\omega C$$

and

$$v_{out} = v_{in} \frac{R}{R_s + R + j\omega C R R_s} \tag{9.15}$$

At very low frequency $\omega C R R_s \ll 1$ and:

$$v_{out} \approx v_{in} \frac{R}{R_s + R} \tag{9.16}$$

This equation describes the normal potential divider action associated with two resistors. The capacitor is assumed to be an open circuit at low frequency. The magnitude of v_{out} is obtained from equation 9.15.

$$v_{out} = v_{in} \frac{R}{R_s + R} \cdot \frac{1}{\{1 + [j\omega C R R_s / (R_s + R)]\}}$$

and

$$|v_{out}| = |v_{in}| \frac{R}{R_s + R} \cdot \frac{1}{\sqrt{1 + [\omega C (R \| R_s)]^2}} \qquad (9.17)$$

The corner frequency occurs when:

$$\omega CR \| R_s = 1$$

or

$$f_c = \frac{1}{2\pi CR \| R_s} \text{ Hz} \qquad (9.18)$$

For the low-pass filter the two resistors behave as if they are in parallel. Another way of viewing the network is to regard the input voltage source v_{in} and the two resistors as a network which can be replaced by its Thévenin equivalent, where the Thévenin resistance is $R \| R_s$. The network then effectively reduces to that shown in Figure 9.6(a) of a single series resistor and shunt capacitor.

▶ **EXAMPLE 9.9** ————————————————————————————————————

For the amplifier shown in Figure 9.14 the mid-band voltage gain $|v_{out}/v_{in}|$ is 45 dB. Determine

(a) the voltage gain $|v_{out}/v_{in}|$
(b) the upper corner frequency
(c) the gain $|v_{out}/v_s|$ at 20 MHz.

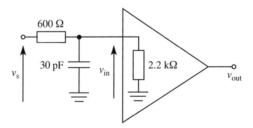

FIGURE 9.14 Amplifier with low-pass filter.

▶ *Solution*

(a) In the mid-band the capacitor is assumed to be open circuit and the gain is:

$$\left|\frac{v_{out}}{v_{in}}\right| = \left|\frac{v_{out}}{v_s}\right| \left|\frac{v_s}{v_{in}}\right| = 20 \log 45 \left|\frac{v_s}{v_{in}}\right|$$

but

$$v_{\text{in}} = v_{\text{s}} \frac{2.2 \text{ k}\Omega}{600 \text{ }\Omega + 2.2 \text{ k}\Omega} = v_{\text{s}}0.786$$

and

$$\left| \frac{v_{\text{out}}}{v_{\text{in}}} \right| = (178)(1/0.786) = 226 = 47 \text{ dB}$$

(b) From equation 9.18 the upper corner frequency is:

$$f_{\text{c}} = \frac{1}{2\pi(30 \times 10^{-12})(600 \text{ }\Omega \,\|\, 2.2 \text{ k}\Omega)} = 11.2 \text{ MHz}$$

(c) The voltage gain is given by:

$$\left| \frac{v_{\text{out}}}{v_{\text{s}}} \right| = \left| \frac{v_{\text{in}}}{v_{\text{s}}} \right| \left| \frac{v_{\text{out}}}{v_{\text{in}}} \right|$$

and from equation 9.17

$$\left| \frac{v_{\text{in}}}{v_{\text{s}}} \right| = \frac{2.2 \text{ k}\Omega}{600 \text{ }\Omega + 2.2 \text{ k}\Omega} \frac{1}{\sqrt{1 + [2\pi(20 \text{ MHz})(30 \text{ pF})(600 \text{ }\Omega \,\|\, 2.2 \text{ k}\Omega)]^2}}$$

$$\left| \frac{v_{\text{in}}}{v_{\text{s}}} \right| = 0.786 \frac{1}{\sqrt{1 + 3.15}} = 0.386$$

and

$$\left| \frac{v_{\text{out}}}{v_{\text{s}}} \right| = (0.386)(226) = 87.2 = 39 \text{ dB}$$

▶ **EXAMPLE 9.10** —————————————————————————————

Use PSpice to determine the upper and lower corner frequencies for the circuit shown in Figure 9.15.

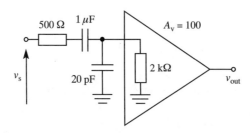

FIGURE 9.15   Example for analysis with PSpice.

▶ Solution

The circuit schematic for PSpice is shown in Figure 9.16.

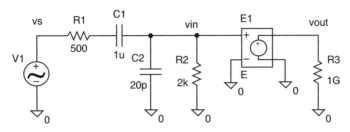

FIGURE 9.16   Schematic for PSpice.

A voltage-controlled voltage source E_1 is used for the amplifier, with the gain set to 100. The frequency response is shown in Figure 9.17.

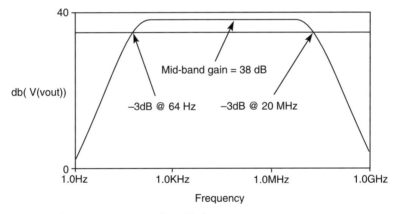

FIGURE 9.17   Frequency response from Probe.

The mid-band gain is determined by the 500 Ω source resistance and the 2 kΩ input resistance of the amplifier. In the mid-band the 1 μF capacitor is a short circuit and the 20 pF capacitor is an open circuit.

▶ EXAMPLE 9.11 _____

Verify the values of mid-band gain and upper and lower corner frequencies for the circuit in Figure 9.15.

▷ *Solution*

The mid-band gain is:

$$A_{\mathrm{v}} = (100)\left(\frac{2\ \mathrm{k\Omega}}{2.5\ \mathrm{k\Omega}}\right) = 38\ \mathrm{dB}$$

The lower corner frequency is:

$$f_{\mathrm{L}} = \frac{1}{2\pi(1\ \mathrm{\mu F})(2.5\ \mathrm{k\Omega})} = 64\ \mathrm{Hz}$$

and the upper corner frequency is:

$$f_{\mathrm{H}} = \frac{1}{2\pi(20\ \mathrm{pF})(500\ \mathrm{\Omega}\,\|\,2\ \mathrm{k\Omega})} = 19.8\ \mathrm{Hz}$$

9.5 ▷ Miller Effect

The stray capacitance associated with a low-pass filter is partly due to the electric field which exists between components and their wiring and ground, and also to the capacitance associated with the transistor (or other amplifying device). For the bipolar transistor there is capacitance between the base and emitter and also between the base and collector associated with the pn junctions. Ideally they should be treated as part of a more complex equivalent circuit known as a hybrid pi (or hybrid π). This circuit is described in Appendix 4, and is mentioned again in Chapter 10 when considering the high-frequency response of bipolar transistors. An alternative approach is to assume that these capacitors exist as discrete components between the leads of the transistor, as shown in Figure 9.18. This approach ignores the effect of the internal base resistance $r_{\mathrm{bb'}}$ of the hybrid pi circuit. Similar interelectrode capacitors exist in the field effect transistor, but without the complication of any internal resistance.

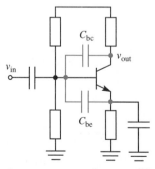

FIGURE 9.18   Stray capacitance in a common-emitter amplifier.

For the common-emitter amplifier the base–collector capacitance C_{bc} forms a link between the input and the output. The base–emitter capacitance C_{be} forms a shunt path to ground.

The presence of C_{bc} (C_{gd} for FETs) complicates the analysis because it provides a feedback path between the output and input. The analysis can be simplified by making use of Miller's theorem. A simplified form of the amplifier, illustrated in Figure 9.18, is shown in Figure 9.19, where C is the capacitance and A_v is the voltage gain of the amplifier.

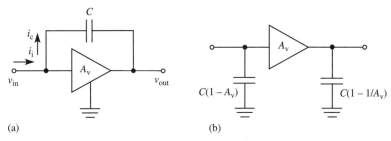

(a)   (b)

FIGURE 9.19   (a) General case of an amplifier with capacitive feedback and (b) the equivalent amplifier with Miller input and output capacitance.

The Miller capacitance is obtained from Figure 9.19(a) as follows:

$$v_{in} = i_c X_c + v_{out} \quad \text{assuming that } i_i = i_c$$

and

$$v_{in} = i_c X_c + A_v v_{in} \quad \text{where } v_{out} = A_v v_{in}$$

Therefore:

$$v_{in} = i_c \frac{X_c}{1 - A_v}$$

$$v_{in} = i_c \frac{1}{2\pi f C (1 - A_v)} = i_c \frac{1}{2\pi f C_{inM}}$$

where

$$C_{inM} = C(1 - A_v) \tag{9.19}$$

This equation shows that as far as the input circuit is concerned the capacitor C can be replaced by the equivalent Miller capacitance C_{inM} connected between the input and ground rather than the input and output.

For the output:

$$v_{out} = v_{in} - i_c X_c$$

$$= \frac{v_{out}}{A_v} - i_c X_c$$

or

$$V_{out} = -i_c \frac{X_c}{1 - (1/A_v)}$$

$$V_{out} = -i_c \frac{1}{2\pi f C_{outM}}$$

where

$$C_{outM} = C\left(1 - \frac{1}{A_v}\right) \tag{9.20}$$

For most amplifiers where the gain is greater than 10 this capacitance approximates to C, because the term $1/A_v \ll 1$. That is, in the equivalent circuit of Figure 9.19(b) the capacitor in the output is simply the feedback capacitor connected between the output and ground.

Miller's theorem provides a means of simplifying a network which contains an impedance connected between the input and output terminals of an amplifier.

▶ **EXAMPLE 9.12** _____

Determine the corner frequencies for the gate to drain capacitor for the JFET amplifier shown in Figure 9.20 if g_m is 3500 μS.

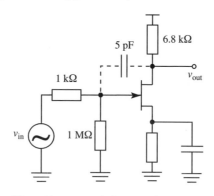

FIGURE 9.20   JFET amplifier with gate to drain capacitance.

▶ *Solution*

The mid-band gain is $-g_m R_D$, that is:

$$A_v = -(3500 \text{ μS})(6.8 \text{ kΩ}) = -23.8$$

The equivalent Miller circuit for the amplifier is shown in Figure 9.21.

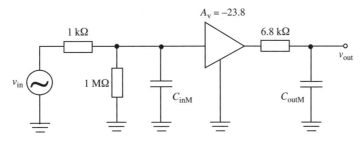

FIGURE 9.21   Miller equivalent of JFET with gate to drain capacitance.

From equation 9.19 the input capacitance is:

$C_{\text{inM}} = 5 \text{ pF}(1 + 23.8) = 124 \text{ pF}$

From equation 9.20 the output capacitance is simply 5 pF.
The 3 dB corner frequency for the input circuit is:

$$f_{\text{cin}} = \frac{1}{2\pi(1\text{k} \| 1\text{M})124 \text{ pF}} = 1.28 \text{ MHz}$$

The 3 dB corner frequency for the output circuit is:

$$f_{\text{cout}} = \frac{1}{2\pi(6.8 \text{ k}\Omega)(5 \text{ pF})} = 46.8 \text{ MHz}$$

The overall 3 dB corner frequency is the lower of these, that is 1.28 MHz.

9.6   Multiple Corner Frequencies

In real circuits there are usually a number of *RC* networks, each of which is likely to have a unique corner frequency. Thus the input coupling capacitor in a transistor amplifier results in one high-pass filter with a corner frequency, while the output coupling capacitor and load resistance results in another high-pass filter with a different corner frequency. Similarly the stray capacitance associated with the input results in a low-pass filter and the capacitance at the output results in another low-pass filter. If the corner frequencies associated with each type of filter are reasonably well separated, then the bandwidth $(f_{\text{H}} - f_{\text{L}})$ is determined by the high-pass filter with the highest corner frequency and the low-pass filter with the lowest corner frequency. The calculation is more difficult if the frequencies are not well separated, and while hand calculations can still be used to obtain an estimate, a far simpler solution is to use a circuit simulator.

▶ **EXAMPLE 9.13** ————————————————————————————

The circuit in Figure 9.22 contains input and output coupling capacitors as well as input and output stray capacitance. Use hand calculations to determine the gain $|v_{\text{out}}/v_{\text{s}}|$ and the upper and lower corner frequencies and verify with PSpice.

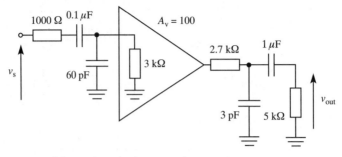

FIGURE 9.22 Amplifier with multiple corner frequencies.

▷ *Solution*

The mid-band gain is obtained by assuming that the series coupling capacitors are short circuits and that the stray shunt capacitances are open circuits. Then the gain is:

$$\left|\frac{v_{out}}{v_{in}}\right| = \left(\frac{3\ k\Omega}{1\ k\Omega + 3\ k\Omega}\right)(100)\left(\frac{5\ k\Omega}{2.7\ k\Omega + 5\ k\Omega}\right) = 48.7$$

or in terms of decibels:

$$\left|\frac{v_{out}}{v_{in}}\right| = 20 \log 48.7 = 33.7\ \text{dB}$$

Low Frequency

For the input the corner frequency is:

$$f_{Linput} = \frac{1}{2\pi(0.1\ \mu F)(1\ k\Omega + 3\ k\Omega)} \approx 400\ \text{Hz}$$

For the output the corner frequency is:

$$f_{Loutput} = \frac{1}{2\pi(1\ \mu F)(2.7\ k\Omega + 5\ k\Omega)} \approx 21\ \text{Hz}$$

High Frequency

For the input the corner frequency is:

$$f_{Hinput} = \frac{1}{2\pi(60\ pF)(1\ k\Omega \| 3\ k\Omega)} = 3.5\ \text{MHz}$$

For the output the corner frequency is:

$$f_{Houtput} = \frac{1}{2\pi(3\ pF)(2.7\ k\Omega \| 5\ k\Omega)} = 30\ \text{MHz}$$

The bandwidth is:

BW = 3.5 MHz − 400 Hz

The circuit schematic for PSpice is shown in Figure 9.23.

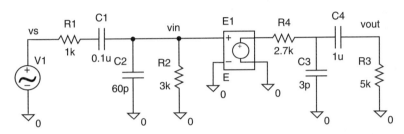

FIGURE 9.23   Schematic for PSpice.

A voltage-controlled voltage source (E) is used to represent the amplifier with a gain of 100. The frequency response is shown in Figure 9.24.

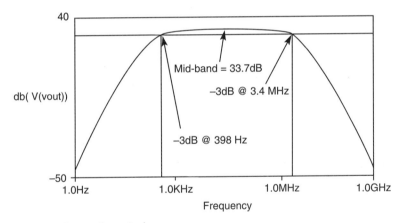

FIGURE 9.24   Output from Probe.

The mid-band voltage gain measured from the Probe output is 33.7 dB and the corner frequencies are as shown on the frequency plot.

Comparing the low-frequency response of Figure 9.24 with that shown in Figure 9.9, it can be seen that in Figure 9.24 the rate of change of gain with frequency below 400 Hz increases as the frequency decreases. In Figure 9.24 the rate of change remains constant at −20 dB per decade. In Figure 9.24 the rate of change increases to 40 dB per decade as the second *RC* filter takes effect. A similar increase in the rate of change takes place above 3.4 MHz.

9.7 Phase Shift

So far only the magnitude of the voltage gain has been considered. Near the corner frequencies (f_L and f_H) and below f_L and above f_H the reactance of the capacitor affects the magnitude of the gain. Because the gain versus frequency relationship in these regions includes both real and complex quantities there is also a phase change as well as a magnitude change.

The phase shift for equation 9.3 is given by (see Appendix 3):

$$\phi = \tan^{-1}\left(\frac{X_C}{R}\right) \tag{9.21}$$

In the mid-band the capacitor is assumed to be a short circuit and $X_C \cong 0$, so

$$\phi = \tan^{-1}\left(\frac{0}{R}\right) = \tan^{-1}(0) = 0°$$

At the corner frequency when $X_C = R$ and the gain is reduced by 3 dB from the mid-band value, the phase is:

$$\phi = \tan^{-1}\left(\frac{R}{R}\right) = \tan^{-1}(1) = 45°$$

A decade below the corner frequency when $X_C = 10R$ the phase is:

$$\phi = \tan^{-1}\left(\frac{10R}{R}\right) = \tan^{-1}(10) = 84.3°$$

These values show that the phase changes from 0° in the mid-band through 45° at the corner frequency and approaches 90° beyond $f_c/10$.

For the low-pass filter described by equation 9.12 the phase angle is given by:

$$\phi = -\tan^{-1}\left(\frac{X_C}{R}\right) \tag{9.22}$$

Notice that the only difference is the change of sign, otherwise the values for the phase are the same as for the high-pass filter with the phase changing from 0° through −45° to −90° as the frequency increases from the mid-band, through f_H and beyond f_H.

In this text phase is not considered in any further detail. The calculation of phase is further complicated by the presence of more than one *RC* network, and the simplest approach is to use PSpice to plot the phase versus frequency.

NB This omission regarding a detailed discussion about phase does not mean that phase changes are unimportant. Far from it, for circuits employing feedback the phase change is very important, but it is difficult to determine, and is considered inappropriate for the level intended for this text. It is left for the reader to study this aspect of analog electronic design when he or she is more experienced.

▷ **EXAMPLE 9.14** ————————————————————————————————

Use PSpice to plot the phase versus frequency for the circuit used for Example 9.12.

▷ *Solution*

In Probe use $p(v_{out})$ where p is for phase to produce the plot shown in Figure 9.25.

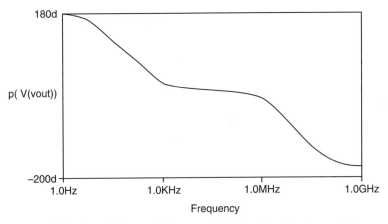

FIGURE 9.25   Variation of phase with frequency for the circuit in Figure 9.23.

Notice that the phase of the output with respect to the input is advanced for low frequencies and retarded for the high frequencies. Note that each low-pass and each high-pass network produces a maximum phase shift of 90° and for two networks the total phase change is 180°. Thus the maximum phase shift at 1 Hz is 180° and at 10 GHz it is $-180°$. Within the mid-band region the phase remains relatively constant, with some departure from 0° near the corner frequencies.

9.8 ▷ Bode Plots

For a single *RC* network, whether it acts as a high-pass or a low-pass filter, the roll-off beyond the 3 dB corner frequency is -20 dB per decade, as illustrated in Figure 9.9. Apart from the region near the corner frequency, the roll-off region can be represented by a straight line with a slope of -20 dB per decade. In the mid-band region where the capacitor is regarded as a short circuit, the gain is constant and the gain versus frequency response is represented by a horizontal straight line. The two lines intersect at the corner frequency as shown in Figure 9.26.

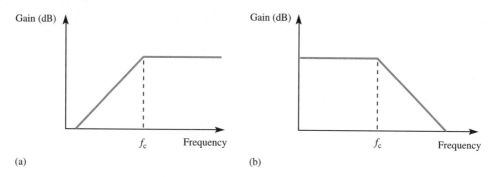

FIGURE 9.26 Straight line approximations of the frequency response for (a) high-pass and (b) low-pass *RC* filters.

These straight line approximations are known as *Bode plots*. They provide a simple means for constructing frequency response graphs for networks which contain *RC* networks.

Multiple RC Networks

The graphs in Figure 9.26 are for single *RC* networks, but the Bode plot can be extended to multiple networks. In Figure 9.27 the input and output *RC* filters are separated from each other by an amplifier, and the networks may be considered to act independently of each other.

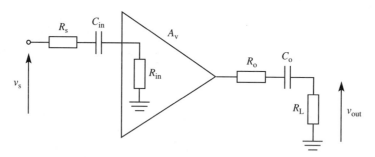

FIGURE 9.27 Amplifier with two *RC* networks.

The input corner frequency is obtained from equation 9.10 and is given by:

$$f_{cin} = \frac{1}{2\pi C_{in}(R_s + R_{in})}$$

and the output corner frequency is:

$$f_{cout} = \frac{1}{2\pi C_o(R_o + R_L)}$$

Notice that each network is a high-pass filter and that the 3 dB frequency is determined by the corner frequency which has the highest value. Below the corner frequency each network introduces a roll-off of −20 dB per decade. Provided that the two corner frequencies are well separated, then there is a distinct change in the slope of the gain versus frequency graph in the roll-off region. Assume that $f_{cin} > f_{cout}$; then the 3 dB frequency is determined by the input network and the frequency response is as shown in Figure 9.28 with a change in the roll-off at point C.

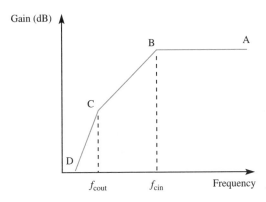

FIGURE 9.28 Frequency response for $f_{cin} > f_{cout}$.

The frequency response can be approximated by a series of straight lines. In the mid-band above f_{cin} the gain is constant and represented by a horizontal line AB at a value of gain which is calculated on the assumption that the coupling capacitors are short circuits. Below f_{cin} the gain begins to decrease and can be approximated by a straight line BC with a slope of −20 dB per decade. At f_{cout} the second network begins to take effect and adds a further −20 dB per decade fall in the gain. Thus below f_{cout} the slope of the straight line section CD is −40 dB per decade.

The roll-off at high frequency as a result of low-pass filters formed by stray capacitance follows a similar pattern with each network contributing to the overall response.

Each RC network contributes a −20 dB per decade rate of change to the gain versus frequency characteristic, that is:

No. of RC networks	Roll-off (dB per decade)
1	− 20
2	− 40
3	− 60
4	− 80

The Bode plot can be used to approximate the frequency response, provided that the corner frequencies are well separated. In practice this may not be the case, but it still provides an indication of the appearance of the frequency response.

To a lesser extent the phase response can also be approximated by noting that each *RC* network produces a maximum phase change of 90°. However, the phase change of 90° does not follow a simple straight line approximation as can be seen by reference to Figure 9.25. In the same way that the gain changes of multiple *RC* networks add, so the phase contributions of each network add to produce an overall phase change. The maximum phase change is as follows:

No. of *RC* networks	Maximum phase change
1	90°
2	180°
3	270°
4	360°

The sign of the phase change depends on whether the filter is low pass or high pass as can be seen from Figure 9.25. At low frequencies the phase change is positive while at high frequencies it is negative.

In practical circuits the corner frequencies are not well separated and it is often not possible to identify the straight line segments which represent the contribution of each network. The changes of gain and phase are likely to be continuous and the frequency response is likely to be a continuous curve beyond the corner frequency.

▷ **EXAMPLE 9.15** ————————————————————————————

Sketch the Bode plot of the gain versus frequency for the circuit used in Example 9.13.

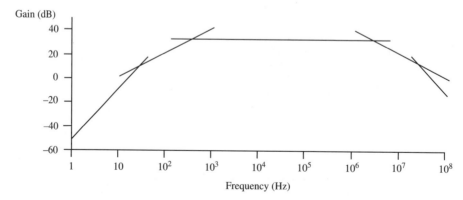

FIGURE 9.29   Bode plot for the gain for the circuit in Figure 9.23.

▷ *Solution*

The 3 dB corner frequencies are:

high pass   21 Hz and 400 Hz

low-pass   3.5 MHz and 30 MHz

The Bode plot for the gain is shown in Figure 9.29.

A reasonable representation of the frequency graph can be created from a series of straight lines with slopes of either -20 dB per decade or -40 dB per decade. However, it becomes more difficult when the corner frequencies are not so well separated. The Bode plot provides a useful means of estimating the likely response, and for circuits which employ negative feedback it can be useful in determining whether the network will be stable when negative feedback is applied. However, this is another topic for a more advanced course.

9.9   Risetime and Bandwidth

The frequency response of an amplifier is obtained by applying a sine wave to the input and measuring the amplitude of the sine wave at the output in order to obtain the gain. This is repeated for different frequencies. In practice the measurement is made with a piece of equipment which automatically scans the frequency over the range of interest, and monitors the input and output voltages while doing so. The result is a set of data, or a plot, of gain versus frequency. Often the only information of interest is the 3 dB corner frequencies and the mid-band gain. For many wide-band applications the upper 3 dB frequency is of more interest than the lower one. The upper 3 dB frequency can be obtained very simply by measuring the *pulse response* of an amplifier.

Risetime

When a voltage step is applied through a resistor to an uncharged capacitor, as shown in Figure 9.30, the voltage across the capacitor rises exponentially from zero to the value of the applied voltage. The voltage across the capacitor is given by:

$$v_c = V\left[1 - \exp\left(\frac{-t}{RC}\right)\right]$$

where v_c is the voltage across the capacitor, V is the applied voltage, RC is the time constant and t is the time after the application of the input voltage step. It is accepted practice to define the time difference $t_2 - t_1$ in Figure 9.30 as the *risetime* (t_R) for the output waveform. These times are associated with the voltage levels which correspond to 10% and 90% of the voltage transition.

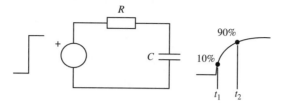

FIGURE 9.30   Variation of voltage across a capacitor.

At the 10% level the voltage across the capacitor is 0.1 V, while at the 90% level the voltage is 0.9 V. Substituting these values for v_c into the above equation and solving for the times t_1 and t_2 gives:

$$t_R = t_2 - t_1 = 2.2RC \tag{9.23}$$

The upper corner frequency for an amplifier is given by equation 9.14 as:

$$f_H = \frac{1}{2\pi RC} \text{ Hz}$$

Substituting for *RC* from equation 9.23 gives:

$$f_H = \frac{0.35}{t_R} \tag{9.24}$$

where f_H is the upper corner frequency and t_R is the risetime of the amplifier output voltage. This equation provides a relationship between the frequency response of an amplifier and its transient response. This relationship is particularly useful for the design of pulse amplifiers as it provides a means of determining the bandwidth required in order to amplify a pulse without affecting the risetime.

▶ **EXAMPLE 9.16** ────────────────────────────────

Determine the bandwidth required for an amplifier which must reproduce pulses with 70 ns risetimes.

▶ *Solution*

From equation 9.24:

$$f_H = \frac{0.35}{70 \times 10^{-9}}$$

and

$$f_H = 5 \text{ MHz}$$

This bandwidth is typical of that required for the video amplifiers in TV receivers, and is necessary to reproduce the step changes in picture intensity in moving from a bright part of a picture to a black part.

The relationship shown in equation 9.24 can also be used to measure the bandwidth of an amplifier. If a square wave is applied to an amplifier, then the bandwidth is obtained by measuring the risetime of the output wave, and obtaining f_H from equation 9.24.

▶ **PROBLEMS** ───────────────────────────

9.1 Determine the voltage gain v_{out}/v_{in} in dB for each of the following:

(a) $v_{in} = 100 \ \mu V_{rms}$ and $v_{out} = 20 \ mV_{rms}$
(b) $v_{in} = 12 \ mV_{rms}$ and $v_{out} = 1$ V peak
(c) $v_{in} = 5 \ mV_{rms}$ and $v_{out} = 2.5$ V peak to
      peak

[(a) 46 dB, (b) 35.4 dB, (c) 45 dB]

9.2 For the circuit shown $v_s = 20 \ mV_{rms}$ and $v_{out} = 1 \ V_{rms}$ Determine

(a) the voltage gain in dB
(b) the power gain in dB.

[(a) 34 dB, (b) 37 dB]

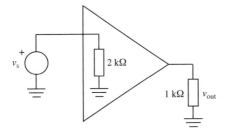

9.3 Repeat Problem 9.2 with the 1 kW load replaced with 2 kW.

9.4 The voltage gain of an amplifier is 25 dB. If the output voltage is 2 V_{rms} and the input resistance is 2.5 kΩ and the output resistance is 1.5 kΩ, determine

(a) the input voltage
(b) the power gain in dB.

[(a) 112 mV, (b) 27 dB]

9.5 An amplifier has two gain stages such that the overall gain is 40 dB. If the output voltage is 2 V_{rms} and the voltage at the output of the first stage is 100 mV, determine

(a) the input voltage
(b) the gain of the first stage.

[(a) 20 mV, (b) 14 dB]

9.6 The input voltage to an amplifier is 2.5 mV$_{rms}$. The amplifier has a gain of 30 dB and drives a network cable which has an attenuation of 20 dB. At the end of the network cable there is a second amplifier. Determine

(a) the voltage at the input to the second amplifier
(b) the overall gain if the final output voltage is 1 V$_{rms}$.

[(a) 7.9 mV, (b) 52 dB]

9.7 A signal processing circuit consists of an amplifier with a gain of 40 dB, followed by a filter with an attenuation of −10 dB and finally an output stage with a gain of 25 dB. Determine

(a) the overall gain
(b) the voltage either side of the filter if the input to the first amplifier is 1 mV$_{rms}$
(c) the output voltage from the second amplifier.

[(a) 55 dB, (b) 100 mV, 31.6 mV, (c) 562 mV]

9.8 For the high-pass filter in Example 9.4 the shunt resistor is 4.7 kΩ and the series capacitor is 1 μF. Determine

(a) the corner frequency
(b) the frequency at which
 $v_{out}/v_{in} = -2$ dB
(c) the attenuation at a frequency of $f_c/2$.

[(a) 33.8 Hz, (b) 44.3 Hz, (c) −7 dB]

9.9 The mid-band gain of the amplifier is 60 dB. Determine

(a) the lower 3 dB frequency
(b) the frequency at which the overall gain is 59 dB.

[(a) 318 Hz, (b) 627 Hz]

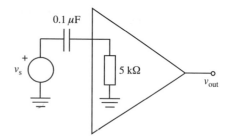

9.10 The mid-band gain of the amplifier is 150. Determine

(a) the lower 3 dB frequency
(b) the overall gain at $f_c/2$.

[(a) 106 Hz, (b) 27 dB]

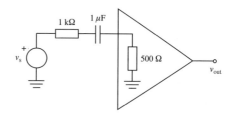

9.11 For the circuit shown an output voltage of 1 V$_{rms}$ is required in the mid-band for an input of 1 mV$_{rms}$. Determine the value of the capacitor to ensure that v_{out} is not less than 0.5 V at 20 Hz.

[4.6 μF]

9.12 For the circuit shown the mid-band gain of the amplifier is 100. Determine

(a) the upper 3 dB frequency
(b) the overall gain v_{out}/v_s in dB
(c) the overall gain at 10 MHz in dB.

[(a) 3.5 MHz, (b) 39 dB, (c) 29.5 dB]

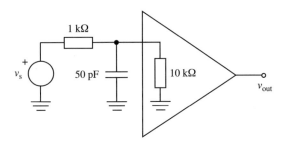

9.13 For the circuit in Problem 9.12
determine the maximum value of the source
resistance to ensure that the upper 3 dB
frequency is not less than 1 MHz.

[4.7 kΩ]

9.14 Determine the lower and upper 3 dB frequencies for the circuit shown.

[34.6 Hz, 5 MHz]

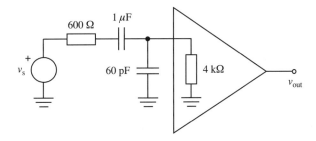

9.15   For the circuit shown determine

(a) the lower and upper 3 dB frequencies.
(b) The phase shift at $f_L/2$ and $2f_H$.

Hint: For $f_L/2$ there are two phase shifts, one for the input and one for the output.

   [(a) 72 Hz, 119 kHz, (b) 114°, −27°]

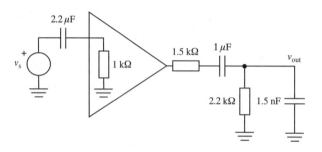

9.16   The amplifier has a mid-band gain of −150. Determine

(a) the upper 3 dB frequency
(b) the overall gain v_{out}/v_s.

   [(a) 306 kHz, (b) 40.7 dB]

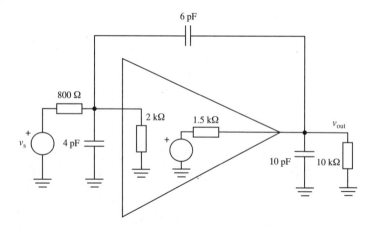

9.17 Use the Bode plot to estimate the gain at $10f_H$.

CHAPTER 10

Frequency Response of Transistor Amplifiers

It was shown in Chapter 9 that the typical frequency response of a transistor amplifier with capacitors for coupling and decoupling, and with stray capacitance associated with wires, components and the transistors, could be described by means of simple low-pass and high-pass RC networks. The corner frequency for a single network was obtained in Chapter 9, equations 9.6 and 9.14, and is reproduced here as:

$$f_c = \frac{1}{2\pi RC} \tag{10.1}$$

For a transistor amplifier there are likely to be a number of such networks, each with a different corner frequency. To determine the complete frequency response it is necessary to identify the networks and the value of the resistance and capacitance for each network. For high-pass RC networks this usually involves obtaining the input or output resistance of the transistor amplifier from the viewpoint of one or more of the coupling capacitors, that is looking into the circuit from one or other of the capacitor terminals to obtain a value of the effective resistance. For low-pass RC filters it is necessary to identify the stray capacitance associated with components and wires, and also the pn junctions of transistors, in addition to calculating the resistance.

Circuit simulators are ideally suited to investigating the frequency response of an amplifier and can be used to replace complex analysis, but in order to select suitable capacitor values to achieve a required bandwidth it is necessary to apply simple design rules. The simulator can then be used to verify the response and if necessary to investigate the effect of changing component values.

The overall frequency response of an amplifier can be divided into three regions, low frequency, mid-band and high frequency. The mid-band conditions have already been considered (Chapters 6 and 8), when the capacitors are considered to be either short circuits or open circuits. In other words they do not enter into the calculations of gain or impedance. In the other two regions the capacitors must be included in the calculations, but only to the extent of determining the cut-off frequencies. The gain beyond cut-off is usually not of importance, and in any case is best observed with the aid of a circuit simulator.

10.1 Low-Frequency Response of BJT Amplifiers

The common-emitter stage shown in Figure 10.1 is typical of a BJT transistor amplifier with frequency-dependent networks resulting from the coupling and decoupling capacitors.

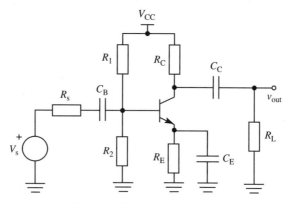

FIGURE 10.1   Typical capacitively coupled transistor amplifier.

The first step is to redraw the circuit as a small-signal circuit with the power supply replaced by a short circuit. This is shown in Figure 10.2.

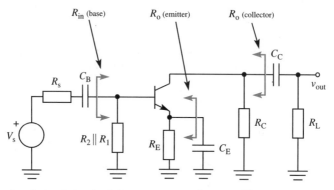

FIGURE 10.2   Small-signal circuit of the common-emitter amplifier.

There are three RC networks. One at the input with C_B, R_S and the input resistance of the transistor as seen from the capacitor; another at the output formed with C_C, R_L and the output resistance seen from the capacitor; and finally the emitter decoupling capacitor with R_E in parallel with the resistance at the emitter terminal looking towards the base. For the circuit shown in Figure 10.2 the value of R in equation 10.1 is likely to be smallest for the decoupling capacitor because the resistance seen by the capacitor at the emitter terminal is the output resistance of an emitter follower, which is small. Therefore, for a given corner frequency the emitter decoupling capacitor is much larger than either of the other two capacitors.

With three RC networks there is the possibility of there being three different corner frequencies. To simplify the calculation of any one corner frequency it is useful to ignore the effect of the other two RC networks, that is to assume that the other two capacitors are either short circuits or open circuits. Depending on the value of each corner frequency this may or may not be true, but it does simplify the calculations, and a circuit simulator can be used to obtain greater accuracy.

The choice as to whether the other capacitors are short circuits or open circuits depends on the value of the other capacitors. In this text the initial calculation is made assuming that the other capacitors are short circuits. Some of the discrepancies which result from this assumption will be examined in the examples below.

Input Coupling Capacitor

For the input circuit the input resistance is the parallel combination of the base potential divider resistors and the input resistance of the transistor. To simplify the calculation it is assumed that C_E acts as a short circuit. Then the input resistance as seen by C_B is:

$$R_{in} = R_1 \| R_2 \| \beta r_e \tag{10.2}$$

and from equation 9.10 the corner frequency is:

$$f_c = \frac{1}{2\pi C_B (R_s + R_{in})} \tag{10.3}$$

▷ **EXAMPLE 10.1** ———————————————————————————————

For the circuit shown in Figure 10.3 estimate the corner frequency produced by the input RC network. Assume that $\beta = 100$ and $r_e = 13.6\ \Omega$.

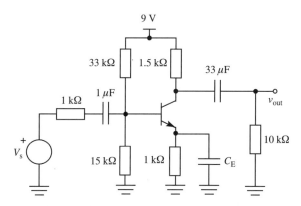

FIGURE 10.3   CE amplifier with capacitive coupling for frequency calculations.

▷ **Solution**

With the emitter decoupling capacitor acting as a short circuit, then:

$$R_{in} = 33\ \text{k}\Omega \| 15\ \text{k}\Omega \| (100)(13.6\ \Omega) = 1.2\ \text{k}\Omega$$

and

$$f_c = \frac{1}{2\pi(1\ \mu F)(1\ k\Omega + 1.2\ k\Omega)} = 72\ Hz$$

Consider now the situation where the decoupling capacitor is not a short circuit at 72 Hz; then the input resistance with C_E assumed to be an open circuit is:

$$R_{in} = 33\ k\Omega \,\|\, 15\ k\Omega \,\|\, (100)(13.6\ \Omega + 1\ k\Omega) = 9.4\ k\Omega$$

and

$$f_c = \frac{1}{2\pi(1\ \mu F)(1\ k\Omega + 9.4\ k\Omega)} = 15.3\ Hz$$

In practice neither approximation is likely to be perfect, and the actual corner frequency will have a value somewhere between these two values. For the purposes of hand-based calculations either value gives a general indication of the possible value, while both provide an indication of the two extremes.

▶ **EXAMPLE 10.2** ————————————————————————————

Use PSpice and the QbreakN transistor from the breakout library to investigate the low-frequency performance of the circuit in Figure 10.3 with $C_E = 1000\ \mu F$ and $C_B = 1\ \mu F$. Assume that $\beta = 100$ and that there is no load resistor.

NB With $C_E = 1000\ \mu F$ the reactance is very small and the approximation that C_E is a short circuit is likely to be true.

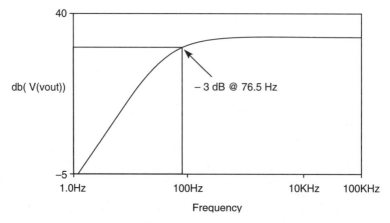

FIGURE 10.4 Probe output of gain versus frequency for a CE amplifier.

▷ *Solution*

The output from Probe is shown in Figure 10.4.

The reactance of C_E at 72 Hz is 2.2 Ω and the assumption that it is a short circuit is quite reasonable and, as can be seen from Figure 10.4, there is good agreement between the calculated value of the corner frequency produced by C_B and the PSpice simulation.

Output Coupling Capacitor

The equivalent circuit for Figure 10.2, assuming that CE is a short circuit, is shown in Figure 10.5. For the output capacitor it does not matter whether C_E is a short circuit or not, as the emitter circuit does not influence the output.

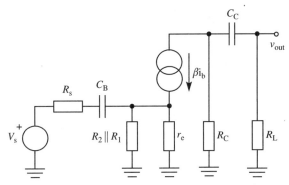

FIGURE 10.5 Equivalent circuit for identifying the output *RC* network.

The current generator and the collector resistance can be replaced by the Thévenin circuit to give the circuit shown in Figure 10.6.

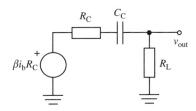

FIGURE 10.6 Thévenin equivalent for the output section of Figure 10.5.

The network in Figure 10.6 is identical to that for the input with R_C acting as the source resistor, and C_C and R_L the basic *RC* network. Equation 9.10 applies. The corner frequency produced by the output coupling capacitor is:

$$f_c = \frac{1}{2\pi C_C(R_C + R_L)} \tag{10.4}$$

where R_C is the collector resistance and R_L is the load resistance. The load resistance could be the input resistance of a second amplifier stage.

Decoupling Capacitor

In considering the effect of the base coupling capacitor it was assumed that the emitter decoupling capacitor was a short circuit. Now in considering the decoupling capacitor it is assumed that the base coupling capacitor is a short circuit.

The equivalent circuit for calculating the effect of C_E is shown in Figure 10.7.

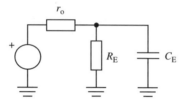

FIGURE 10.7   Equivalent circuit for the decoupling capacitor.

The resistor r_o is the output resistance as seen by the capacitor at the emitter of the transistor, that is the output resistance of an emitter follower and is:

$$r_o = \left(\frac{R_s \| R_1 \| R_2}{\beta} + r_e \right) \tag{10.5}$$

and the corner frequency is given by:

$$f_c = \frac{1}{2\pi C_E (r_o \| R_E)} \tag{10.6}$$

The effect of the emitter decoupling capacitor on the frequency response is different to that of the coupling capacitors. In the mid-band the emitter resistor R_E is fully decoupled (shorted out) and does not appear in the voltage gain equations for the amplifier. However, well below the cut-off frequency as determined by equation 10.6 the emitter resistor is no longer decoupled and forms part of the equivalent circuit for the calculation of gain. The gain with R_E present is given by:

$$A_v = - \frac{R_C \| R_L}{(r_e + R_E)}$$

The gain is reduced from the mid-band value of $-(R_C \| R_L)/r_e$. The gain versus frequency curve may contain a plateau at an intermediate value of gain, as given above, when C_E is no longer acting as a short circuit, before finally decreasing to zero gain for very low frequency, because of the coupling capacitors. An example is shown in Figure 10.9 below. Normally this part of the gain versus frequency curve is of little interest.

▷ **EXAMPLE 10.3** ──

For the circuit shown in Figure 10.3 determine the value of C_E to provide a 3 dB corner frequency of 50 Hz. Assume that C_B is a short circuit.

▷ *Solution*

The resistance seen from the emitter looking back into the base is given by equation 10.5 and is:

$$r_o = \frac{1\ \mathrm{k\Omega}\,\|\,33\ \mathrm{k\Omega}\,\|\,15\ \mathrm{k\Omega}}{100} + 13.6\ \Omega = 22.7\ \Omega$$

For a 3 dB frequency of 50 Hz the required value of C_E is obtained from equation 10.6 as:

$$C_E = \frac{1}{2\pi(50)(22.7\ \Omega\,\|\,1\ \mathrm{k\Omega})} = 143\ \mu\mathrm{F}$$

───

This value of C_E is fairly typical for an emitter decoupling capacitor, because of the small value of the emitter follower output resistance. An important point to note is that the dc working voltage for these capacitors is generally less than 5 V, because the dc bias voltage across R_E is usually no greater than $V_{CC}/10$ (see Chapter 5). This fact has important cost implications since large-value capacitors can be expensive if the working voltage is also large.

▷ **EXAMPLE 10.4** ──

Use PSpice to evaluate the low-frequency performance of the circuit in Figure 10.3 with $C_B = 1.5\ \mu\mathrm{F}$, $C_E = 0.72\ \mu\mathrm{F}$ and $C_C = 14\ \mu\mathrm{F}$.

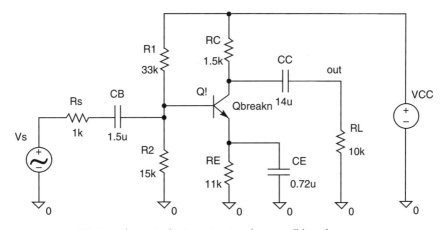

FIGURE 10.8   PSpice schematic for investigating the overall low-frequency response.

▶ *Solution*

The PSpice schematic is shown in Figure 10.8 and the Probe output in Figure 10.9.

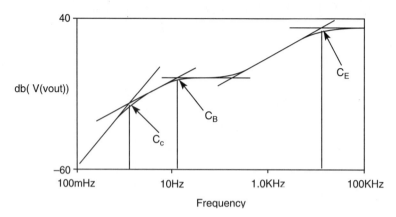

FIGURE 10.9   Low-frequency response from Probe with straight line approximations to show the corner frequencies produced by the different capacitors.

The frequency response shown in Figure 10.9 is for illustrative purposes only and is not intended to represent the ideal response of a practical circuit. The capacitor values are chosen to ensure that the corner frequencies for each *RC* network are well separated. The frequency which determines the lower 3 dB point is 10 kHz and is determined by C_E. As the frequency decreases below 10 kHz the decoupling capacitor becomes less effective until at about 100 Hz it is effectively an open circuit and the gain is reduced to approximately 2 dB. As the frequency is reduced further the base coupling capacitor reactance starts to increase and a further corner frequency occurs at about 10 Hz. Finally at about 1 Hz the collector coupling capacitor produces a final corner frequency. Notice that the slope of the curve increases below 1 Hz. The slope changes from −20 dB per decade to −40 dB per decade as the actions of the *RC* networks formed by the two coupling capacitors combine.

Partially Decoupled Emitter Resistor

A very common circuit configuration for a common-emitter amplifier is one with a partially decoupled emitter resistor as shown in Figure 10.10.

The output resistance as seen from the decoupling capacitor C_E is:

$$r_o = R_{E1} + r_e + \frac{R_s \| R_1 \| R_2}{\beta} \tag{10.7}$$

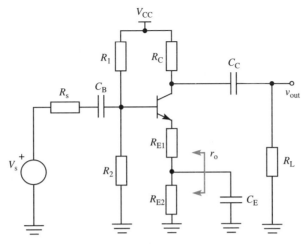

FIGURE 10.10   Partially decoupled emitter resistor.

and

$$f_c = \frac{1}{2\pi C_E(r_o \| R_{E2})} \tag{10.8}$$

The main difference is in the value of r_o which is increased by the presence of R_{E1}. The input resistance is also affected and is increased by the presence of R_{E1}. This will have the effect of reducing the value of the base decoupling capacitor for a given corner frequency, in comparison with the previous circuit with a fully decoupled emitter resistor.

▶ **EXAMPLE 10.5** ─────────────────────────────────

For the circuit shown in Figure 10.3 the emitter resistor is divided into two with values of $R_{E1} = 150\ \Omega$ and $R_{E2} = 850\ \Omega$. Determine the value of C_E to produce a corner frequency of 50 Hz.

▶ *Solution*

The output resistance is obtained from equation 10.7:

$$r_o = 150 + 13.6 + \frac{1\ \text{k}\Omega \| 33\ \text{k}\Omega \| 15\ \text{k}\Omega}{100} = 173\ \Omega$$

and

$$C_E = \frac{1}{2\pi(50)(850\ \Omega \| 173\ \Omega)} = 25\ \mu\text{F}$$

The value of emitter decoupling capacitance is decreased by the presence of a series emitter resistance as compared with the fully decoupled emitter (143 µF).

Selection of Coupling and Decoupling Capacitors

An important point to note in the selection of the capacitors is that the corner frequency should be determined by only one of the *RC* networks. Since the largest value of capacitance is likely to be the emitter decoupling capacitor, then it is usual for this capacitor to determine the corner frequency. The frequencies used to determine values for the base and collector coupling capacitors are then selected to be 1/10th of the frequency used for the emitter decoupling capacitor. This means that when calculating the value of C_E the base coupling capacitor can be assumed to be a short circuit.

When calculating the value of the base coupling capacitor the emitter decoupling capacitor is more likely to have a finite impedance at a frequency which is 1/10th of the frequency for which C_E was chosen to establish the corner frequency. In practice, however, assuming it to be a short circuit will result in a value of C_B that is larger than necessary, but this will not affect the 3 dB frequency. The final optimization of capacitor value can be made with the aid of a circuit simulator.

▶ **EXAMPLE 10.6** ──────────────────────────────────────

For the circuit shown in Figure 10.11 determine suitable values for the capacitors to produce a 3 dB frequency of 20 Hz. Assume that $r_e = 26\ \Omega$ and $\beta = 100$.

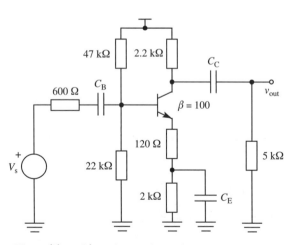

FIGURE 10.11   CE amplifier with emitter series resistance.

▷ Solution

Decoupling Capacitor

From equation 10.7 the output resistance as seen by the decoupling capacitor is:

$$r_o = \frac{600\ \Omega \| 47\ \text{k}\Omega \| 22\ \text{k}\Omega}{100} + 26\ \Omega + 120\ \Omega = 152\ \Omega$$

and

$$C_E = \frac{1}{2\pi(20\ \text{Hz})(850\ \Omega \| 152\ \Omega)} = 62\ \mu\text{F}$$

Base Coupling Capacitor

The calculation for the coupling capacitors is made at 2 Hz. The input resistance as seen by C_B with C_E assumed to be a short circuit is:

$$R_{in} = 47\ \text{k}\Omega \| 22\ \text{k}\Omega \| \beta(26\ \Omega + 120\ \Omega) = 7{,}4\ \text{k}\Omega$$

and

$$C_B = \frac{1}{2\pi(2\ \text{Hz})(600\ \Omega + 7.4\ \text{k}\Omega)} = 10\ \mu\text{F}$$

Collector Coupling Capacitor

The collector capacitance is obtained with the aid of equation 10.4:

$$C_C = \frac{1}{2\pi(2\ \text{Hz})(2.2\ \text{k}\Omega + 5\ \text{k}\Omega)} = 11\ \mu\text{F}$$

▷ **EXAMPLE 10.7** ──────────────────────────────

With $V_{CC} = 9$ V simulate the circuit in Figure 10.11 with PSpice and obtain the low-frequency gain versus frequency response.

▷ Solution

The circuit schematic for PSpice is shown in Figure 10.12 and the gain versus frequency plot for the circuit is shown in Figure 10.13.

It can be seen that the corner frequency is 20 Hz as required and that the gain in the mid-band is 20 dB. Notice also that the roll-off is continuous below 20 Hz and not the stepped response shown in Figure 10.9.

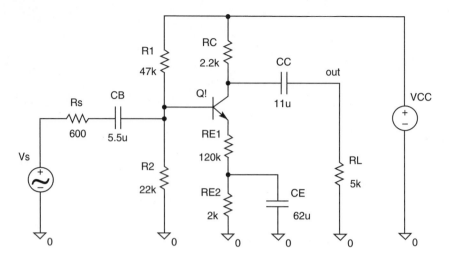

FIGURE 10.12 PSpice schematic.

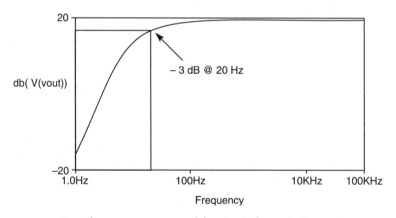

FIGURE 10.13 Low-frequency response of the circuit shown in Figure 10.12.

NB It is important to note that the choice of capacitor may not always follow these rules and that the corner frequency may in fact be determined by one of the coupling capacitors rather than the emitter decoupling capacitor. For the purposes of hand calculations the highest corner frequency is the one which governs the overall low-frequency 3 dB roll-off point. A circuit simulator will obviously provide a more accurate picture which takes account of the effects of all capacitors acting together. Hand calculations are only really accurate if the corner frequencies produced by each *RC* network are well separated. If they are close together then it is no longer possible to ignore one capacitor while determining the effect of another, and it is also more difficult to predict the actual value of the 3 dB corner frequency.

Common-Base Input Coupling Capacitor

The important capacitor for the common-base (CB) stage is the input capacitor (C_1) in Figure 10.14, because the input impedance of the CB stage is small (equal to r_e).

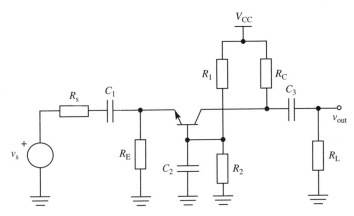

FIGURE 10.14   Capacitors in the common-base stage.

The corner frequency for the input stage is given by:

$$f_c = \frac{1}{2\pi C_1 (R_s + r_e \| R_E)} \tag{10.9}$$

and the corner frequency for the output coupling capacitor is:

$$f_c = \frac{1}{2\pi C_3 (R_C + R_L)} \tag{10.10}$$

For the base decoupling capacitor (C_2) the two bias resistors (R_1 and R_2) appear in parallel with this capacitor, but the output impedance of the transistor as seen from the base is also in parallel with the capacitor. In Chapter 6 the concept of impedance reflection was introduced (Section 6.5). The impedance of the emitter circuit as seen from the base terminal is:

$$r_o = \beta(r_e + R_s \| R_E) \tag{10.11}$$

and the corner frequency for the base decoupling capacitor is:

$$f_c = \frac{1}{2\pi C_2 (r_o \| R_1 \| R_2)} \tag{10.12}$$

To obtain suitable values of capacitance for a CB stage the input coupling capacitor is calculated for the required corner frequency. The other two capacitors are then obtained for a frequency that is 1/10th of this frequency.

▶ **EXAMPLE 10.8** ────────────────────────────────

For the circuit shown in Figure 10.14 the components have the following values:

$R_E = 470\ \Omega$ $R_C = 1.8\ k\Omega$
$R_1 = 22\ k\Omega$ $R_2 = 6.8\ k\Omega$
$R_S = 50\ \Omega$ $R_L = 3\ k\Omega$
$C_1 = 100\ \mu F$ $C_2 = 47\ \mu F$
$C_3 = 4.7\ \mu F$

The transistor β is 100 and $r_e = 9.5\ \Omega$. Determine the low-frequency 3 dB corner frequency.

▶ *Solution*

Input RC Network

The input resistance for a CB stage is:

$$R_{in} = 470\ \Omega\,||\,9.5\ \Omega \approx 9.5\ \Omega$$

and

$$f_c = \frac{1}{2\pi(100\ \mu F)(50\ \Omega + 9.5\ \Omega)} \approx 27\ Hz$$

Base Capacitor

The reflected impedance from the emitter is:

$$r_o = (100)(9.5\ \Omega + 50\ \Omega\,||\,470\ \Omega) \approx 5.5\ k\Omega$$

and from equation 10.12 the corner frequency is:

$$f_c = \frac{1}{2\pi(47\ \mu F)(5.5\ k\Omega\,||\,22\ k\Omega\,||\,6.8\ k\Omega)} \approx 1.3\ Hz$$

Output RC Network

For the output network the corner frequency is obtained from equation 10.10:

$$f_c = \frac{1}{2\pi(4.7\ \mu F)(1.8\ k\Omega + 3\ k\Omega)} \approx 7\ Hz$$

The actual 3 dB corner frequency for the complete amplifier is the highest of the three and is that produced by the input *RC* network at 27 Hz.

EXAMPLE 10.9

For the same resistor values and β select suitable capacitors to produce a 3 dB corner frequency of 100 Hz. Verify that the gain is 25 dB and use PSpice to check the frequency response. Assume a supply voltage of 9 V.

▷ Solution

With the aid of equations 10.9, 10.10 and 10.12 the capacitors are:

$C_1 = 27 \ \mu F$

$C_2 = 6 \ \mu F$

$C_3 = 3.3 \ \mu F$

The gain versus frequency plot produced by Probe is shown in Figure 10.15.

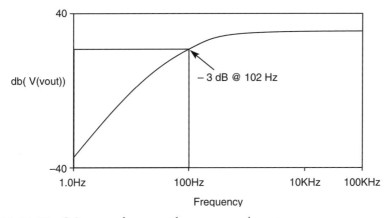

FIGURE 10.15   Gain versus frequency for a common-base stage.

Emitter Follower

The emitter follower is shown in Figure 10.16.

The important features of the emitter follower are high input resistance and low output resistance, which will result in a relatively small value of capacitance for the input coupling capacitor and a large value for the output coupling capacitor. The corner frequency should be determined by the output capacitor and it is given by:

$$f_c = \frac{1}{2\pi C_O\{R_E \| [(R_s \| R_1 \| R_2/\beta) + r_e] + R_L\}} \tag{10.13}$$

where the term in the square brackets is the output resistance of the emitter follower.

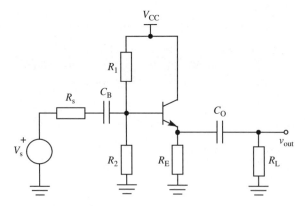

FIGURE 10.16   Emitter follower with coupling capacitors.

For the input coupling capacitor the corner frequency is given by:

$$f_c = \frac{1}{2\pi C_B (R_s + R_{in})} \tag{10.14}$$

where

$$R_{in} = R_1 \,\|\, R_2 \,\|\, \beta(r_e + R_E \,\|\, R_L)$$

To select suitable capacitors for an emitter follower the value of C_O is obtained for the required corner frequency and then the value of C_B is calculated for a frequency of 1/10th of this value.

▷ **EXAMPLE 10.10** ──

Select capacitors for the circuit shown in Figure 10.17 to provide a low-frequency corner frequency of 30 Hz. Assume that $\beta = 100$.

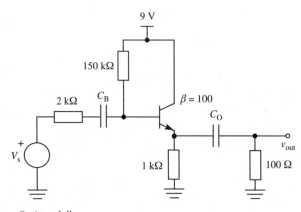

FIGURE 10.17   Emitter follower.

▶ *Solution*

DC Conditions

The collector current is given by:

$$I_C = \frac{9\text{ V} - 0.7\text{ V}}{1\text{ k}\Omega + (150\text{ k}\Omega/100)} = 3.3\text{ mA}$$

and

$$r_e = 7.8\ \Omega$$

AC Conditions

The output resistance of the emitter follower as seen by the capacitor C_O is:

$$r_o = 1\text{ k}\Omega \left\| \left(\frac{2\text{ k}\Omega \,\|\, 150\text{ k}\Omega}{100} + 7.8\ \Omega \right) \approx 27\ \Omega \right.$$

and the value of the output capacitance for a corner frequency of 30 Hz is:

$$C_O = \frac{1}{2\pi(30\text{ Hz})(27\ \Omega + 100\ \Omega)} = 42\ \mu\text{F}$$

The base coupling capacitor is determined for a frequency of 3 Hz and the *RC* network comprises the 2 kΩ source resistance plus the input resistance of the emitter follower. At 3 Hz the reactance of C_O is very large and for simplicity it may be regarded as an open circuit. Thus the capacitance is:

$$C_B = \frac{1}{2\pi(3\text{ Hz})[2\text{ k}\Omega + 150\text{ k}\Omega \,\|\, 100(7.8\ \Omega + 1\text{ k}\Omega)]} \approx 1\ \mu\text{F}$$

▶ **EXAMPLE 10.11** ──────────────────────────────

Use PSpice to verify the accuracy of the capacitor calculations in example 10.9.

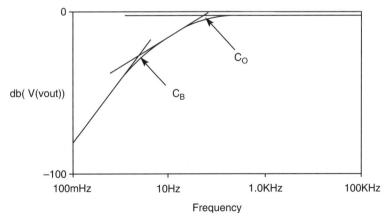

FIGURE 10.18 Frequency versus gain for an emitter follower.

▶ *Solution*

The gain versus frequency plot from Probe is shown in Figure 10.18.

High-Frequency Response of BJT Amplifiers

The high-frequency performance of a BJT amplifier may be limited by the cut-off frequency of the transistor. For small-signal transistors this is typically tens or even hundreds of GHz which far exceeds the corner frequency of practical amplifiers. A more likely cause of the high frequency limitation of real circuits is stray capacitance associated with the components, the wiring, and the pn junctions of the base–emitter and particularly the base–collector which provides a feedback path from the collector back to the base. The combined effect of these components of capacitance results in signal paths to ground which eventually short the signal to ground at sufficiently high frequency. The point at which this first occurs in the circuit depends on the *RC* network formed by these capacitors and resistors in the circuit. This frequency establishes the high-frequency 3 dB corner of the gain versus frequency response. The most important components of stray capacitance are those associated with the pn junctions and these form the basis of the hybrid pi equivalent circuit (see Appendix 4, and the section below). A simplification of the hybrid pi allows these capacitors to be considered as discrete components connected between the external leads of the transistor, as shown in Figure 10.19.

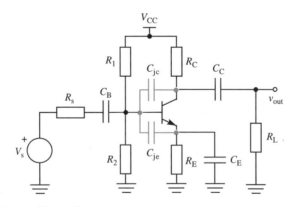

FIGURE 10.19   CE amplifier with junction capacitance.

At high frequency the decoupling capacitor is a short circuit and the base–emitter junction capacitance C_{je} provides a signal path to ground through C_E. There is an additional component of capacitance which contributes to this signal path, which results from the base–collector capacitor C_{jc}. This capacitor is connected between the base and collector, or the input and output of an amplifier. In Chapter 9 it was shown that such a capacitor can be reflected to the input and output as two separate capacitors by application of Miller's theorem. Thus the total capacitance from base to ground is:

$$C_{in} = C_{je} + C_{jc}(1 - A_v) \qquad (10.15)$$

and at the output:

$$C_{out} = C_{jc}\left(1 - \frac{1}{A_v}\right) \qquad (10.16)$$

There are two important *RC* networks which contribute to the overall high-frequency performance of a CE stage: the input network and the output network.

Input RC Network

The equivalent circuit for the input is shown in Figure 10.20. Notice the presence of the source resistance R_s. If this were not present then the corner frequency would be infinity, since the Thévenin resistance of the source would be 0 Ω, and based on equation 10.1 the corner frequency would be infinity.

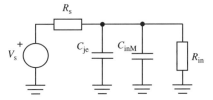

FIGURE 10.20   Equivalent high-frequency circuit for the input.

The capacitance is obtained from equation 10.15 and represents the effect of the junction capacitance associated with the transistor. The corner frequency for the network in Figure 10.20 is:

$$f_c = \frac{1}{2\pi(C_{je} + C_{inM})(R_s \| R_{in})} \qquad (10.17)$$

where

$$C_{inM} = C_{jc}(1 - A_v)$$

and

$$R_{in} = R_1 \| R_2 \| \beta r_e$$

Any capacitance associated with the bias resistors and the electrical wiring would be represented as a lumped component in parallel with the two capacitors in Figure 10.20.

▶ **EXAMPLE 10.12**

In the circuit shown in Figure 10.21 estimate the corner frequency produced by the low-pass filter at the input if the capacitors associated with the transistor are:

$C_{je} = 10$ pF

$C_{jc} = 4$ pF

Assume that $r_e = 43$ Ω.

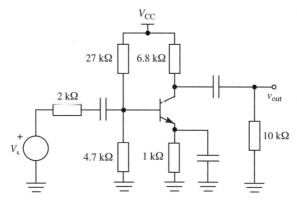

FIGURE 10.21 CE amplifier for high-frequency calculations.

▶ *Solution*

The voltage gain is:

$$A_v = -\frac{R_C \| R_L}{r_e} = -\frac{6.8 \text{ k}\Omega \| 10 \text{ k}\Omega}{43} = -94$$

The input Miller capacitance is:

$$C_{inM}|_{input} = (4 \text{ pF})(1 + 94) = 380 \text{ pF}$$

and the corner frequency for the low-pass filter is:

$$f_c|_{input} = \frac{1}{2\pi(10 \text{ pF} + 380 \text{ pF})(2 \text{ k}\Omega \| R_{in})}$$

where

$$R_{in} = 27 \text{ k}\Omega \| 4.7 \text{ k}\Omega \| (100 \times 43 \text{ }\Omega) \approx 2 \text{ k}\Omega$$

Therefore substituting for R_{in} gives:

$$f_c|_{input} = \frac{1}{2\pi(10 \text{ pF} + 380 \text{ pF})(2 \text{ k}\Omega \| 2 \text{ k}\Omega)} \approx 400 \text{ kHz}$$

Notice the importance of the base–collector capacitance. Although this component is small it is multiplied by the gain of the stage so that it dominates the input RC network.

Output RC Network

The output equivalent circuit is shown in Figure 10.22.

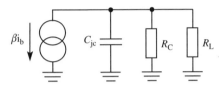

FIGURE 10.22   High-frequency equivalent circuit for the output of the CE amplifier.

The capacitance C_{jc} results from the Miller capacitance C_{out} given by equation 10.16 where A_v is assumed to be $\gg 1$. The corner frequency is:

$$f_c = \frac{1}{2\pi C_{jc}(R_C \| R_L)} \tag{10.18}$$

In some instances it may be desirable to limit the high-frequency performance of an amplifier. This may happen in an amplifier designed for speech channels where the maximum frequency is generally not greater than 4 kHz. There is no point in having a bandwidth of hundreds of kHz if the maximum frequency of the signal is not going to exceed 4 kHz. From equation 10.18 it is obvious that the bandwidth can be limited by increasing the value of the capacitance, that is by placing additional capacitance in parallel with C_{jc}. This can be achieved by placing a suitable capacitor from the collector to ground to establish the required value of f_c as defined in equation 10.18. The same effect could also be achieved by placing a capacitor across the input, but because the resistance in the output circuit is generally greater than in the input and more easily identified, it is usually simpler to place such a capacitor across the output.

▶ **EXAMPLE 10.13** ─────────────────────────────────

For the circuit shown in Figure 10.21 determine the corner frequency formed by the output low-pass filter.

▶ *Solution*

The capacitance in the output circuit is simply C_{jc} in parallel with the collector and output load resistors, that is:

$$f_c \big|_{output} = \frac{1}{2\pi(4 \text{ pF})(6.8 \text{ k}\Omega \| 10 \text{ k}\Omega)} = 9.8 \text{ MHz}$$

Combined High-Frequency Response

The combined frequency response which includes both the input and output RC networks should produce a response as shown in Figure 10.23.

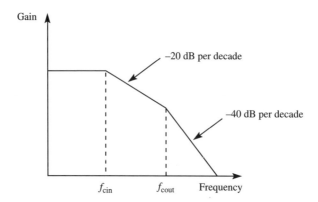

FIGURE 10.23   Combined high-frequency response of the CE amplifier due to C_{je} and C_{jc}.

The 3 dB corner frequency is determined by the input RC network because of the large increase in the effective capacitance produced by the Miller effect. In practice the clearly identified break in the rate of change of gain beyond f_{cin} may not be so obvious because of the approximations necessary to obtain a simple expression for the Miller capacitance.

The greatest contribution to the input capacitance is the result of the Miller effect, but in order to simplify the calculation it is assumed that the input resistance of the amplifier is very large compared with the impedance of the current path through the capacitor linking the input and output (C_{jc}). The relationship between corner frequency, Miller effect and input resistance is complex, but it is easily resolved with a circuit simulator which is not confined to the simplifications imposed on the circuit for the purposes of hand calculations. The hand calculations are useful in predicting the approximate position of the 3 dB corner frequency, but not in predicting the exact nature of the roll-off beyond the corner frequency. An added complication is that the junction capacitances are voltage dependent, and the values given in the manufacturer's data are zero-voltage values. For most applications this degree of accuracy is not generally important for hand-based calculations and an approximate knowledge of the corner frequency is all that is required. A circuit simulator can provide more detail.

▶ **EXAMPLE 10.14** ——————————————————————————————

Use PSpice to simulate the effect of the low-pass filters formed by placing a 380 pF capacitor from the base to ground and 4 pF from the collector to ground. Use the default npn transistor QbreakN.

▶ Solution

The circuit schematic from PSpice is shown in Figure 10.24.

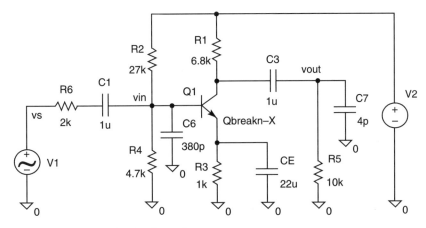

FIGURE 10.24   PSpice circuit schematic.

The junction capacitors are represented by the lumped capacitors C6 (C_{je} plus the Miller capacitance) and C7 (C_{jc}). The default values of junction capacitance in the breakout transistor are 0 pF. For the AC analysis the frequency is swept from 10 kHz to 100 MHz and the resulting frequency response is shown in Figure 10.25.

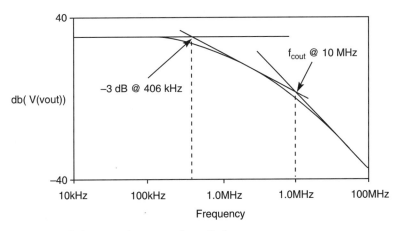

FIGURE 10.25   Gain versus frequency from Probe.

The response produced by Probe for Example 10.14 is very similar to that predicted in Figure 10.23, but it does not represent what happens in a real circuit, because in practice the junction capacitors have finite values. The generic model for the QbreakN transistor does not include values for the junction capacitance. To obtain a more realistic frequency response curve it is necessary to model the transistor more accurately by including in the model statement values for the junction capacitors. These are used in place of the external lumped capacitors used in Figure 10.24.

▶ **EXAMPLE 10.15** ——

Use PSpice again but remove the discrete capacitors C6 and C7 from the circuit in Example 10.14 and from **Edit** select **Model...** and add values for the two junction capacitors as:

.model QbreakN npn cjc = 4p cje = 10p

▶ *Solution*

The output from Probe is shown in Figure 10.26.

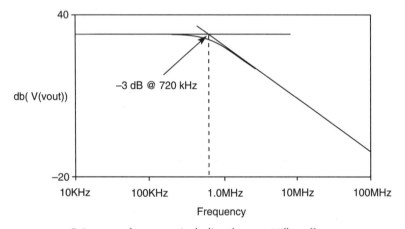

FIGURE 10.26   Gain versus frequency including the true Miller effect.

It can be seen from Figure 10.26 that there is only one break point in the frequency response at 720 kHz. This is higher than the predicted 400 kHz, but the capacitor values for $C_{je} = 10$ pF and $C_{jc} = 4$ pF are zero-voltage values. When the transistor is biased the emitter–base junction is forward biased, and therefore the capacitance will be greater than 10 pF. Similarly the collector–base junction is reverse biased and the capacitance of this junction will be less than 4 pF. Actual values for the two capacitors may be obtained by examining the PSpice output data file as:

$C_{je} = 16.8$ pF

$C_{jc} = 2.2$ pF

If these values are used to calculate the corner frequency of the input low-pass filter then the value is:

$C_M|_{input} = (2.2 \text{ pF})(1 + 94) = 209 \text{ pF}$

where 94 represents the measured voltage gain.

$$f_c \big|_{\text{input}} = \frac{1}{2\pi(16.8 \text{ pF} + 209 \text{ pF})(2 \text{ k}\Omega \,\|\, 2 \text{ k}\Omega)} \approx 690 \text{ kHz}$$

This value is close to that observed from the Probe output in Figure 10.26.

The second break point predicted by the calculation of the corner frequency of the output low-pass filter is masked by the complex nature of the relationship between the Miller effect and the input impedance of the transistor.

▶ **EXAMPLE 10.16** ——————————————————————

Repeat Example 10.15 with a capacitor of 10 nF connected between the collector and the ground and observe the value of the high-frequency response.

▶ *Solution*

The output produced with Probe should show a −3 dB point of about 4.3 kHz with a roll-off beyond this frequency of −20 dB per decade.

The Hybrid Pi Approach

In Figure 10.19 the high-frequency limitations of the bipolar transistor were described in terms of capacitance associated with the pn junctions of the transistor, and were shown as external components. In reality these capacitors are an integral part of the transistor. A more accurate representation may be achieved by adopting a more suitable high-frequency equivalent circuit for the transistor in place of the very simple r parameter circuit. The most suitable equivalent circuit is the hybrid pi, as described in Chapter 4 and Appendix 4. The capacitors C_{jc} and C_{je} do not appear in the discrete component circuit diagram, as shown in Figure 10.19, but now form an integral part of the equivalent circuit of the transistor. The circuit with the transistor replaced by the hybrid pi equivalent circuit is shown in Figure 10.27.

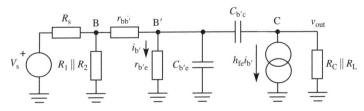

FIGURE 10.27   CE amplifier with hybrid pi equivalent circuit.

The analysis is exactly the same with Miller's theorem being used to transfer $C_{b'c}$ to the input or output as appropriate. For the purposes of hand-based calculations the base resistance $r_{bb'}$ can be ignored.

▶ **EXAMPLE 10.17** ───

Determine the high-frequency 3 dB point for the circuit shown in Figure 10.28 for which V_{CC} is 9 V. For the 2N2222A h_{fe} is typically 200, the depletion-layer capacitance for the collector, $C_{b'c}$, is 6 pF, and for the emitter is 20 pF and f_T is quoted as 300 MHz.

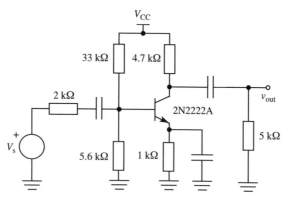

FIGURE 10.28   CE amplifier.

▶ *Solution*

DC Conditions

$R_T = 33k \parallel 5.6k = 4.78\ k\Omega$

$V_T = 1.3\ V$

and

$$I_C = \frac{1.3\ V - 0.7\ V}{1k + (4.78k/200)} = 0.58\ mA$$

AC Conditions

From Appendix 4, equation A4.7, the delay capacitance for the emitter−base junction, $C_{b'e}$, is given by

$$C_{b'e}\big|_{delay} = \frac{g_m}{2\pi f_T} = \frac{0.58\ mA}{0.026}\ \frac{1}{2\pi(300\ MHz)} = 11.8\ pF$$

This capacitance must be added to the depletion-layer capacitance, that is:

$$C_{b'e}\big|_{total} = C_{b'c}\big|_{depletion} + C_{b'c}\big|_{delay} = 20 \text{ pF} + 11.8 \text{ pF} = 31.8 \text{ pF}$$

The emitter-base junction small-signal resistance is:

$$r_{b'e} = \frac{h_{fe}0.026}{I_E} = \frac{200 \times 0.026}{0.58 \text{ mA}} \approx 9 \text{ k}\Omega$$

The complete small-signal circuit, with the exception of $r_{bb'}$ which has been ignored ($r_{bb'} \ll R_s$), is shown in Figure 10.29.

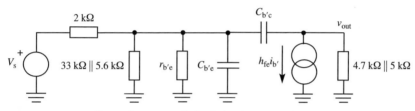

FIGURE 10.29   Small-signal circuit with the hybrid pi equivalent circuit.

The analysis now proceeds as before. The capacitor $C_{b'c}$ (6 pF) is reflected to the input by means of Miller's theorem as

$$C_{inM} = C_{b'c}(1 - A_v)$$

where

$$A_v = -\frac{R_c \| R_L}{r_e} = -\frac{R_c \| R_L}{r_{b'e}/h_{fe}} = -\frac{2.4k}{9k/200} \approx -54$$

and

$$C_{inM} = (6 \text{ pF})(1 + 55) = 330 \text{ pF}$$

The corner frequency for the input is:

$$f_c\big|_{input} = \frac{1}{2\pi(3.18 \text{ pF} + 330 \text{ pF})(2 \text{ k}\Omega \| R_{in})}$$

where

$$R_{in} = R_1 \| R_2 \| r_{b'e} = 33 \text{ k}\Omega \| 5.6 \text{ k}\Omega \| 9 \text{ k}\Omega = 3.1 \text{ 1}\Omega$$

and

$$f_c\big|_{input} = \frac{1}{2\pi(361.8 \text{ pF})(2 \text{ k}\Omega \| 3.1 \text{ k}\Omega)} = 362 \text{ kHz}$$

For the output circuit the corner frequency is:

$$f_c\big|_{output} = \frac{1}{2\pi C_{b'c}(R_c \| R_L)} = \frac{1}{2\pi(6\text{ pF})(2.4\text{ k}\Omega)} = 11\text{ MHz}$$

Thus the upper 3 dB frequency for the circuit is 362 kHz.

The hybrid pi equivalent circuit provides a more accurate representation of the transistor at high frequencies. Its most important function is to provide a means for estimating the value of the emitter–base capacitance which is associated with the signal delay imposed by the transport of charge carriers through the base of the transistor (see Appendix 4). Knowledge of the unity gain frequency f_T allows this capacitance to be determined. The base resistance $r_{bb'}$ does complicate the analysis, but it could be included by repeated application of Thévenin's theorem. For many applications it can be ignored.

High-frequency analysis of bipolar transistor circuits is difficult. The hybrid pi circuit provides a good approximation for hand-based calculations, but it is usually adequate to simply include the junction capacitors as external components, as shown in Figure 10.19, and to ignore the effect of $r_{bb'}$.

Common-Base Circuit

The CB configuration is widely used for high-frequency applications and it is of interest to understand how the high-frequency 3 dB frequency differs from that of the CE amplifier. A typical CB circuit is shown in Figure 10.30.

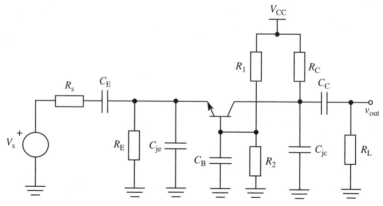

FIGURE 10.30   CB circuit with interelectrode capacitors present.

Notice that the two junction capacitors C_{je} and C_{jc} now appear directly across the input and output respectively, and that there is no junction capacitor shown between the emitter and the collector. A small interelectrode capacitance will exist between the leads, both inside the case of the transistor and outside, but it is very much smaller than the junction capacitors. Thus the Miller capacitance, which is a major limitation of the high-frequency response of the CE amplifier, is not a serious problem for the CB amplifier. The equivalent circuits for the input and output are shown in Figure 10.31.

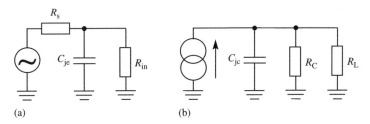

(a)   (b)

FIGURE 10.31 High-frequency equivalent circuits for (a) the input and (b) the output.

For the input circuit the cut-off frequency is given by:

$$f_c = \frac{1}{2\pi C_{je}(R_s + R_{in})} \tag{10.19}$$

where the input resistance is:

$$R_{in} = R_E \parallel r_e \approx r_e$$

The input resistance of the CB stage is very small and if the source resistance is also small then the corner frequency is large. The corner frequency for the output is:

$$f_c = \frac{1}{2\pi C_{jc} R_C \parallel R_L} \tag{10.20}$$

This equation is the same as that for the CE stage (equation 10.18).

▶ **EXAMPLE 10.18** ───────────────────────────────────────

Determine the 3 dB frequency for the CB circuit shown in Figure 10.32. Assume that $r_e = 26\ \Omega$, $C_{je} = 15\ pF$ and $C_{jc} = 2\ pF$.

▶ *Solution*

For the input circuit the corner frequency is obtained from equation 10.19 as:

$$f_c|_{input} = \frac{1}{2\pi(16\ pF)(50\ \Omega + 26\ \Omega)} \approx 130\ MHz$$

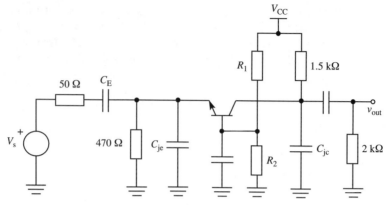

FIGURE 10.32 CB circuit.

For the output from equation 10.20 the corner frequency is:

$$f_c|_{output} = \frac{1}{2\pi(2\text{ pF})(1.5\text{ k}\Omega \| 2\text{ k}\Omega)} \approx 93\text{ MHz}$$

The 3 dB frequency is determined by the lower of the two, that is 93 MHz.

The corner frequency for the CB stage in Example 10.17 is considerably higher than the CE stage, but it should be treated with some caution because the equivalent circuit which is being used for the transistor is reaching a high-frequency limit. The more complex equivalent circuit described in Chapter 4 involving the Gummel–Poon configuration is more appropriate. This circuit is too complex to use for hand calculations and it is better to use a circuit simulator, provided that the transistors required are modelled.

▷ **EXAMPLE 10.19** ————————————————————————

Use PSpice to determine the 3 dB corner frequency for a CB stage using a 2N2222 transistor with a collector current of 1 mA. The schematic is shown in Figure 10.33.

▷ *Solution*

The output from Probe is shown in Figure 10.34.
Referring to the output file produced by PSpice yields the following values of capacitance for the junctions:

$C_{BE} = C_{je}$ at the operating current and voltage $= 53$ pF
$C_{BC} = C_{jc}$ at the operating current and voltage $= 3.4$ pF

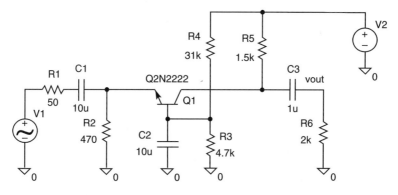

FIGURE 10.33   PSpice schematic for CB stage.

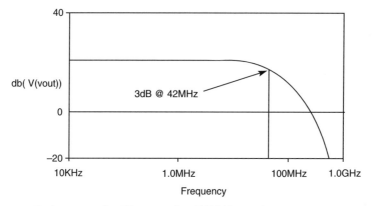

FIGURE 10.34   Probe output for CB stage using 2N2222 transistor.

These values yield theoretical corner frequencies based on equations 10.19 and 10.20 of:

$$f_c\big|_{input} = \frac{1}{2\pi(53\ \text{pF})(76\ \Omega)} = 40\ \text{MHz}$$

and

$$f_c\big|_{output} = \frac{1}{2\pi(3.4\ \text{pF})(857\ \Omega)} = 55\ \text{MHz}$$

The 3 dB frequency is 40 MHz which agrees surprisingly well with the simulated value of 42 MHz. The 2N2222 is a general purpose npn transistor and is not particularly suited to very high-frequency use. Higher-frequency transistors with smaller junction capacitance are available, but the accuracy of the above calculations based on simple models will be limited and it would be better to use a simulator, provided that the transistor model is available

The Emitter Follower

The emitter follower is used extensively as a buffer amplifier for low-impedance loads for a wide range of frequencies. Typically the load is a coaxial cable and the signal may extend to hundreds of MHz. The emitter follower with stray capacitance is shown in Figure 10.35.

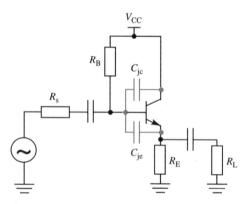

FIGURE 10.35   Emitter follower amplifier with interelectrode capacitors.

One terminal of the base–collector capacitor, C_{jc}, is grounded through V_{CC} while the other is attached to the base. Miller's theorem can be applied to the base-emitter capacitor, C_{je}, to give a capacitor of $C_{inM} = C_{je}(1 - A_v)$ from base to ground. A second capacitor will exist at the output, but because it is in parallel with the low output impedance of the emitter follower, its effect on the high-frequency corner frequency is not important. The small-signal circuit of the low-pass input network is shown in Figure 10.36.

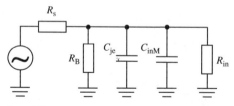

FIGURE 10.36   Low-pass input filter.

The corner frequency is given by:

$$f_c = \frac{1}{2\pi(C_{jc} + C_{inM})(R_s \| R_B \| R_{in})} \tag{10.21}$$

where

$$C_{\text{inM}} = C_{\text{jw}}(1 - A_{\text{v}})$$

and

$$R_{\text{in}} = \beta(r_{\text{e}} + R_{\text{e}} \parallel R_{\text{L}})$$

In comparison with the CE stage the Miller capacitance is small, because the voltage gain is almost unity, and hence $(1 - A_{\text{v}})$ approaches zero. The corner frequency for the emitter follower is much higher than that of the common-emitter amplifier.

▶ **EXAMPLE 10.20** ────────────────────────────────────

A PSpice schematic of an emitter follower using a 2N3904 transistor is shown in Figure 10.37. Examine the high-frequency response.

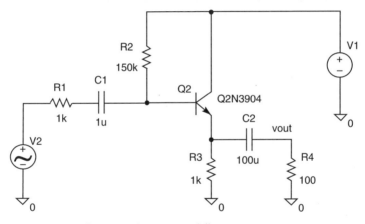

FIGURE 10.37   PSpice schematic of an emitter follower.

▶ *Solution*

The frequency response without a capacitive load is shown in Figure 10.38. From the PSpice output file the following information is obtained:

$$I_{\text{C}} = 4.3 \text{ mA} \qquad C_{\text{BE}} = 54 \text{ pF} \qquad C_{\text{BC}} = 2 \text{ pF}$$
$$A_{\text{v}} = -1.14 \text{ dB} = 0.877 \qquad \beta = 162$$

Utilizing equation 10.21:

$$f_{\text{c}} = \frac{1}{2\pi(2 \text{ pF} + 6.6 \text{ pF})(1 \text{ k}\Omega \parallel 150 \text{ k}\Omega \parallel 15.7 \text{ k}\Omega)} \approx 20 \text{ MHz}$$

This should be compared with the 29 MHz predicted by PSpice.

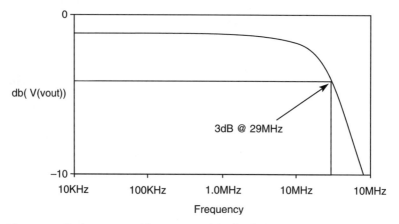

FIGURE 10.38   Probe output without a capacitive load.

Emitter followers are often used to drive a coaxial cable which has capacitance associated with it. The addition of 200 pF to the 100 Ω load in Figure 10.37 produces the frequency response shown in Figure 10.39.

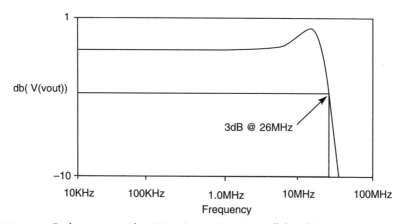

FIGURE 10.39   Probe output with a 200 pF capacitor in parallel with the 100 Ω load.

The corner frequency has not changed greatly, but the significant feature is the increase in gain just prior to the fall-off in gain. The peak is dependent on the load capacitor, the transistor parameters and the collector current. A much more detailed analysis using the Gummel–Poon equivalent circuit shows that under certain circumstances the output impedance of the emitter follower can exhibit an increasing value with increasing frequency. This represents inductive behaviour which can interact with a capacitive load to give frequency characteristics which are reminiscent of a parallel *RLC* circuit. The effect diminishes at lower values of collector current, but also depends on the base resistance in the Gummel–Poon equivalent circuit and also the source resistance. Some experimentation with a circuit simulator is necessary to optimize a design and to determine whether the effect is a problem.

10.3 Low-Frequency Response of FET Amplifiers

The frequency response of an FET amplifier stage is affected by capacitance in a similar manner to that of a bipolar transistor amplifier. The low-frequency response is determined by the coupling and decoupling capacitors, while the high-frequency response is affected by the stray capacitance in the transistor and the circuit wiring and components. Stray capacitance is a particular problem for FET amplifier stages because of the high impedance levels associated with FET devices, and usually results in the corner frequency of the low-pass filters being much lower than the equivalent corner frequency in a low-impedance bipolar circuit.

A typical common-source amplifier is shown in Figure 10.40.

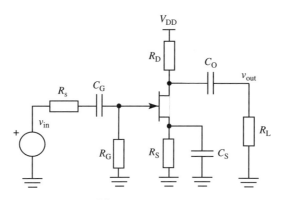

FIGURE 10.40   Common-source amplifier.

The capacitors C_G and C_O form high-pass filters with the input resistance and output resistance respectively, while C_S influences the negative feedback produced by the source resistor R_s.

Input RC Network

The input high-pass filter comprises C_G, R_s and the input resistance of the JFET as shown in Figure 10.41.

The input resistance is simply R_G and the corner frequency is:

$$f_c = \frac{1}{2\pi C_G(R_s + R_G)} \tag{10.22}$$

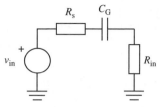

FIGURE 10.41   Input high-pass filter.

▶ **EXAMPLE 10.21** ——

Determine the value of C_G required to produce a 3 dB corner frequency of 10 Hz
if $R_s = 10$ kΩ and $R_G = 1$ MΩ.

Solution

$$C_G = \frac{1}{2\pi(10 \text{ Hz})(10 \text{ k}\Omega + 1 \text{ M}\Omega)} = 15.7 \text{ nF}$$

Because of the very large input resistance of the FET amplifier, the value of the input
coupling capacitor required to achieve a low corner frequency is relatively small in
comparison with what is required for an equivalent bipolar amplifier.

Output RC Network

The equivalent circuit for the output circuit is shown in Figure 10.42 with the transistor
replaced by the current generator $g_m v_{gs}$.

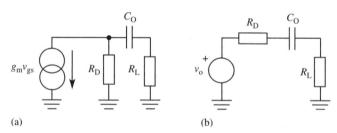

FIGURE 10.42   Output high-pass filter for the CS JFET amplifier with the current generator in
(a) and the Thévenin equivalent in (b).

The Norton equivalent circuit formed by $g_m v_{gs}$ and R_D in Figure 10.42(a) is converted
into the Thévenin equivalent in Figure 10.42(b). By inspection the 3 dB corner
frequency is:

$$f_c = \frac{1}{2\pi C_O(R_D + R_L)} \tag{10.23}$$

▷ **EXAMPLE 10.22** ——————————————————————————————————

Determine the 3 dB corner frequency if the drain resistor $R_D = 6.8$ kΩ, the load resistor $R_L = 10$ kΩ and the coupling capacitor is 1 μF.

▷ *Solution*

$$f_c = \frac{1}{2\pi(1 \ \mu\text{F})(6.8 \ \text{k}\Omega + 10 \ \text{k}\Omega)} = 9.5 \ \text{Hz}$$

Notice that although the corner frequency is similar to that in Example 10.21 the value of the capacitor is considerably larger. While the input impedance of the JFET is very high the output circuit has a similar impedance level to that of a CE bipolar transistor amplifier.

Decoupling Capacitor

In Figure 10.40 the source resistor R_S is decoupled with a source capacitor C_S in the same manner that the emitter resistor is decoupled in a bipolar circuit. In Section 8.2 it is shown that the gain of the common-source amplifier with a source resistor is given by:

$$A_v = -\frac{g_m(R_D \| R_L)}{1 + g_m R_S} \tag{10.24}$$

That is, the gain is reduced from $-g_m(R_D \| R_L)$, when the source resistor is fully decoupled, to the value given above. As the frequency of the input signal decreases the impedance of C_S increases and the gain decreases as the effect of R_S becomes more significant.

To determine the effect of C_S it is necessary to consider the impedance seen by C_S as illustrated in Figure 10.43.

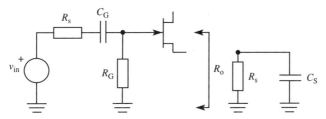

FIGURE 10.43   Output resistance at the source as seen by the decoupling capacitor.

The output resistance R_o is the same as the output resistance of the common-drain amplifier as determined in Section 8.3 and is equal to $1/g_m$. Thus the *RC* network formed by C_S is R_S in parallel with $1/g_m$, and the corner frequency is:

$$f_c = \frac{1}{2\pi C_S(1/g_m \| R_s)} \tag{10.25}$$

▶ **EXAMPLE 10.23** ─────────────────────────

For the circuit in Figure 10.40 $R_S = 1 \text{ k}\Omega$, $C_S = 22 \text{ µF}$ and $g_m = 3500 \text{ µS}$. Determine the corner frequency assuming that the coupling capacitors act as short circuits.

▶ *Solution*

The output resistance of the FET at the source terminal is:

$$R_o = \frac{1}{g_m} = \frac{1}{3500 \times 10^{-6}} = 285.7 \ \Omega$$

and the corner frequency is:

$$f_c = \frac{1}{2\pi(22 \times 10^{-6})(285.7 \ \Omega \| 1 \text{ k}\Omega)} = 32.5 \text{ Hz}$$

Choice of Capacitor

The designer of a common-source amplifier must make a suitable choice of capacitors for coupling and decoupling. Since the source decoupling capacitor is the largest it is usually sensible to allow this capacitor to determine the overall frequency response. The coupling capacitors are then selected for a frequency that is one-tenth of the frequency used for determining C_S.

▶ **EXAMPLE 10.24** ─────────────────────────

Select coupling and decoupling capacitors for the circuit in Figure 10.40 to provide a 3 dB corner frequency of 20 Hz if $R_s = 4 \text{ k}\Omega$, $R_G = 1 \text{ M}\Omega$, $R_S = 1 \text{ k}\Omega$, $R_D = 6.8 \text{ k}\Omega$, $R_L = 12 \text{ k}\Omega$ and $g_m = 2500 \text{ µS}$.

▶ *Solution*

Let the 3 dB frequency be determined by the decoupling capacitor. Then:

$$C_S = \frac{1}{2\pi(20 \text{ Hz})(1/2500 \times 10^{-6} \| 1 \text{ k}\Omega)} \approx 28 \text{ µF}$$

The values of the coupling capacitors are determined for a frequency of 2 Hz, that is for the input capacitor:

$$C_G = \frac{1}{2\pi(2 \text{ Hz})(4 \text{ k}\Omega + 1 \text{ M}\Omega)} \approx 80 \text{ nF}$$

and for the output capacitor:

$$C_o = \frac{1}{2\pi(2 \text{ Hz})(6.8 \text{ k}\Omega + 12 \text{ k}\Omega)} \approx 4.4 \text{ }\mu\text{F}$$

▷ **EXAMPLE 10.25** ──────────────────────────────

Use PSpice to determine the lower 3 dB frequency for the JFET amplifier shown in Figure 10.44. The power supply is 30 V and the BETA of the transistor is 35×10^{-4}. From **Analysis** select **AC Sweep...** and set the start frequency to 1 Hz and the stop frequency to 100 kHz with 10 points per decade.

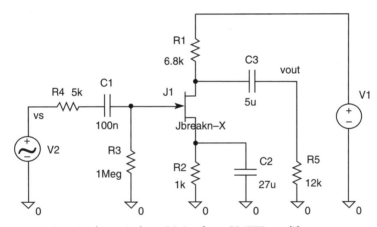

FIGURE 10.44   Circuit schematic from PSpice for a CS JFET amplifier.

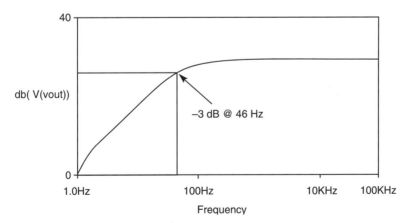

FIGURE 10.45   Frequency response produced by Probe.

▶ *Solution*

The frequency response produced by Probe is shown in Figure 10.45.

From the output file produced by PSpice g_m may be observed as 7000 µS. From equation 10.25 the corner frequency may be calculated as:

$$f_c = \frac{1}{2\pi(27\ \mu\mathrm{F})(1/7000\ \mu\mathrm{S}\ \|\ 1\ \mathrm{k}\Omega)} = 47\ \mathrm{Hz}$$

which corresponds to the value predicted by PSpice.

10.4   High-Frequency Response of FET Amplifiers

The high-frequency performance is affected by the low-pass filters formed by the stray capacitance associated with components, wire or pcb interconnections and in particular the junction capacitors associated with the transistor. Because of the high impedance levels of the JFET, relatively small values of capacitance can have a significant effect on the high-frequency response, particularly for the input low-pass filter. The CS amplifier with the transistor junction capacitors added is shown in Figure 10.46.

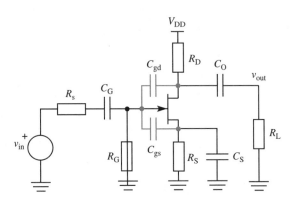

FIGURE 10.46   CS amplifier with gate to source and gate to drain capacitors.

These capacitors are formed by the reverse-biased pn junctions of the gate and source and the gate and drain. As for any reverse-biased pn junction the value of the capacitance is voltage dependent (see Chapter 2), and decreases with increasing reverse bias.

Input RC Network

For the input circuit the coupling capacitor C_G and the source decoupling capacitor C_S may be regarded as short circuits at the frequencies at which the stray capacitance is likely to have an effect. The most significant capacitor for the input low-pass filter is C_{gd}

because of the Miller effect which multiplies the capacitance by the voltage gain. The equivalent small-signal circuit for the input is shown in Figure 10.47.

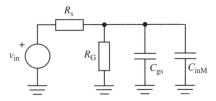

FIGURE 10.47   Small-signal circuit for the input of the CS JFET amplifier.

In the diagram the Miller capacitance C_{inM} is given by:

$$C_{\text{inM}} = C_{\text{gd}}(1 - A_{\text{v}})$$

where A_{v} is the voltage gain $(-g_{\text{m}}R_{\text{D}} \| R_{\text{L}})$. The 3 dB corner frequency for the low-pass input filter is:

$$f_{\text{c}} = \frac{1}{2\pi(R_{\text{s}} \| R_{\text{G}})(C_{\text{gs}} + C_{\text{inM}})} \tag{10.26}$$

Notice that the Thévenin equivalent resistance of R_{s} and R_{G} results in the parallel combination of these two resistors rather than a series combination. Therefore, the corner frequency is likely to be determined by the value of the source resistance, rather than R_{G}.

An example of a circuit which may contain a large source resistance and which, as a result, may be subject to high-frequency limitations is shown in Figure 10.48. Modulated optical radiation is directed onto the reverse-biased photodetector. The variation in the diode saturation current produces a variable voltage across the load resistor. To maximize this voltage the load resistor should be as large as possible and the use of the FET with its very high input impedance minimizes the loss of signal. To detect very small optical signals requires a large value of load resistor, but the larger the load resistor the lower will be the upper 3 dB corner frequency, as given by equation 10.26, and therefore the smaller the bandwidth of the modulated optical signal.

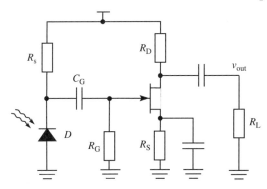

FIGURE 10.48   Photodetector circuit with FET amplifier.

NB This is not the only circuit for detecting optical signals and is only presented here as an example.

Output RC Network

The output low-pass filter comprises the drain resistor, any load resistance and the Miller capacitance $C_{gd}(1 - 1/A_v)$ which is approximately C_{gd} provided that $A_v \gg 1$. Thus the 3 dB corner frequency associated with the output is:

$$f_c = \frac{1}{2\pi C_{gd}(R_D \| R_L)} \tag{10.27}$$

This corner frequency is usually much higher than that for the input low-pass filter, and consequently the 3 dB corner frequency is usually determined by the input low-pass filter.

▶ **EXAMPLE 10.26** ─────────────────────────────────

Determine the upper 3 dB frequency for a CS JFET amplifier if $R_D = 5.6$ kΩ, $R_s = 47$ kΩ, $R_G = 1$ MΩ, $R_L = 5$ kΩ, $g_m = 3000$ μS, $C_{gs} = 4$ pF and $C_{gd} = 2.5$ pF.

▶ *Solution*

The voltage gain is:

$$A_v = -(3000 \times 10^{-6})(5.6 \text{ k}\Omega \| 5 \text{ k}\Omega) \approx -8$$

and the input Miller capacitance is:

$$C_{inM} = (2.5 \text{ pF})(1 + 8) = 22.5 \text{ pF}$$

and the corner frequency for the input from equation 10.26 is:

$$f_c = \frac{1}{2\pi(47 \text{ k}\Omega \| 1 \text{ M}\Omega)(4 \text{ pF} + 22.5 \text{ pF})} \approx 134 \text{ kHz}$$

From equation 10.27 the corner frequency for the output is:

$$f_c = \frac{1}{2\pi(2.5 \text{ pF})(5.6 \text{ k}\Omega \| 5 \text{ k}\Omega)} \approx 24 \text{ MHz}$$

The 3 dB frequency for the combined networks is 134 kHz.

Combined lf and hf Frequency Response

PSpice can be used to observe the combined effects of the coupling and decoupling capacitors on the low-frequency response and the stray capacitance on the high-frequency response of an amplifier.

▷ **EXAMPLE 10.27** ——————————————————————————————

For the circuit shown in Figure 10.44 increase the source resistance to 47 kΩ, and add to the **.model…** information for the transistor the values of the gate to source and gate to drain capacitors of 4 pF and 2.5 pF respectively.

▷ *Solution*

The output from Probe is shown in Figure 10.49.

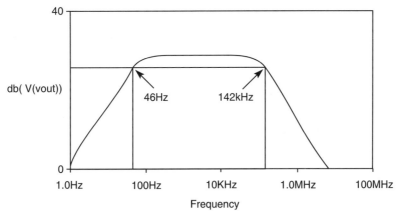

FIGURE 10.49 Combined frequency response from Probe.

The low-frequency 3 dB corner is at 46 Hz as observed in Example 10.25 and is determined by the source decoupling capacitor. The actual value of g_m for the transistor obtained from the PSpice **output file** is 7000 μS and hence the gain is:

$$A_v = -(7000 \times 10^{-6})(6.87 \text{ k}\Omega \,\|\, 12 \text{ k}\Omega) = -30 = -29.5 \text{ dB}$$

The values for the gate to source and gate to drain capacitors also obtained from the PSpice **output file** are:

$$C_{gs} \approx 2 \text{ pF and } C_{gd} \approx 0.75 \text{ pF}$$

These values are smaller than those entered into the **.model…** statement as a result of the voltage dependence of the junction capacitance.

The Miller capacitance for the input is:

$$C_{inM} = (0.75 \text{ pF})(1 + 30) \approx 23 \text{ pF}$$

The 3 dB frequency formed by the input low-pass filter comprising the source resistor, the gate resistor, the gate to source capacitance and the Miller capacitance is:

$$f_c = \frac{1}{2\pi(47 \text{ k}\Omega \,\|\, 1 \text{ M}\Omega)(2 \text{ pF} + 23 \text{ pF})} \approx 142 \text{ kHz}$$

This value corresponds well with that obtained from Probe in Figure 10.49.

It is worth noting that the upper corner frequency is not very high, and this is an example of the problem in using FETs for high-frequency applications. The low-pass filters formed by stray capacitance, and particularly the Miller capacitance, are a major limitation. A common solution is the FET–BJT *cascode*, as shown in Figure 8.26. With a cascoded stage the load which is presented to the FET is the input resistance of a CB stage r_e, and consequently the gain is small. The gain of the FET is $-g_m r_e$.

In Example 10.27 the drain current is approximately 3 mA, which when reproduced in a cascode is equal to the collector current. Thus r_e is $0.026/3$ mA $\approx 9\ \Omega$. Using the value of g_m observed in the PSpice output file of 7000 μS, then the gain of the FET is:

$$A_v = -(7000\ \mu S)(9\ \Omega) = -0.063$$

and the Miller capacitance, again using the actual value for C_{gd} from the PSpice output file of 0.75 pF, is:

$$C_{inM} = (0.75\ \text{pF})(1 + 0.063) \approx 0.8\ \text{pF}$$

and

$$f_c = \frac{1}{2\pi(47\ \text{k}\Omega \,\|\, 1\ \text{M}\Omega)(2\ \text{pF} + 0.8\ \text{pF})} \approx 1.26\ \text{MHz}$$

Again the actual value for C_{gs} of 2 pF is used, as obtained from the PSpice file. There is a considerable increase in the upper corner frequency. The corner frequency for the output low-pass filter does not change.

▶ **EXAMPLE 10.28** ─────────────────────────────────────

Use PSpice to examine the frequency response of the cascode circuit shown in Figure 8.26 and shown in Figure 10.50 with suitably modified resistor values to

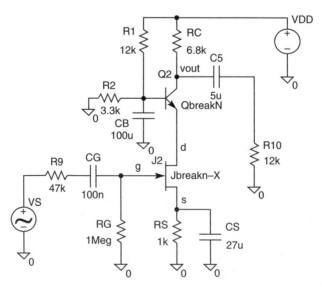

FIGURE 10.50   PSpice schematic for JFET cascode.

ensure a collector/drain current of approximately 3 mA. In the **.model...** information for the generic JFET add the two capacitors cgs = 4 pF and cgd = 2.5 pF and vto = −4.5 V, BETA = 35e-4.

▷ *Solution*

The output from Probe is shown in Figure 10.51.

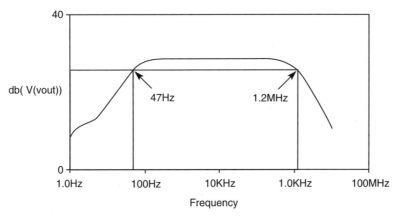

FIGURE 10.51   Probe output for the JFET cascode.

The observed results are in good agreement with the calculated values. The overall gain is also similar at 30 dB as the circuit described in Example 10.27.

▷ **PROBLEMS** ————————————————————————

10.1   For the circuit shown $r_e = 18\ \Omega$ and $\beta = 120$. Determine the lower corner frequency.

[148 Hz]

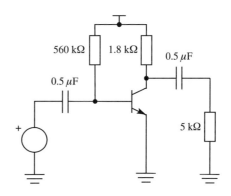

10.2 For the circuit shown $r_e = 13 \ \Omega$ and $\beta = 150$. Determine the lower 3 dB frequency.

[432 Hz]

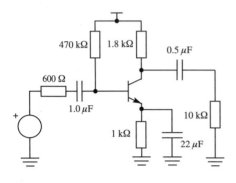

10.3 For the circuit shown $r_e = 11 \ \Omega$ and $\beta = 120$. Determine the three corner frequencies and identify the 3 dB frequency.

[30 Hz, 12 Hz, 398 Hz]

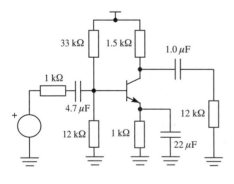

10.4 For the circuit shown in Problem 10.2, determine suitable values for the three capacitors to provide a lower 3 dB corner frequency of 100 Hz.

$[C_E = 95 \ \mu\text{F}, \ C_B = 6 \ \mu\text{F}, \ C_O = 1.3 \ \mu\text{F}]$

10.5 For the circuit shown in Problem 10.3, determine suitable values for the three capacitors to provide a 3 dB corner frequency of 200 Hz.

$[C_E = 44 \ \mu\text{F}, \ C_B \approx 4 \ \mu\text{F}, \ C_O = 0.6 \ \mu\text{F}]$

10.6 For the circuit shown $r_e = 18.6 \ \Omega$ and $\beta = 100$. Determine the lower 3 dB frequency.

[144 Hz]

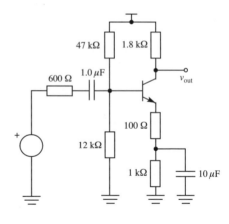

10.7 For the circuit in Problem 10.6 select suitable values for the capacitors to provide a lower 3 dB frequency of not less than 70 Hz.

$[C_E = 20 \ \mu\text{F}, \ C_B \approx 4 \ \mu\text{F}]$

10.8 For the circuit shown $r_e = 10.4$ Ω and $\beta = 180$. Determine the corner frequency for each capacitor.

[120 Hz, 58 Hz, 5 Hz]

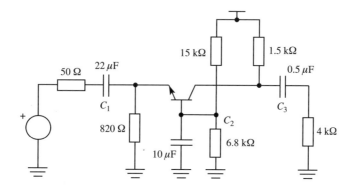

10.9 For the circuit shown in Problem 10.8 select suitable values for the capacitors to provide a corner frequency of 50 Hz.

$[C_1 = 53$ µF, $C_2 = 10$ µF, $C_3 = 6$ µF]

10.10 For the circuit shown $r_e = 7.4$ Ω and $\beta = 150$. Determine suitable values for the two capacitors to provide a 3 dB corner frequency of 20 Hz.

$[C_B = 8$ µF, $C_O = 113$ µF]

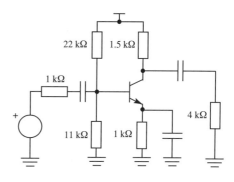

10.11 For the circuit shown $r_e = 12.4$ Ω, $\beta = 100$, $C_{bc} = 4$ pF, $C_{be} = 20$ pF and $C_{ce} = 6$ pF. Determine the upper corner frequencies.

[822 kHz, 14.6 MHz]

10.12 Determine the corner frequency for the circuit shown in Problem 10.11 if the cut-off frequency (f_T) for the transistor is 10 MHz and sketch the gain versus frequency Bode plot showing clearly the rate of roll-off beyond the 3 dB frequency.

[100 kHz]

10.13 For the CB circuit shown determine the upper 3 dB corner frequency if $I_C = 2$ mA, $C_{jc} = 1.5$ pF, $C_{je} = 40$ pF.

[45 MHz]

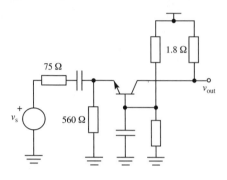

10.14 Use PSpice to simulate the circuit shown in Figure 10.33 with the 2N2222 transistor replaced with the 2N3904. Use a dc supply of 9 V. Examine the output file to obtain the junction capacitance and the collector current, and compare the theoretical corner frequency with that observed with Probe.

[calculated $f_c \approx 100$ MHz]

10.15 For the emitter follower shown $I_C = 1$ mA, $C_{jc} = 2$ pF, $C_{je} = 20$ pF and $\beta = 100$. Determine the upper 3 dB corner frequency.

[13 MHz]

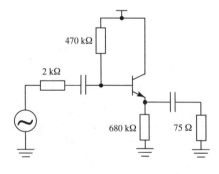

10.16 If $g_m = 3000$ μS

(a) determine the mid-band gain
(b) determine the three corner frequencies which result form the two coupling capacitors and the decoupling capacitor
(c) sketch the Bode plot.

[(a) 21 dB, (b) 1.6 Hz, 10 Hz, 64 Hz]

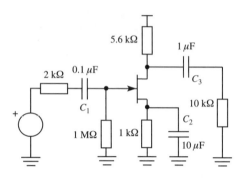

10.17 For the circuit shown $g_m = 4500$ μS. Determine suitable values for the three capacitors to provide a low-frequency 3 dB point of 100 Hz.

[$C_1 = 16$ nF, $C_2 = 9$ μF, $C_3 = 0.9$ μF]

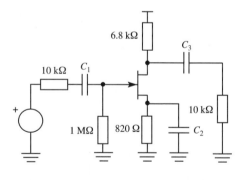

10.18 For the circuit shown in Problem 10.17 determine the upper 3 dB frequency and sketch the Bode plot if $C_{gs} = 3$ pF, $C_{gd} = 1$ pF and $C_{ds} = 1.5$ pF.

[724 kHz]

10.19   The circuit shown is a photodetector circuit which can convert a modulated light signal into an electrical signal. Determine the maximum value of the diode load resistor which would allow the circuit to amplify an optical signal with a modulation frequency of 500 kHz, if $g_m = 3500$ µS, $C_{gs} = 3.5$ pF, $C_{gd} = 1.5$ pF, $C_{ds} = 2$ pF and the junction capacitance of the diode is 2 pF.

[~10 kΩ]

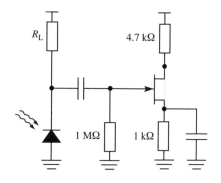

CHAPTER 11

Differential and Operational Amplifiers

The *single-ended* amplifiers considered so far have involved one signal source followed by one or more transistor gain stages which produces an output which is proportional to the input. The *differential amplifier* is an arrangement of transistors which allows the difference between *two* signal sources to be amplified and the output is proportional to the difference between these two inputs. The differential amplifier is used extensively in integrated circuit operational amplifiers. This chapter deals with the differential amplifier and introduces the operational amplifier. The operational amplifier is considered further in Chapter 13.

11.1 Differential or Difference Amplifier

A simple symbolic representation of the differential or difference amplifier is shown in Figure 11.1.

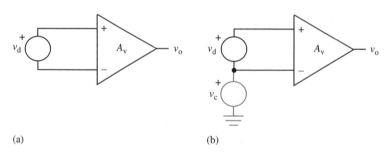

(a)   (b)

FIGURE 11.1   Simple difference amplifier with (a) one signal source applied between the inputs and (b) an additional common-mode signal source common to each input.

In Figure 11.1(a) a single source is connected between the two inputs and the output may be expressed as:

$$v_o = v_d A_v \tag{11.1}$$

Figure 11.1(a) could have been provided with two separate sources, v_1 and v_2, connected to each of the inputs and ground. The actual input would then equal the difference between these two sources, that is $v_d = v_1 - v_2$, and this provides the reason for the term *difference amplifier*.

It is also possible to connect another source (v_c) which is common to both inputs.

NB that when two voltage sources are present in a circuit, the effect of one is determined by short-circuiting the other. Thus in considering the effect of v_c the other source is replaced by a short circuit, which means that v_c becomes common to both inputs.

This *common-mode* signal also produces an output which may be expressed as:

$$v_o = v_c A_c \tag{11.2}$$

Notice that this is not possible with the amplifiers which only have a single input.

A common example of how a differential amplifier has a clear advantage over a single-ended amplifier is in the amplification of the signal from a strain gauge. A strain gauge is a transducer which produces a voltage output when it is subjected to mechanical force. It often takes the form of four thin-film resistors connected together to form a Wheatstone bridge. In use the transducer is bonded to a mechanical structure which is being examined and any deflection of the structure causes the resistors to be subjected to a force which changes their value. Two corners of the Wheatstone bridge are connected to the inputs of the differential amplifier, with the dc supply and ground being connected to the other two corners. The amplifier may be some distance from the transducers, and as a result the wires connecting the bridge to the amplifier are likely to be subjected to electromagnetic interference, particularly from the 50 Hz (or 60 Hz) line voltage. Both inputs are subjected to the same interference. This is the situation which is represented in Figure 11.1(b), where the difference-mode signal from the strain gauge bridge is represented by v_d and the signal from the interference is represented by the common-mode signal source v_c. The output voltage is obtained by considering each input separately and then applying superposition.

The output voltage with v_c short-circuited is:

$$v_o = v_d A_d$$

where A_d is the difference-mode gain. The output with v_d short-circuited is:

$$v_o = v_c A_c$$

where A_c is the common-mode gain. The combined output is:

$$v_o = v_d A_d + v_c A_c \tag{11.3}$$

The required output is $v_d A_d$ and the component $v_c A_c$ is an unwanted source of interference. Rearranging equation 11.3 it is possible to identify how the source of interference, or error, can be minimized. Thus:

$$v_o = v_d A_d \left(1 + \frac{v_c/v_d}{\text{CMRR}} \right) \tag{11.4}$$

where $\text{CMRR} = A_d/A_c$ and is the *common-mode rejection ratio*. It is apparent from equation 11.4 that if $v_d A_d$ is considered to be the required signal then the second term in parentheses represents an error signal. This error is minimized if CMRR is very large. Since CMRR is the ratio of the difference-mode gain to the common-mode gain, then in order to minimize the effect of the common-mode signal it is necessary for $A_d \gg A_c$.

▷ **EXAMPLE 11.1** ————————————————————————————

Consider a strain gauge amplifier in which $v_d = 1$ mV and the common-mode interference is $v_c = 1$ V. If the difference-mode gain $A_d = 2500$, and CMRR = 30000, determine the output due to the difference-mode signal and the error due to the interference.

▷ *Solution*

The schematic of the arrangement for the measurement of gain is shown in Figure 11.2.

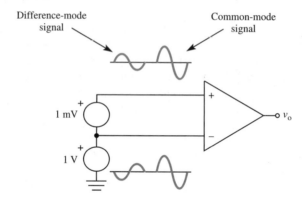

FIGURE 11.2 Instrumentation setup with difference-mode and common-mode sources.

If one assumes that the signals are sinusoidal then the phase relationships and amplitudes are as shown in Figure 11.2. Notice that the difference-mode signals are in anti-phase, while the common-mode signals are in phase.

By superposition the output voltage is:

$$v_o = (1 \text{ mV})(2500)\left(1 + \frac{1 \text{ V}/1 \text{ mV}}{30000}\right)$$

$$= 2.5 \text{ V} + 0.083 \text{ V}$$

That is, the required signal is 2.5 V, while the error caused by the common-mode signal is 83 mV.

NB that the difference-mode signal does not have to be a sine wave, and is only used here to illustrate the phase relationships between the different signals. The difference signal could equally well be a dc voltage which would change polarity or magnitude as strain is applied.

11.2 ▷ The Emitter-Couple for the Amplifier

The symbols used in Figures 11.1 and 11.2 are those associated with an operational amplifier which has two inputs and a single output. A simplified discrete component version of the difference amplifier is the emitter-coupled transistor pair shown in Figure 11.3.

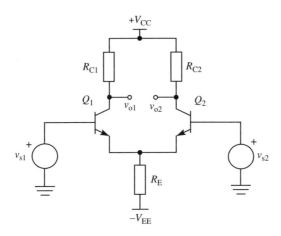

FIGURE 11.3 Circuit schematic of a differential amplifier.

Each transistor is in fact a common-emitter amplifier with a signal applied to the base, and an output created at the collector. However, the fact that both emitters are connected together through a single resistor R_E means that the base–emitter voltages of the two transistors interact and it is this interaction which establishes the unique properties of the differential amplifier.

The analysis can be greatly simplified by restricting it to those situations which are likely to exist in practice. These involve single-ended mode of operation where the signal is applied to only one of the inputs, with the other input grounded. Alternatively the signal is applied between the two inputs so that each input is in anti-phase to the other, that is when one input is moving in a positive direction the other is moving in a negative direction; this is the differential-input mode.

Single-Ended Input Mode

In this mode the signal is only applied to one of the inputs as shown in Figure 11.4. A sine wave is applied to the base of Q_1, while the base of Q_2 is grounded. Transistor Q_1 acts as a common-emitter amplifier with an inverted output at its collector (v_{o1}). At the same time an in-phase voltage appears across R_E by emitter follower action. Usually the load resistance of an emitter follower is much larger than the internal resistance of the transistor (r_e) and the voltage gain is unity (see Section 6.5), but the load

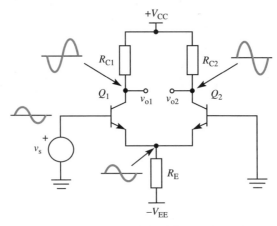

FIGURE 11.4   Single-ended input operation of the differential amplifier.

presented to the emitter of Q_1 is the input resistance of Q_2. With the base of Q_2 grounded this transistor acts as a common-base stage with an input resistance equal to r_e. The gain from the base of Q_1 to the emitter of Q_1 is (from equation 6.26):

$$A = \frac{v_e}{v_s} = \frac{R_E \| r_e}{r_e + R_E \| r_e}$$

$$\approx \frac{r_e}{r_e + r_e} = 0.5 \quad \text{since } r_e \ll R_E$$

$$v_e = 0.5 \times v_s \tag{11.5}$$

where v_e is the voltage at the emitter and v_s is the input voltage at the base of Q_1.

The voltage applied to the emitter of Q_2 is $0.5v_s$. With Q_2 acting as a common-base stage this signal is amplified and appears non-inverted (refer to Section 6.3) at its collector (v_{o2}). Because of the symmetry of the circuit the same conditions apply when the signal is applied to Q_2 and the base of Q_1 is grounded.

The gain of the circuit may be obtained by considering each transistor in turn and using superposition to determine the final output. With a signal applied to the base of Q_1 this transistor acts as a CE amplifier with a resistor R_E in series with the emitter. In addition Q_2 acts as a CB stage with its emitter–base junction appearing directly across R_E. Thus when considering the gain of Q_1 the effect of Q_2 may be included by replacing Q_2 with its equivalent circuit, as shown in Figure 11.5.

The input resistance of the CB stage is simply r_e and since this is $\ll R_E$ the gain of the CE amplifier stage based on Q_1 is:

$$\frac{v_{o1}}{v_s} = -\frac{R_{C1}}{(r_{e1} + r_{e2})}$$

and since the transistors are identical, and both are supplied by the same power supply

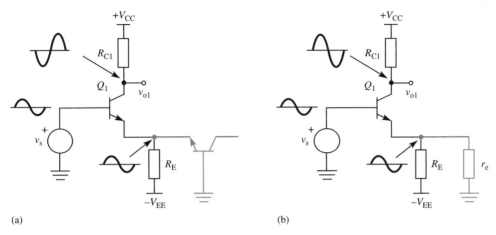

FIGURE 11.5   Simplified circuit to use when considering the gain of Q_1 with (a) showing Q_2 as a CB amplifier and (b) with Q_2 replaced by its equivalent input resistance.

V_{EE} which results in identical emitter currents, then $r_{e1} = r_{e2} = r_e$ and:

$$\frac{v_{o1}}{v_s} = -\frac{R_{C1}}{2r_e} \tag{11.6}$$

The gain of Q_2, which acts as a CB stage is simply:

$$\frac{v_{o2}}{v_e} = \frac{R_{C2}}{r_e}$$

and from equation 11.5 $v_e = 0.5v_s$. Then

$$\frac{v_{o2}}{v_s} = -\frac{R_{C2}}{2r_e} \tag{11.7}$$

The output between the two collectors is:

$$v_{o1} - v_{o2} = -\frac{v_s R_{C1}}{2r_e} - \frac{v_s R_{C2}}{2r_e}$$

$$= -\frac{v_s R_C}{r_e} \quad \text{if } R_{C1} = R_{C2} = R_C$$

The more usual form of this equation is:

$$\frac{v_{o1-2}}{v_{in}} = -\frac{R_C}{r_e} \tag{11.8}$$

An important point to note is that the emitter resistor (R_E) does not appear in this equation. This has important consequences for the design of integrated circuit operational amplifiers, because unlike the single-ended CE amplifier a capacitor is not required to decouple the R_E of an emitter-coupled pair in order to achieve large values of gain.

If the output is only taken from one of the outputs, for example the collector of Q_2, then from equation 11.7 with $v_o = v_{o2}$:

$$\frac{v_{o1}}{v_{in}} = -\frac{v_{o2}}{v_{in}} = -\frac{R_C}{2r_e} \tag{11.9}$$

and the gain is reduced by a factor of 2.

Equations 11.8 and 11.9 describe the gain for a single-input version of the differential amplifier. For the differential-input mode the signal is applied between both inputs.

Differential-Input Mode

The differential mode is illustrated in Figure 11.6.

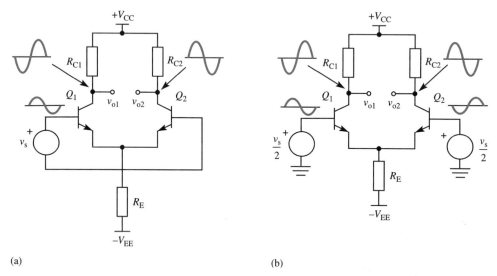

(a)   (b)

FIGURE 11.6   (a) Differential amplifier with differential input and (b) amplifier with individual inputs equal to half of the single input.

In Figure 11.6(a) a single sine wave source is connected between the two inputs. This means that when the voltage on the base of Q_1 is rising the voltage on the base of Q_2 is falling, and vice versa. In order to simplify the analysis the single source is split into two, each equal to half of the original source ($v_s/2$), as shown in Figure 11.6(b). Notice the phase relationship between the two input waveforms. With the two transistors acting as CE amplifiers the two collector waveforms are inverted with respect to their respective inputs, and are out of phase with each other.

The analysis follows the same procedure as for the single-input mode of operation except that instead of an input of v_s the input is now $v_s/2$. Each source is considered separately and the final output obtained using superposition.

The two stages in the analysis are illustrated in Figure 11.7 where in 11.7(a) the source is applied to Q_1, while in 11.7(b) an anti-phase source is applied to Q_2. Notice that the output at each collector is the same for both cases: for 11.7(a) there is a 180° phase shift between the collector of Q_1 and its input, and for 11.7(b) there is a similar phase shift between its collector and its input. Because of the 180° phase shift between the two inputs the resultant collector waveforms are the same for each case.

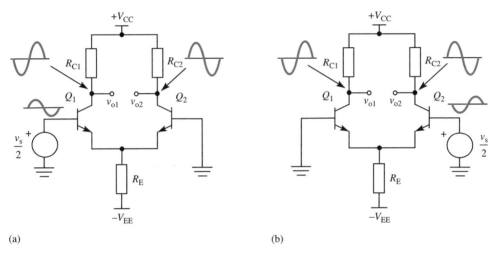

(a)   (b)

FIGURE 11.7   The differential amplifier shown with (a) a source applied to Q_1 and (b) a source with the opposite phase applied to Q_2

From equation 11.7 the output at the collector of Q_1 with the base of Q_2 grounded is:

$$v'_{o1} = -v_{s1} \frac{R_{C1}}{2r_e} \tag{11.10}$$

where $v_{s1} = v_s/2$.

The output at the collector of Q_2 is:

$$v'_{o2} = v_{s1} \frac{R_{C2}}{2r_e} \tag{11.11}$$

An identical pair of expressions are obtained when the base of Q_1 is grounded and the signal attached to Q_2 and are:

$$v''_{o1} = v_{s2} \frac{R_{C1}}{2r_e} \tag{11.12}$$

where $v_{s2} = v_s/2$ and is $180°$ out of phase with v_{s1}, that is:

$$v''_{o1} = -v_{s1} \frac{R_{C1}}{2r_e} \tag{11.13}$$

Similarly for the collector of Q_2:

$$v''_{o2} = v_{s1} \frac{R_{C2}}{2r_e} \tag{11.14}$$

The signs represent the phase changes with respect to each input. Remember that in each case one transistor acts as a CE amplifier with a $180°$ phase shift while the other acts as a CB amplifier with no phase shift.

The total output voltage at each collector is obtained by superposition as:

$$v'_{o1} + v''_{o1} = -v_{s1} \frac{R_{C1}}{r_e}$$

$$v_{o1} = -\frac{v_s}{2} \frac{R_{C1}}{r_e}$$

and for Q_2:

$$v'_{o2} + v''_{o2} = -v_{s2} \frac{R_{C2}}{r_e}$$

$$v_{o2} = -\frac{v_s}{2} \frac{R_{C2}}{r_e}$$

The combined signal from between the two collectors is:

$$v_{o1} - v_{o2} = -\left(\frac{v_s}{2} + \frac{v_s}{2}\right)\left(\frac{R_{C1}}{r_e} + \frac{R_{C2}}{r_e}\right) \tag{11.15}$$

and since $R_{C1} = R_{C2} = R_C$ then:

$$\frac{v_{o1-2}}{v_{in}} = -\frac{R_C}{r_e} \tag{11.16}$$

If the output is taken from only one of the collectors then:

$$\frac{v_{o1}}{v_{in}} = -\frac{v_{o2}}{v_{in}} = -\frac{R_C}{2r_e} \tag{11.17}$$

The different configurations are summarized in Figure 11.8. Notice the phase relationships which exist between the inputs and their respective outputs. Notice also that the differential input v_d is equivalent to two signal sources, v_{in1} and v_{in2}, which are each equal to $v_d/2$ and are in anti-phase to each other. Notice that the amplitude of the signals at each output for the differential mode, Figure 11.8(d), is twice that for the single-input situations in 11.8(a) and 11.8(b), provided that $|v_{in1}| = |v_{in2}| = |v_d|$.

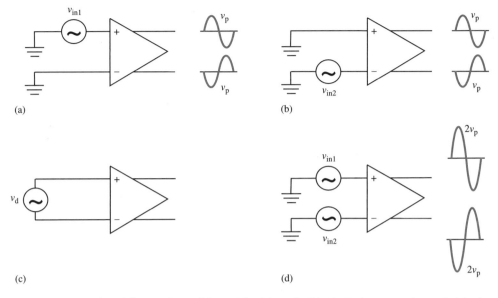

FIGURE 11.8 The differential amplifier with (a) and (b) single-input mode and (c) the differential mode with (d) its equivalent two-generator representation.

▶ **EXAMPLE 11.2** _____

For the differential amplifier shown in Figure 11.9 determine

(a) the value of R_E to provide an emitter current I_E of 1 mA
(b) the dc output voltages V_{o1} and V_{o2}
(c) the double-ended output voltage gain v_{o1-2}/v_{in}
(d) the single-ended output voltage gain v_{o2}/v_{in}.

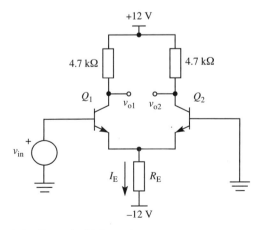

FIGURE 11.9 Circuit for Example 11.2.

▶ *Solution*

(a) Applying KVL to the base–emitter circuit of Q_1 gives:

$$-12 \text{ V} = (1 \text{ mA})R_E + 0.7 \text{ V}$$

and

$$R_E = 11.3 \text{ k}\Omega$$

The emitter resistance of each transistor is:

$$r_e = \frac{0.026}{0.5 \text{ mA}} = 52 \ \Omega$$

(b) The current flowing in R_E divides equally between the two transistors, assuming that they are identical. Thus $I_{C1} = I_{C2} = 0.5$ mA, and applying KVL to the collector circuit gives:

$$12 \text{ V} = (0.5 \text{ mA})(4.7 \text{ k}\Omega) + v_{o1}$$

and

$$v_{o1} = 9.65 \text{ V}$$

(c) The double-ended voltage gain is:

$$\frac{v_{o1-2}}{v_{in}} = -\frac{4.7 \text{ k}\Omega}{52} = -90.4$$

(d) The single-ended voltage gain is:

$$\frac{v_{o2}}{v_{in}} = \frac{4.7 \text{ k}\Omega}{2 \times 52} = 45.2$$

▶ **EXAMPLE 11.3** ─────────────────────────────────

Use PSpice, with a value of $R_E = 11.3$ kΩ, to verify the values obtained in Example 11.2. Select a 2N2222 from the **eval.lib**. From **Analysis** and **setup...** select **Transient**. For a 1 kHz input sine wave set the **Print step** to 10 µs and the **Final time** to 2 ms.

▶ *Solution*

The schematic used for PSpice is shown in Figure 11.10. The sine wave input signal has an amplitude of 1 mV and a frequency of 1 kHz. The dc voltage sources have dc values of 12 V. Notice that it is convenient to name the outputs (out1 and out2) in order to identify them more clearly when using Probe, or examining the text output file.

The sine wave output may be examined with Probe (Figure 11.11), and the dc values at the collectors may be examined by selecting **Examine output** from **Analysis**.

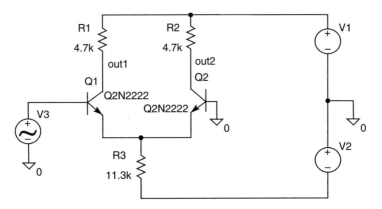

FIGURE 11.10   Schematic from PSpice for Example 11.3.

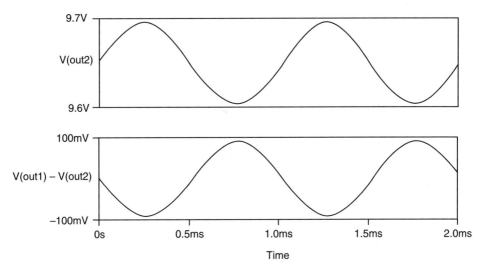

FIGURE 11.11   Single-ended and double-ended outputs from Probe.

For part (b) of example 11.2 the dc voltage at the collectors is:

$v_{out1} = v_{out2} = 9.65$ V

For part (c) the double-ended voltage gain as obtained from Probe is:

$$A_{v1-2} = \frac{175.4 \text{ mV}}{2 \text{ mV}} = 87.7$$

For part (d) the single-ended voltage gain is:

$$A_{v2} = \frac{87.9 \text{ mV}}{2 \text{ mV}} = 43.9$$

These values are in close agreement with those values obtained by calculation in Example 11.2.

Common-Mode Gain

An important feature of the differential amplifier is its ability to reject common-mode signals. The common-mode signal results if the same signal, both amplitude and phase, is applied to the two inputs, as shown in Figure 11.12.

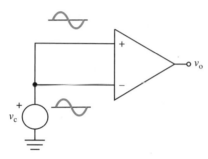

FIGURE 11.12   Amplifier with common-mode signal.

For the ideal differential amplifier the output voltage is zero, because when applying superposition to the output for each of the inputs, the resultant output voltages at each collector are in anti-phase to each other and cancel. This would appear to imply that the common-mode gain (A_c) is zero, and that the CMRR is infinite. In practice the transistors which form the differential amplifier are not identical and the small differences in their parameters result in an output.

Even with perfectly matched components and, therefore, zero output, each transistor responds to the input voltage applied to its base, and consequently there is an output at each collector. The effective output resulting from superposition is zero, but each transistor exhibits common-mode gain. The analysis is based on the circuit shown in Figure 11.13(a).

Notice that the signals applied to each input are in phase with each other. This means that with a positive-going input signal the emitter current in both transistors increases, while when the inputs decrease, the emitter currents decrease. There is, therefore, a rising and falling voltage across R_E with both emitters at the same voltage level because of the common connection. This is in contrast to the case for the difference amplifier with a source connected between the two inputs when each input moves in opposite directions.

In Figure 11.13(b) the emitter resistor has been replaced with two resistors each with a value of $2R_E$. The parallel value remains the same so that the dc conditions remain unchanged. However, the small-signal circuit can now be simplified because the voltage at the two emitters, and therefore at each end of the resistors, remains the same.

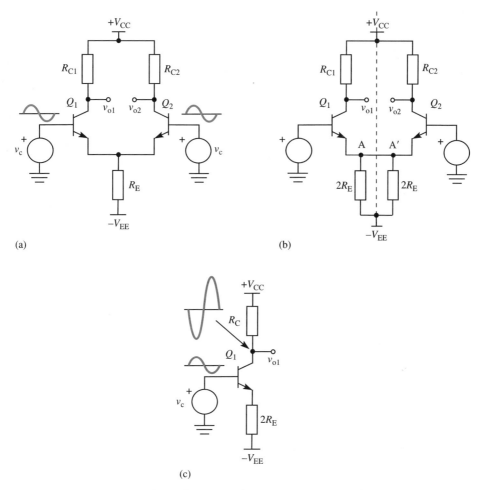

(a)

(b)

(c)

FIGURE 11.13   Circuit used to determine the common-mode voltage gain with (a) the differential amplifier, (b) the symmetry in the circuit and (c) one-half of the circuit.

Therefore, no current flows in the wire AA'. This connection can, therefore, be removed without affecting the small-signal conditions of each half of the circuit. The gain of half of the circuit, as shown in Figure 11.13(c), is:

$$\frac{v_{o1}}{v_c} = -\frac{R_C}{r_e + 2R_E} \approx -\frac{R_C}{2R_E} \tag{11.18}$$

This equation represents the gain of a CE amplifier with a resistor in series with the emitter. It is usually possible to apply the approximation that $R_E \gg r_e$ as shown. To provide a large value of CMRR the common-mode gain should be as small as possible, that is R_E should be large.

▶ **EXAMPLE 11.4** ————————————————————————————————————

For the circuit shown in Figure 11.14 determine the difference-mode gain for a single-ended output, the common-mode gain and the CMRR.

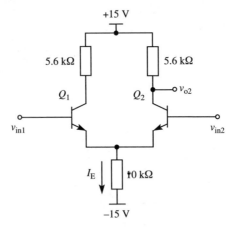

FIGURE 11.14   Circuit schematic for Example 11.4.

▶ *Solution*

The current I_E is:

$$I_E = \frac{15\ V - 0.7\ V}{10\ k\Omega} = 1.43\ mA$$

The current in each transistor is 0.71 mA and the internal emitter resistance is:

$$r_e = \frac{0.026}{0.71\ mA} = 36.6\ \Omega$$

The difference-mode gain for a single-ended output is:

$$A_d = \frac{5.6\ k\Omega}{2 \times 36.6\ \Omega} = 76.5$$

and the common-mode gain is:

$$A_c = -\frac{5.6\ k\Omega}{36.6\ \Omega + (2 \times 10\ k\Omega)} = -0.28$$

The common mode rejection ratio is:

$$CMRR = \frac{A_d}{A_c} = \frac{76.5}{0.28} = 48.7\ dB$$

The most important parameter for the differential amplifier is the CMRR, because the ability to discriminate between the required signal and the common-mode interference depends on its value. For good discrimination it should be large, typically >80 dB. This requires a large value of difference-mode gain (A_d) and a small value of common-mode gain (A_c). It can be seen from equation 11.18 for the common-mode gain that R_E needs to be large for A_c to be small. For the difference-mode gain the value of R_c needs to be large to obtain large values of A_d. Thus to achieve the required conditions for large CMRR it is necessary to use large-value resistors in both the emitter and collector circuits. One of the difficulties in using large value resistors is that the dc voltage drops caused by the collector and emitter currents are large. These problems can be overcome by using *current sources* and *active loads*.

▶ **EXAMPLE 11.5** ————————————————————————————————

Use PSpice with a 2N2222 transistor to measure the difference-mode and common-mode gains of the amplifier shown in Figure 11.14. Use transient analysis with a 1 kHz sine wave having an amplitude of 1 mV.

▶ *Solution*

The schematic for the measurement of the difference mode is shown in Figure 11.15.

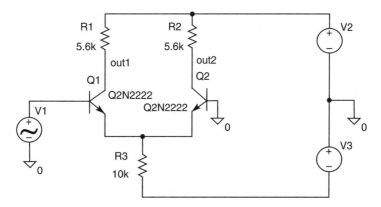

FIGURE 11.15   Schematic for measurement of the difference-mode gain.

The difference-mode gain is measured as:

$$A_{vd} = \frac{146\,\text{mV}}{2\,\text{mV}} = 73$$

The schematic for the measurement of the common-mode gain is shown in Figure 11.16.

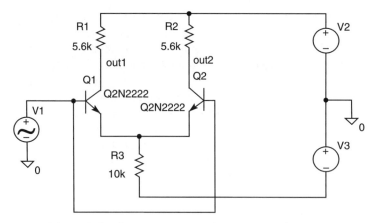

FIGURE 11.16   Schematic for the measurement of common-mode gain.

The common-mode gain is measured as:

$$A_{vc} = \frac{549\ \mu V}{2\ mV} = 0.27$$

The CMRR is:

$$CMRR = \frac{73}{0.27} = 270 = 48.6\ dB$$

11.4 ▷ Current Mirror

The first step to developing a more useful difference amplifier is to replace the emitter resistor with a current source. The main purpose of the emitter resistor (R_E) is to establish the emitter current for the two transistors, but this can be achieved more precisely, and without the dc voltage drop associated with resistors, by using a current source. Current sources are particularly important for integrated circuits where large-value resistors are difficult to manufacture, and where transistors are much easier to manufacture than resistors.

The current source which is used extensively in integrated circuits is the current mirror. It is based on the collector current of a transistor, and the fact that the current remains substantially constant for large variations of the collector to emitter voltage, as shown in Figure 11.17.

The characteristics shown in Figure 11.17 are copied from Chapter 4, and were obtained by means of the dc sweep mode within PSpice for a generic transistor. They show that for a large variation of collector to emitter voltage the collector current remains substantially constant. The value of the current is established by means of the base current, and the base current is established by means of V_{BE}.

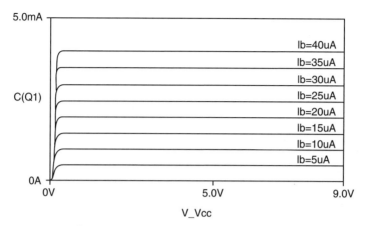

FIGURE 11.17   Output characteristics of a transistor which are utilized in a constant current source.

In the current mirror two transistors are used: one acts as the constant current source while the other establishes the correct value of V_{BE}. The circuit is shown in Figure 11.18.

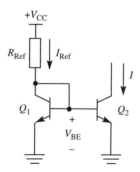

FIGURE 11.18   Current mirror.

The collector and base of Q_1 are shorted together so that they effectively operate as a diode. The relationship between the current flowing in the diode and the voltage across it is given by the diode equation as:

$$I_F = I_S A_1 \exp\left(\frac{V_{BE}}{V_T}\right) \tag{11.19}$$

where I_S is the saturation current which is determined by the properties of the semiconductor, A_1 is the area of the emitter of Q_1 and V_T is the thermal voltage (0.026 V at 300 K). Establishing the current through the diode establishes the value of V_{BE}, and if the transistor Q_2 is made from identical semiconductor material then establishing the V_{BE} for Q_1 also establishes the V_{BE} for Q_2 which determines the current (I) which flows in Q_2.

The current in Q_1 is obtained by applying KVL as:

$$V_{CC} = I_{Ref}R_{Ref} + V_{BE}$$

and

$$I_{Ref} \approx I_F = \frac{V_{CC} - V_{BE}}{R_{Ref}} \qquad (11.20)$$

The relationship between I_{Ref} and I_F is approximate because it ignores the base current of Q_2 which flows in R_{Ref}. However, provided the emitter gain (β) is large this approximation is valid for most applications. In making the above calculation a value of between 0.6 and 0.7 V is assumed for V_{BE}.

Provided that the two transistors are made from the same material, are at the same temperature and are of the same area, then the currents flowing in each transistor are the same. A more general relationship involves the area, with the current in each transistor being proportional to the area of the emitter. That is:

$$I = I_{Ref}\frac{A_2}{A_1} \qquad (11.21)$$

This is an important relationship for integrated circuits because it provides a means for scaling the constant current sources by scaling the areas, as illustrated in Figure 11.19.

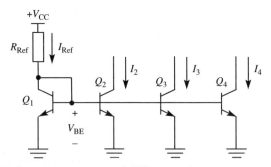

FIGURE 11.19   Multiple current sources with different areas.

If the areas of Q_2, Q_3 and Q_4 are scaled 2, 4 and 8, then the currents I_2, I_3 and I_4 are also scaled 2, 4 and 8. Current sources such as this find uses in digital-to-analog converters.

▶ **EXAMPLE 11.6** ————————————————————————

For the circuit shown in Figure 11.20 the area of Q_4 is half the area of Q_3.

Determine

(a) the current in Q_4
(b) the difference-mode gain.

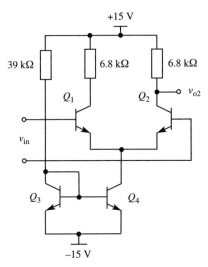

FIGURE 11.20 Differential amplifier with current source.

▷ *Solution*

(a) The current in Q_3 is:

$$I_{Ref} = \frac{15\ V + 15\ V - 0.7\ V}{39\ k\Omega} = 750\ \mu A$$

and the current in Q_4 is:

$$I_4 = 750\ \mu A\ \frac{0.5A_3}{A_3} = 375\ \mu A$$

(b) The current provided by the constant current source divides equally between Q_1 and Q_2 so that:

$$r_{e1} = r_{e2} = \frac{0.026}{187.5\ \mu A} = 139\ \Omega$$

With a differential input and single-ended output the difference gain is obtained from equation 11.16:

$$A_d = -\frac{6.8\ k\Omega}{2 \times 139\ \Omega} = -24.5$$

CMRR with a Current Mirror

An ideal current source has infinite resistance and based on equation 11.18 the common-mode gain would be zero, and the CMRR would be infinite. In practice the current source which uses a transistor is not ideal. In Chapter 4 it was seen that the non-ideal properties of the output characteristics of a transistor can be described in terms of the Early voltage as illustrated in Figure 11.21, which is reproduced from Chapter 4.

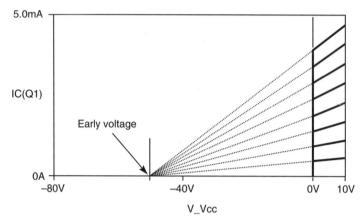

FIGURE 11.21   Finite output resistance described in terms of the Early voltage.

The horizontal lines of the output characteristics shown in Figure 11.17, which would suggest that the current remains constant when V_{CE} varies, do in fact have a finite slope which can be projected back to the Early voltage as shown in Figure 11.21. The practical current generator must be represented by a current source with a resistor in parallel with it. The resistor represents the output resistance of the transistor and is obtained from the Early voltage as:

$$R_o = \frac{V_A}{I_C}$$

where V_A is the Early voltage.

The common-mode voltage gain is obtained from equation 11.18 where R_E is replaced by the output resistance of the transistor used in the current mirror. In practice the Early voltage for all of the transistors of one conductivity type will be the same. The Early voltage for pnp transistors is generally less than that for npn transistors.

▶ **EXAMPLE 11.7** ───────────────────────────────────

For the differential amplifier shown in Figure 11.20 the Early voltage for the transistors is 110 V. Determine the CMRR.

▷ *Solution*

The current in Q_4 of the circuit in Figure 11.20 is 375 µA and the output resistance is:

$$R_o = \frac{110 \text{ V}}{375 \text{ µA}} = 293 \text{ k}\Omega$$

The common-mode gain is:

$$A_c = - \frac{6.8 \text{ k}\Omega}{139 \text{ }\Omega + 2 \times 293 \text{ k}\Omega} = -0.012$$

and the CMRR is:

$$\text{CMRR} = \frac{24.5}{0.012} = 2042 = 66 \text{ dB}$$

An important point to note is the effective value of resistance which has been achieved by using a current source. If a discrete resistor of 293 kΩ were used in place of the current mirror, then with a direct current of 375 µA the dc voltage drop across such a resistor is 110 V. This would require dc supplies in excess of 100 V being required to operate such amplifiers!

▷ **EXAMPLE 11.8** ─────────────────────────────────

Use PSpice to measure the difference-mode and common-mode gains and hence the CMRR for the circuit in Figure 11.20. Use the transistors from the breakout library and set the Early voltage (vaf) to 110 V. Set the area multiplier for Q_3 to 0.5.

▷ *Solution*

The schematic for measuring the differential-mode gain is shown in Figure 11.22. Notice that the signal source is divided into two parts.

NB For the source attached to Q_2 there is a time delay (t_d) of half a cycle. For a 1 kHz waveform this is 0.5 ms. Each source has an amplitude of 0.5 mV.

The dc collector current observed in Q_1 and Q_2 is 205 µA (cf. a calculated value of 187 µA) and the difference-mode gain is:

$$A_{vd} = 26.5 \quad \text{(cf. a calculated value of 24.5)}$$

To measure the common-mode gain remove the sine wave source from the base of Q_2, and connect the base to that of Q_1. It is also worth while increasing the amplitude of the sine wave source to 100 mV. With an increased CMRR the signal at the output is very small, and a larger input provides more realistic outputs for the simulation.

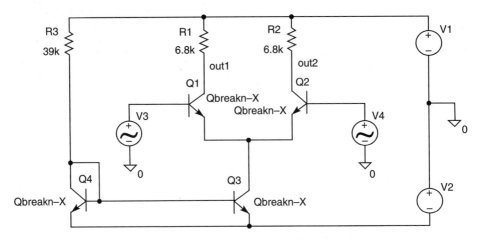

FIGURE 11.22 Schematic for measuring the differential-mode voltage gain.

The common-mode gain is:

$A_{cm} = 0.011$ (cf. a calculated value of 0.012)

The CMRR is:

$$CMRR = 20 \log\left(\frac{26.5}{0.011}\right) = 67.6 \text{ dB}$$

Current Mirror for Very Small Current

The current mirror is very important for the design of analog integrated circuits, for example operational amplifiers. To satisfy the design requirements, particularly for the input stages, it is usually necessary to design current mirrors to operate at a few microamps. For the current mirror shown in Figure 11.10 this would require either that the reference current be very small, or that the ratio of the areas be large for equation 11.21 to be satisfied. A small reference current implies a large reference resistor, which is not always possible with conventional integrated circuit technology. A large ratio of the areas decreases the accuracy of the current because of manufacturing limitations in accurately defining greatly different areas in an integrated circuit. An alternative form of current mirror removes these limitations. The current mirror is the *Widlar current mirror* and is shown in Figure 11.23.

From equation 11.19 the current for Q_1 is:

$$I_F \approx I_{ref} = I_S A \exp\left(\frac{V_{BE1}}{V_T}\right) \tag{11.22}$$

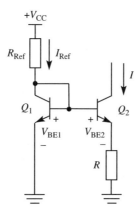

FIGURE 11.23 Widlar current mirror.

The current for Q_2 is:

$$I_F = I = I_S A \, \exp\!\left(\frac{V_{BE2}}{V_T}\right) \tag{11.23}$$

In these equations it is assumed that the areas are the same. Applying KVL to the base–emitter loop gives:

$$V_{BE1} = V_{BE2} + IR$$

and

$$I = \frac{V_{BE1} - V_{BE2}}{R} \tag{11.24}$$

From equations 11.22 and 11.23 V_{BE1} and V_{BE2} are:

$$V_{BE1} = V_T \ln \frac{I_{ref}}{I_S A}$$

and

$$V_{BE2} = V_T \ln \frac{I}{I_S A}$$

Substituting into equation 11.24 gives:

$$I = \frac{V_T}{R} \ln \frac{I_{ref}}{I} \tag{11.25}$$

▶ EXAMPLE 11.9 _____

Determine the value of the resistor required for the Widlar current mirror shown in Figure 11.23 if the reference current is 1 mA and the control current (I) is 10 μA.

▶ *Solution*

From equation 11.25 the resistor is given by:

$$R = \frac{V_T}{I} \ln \frac{I_{ref}}{I} = \frac{0.026}{10\,\mu A} \ln \frac{1\,mA}{10\,\mu A} \approx 12\,k\Omega$$

This value of resistance is well within the capabilities of integrated circuit technology. If the simple current mirror of Figure 11.18 were used then from equation 11.21 the area ratio would need to be 100. Accurate ratios of this magnitude are difficult to achieve in an integrated circuit.

Active Loads

A form of current mirror can also be used to replace the collector resistors in the emitter-coupled transistors of the differential amplifier. In Figure 11.24 pnp transistors Q_3 and Q_4 form a current mirror in which the diode-connected Q_3 reflects the current in Q_3, which is equal to I_{C1} into Q_4.

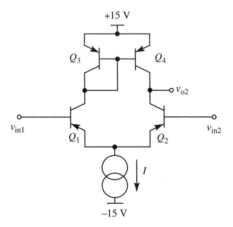

FIGURE 11.24 Active loads formed by pnp current mirrors.

Since Q_3 simply acts as a forward-biased diode the resistance in the collector of Q_1 is very low, and the gain of Q_1 is small. For Q_2 the load is the output resistance of Q_4 which is very large, and the gain for Q_2 is large. Because of this asymmetry in the voltage gains it is usual to operate this circuit with a single-ended output v_{o2}. The analysis of the circuit is more complex than the symmetrical circuits considered above, and it is not reproduced here, but is presented in Appendix 2. However, it is worth while presenting the expression for the gain, because it demonstrates how the replacement of resistors with transistors can greatly improve the performance of the differential amplifier. It can be shown that the single-ended gain is:

$$A_d = \frac{r_{o2} \| r_{o4}}{r_e} \qquad (11.26)$$

where r_{o2} and r_{o4} are the output resistances of Q_2 and Q_4 respectively. The values are obtained from the knowledge of the Early voltages for each of the transistors.

▶ **EXAMPLE 11.10** ——————————————————————————

For the circuit shown in Figure 11.24 the current I is 20 μA, the Early voltage for the npn transistors is 100 V and for the pnp transistors 80 V. Determine the difference-mode gain.

▶ *Solution*

The current through Q_1 and Q_2 is 10 μA and the internal resistance is:

$$r_e = \frac{0.026 \text{ V}}{10 \text{ μA}} = 2600 \text{ } \Omega$$

The small-signal output resistances for Q_2 and Q_4 are:

$$r_{o2} = \frac{100 \text{ V}}{10 \text{ μA}} = 10 \text{ M}\Omega$$

$$r_{o4} = \frac{80 \text{ V}}{10 \text{ μA}} = 8 \text{ M}\Omega$$

and the difference-mode gain is:

$$A_d = \frac{10 \text{ M}\Omega \| 8 \text{ M}\Omega}{2600} = 1709 = 64.6 \text{ dB}$$

It is again worth noting that the use of transistors in place of discrete resistors for the collector loads has greatly increased the voltage gain by providing very large effective resistor values. This has been achieved without the need for large-voltage power supplies (if resistors were to be used then 10 μA × 10 MΩ = 100 V).

▷ **11.5** **Input Resistance**

With two inputs there are two values of input resistance for the differential amplifier: the difference-mode resistance, which exists between the two inputs, and the common-mode resistance, which exists between each input and ground, as shown in Figure 11.25.

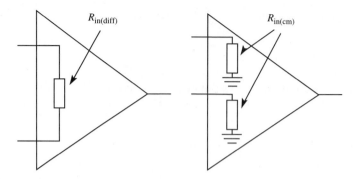

FIGURE 11.25   Difference amplifier input resistances.

The difference-mode input resistance is:

$$R_{\text{in(diff)}} = 2\beta r_e \tag{11.27}$$

Notice that this is similar to the value obtained for the CE amplifier, with the addition of the factor of 2.

The common-mode input resistance is approximately given by:

$$R_{\text{in(cm)}} = 2\beta R_E \tag{11.28}$$

where R_E is the emitter resistor, or when R_E is replaced with a current mirror, the output resistance of the transistor associated with the current source. The approximation arises because the value is very large, and a more detailed analysis would show that the output resistance of the emitter-coupled input transistors should also be included in parallel with R_E.

▶ **EXAMPLE 11.11** ───

For the circuit shown in Figure 11.26 determine the differential- and common-mode input resistances. The Early voltage is 90 V and the common-emitter gain (β) is 200. Assume that the area of Q_4 is one-fifth that of Q_3.

▶ *Solution*

DC Conditions

The reference current in Q_3 is:

$$I_{\text{Ref}} = \frac{(24\text{ V} - 0.7\text{ V})}{30\text{ k}\Omega} = 776\ \mu\text{A}$$

The collector current in Q_4 by current mirror action is:

$$I_4 = \frac{I_{\text{Ref}}}{5} = 155\ \mu\text{A}$$

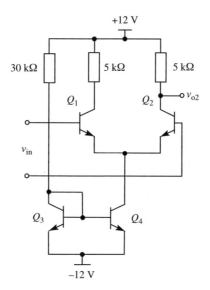

FIGURE 11.26 Circuit schematic for Example 11.7.

Small-Signal Conditions

The small-signal emitter resistance of Q_1 and Q_2 is:

$$r_e = \frac{0.026}{77.5 \ \mu A} = 335 \ \Omega$$

The differential-input resistance is:

$$R_{in\,(diff)} = 2 \times 200 \times 335 \ \Omega = 134 \ k\Omega$$

The output resistance of Q_4 is:

$$R_{o4} = \frac{90 \ V}{155 \ \mu A} = 580 \ k\Omega$$

The common-mode input resistance is:

$$R_{in\,(cm)} = 2 \times 200 \times 580 \ 1\Omega = 232 \ M\Omega$$

11.6 Operational Amplifier

The most important application for the differential amplifier is as the input stage of the integrated circuit operational amplifier. The differential amplifier with active loads and constant current source to supply the current to the emitter-coupled pair of input transistors, provides a large value of gain with large CMRR. In order that the amplifier can drive low-resistance loads, it is necessary to provide some form of output stage. Between the differential stage and the output stage there is likely to be a dc level-shifting stage and a single-ended gain stage. A simplified schematic of the complete amplifier is shown in Figure 11.27.

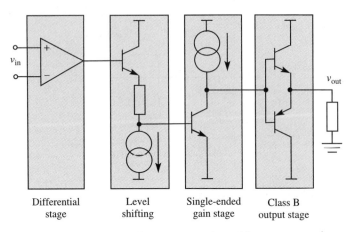

| Differential | Level | Single-ended | Class B |
| stage | shifting | gain stage | output stage |

FIGURE 11.27   Simplified schematic of an operational amplifier.

The level-shifting circuit is necessary because coupling capacitors cannot be used in an integrated circuit. The level-shifter provides a means of connecting together two nodes within a circuit which are at different dc voltage levels, without loss of gain. In the circuit in Figure 11.27 the level-shifter uses an emitter follower. This circuit changes the dc voltage level from a value which may be close to the positive supply (~12 V) at its input to a value close to the negative voltage supply ($-V_{EE} + 0.7$ V) at its output. The single-ended gain stage uses an active load, which is represented by the current source, and provides additional gain. It may also have internal feedback for gain stabilization. The output is a push–pull class B stage which has a low output resistance, and is capable of producing a large voltage swing across low values of load resistance.

It is not the intention of this text to describe these circuits in any more detail, and they have only been included here to show how the differential amplifier is incorporated into an operational amplifier. In practice the differential stage is more complex than those described above, and similarly the level-shifter, single-stage amplifier and the output stage are all much more complex than suggested in Figure 11.27. This text treats the operational amplifier as a black box and considers those parameters which may be obtained by making measurements from its terminals.

There are a number of important parameters which must be considered when selecting an operational amplifier. The more important of these are concerned with the input bias current, the input offset voltage, the small-signal response, the rate at which the output responds to a step response, and the large-signal frequency response.

Input Bias Current

The inputs to an operational amplifier are the inputs to the differential amplifier. For the amplifiers considered so far, these inputs are the base terminals of bipolar transistors. For correct dc operation it is necessary that there is a dc path for the base current. This current must flow in the components which are external to the operational amplifier, that is the resistors which form part of the external circuit. The voltage drop resulting from these currents may affect the dc voltage at the input of the amplifier, and also the output voltage because of the large gain of the internal amplifiers. It is important to know the value of the input bias current in order to take account of its presence.

The *input bias current* is defined as the average value for the two inputs, that is:

$$I_{\text{Bias}} = \frac{I_{B1} + I_{B2}}{2} \tag{11.29}$$

where I_{B1} and I_{B2} are the currents for the emitter-coupled pair of transistors in the differential amplifier.

Ideally the currents should be equal, but in practice they will vary as a result of variation in dimensions of the transistors and the variations in the semiconductor material. To minimize the effect that the bias current has when passing through the external components, the value should be as small as possible. The magnitude of the bias current is a measure of the quality of the operational amplifier. In a bipolar transistor circuit the current is minimized by operating the emitter-coupled transistors at very low currents, typically $10-20 \, \mu A$. The base current is reduced by the value of the common-emitter current gain, which may be increased by the inclusion of a Darlington pair of transistors for each of the emitter-coupled transistors. For a 741 amplifier the bias current is typically $50-80$ nA.

Greatly reduced input current is achieved by replacing the bipolar transistors with field effect transistors, either JFETs or MOSFETs. The bias current with these devices, particularly for the MOSFETs, is reduced to extremely low values. For a JFET the input current may be $5-20$ pA.

Input Offset Current

The *input offset current* is a measure of how well the two emitter-coupled transistors are matched. It is given by:

$$I_{\text{Offset}} = |I_{B1} - I_{B2}| \tag{11.30}$$

For good-quality amplifiers the value should be as small as possible. For a 741 it may typically be about $10-40$ nA. For a JFET the offset current may be $1-5$ pA.

▷ **EXAMPLE 11.12** ────────────────────────────────

If the base currents for the emitter-coupled transistors of a differential amplifier are $20 \, \mu A$ and $24 \, \mu A$, determine I_{Bias} and I_{Offset}.

▶ *Solution*

$$I_{Bias} = \frac{20 \ \mu A + 24 \ \mu A}{2} = 22 \ \mu A$$

$$I_{Offset} = |20 \ \mu A - 24 \ \mu A| = 4 \ \mu A$$

▶ **EXAMPLE 11.13**

For the circuit shown in Figure 11.28 it is assumed that the common-emitter current gain differs for each of the transistors, with $\beta_1 = 100$ and $\beta_2 = 120$. Determine I_{Bias} and I_{Offset}.

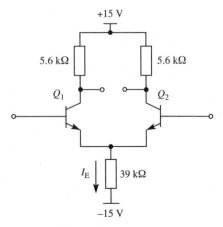

FIGURE 11.28   Circuit for Example 11.13.

▶ *Solution*

$$I_E = \frac{15 \ V - 0.7 \ V}{39 \ k\Omega} = 367 \ \mu A$$

$$I_{E1} = I_{E2} = \frac{367 \ \mu A}{2} = 183.5 \ \mu A$$

$$I_{B1} = \frac{183.5 \ \mu A}{100} = 1.84 \ \mu A$$

$$I_{B2} = \frac{183.5 \ \mu A}{120} = 1.53 \ \mu A$$

and:

$$I_{Bias} = \frac{1.84 \ \mu A + 1.53 \ \mu A}{2} = 1.69 \ \mu A$$

$$I_{Offset} = |184 \ \mu A - 1.53 \ \mu A| = 0.31 \ \mu A$$

Input Offset Voltage

For an ideal amplifier the output voltage is 0 V if both of the inputs are grounded (at 0 V). In practice, because of variations in device geometry and variations in the semiconductor materials, V_{BE} and the current in the transistors of the emitter-coupled pair are not perfectly matched and the output voltage will not be 0 V. The *offset voltage* is the input voltage which is required to make the output zero. In Figure 11.29 it is shown as a positive voltage connected to the non-inverting input terminal (+), but it could equally well have been a negative voltage, or it could have been connected to the inverting terminal (−). It is typically a few mV.

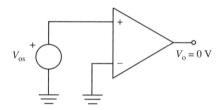

FIGURE 11.29   Example of the voltage required to set the output to 0 V.

For those applications where the output must be at 0 V when the inputs are at 0 V, then it may be necessary to include some form of dc bias at the input which is capable of being adjusted in order to compensate for any offset. This adjustment will take account of the offset voltage and also any offsets resulting from the input bias current flowing in the external resistors which are attached to the input. Alternatively, many operational amplifiers have provision for the attachment of an external potentiometer which is used to cancel any offset, as shown in Figure 11.30.

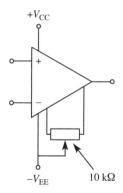

FIGURE 11.30   Input offset voltage compensation.

The terminals to which the potentiometer is attached are connected to the emitter-coupled transistors in the differential stage (which in an actual amplifier is more complex than the differential circuits described above) and a voltage applied from the negative supply generates small values of voltage within the differential amplifier. With the inputs grounded it is a simple matter to adjust the potentiometer to zero the output voltage. However, for good operational amplifier design it is better to choose an amplifier with low offset figures rather than use a potentiometer which may change in value with time or temperature.

Small-Signal Response

The typical gain versus frequency response of a dc-coupled operational amplifier is shown in Figure 11.31.

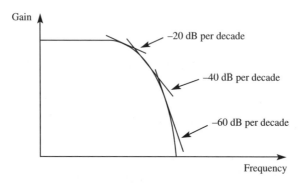

FIGURE 11.31   Small-signal frequency response.

Remember that for each *RC* filter section the rate of change of the gain with frequency increases by −20 dB per decade. The capacitance is associated with the pn junctions of the different transistors, and it is unlikely that the *RC* products for the different low-pass filters will be the same. Hence the shape of the curve shown in Figure 11.31, where the rate of change of gain with frequency increases with increasing frequency. The overall frequency response can be approximated with the Bode straight line plots (Section 9.8) which are shown in Figure 11.31.

For many applications this type of frequency response is undesirable, because for frequencies where the rate of change is −60 dB per decade or greater (which corresponds to three or more *RC* networks), there is also a very large phase shift (90° for each *RC* section) between input and output. If negative feedback is employed to control the gain, then at these frequencies the signal which is fed back to the input may have undergone such a large phase shift that the feedback is no longer negative, but positive. Under these conditions the amplifier will become unstable and may oscillate. A much more suitable response is one where the gain decreases uniformly with frequency at a rate of −20 dB per decade until the gain is reduced to unity, as shown in Figure 11.32.

The response shown in Figure 11.32 is typical of a class of operational amplifiers known as *compensated amplifiers*, and the 741 is an example of a compensated amplifier. The gain A_o is the open-loop gain, f_c is the 3 dB corner frequency and f_t is the frequency at which the gain is reduced to unity. The constant rate of change of gain

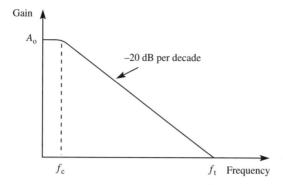

FIGURE 11.32   Frequency response of a compensated amplifier.

with frequency implies that there is only one dominant RC network. The phase change associated with one RC network cannot exceed 90° and, therefore, the amplifier is unconditionally stable for any amount of negative feedback.

For the response shown in Figure 11.32 there is a simple relationship between A_o, f_c and f_t given by:

$$f_t = A_o f_c \qquad\qquad (11.31)$$

▶ **EXAMPLE 11.14** ──────────────────────────

For a 741 operational amplifier the open-loop gain is typically 2×10^5 and the unity gain frequency is 1.5 MHz. Determine the 3 dB corner frequency.

▶ *Solution*

$$f_c = \frac{f_t}{A_o} = \frac{1.6\ \text{MHz}}{2 \times 10^5} = 8\ \text{Hz}$$

▶ **EXAMPLE 11.15** ──────────────────────────

Use PSpice to examine the open-loop frequency of the UA741 operational amplifier which is to be found in **eval.lib**. From **AC sweep** in **Setup...** select **decade**, **10 points/decade**, **start frequency** of 1 Hz and a **stop frequency** of 10 MHz.

▶ *Solution*

The schematic is shown in Figure 11.33 with the Probe output in Figure 11.34.

From the Probe output the open-loop gain is 104 dB, the 3 dB corner frequency is 5 Hz and the unity gain frequency is 857 kHz.

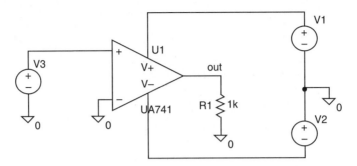

FIGURE 11.33   Schematic for the frequency response of a 741 op-amp.

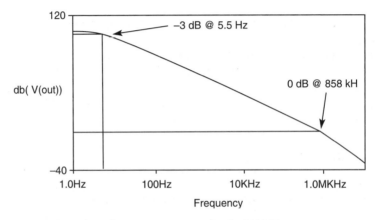

FIGURE 11.34   Open-loop frequency response for the UA741 op-amp.

Frequency Compensation

The frequency compensation which is required to produce the frequency response shown in Figure 11.32 involves a low-pass *RC* filter. The resistive part of the filter is the output resistance of the differential amplifier. For the differential amplifier with active loads as shown in Figure 11.24, the output resistance is:

$$r_o = r_{o2} \parallel r_{o4} \tag{11.32}$$

If a capacitor is attached from the collectors of Q_2 and Q_4, in Figure 11.24, then the resultant *RC* network is as shown in Figure 11.35

The corner frequency is:

$$f_c = \frac{1}{2\pi r_o C}$$

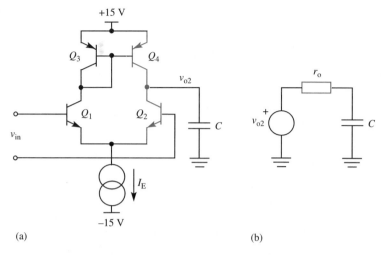

(a) (b)

FIGURE 11.35 Frequency-compensated amplifier with (a) a capacitor at the collector of the differential amplifier and (b) the equivalent RC low-pass filter.

▷ EXAMPLE 11.16

For the circuit shown in Figure 11.35 the current I_E is 30 μA and the Early voltage of the npn transistors is 120 V and for the pnp transistors it is 75 V. Determine

(a) the 3 dB corner frequency if the compensation capacitor is 3.5 nF
(b) the difference-mode gain A_{vd}
(c) the unity gain frequency

▷ Solution

(a) The output resistances of the transistors Q_2 and Q_4 are:

$$r_{o2} = \frac{120 \text{ V}}{15 \text{ μA}} = 8 \text{ MΩ} \quad \text{and} \quad r_{o4} = \frac{75 \text{ V}}{15 \text{ μA}} = 5 \text{ MΩ}$$

The corner frequency is:

$$f_c = \frac{1}{2\pi(8 \text{ MΩ} \| 5 \text{ MΩ})(3.5 \text{ nF})} \approx 15 \text{ Hz}$$

(b) The small-signal emitter resistance is:

$$r_e = \frac{0.026}{15 \text{ μA}} = 1.73 \text{ kΩ}$$

From equation 11.21 the gain is:

$$A_{vd} = \frac{r_o}{r_e} = \frac{3.08 \text{ MΩ}}{1.73 \text{ kΩ}} = 1780$$

(c) The unity gain frequency is:

$$f_t = A_{vd} \times f_c = 1780 \times 15 = 26.7 \text{ kHz}$$

Capacitor values of 3.5 nF cannot be achieved directly with semiconductor integrated circuit technology, and such values must be obtained indirectly if this form of frequency compensation is to be employed. In Chapter 9 (Section 9.5) it was shown that the Miller effect could result in very large values of capacitance being achieved as a result of very small capacitors which appear between the input and output terminals of a high-gain amplifier. This effect is used to advantage for frequency compensation in operational amplifiers by placing a small capacitor across a high-gain stage. From Figure 11.27 an appropriate place for the capacitor is from the collector of the single-ended gain stage to the output of the differential stage. The gain of the single-ended stage is typically of the order of 200. If the required capacitor is 3.5 nF then the actual compensation capacitor need only be 17.5 pF. This is more readily achieved with integrated circuit technology.

▶ **EXAMPLE 11.17** ————————————————————————

Use PSpice to examine the circuit shown in Figure 11.36. Use transistors from the breakout library and set the Early voltage to the correct values (vaf = 120 and vaf = 75). From **Analysis** and **setup...** choose **AC Sweep** and set the frequency sweep for **decade**, 10 points per decade, from 1 Hz to 1 MHz.

▶ *Solution*

The schematic from PSpice is shown in Figure 11.36.
The output from probe is shown in Figure 11.37

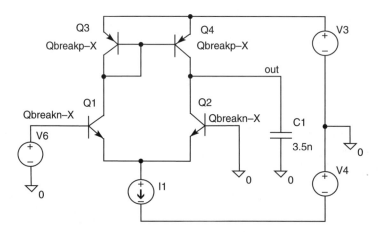

FIGURE 11.36   PSpice schematic.

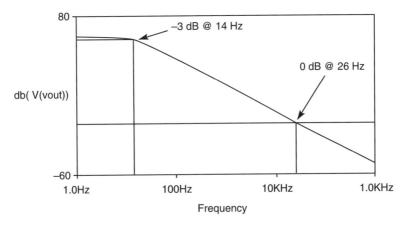

FIGURE 11.37   Gain versus frequency for the circuit shown in Figure 11.26.

From Probe it is seen that the 3 dB corner frequency is 14 Hz and the gain is 65 dB. The unity gain frequency is 26 kHz.

For compensated amplifiers such as the 741, the compensating capacitor is formed on the IC chip during the manufacturing process. This results in an easy to use amplifier which is unconditionally stable for any amount of negative feedback. However, it does limit the frequency response and for applications requiring very large bandwidth, then fixed internal compensation is not ideal. For these applications it is common practice for the operational amplifier to have an additional pair of terminals to which an external capacitor may be attached. This capacitor replaces the built-in capacitor and enables the user to optimize the frequency response by varying its value to satisfy the particular application.

Slew Rate

The capacitors which are used for frequency compensation limit the frequency response of an operational amplifier, but they also limit the amplifier in another way. When a voltage step is applied to the input, the capacitors require a finite time to charge and discharge. Since the voltage across the capacitors cannot change instantaneously, there is a limit to the rate at which the output voltage can change. This limit is known as the *slew rate*. It is measured in volts per second.

The charging current in a capacitor is given by:

$$i = C \frac{\mathrm{d}v}{\mathrm{d}t}$$

where $\mathrm{d}v/\mathrm{d}t$ is the rate of change of the voltage across the capacitor. The equation can be rearranged so that this rate of change is expressed in terms of the current:

$$\frac{dv}{dt} = \frac{i}{C}$$

In order that the output of an operational amplifier accurately follows the input, then the rate of change should be as large as possible, that is the current should be large for a given capacitor. This is illustrated in Figure 11.38 where the current in the current generator is increased from 0 A to 50 µA.

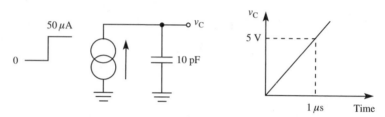

FIGURE 11.38 Maximum slew rate is determined by the charging current.

The voltage across the capacitor increases linearly with time until after 1 µs the voltage is 5 V. The rate of change is:

$$\frac{dv_C}{dt} = \frac{50\ \mu A}{10\ pF} = 5000000\ \text{V s}^{-1}$$

This figure represents the maximum rate of change of the voltage across the capacitor. A more convenient way of expressing the slew rate is volts per microsecond. For the example above this would be 5 V µs^{-1}.

The slew rate S_R is defined as the maximum rate of change of the output voltage and is defined as:

$$S_R = \frac{I_{max}}{C} \tag{11.33}$$

▷ **EXAMPLE 11.18** ──────────────────────────────

Determine the slew rate for the differential amplifier shown in Figure 11.39.

▷ *Solution*

The maximum current which can be diverted to the capacitor is 100 µA. In the figure this would occur when Q_1 is turned completely OFF, and Q_2 is turned ON. The slew rate is:

$$S_R = \frac{100\ \mu A}{200\ pF} = 0.5\ \text{V µs}^{-1}$$

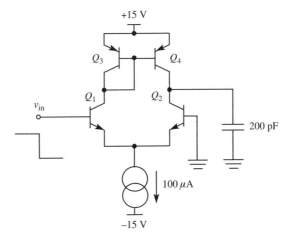

FIGURE 11.39 Differential amplifier for slew rate calculation.

Slew Rate and Distortion

In addition to the distortion which is produced in a square wave or pulse waveform, distortion also occurs in a sine wave. Slew rate results in a limit being placed on the rate at which the output voltage changes. For a sine wave the maximum rate of change occurs when the signal is crossing the zero-voltage axis. The rate of change at this point is a function of the amplitude and the frequency, as shown in Figure 11.40.

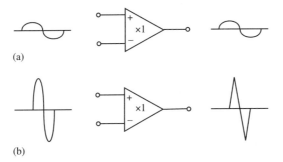

FIGURE 11.40 Distortion resulting from slew rate limitations for (a) a small-amplitude low-frequency wave and (b) a large-amplitude higher-frequency wave.

For an amplifier with a gain of unity, the output for a low-frequency small-amplitude signal is a faithful reproduction of the input. However, as the amplitude and frequency increase the rate at which the voltage changes increases until the slew rate limit is reached. At this point the rate of change is limited by the internal charging and discharging of the internal capacitance, and the output progressively becomes more and more triangular in shape with increasing amplitude and/or frequency, as shown in Figure 11.40(b).

A relationship between amplitude, frequency and slew rate for a sine wave can be obtained as follows:

$$v_s = V_p \sin(2\pi f t)$$

where v_s is the input described as a sine wave, V_p is the peak amplitude and f is the frequency. The rate of change is obtained by taking the differential of v_s with respect to time. Thus:

$$\frac{dv_s}{dt} = \frac{d}{dt}(V_p \sin(2\pi f t))$$

$$= 2\pi f V_p \cos(2\pi f t))$$

The maximum rate of change occurs at the zero crossing point, that is when $2\pi f t = 0$, or:

$$\left.\frac{dv_s}{dt}\right|_{max} = 2\pi f V_p \cos(0) = 2\pi f V_p$$

The slew rate is defined as:

$$S_R = \frac{dV}{dt}$$

Thus:

$$f_{max} = \frac{S_R}{2\pi V_p} \tag{11.34}$$

where f_{max} is the maximum frequency of a sine wave with a peak amplitude of V_p. To avoid distortion the frequency of the applied signal must be less than that predicted by equation 11.34.

▶ **EXAMPLE 11.19** ───────────────────────────

A 741 has a slew rate of 0.5 Vμs^{-1}. Determine the maximum frequency which can be applied to produce an undistorted output which has a peak amplitude of 2 V, 5 V and 10 V.

▶ *Solution*

From equation 11.34:

$$f_{max}|_{2V} = \frac{0.5}{2\pi(2 \text{ V} \times 10^{-6})} \approx 40 \text{ kHz}$$

$$f_{max}|_{5V} = \frac{0.5}{2\pi(5 \text{ V} \times 10^{-6})} \approx 16 \text{ kHz}$$

$$f_{max}|_{10V} = \frac{0.5}{2\pi(10 \text{ V} \times 10^{-6})} \approx 8 \text{ kHz}$$

Notice that for a given slew rate the maximum frequency which can be applied for an undistorted output decreases as the amplitude of the output increases.

▶ EXAMPLE 11.20

Connect the UA741 operational amplifier as a unity gain amplifier as shown in Figure 11.41. The slew rate for a 741 is 0.5 V μs^{-1}. The maximum frequency which can be applied to provide an 8 V peak sine wave output is 10 kHz. Observe the output at 10 kHz and 8 V, 10 kHz and 10 V and 12 kHz and 8 V.

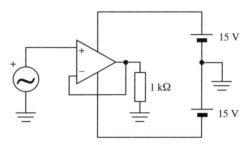

FIGURE 11.41   Unity gain amplifier for Example 11.20.

▶ *Solution*

The output from Probe for each of the conditions is shown in Figure 11.42. It can be clearly seen that for 12 kHz or 10 V the output is distorted and becomes triangular. For 10 kHz and 8 V the output remains is a sine wave. Notice that 12 kHz does not represent the maximum small-signal response of the 741 amplifier. It represents a large-signal limitation.

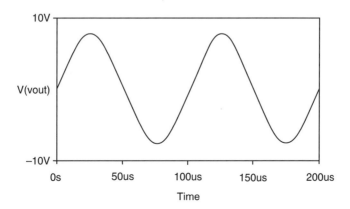

FIGURE 11.42   *Continued.*

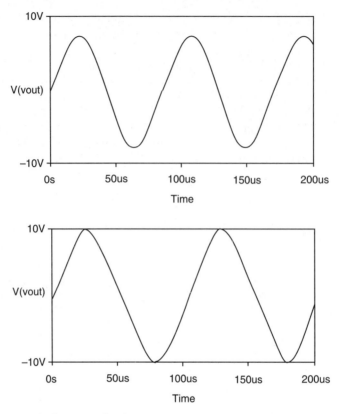

FIGURE 11.42   Probe output for the circuit in Figure 11.41.

► PROBLEMS

11.1 The input to a differential amplifier comprises 500 μV from a transducer and 2 V from a common-mode source of interference. The amplifier has a difference-mode gain of 2000 and a common-mode gain of 0.25. Determine the difference-mode output and the error signal.

[1.25 V, 0.25 V]

11.2 A transducer generates an output of 2 mV. Determine the differential gain and CMRR to provide an output of 2 V with an error of less than 1% if there is a common-mode signal of 5 V.

[1000, 108 dB]

11.3 Determine

(a) the voltages at the collector of each transistor
(b) the double-ended gain v_{01-2}/v_{in}
(c) the single-ended gain v_{o2}/v_{in}

[(a) 13.2 V, (b) −69.4, (c)***]

11.4 When a differential input of 1 mV sin ωt is applied to a differential amplifier the differential output is 0.5 V sin ωt. When both inputs are joined together and 1 V sin ωt applied the output is 200 mV sin ωt. Determine the CMRR.

[68 dB]

11.5 If v_{in} is 1 mV sin ωt determine the output v_{out}.

[53.8 mV sin ωt]

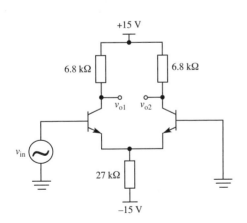

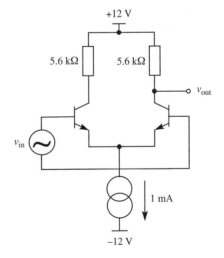

11.6 Determine the single-ended gain and the CMRR.

[65, 44 dB]

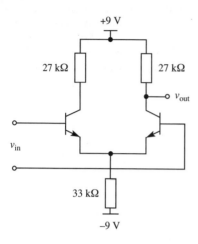

11.7 Determine the double-ended differential gain and the CMRR.

[−275, 55 dB]

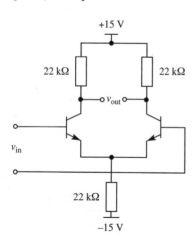

11.8 For the current mirror shown determine the current I if the emitter area of Q_2 is half the area of Q_1.

[~0.5 mA]

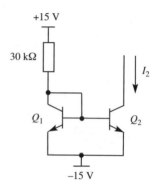

11.9 Determine the currents I_2, I_3 and I_4 if the areas of Q_2, Q_3 and Q_4 are $0.5A_1$, $0.25A_1$ and $0.125A_1$ where A_1 is the area of Q_1.

[0.357 mA, 0.179 mA, 0.089 mA]

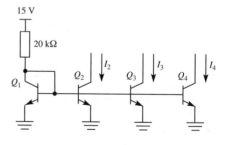

11.10   Show that the collector current I_{C2} is given by:

$$I_{C2} = I_{ref} \frac{A_2}{A_1} \left(1 + \frac{1}{\beta_1} + \frac{A_2}{A_1} \frac{1}{\beta_2} \right)^{-1}$$

where the subscripts relate to the transistors Q_1 and Q_2.

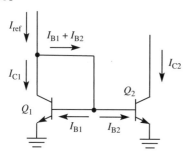

11.11   For the current mirror in Problem 11.10, $A_1 = 2000 \ \mu m^2$, $A_2 = 200 \ \mu m^2$, $\beta_1 = 150$ and $\beta_2 = 100$. Determine the collector current I_{C2} as given by the above relationship if $I_{ref} = 1$ mA and compare this value with that given by the approximate relationship given by equation 11.21.

11.12   For the circuit shown the emitter area of Q_4 is 0.25 that of Q_3. Determine

(a) the collector currents in Q_1 and Q_2
(b) the voltage gain v_{out}/v_{in}.

[(a) ~0.12 mA, (b) 117]

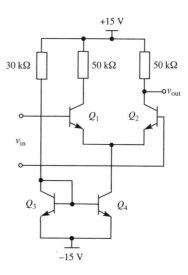

11.13   For the circuit shown $A_4 = 0.4A_3$ and the Early voltage for the npn transistors is 120 V. Determine the gain and the CMRR.

[56, 67 dB]

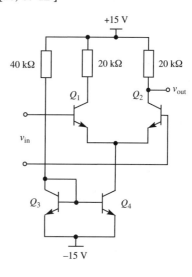

11.14   Use PSpice to verify the values of gain and CMRR for the circuit shown in Problem 11.13 based on the QbreakN transistors from the breakout library. Set vaf = 120 V.

11.15   For the circuit shown in Figure 11.24 the gain is given by equation 11.26. Show that if V_N and V_P are the Early voltages for the npn and pnp transistors respectively, then the gain can be expressed as:

$$A_d = \frac{1}{V_T}\left(\frac{V_N V_P}{V_N + V_P}\right)$$

where V_T is the thermal voltage (0.026 V).

11.16   The Early voltages are 110 V for the npn transistor and 80 V for the pnp transistor. Determine the gain for the circuit shown.

[65 dB]

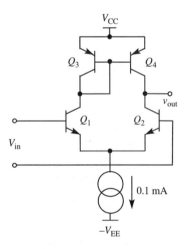

11.17   For the circuit shown the Early voltages are 100 V for the npn transistor and 75 V for the pnp transistor. If the emitter area of Q_4 is 0.1 that of Q_3 determine the gain and the CMRR.

[64 dB, 72 dB]

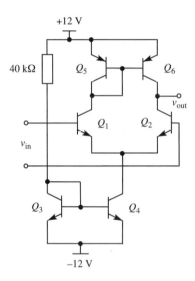

11.18   For the circuit shown in Problem 11.17 determine the differential and common-mode input resistances.

[~180 kΩ, ~350 MΩ]

11.19   If for the circuit shown in Problem 11.17 the current gain of Q_1 is 250 and that of Q_2 is 260, determine the input bias current and the input offset current.

[114 nA, 4.5 nA]

11.20 Identify the different stages of an operational amplifier and explain the purpose of each stage.

11.21 Determine the maximum phase shift for an operational amplifier which has four clearly identifiable RC networks between the input terminal and the output. If the corner frequencies for each network are 100 Hz, 1 kHz, 12 kHz and 64 kHz, identify the frequency range in which the closed-loop gain will become unstable.

11.22 Sketch the frequency response for a compensated amplifier where the open-loop gain is 150000 and the open-loop 3 dB frequency is 20 Hz. Determine f_t.

11.23 The Early voltages for the transistors in the circuit shown are 120 V for the npn and 65 V for the pnp. Determine

(a) the difference-mode gain
(b) the value of the capacitor C required to give a 3 dB frequency of 15 Hz
(c) the unity gain frequency f_t.

[(a) 64 dB, (b) 6.3 nF, .(c) 24.3 kHz]

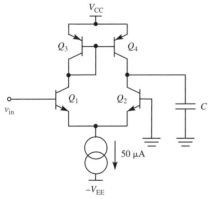

11.24 For the circuit shown determine the value of the capacitor C to create a 3 dB corner frequency of 15 Hz if the gain of the amplifier A is -250. Assume that the area of Q_4 is 0.1 of Q_3 and that the Early voltages of the npn transistors are 125 V and 85 V for the pnp transistors.

[42 pF]

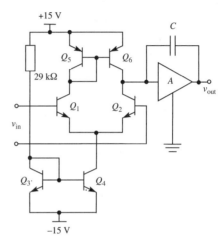

11.25 Use PSpice to simulate the circuit shown in Problem 11.24 using QbreakN and QbreakP transistors with appropriate values for the Early voltages, vaf = 125 and vaf = 85. Use a voltage-controlled voltage source (E) with a gain of -250 for the gain block A. In **setup...** under **Analysis** select **AC sweep...** and a sweep frequency from 1 Hz to 10 MHz. Measure f_c and f_t.

[14.7 Hz, 16 MHz]

11.26 Estimate the slew rate for the circuit shown in Problem 11.24.

[0.01 V µs^{-1}]

11.27   An operational amplifier is required to amplify pulses which have a risetime of 2 μs and an amplitude of 1 V. Determine the minimum value of the slew rate for the amplifier if the pulses are not to be distorted.

[0.5 V μs^{-1}]

11.28   An operational amplifier has a slew rate of 5 V μs^{-1}. Determine the maximum value of the frequency for a sine wave if the output amplitude is to be 8 V. What is the frequency if the slew rate is 0.2 V μs^{-1}?

[100 kHz, 4 kHz]

11.29   An operational amplifier is required to amplify a sine wave with a frequency of 15 kHz and to produce an output with an amplitude of 12 V. Determine the minimum slew rate for the amplifier. Would a 741 operational amplifier be suitable?

[1 V μs^{-1}]

CHAPTER 12

Negative Feedback

In the previous chapters the analysis of a transistor circuit has been based on the simple application of KVL without any regard to the possible existence of negative feedback. In considering a single-stage common-emitter amplifier with a resistor in series with the emitter, and without any decoupling capacitor, the analysis was based on the resultant equivalent circuit, again, without any reference to feedback. However, this circuit is an example of the application of negative feedback by means of a series resistor. The analysis could have been developed in a different manner based on basic negative feedback relationships. For a simple circuit, such as the CE amplifier, there is no advantage in doing this, but for more complex circuits there is an advantage in being able to identify the type of feedback in order to determine the effect that feedback has on the gain, impedance levels and frequency response.

Negative feedback offers a number of advantages, and some of the more important are:

1 Stabilize the gain.
2 Reduce electrical noise which is generated within the feedback loop.
3 Reduce distortion.
4 Improve the frequency response.
5 Modify the input and output impedance levels.

These advantages are obtained at the expense of a reduction in the overall gain. However, amplifiers are often designed to have a gain which is considerably higher than is required, and negative feedback is used to reduce the gain to the required level with a subsequent improvement in the overall performance as listed above. This chapter identifies the different methods of applying negative feedback and the effects which these methods have on the gain and impedance levels. It is not intended to provide a rigorous treatment of negative feedback, but rather a review of the basic principles and methods of analysis. Analysis of complex negative feedback circuits is difficult, but being able to identify the topology and to make some approximations can provide an indication of the gain and impedance values. Simulation can then confirm the results and provide greater accuracy.

12.1 Basic Theory

The circuit shown in Figure 12.1 illustrates a generalized view of feedback, with an amplifier with gain A_o and a feedback circuit which returns a fraction of the output back to the input. The fraction returned is determined by the factor B in the feedback network. The returned signal (v_f) is summed with the input signal (v_s) to produce the input to the amplifier (v_i). The input v_i is often referred to as the error signal.

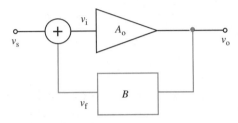

FIGURE 12.1   Basic arrangement for applying feedback.

The relationship between the different parameters is obtained as follows:

$$v_i = v_s - v_f$$

But

$$v_o = v_i A_o \quad \text{and} \quad v_f = v_o B$$

Substituting and rearranging gives:

$$v_s = \frac{v_o}{A_o} + B v_o$$

The overall gain is defined as:

$$A_f = \frac{v_o}{v_s} = \frac{A_o}{1 + B A_o} \tag{12.1}$$

This represents a fundamental relationship[*] for the gain of an amplifier which incorporates feedback. Provided that $1 + BA_o$ is >1 then A_f is $<A_o$ and the feedback is said to be *negative*. Some basic definitions which apply are:

A_o is the open-loop gain, that is without feedback.
BA_o is the loop gain, the gain around the feedback loop.
A_f is the closed-loop gain, that is with feedback applied.

The loop gain is usually arranged to be much greater than unity. Then:

$$A_f \approx \frac{1}{B} \tag{12.2}$$

---

[*]Note that with a different sign convention the denominator becomes $1 - BA_o$. This arises when the gain is defined as $-A_o$ rather than $+A_o$, that is $v_o = -v_i A_o$.

When equation 12.2 applies the gain of the complete amplifier is independent of the open-loop gain. This is an important advantage of negative feedback because the open-loop gain may vary with temperature and also rely heavily on the gain of the active devices, which may vary from one manufactured batch to another. The feedback factor B, on the other hand, is determined by passive components, for example resistors, which are much more stable than active devices.

Provided BA_o is greater than unity and is positive then from equation 12.1 the overall gain (A_f) is less than the open-loop gain and the feedback is said to be negative. If, however, the product BA_o is negative and also has a value which is close to unity, then the denominator in equation 12.1 ($1 - BA_o$) can approach zero, and the closed-loop gain tends towards infinity. Under these circumstances the feedback is said to be *positive*. Positive feedback results in the overall gain being larger than the open-loop gain. The gain may be so large that the circuit becomes unstable and oscillates.

The simple definition given above for negative feedback applies only to the mid-band of the open-loop gain versus frequency response of the amplifier. Typically the phase relationship between the input and output will vary and this variation can be considerable beyond the mid-band region. It was noted in Chapter 9 that each *RC* network contributes a maximum of 90° of phase shift beyond the corner frequencies. When this happens it is possible for the feedback to change from being negative to positive, and the amplifier with feedback may oscillate at the frequency corresponding to this change of phase. This can be a particular problem with wide-band operational amplifier circuits and special frequency compensation methods must be applied to prevent this happening, but these will not be considered in this text.

▶ **EXAMPLE 12.1**

Two amplifiers each with a gain of 100 are connected in cascade and negative feedback applied to provide an overall gain of 20. Determine:

(a) the required feedback factor
(b) the percentage increase in the overall gain if the gain of each amplifier increases by 100%.

▶ *Solution*

(a) $A_f = \dfrac{A_o^2}{(1 + BA_o^2)} = \dfrac{100^2}{(1 + B \times 100^2)} = 20$

$\therefore \quad B = 0.0499$

(b) $A_f = \dfrac{200^2}{(1 + 0.0499 \times 200^2)} = 20.03$

The increase in the overall gain is $(20.03 - 20.00)/(20) \times 100 = 0.15\%$.

Notice how negative feedback has stabilized the gain against variations in the open-loop gain. A 100% variation has been reduced to 0.15%. If the amplifier were designed to have a gain of 20 without negative feedback then there could be large variations of gain as a result of differences in the active devices. Negative feedback reduces the effect of such differences.

12.2 Types of Feedback

Feedback exists when some fraction of the output signal is returned to the input. Since both the output and the input can be characterized by either voltage or current, then four possible types of feedback exist. The output voltage or current may be sampled and either a voltage or current may be returned to the input.

The two different types of output sampling are shown in Figure 12.2. In 12.2(a) the output voltage is sampled by connecting the feedback network across the output load. This type of connection is referred to as *voltage sampling*. In 12.2(b) the feedback network is connected in series with the load so that the output current is sampled. This type of connection is referred to as *current sampling*.

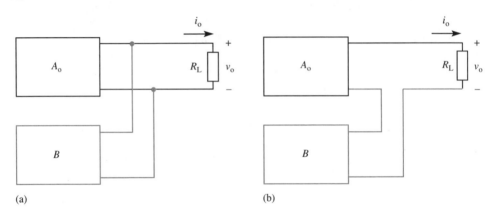

(a)   (b)

FIGURE 12.2 Feedback connections for the output of a feedback amplifier with (a) voltage sampling and (b) current sampling.

Two different types of input connection are shown in Figure 12.3. In 12.3(a) the source voltage, the input voltage and the feedback voltage are all in series, while in 12.3(b) a shunt connection is shown. Figure 12.3(a) represents *series* comparison and 12.3(b) represents *shunt* comparison.

The four feedback configurations are shown in Figure 12.4. The identity of each configuration is determined by the type of sampling at the output and the manner by which the signal is returned to the input. Thus there is (a) *voltage–series*, (b) *voltage–shunt*, (c) *current–series* and (d) *current–shunt*.

The gain factor for each configuration is simply identified by means of A_o, but this parameter has different units depending on the type of feedback. For the voltage–series type of feedback both output and input signals are voltages, and the

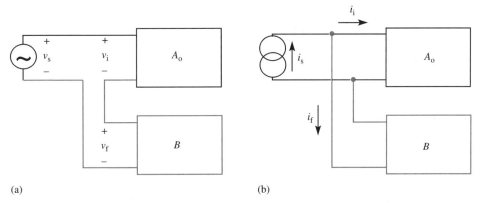

FIGURE 12.3   Feedback connections at the input of a feedback amplifier with (a) series comparison and (b) shunt comparison.

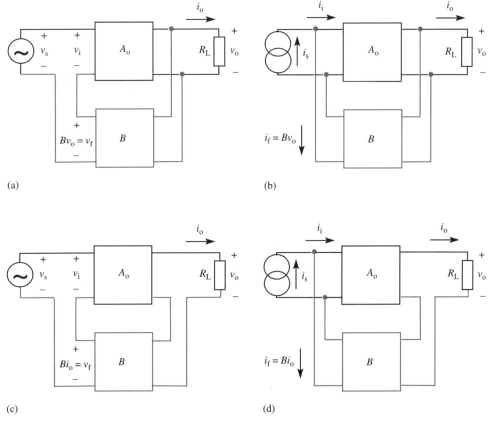

FIGURE 12.4   Feedback configurations with (a) voltage–series, (b) voltage–shunt, (c) current–series and (d) current–shunt.

ratio v_o/v_i is the voltage gain A_v which is dimensionless. For the current–shunt configuration the signals are the output and input currents, so that the ratio i_o/i_i describes the current gain A_i, which again is dimensionless. For the other two configurations the units of the output and input signals are different. Thus for the voltage–shunt configuration the output is a voltage while the input is a current. The gain factor is defined as the ratio of v_o/i_i which describes a resistance. However, the voltage is from the output and the current is from the input and, therefore, the gain factor is defined as a *transresistance* R_m. For the current–series configuration the output is a current and the input a voltage. The transfer parameter in this case is a *transconductance* G_m which describes the ratio i_o/v_i.

The transfer parameter of the feedback network is also similarly described, as is the overall transfer parameter of the complete circuit.. For convenience the general expression for gain uses A_o to represent the transfer parameters for the amplifier without feedback, B for the feedback network and A_f for the overall transfer parameter.

12.3 ▷ Input Resistance

The input resistance is affected by the application of negative feedback. For the case of series feedback, then regardless of whether the signal is proportional to the output voltage or the output current, the *input resistance with series feedback is increased*. The signal returned to the input v_f opposes the input v_s and the input current i_i is less than it would be if v_f were absent. Therefore, the input resistance with feedback $R_{if} = v_s/i_i$ is greater than the input resistance without feedback R_i.

When the negative feedback signal is returned to the input in parallel (shunt) with the applied signal then, regardless of whether the signal is proportional to the output voltage or the output current, *the input resistance with shunt feedback is reduced*. With shunt feedback there is an additional path for the source current and $i_s = i_i + i_f$. The source current is greater with feedback than without it. Therefore, the input resistance $R_{if} = v_i/i_s$ is decreased with feedback.

Voltage–Series

For voltage–series feedback the relationship between the three components of voltage in the input network is shown in Figure 12.5(a) and is:

$$v_s = v_i + v_f$$

The feedback component is:

$$v_f = Bv_o = Bv_i A_{vo}$$

where A_{vo} is the open-loop voltage gain. Substituting for v_f and dividing through by i_i gives:

$$\frac{v_s}{i_i} = \frac{v_i}{i_i} + \frac{Bv_i A_{vo}}{i_i}$$

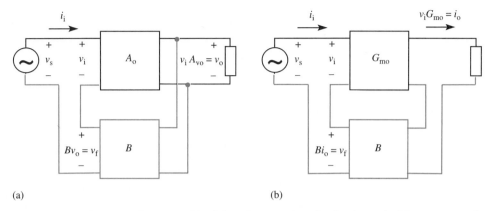

FIGURE 12.5   Input resistance with (a) voltage–series feedback and (b) current–series feedback.

or

$$R_{if} = R_i(1 + BA_{vo}) \tag{12.3}$$

where R_{if} is the input resistance with feedback and R_i is the input resistance without feedback. The input resistance is increased by the factor $(1 + BA_{vo})$.

Current–Series

For current–series shown in Figure 12.5(b) the relationship between the three input voltages is the same as for the voltage–series, but the feedback component is different. It is given by:

$$v_f = Bi_o = Bv_iG_{mo}$$

where G_{mo} is the transconductance for the amplifier without feedback. Note that B is a transresistance, so that the product BG_{mo} is dimensionless. Dividing through by i_i gives:

$$R_{if} = R_i(1 + BG_{mo}) \tag{12.4}$$

This equation is of the same form as equation 12.3 where the expression in parentheses is typically >10, so that the input resistance with feedback is greater than the resistance without feedback.

Voltage–Shunt

For voltage–shunt the relationship between the three input currents is shown in Figure 12.6 and is given by:

$$i_s = i_i + i_f$$

The feedback component is:

$$i_f = Bv_o = Bi_iR_{mo}$$

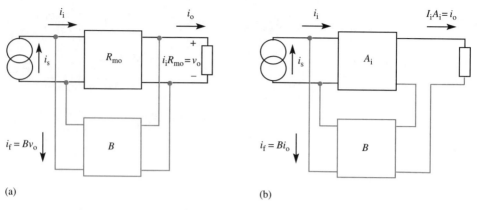

(a)   (b)

FIGURE 12.6 Input resistance with (a) voltage–shunt feedback and (b) current–shunt feedback.

where R_{mo} is the open-loop transresistance. Substituting for i_f and dividing through by v_s gives:

$$\frac{i_s}{v_s} = \frac{i_i}{v_s} + \frac{i_i}{v_s} BR_{mo}$$

or

$$R_{if} = \frac{R_i}{(1 + BR_{mo})} \tag{12.5}$$

where $R_{if} = v_s/i_s$ and $R_i = v_s/i_i$. The term in parentheses is typically >10 so that the resistance with feedback is less than the resistance without feedback.

Current–Shunt

For the circuit shown in Figure 12.6(b) the feedback component is:

$$i_f = Bi_o = Bi_i A_{io}$$

where A_{io} is the current gain without feedback. Substituting i_f in the equation for i_s and dividing through by v_s gives:

$$\frac{i_s}{v_s} = \frac{i_i}{v_s} + \frac{i_i}{v_s} BA_{io}$$

or

$$R_{if} = \frac{R_i}{(1 + BA_{io})} \tag{12.6}$$

The term in the denominator is typically >10 so that the resistance with feedback is less than without feedback.

In summary the input resistance is increased when the signal which is returned to the input is in series with the source, and it is decreased when the returned signal is in parallel with the source. Negative feedback provides a useful means for changing the impedance level of the input to an amplifier. For a voltage source such as a crystal microphone, it is better to have a high input resistance, while for a current source, such as a photodiode, a low input resistance acts as a current sink.

▶ **EXAMPLE 12.2** ——————————————————————

An amplifier has an input resistance of 1 kΩ and a gain of 100. Voltage–series feedback is applied to provide an overall gain of 10. Determine

(a) the feedback factor
(b) the input resistance with feedback.

▶ *Solution*

(a) $A_{vf} = \dfrac{A_{vo}}{(1 + BA_{vo})}$

$\quad 10 = \dfrac{100}{(1 + B \times 100)}$

$\quad \therefore \quad B = 0.09$

(b) $R_{if} = R_i(1 + BA_{vo})$

$\quad R_{if} = (1 \text{ k}\Omega)(1 + 0.09 \times 100) = 10 \text{ k}\Omega$

▶ **EXAMPLE 12.3** ——————————————————————

An amplifier has a current gain of 200 and an input resistance of 2.5 kΩ. Current–shunt feedback is applied to provide a loop gain (BA_{io}) of 9. Determine

(a) the current gain with feedback
(b) the input resistance with feedback.

▶ *Solution*

(a) $A_{if} = \dfrac{A_{io}}{(1 + BA_{io})} = \dfrac{200}{(1 + 9)} = 20$

(b) $R_{if} = \dfrac{R_{io}}{(1 + BR_{io})} = \dfrac{2.5 \text{ k}\Omega}{(1 + 9)} = 250 \ \Omega$

12.4 ▸ Output Resistance

The output resistance is also affected by the application of negative feedback. For negative feedback which samples the output voltage, regardless of how this signal is applied to the input, *the output resistance is reduced for voltage sampling*. There is a tendency for the voltage to remain constant irrespective of what happens to the output current or the load resistance. This is illustrated in Figure 12.7(a) where v_{out} is approximately equal to v_o if $R_{of} \ll R_L$.

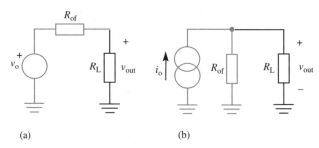

(a)   (b)

FIGURE 12.7   Equivalent circuits for the output with (a) voltage source for voltage sampling and (b) a current source for current sampling.

For negative feedback which samples the output current there is a tendency for the output current to remain constant provided $R_{of} \gg R_L$. This is illustrated in Figure 12.7(b) where the constant current source provides a constant output current if R_{of} is very large, that is *the output resistance is increased for current sampling*.

Voltage Sampling

The output resistance for voltage sampling with either series or shunt comparison at the input is:

$$R_{of} = \frac{R_o}{(1 + BA_o)} \tag{12.7}$$

where A_o is a voltage gain for voltage–series feedback, and a transresistance for voltage–shunt feedback. In both cases the denominator is likely to have a value which is >10, so that the output resistance with feedback is reduced.

Current Sampling

The output resistance for current sampling with either series or shunt comparison at the input is:

$$R_{of} = R_o(1 + BA_o) \tag{12.8}$$

where A_o is a transconductance for current–series, and a current gain for current–shunt. As for the previous equations the term in parentheses is likely to be >10 so that the output resistance with current sampling is greater with feedback than without it.

In summary the output resistance is decreased when the output voltage is sampled, which effectively stabilizes the output voltage, and it is increased when the output current is sampled. A low output resistance voltage source is useful for many amplifiers for driving a loudspeaker or a coaxial cable, while a high-impedance current source may be useful for driving an electromagnetic device, such as a motor or laser diode.

▶ **EXAMPLE 12.4** ——————————————————————————————

A voltage–series feedback amplifier is required to have an output resistance of 100 Ω and a voltage gain of 10. The basic amplifier without feedback has an output resistance of 5 kΩ.

(a) Determine the minimum value of gain required for the basic amplifier and the feedback factor.
(b) Determine the gain with feedback if the gain of the basic amplifier doubles.

▶ *Solution*

(a) $R_{of} = \dfrac{R_o}{(1 + BA_{vo})}$

$100 = \dfrac{5 \text{ k}\Omega}{(1 + BA_{vo})}$

and

$BA_{vo} = 49$

$A_{vf} = \dfrac{A_{vo}}{(1 + BA_{vo})}$

so

$A_{vo} = A_{vf}(1 + BA_{vo}) = 10(1 + 49) = 500$

and

$B = \dfrac{49}{500} = 0.098$

(b) $A_f = \dfrac{1000}{(1 + 0.098 \times 1000)} = 10.1$

12.5 ▷ Summary of Feedback Topologies

A summary of the effects of the different negative feedback topologies is shown in Table 12.1. Note that the transfer parameter differs for each of the four configurations. This does not mean that voltage–series feedback is the only configuration which exhibits voltage gain. *All* of the configurations exhibit voltage gain, although for current–shunt, current gain may be more appropriate, but even for this configuration voltage gain can be specified as:

$$A_{if} = \frac{i_o}{i_i} = \frac{i_o}{i_i} \frac{v_o}{v_o} \frac{v_i}{v_i} = A_{vf} \frac{R_{if}}{R_{of}} \tag{12.9}$$

or

$$A_{vf} = A_{if} \frac{R_{of}}{R_{if}}$$

TABLE 12.1   Negative feedback summary

Property	Voltage–series	Voltage–shunt	Current–series	Current–shunt
Feedback signal	Voltage	Current	Voltage	Current
Sampled signal	Voltage	Voltage	Current	Current
Input signal source	Thévenin	Norton	Thévenin	Norton
B	v_f/v_o	i_f/v_o	v_f/i_o	i_f/i_o
A_o	$A_{vo} = v_o/v_i$	$R_{mo} = v_o/i_i$	$G_{mo} = i_o/v_i$	$A_{io} = i_o/i_i$
A_f	$A_{vo}/(1 + BA_{vo})$	$R_{mo}/(1 + BR_{mo})$	$G_{mo}/(1 + BG_{mo})$	$A_{io}/(1 + BA_{io})$
R_{if}	$R_i(1 + BA_{vo})$	$R_i/(1 + BR_{mo})$	$R_i(1 + BG_{mo})$	$R_i/(1 + BA_{io})$
R_{of}	$R_o/(1 + BA_{vo})$	$R_o/(1 + BR_{mo})$	$R_o(1 + BG_{mo})$	$R_o(1 + BA_{io})$

Similar expressions may be obtained for voltage–shunt and current–series configurations, for example:

$$G_{mf} = \frac{i_o}{v_i} = \frac{i_o}{v_i} \frac{v_o}{v_o} = \frac{A_{vf}}{R_{of}} \tag{12.10}$$

or

$$A_{vf} = G_{mf} R_{of}$$

$$R_{mf} = \frac{v_o}{i_i} = \frac{v_o}{i_i} \frac{v_i}{v_i} = A_{vf} R_{if} \tag{12.11}$$

or

$$A_{vf} = \frac{R_{mf}}{R_{if}}$$

▶ **EXAMPLE 12.5** ─────────────────────────────────────

A voltage–shunt amplifier has an open-loop transresistance of 100 kΩ, an input resistance of 1.5 kΩ and an output resistance of 2 kΩ. If negative feedback is applied to give a loop gain of 4, determine the transresistance with feedback, the input resistance, the output resistance and the voltage gain with feedback.

▶ *Solution*

The transresistance with feedback is:

$$R_{mf} = \frac{-R_{mo}}{(1 + BR_{mo})} = \frac{-100 \text{ k}\Omega}{(1+4)} = -20 \text{ k}\Omega$$

The input and output resistances are:

$$R_{if} = \frac{R_i}{(1 + BR_{mo})} = \frac{1.5 \text{ k}\Omega}{5} = 300 \ \Omega$$

$$R_{of} = \frac{R_o}{(1 + BR_{mo})} = \frac{2.0 \text{ k}\Omega}{5} = 400 \ \Omega$$

The voltage gain is:

$$A_{vf} = \frac{-R_{mf}}{R_{if}} = \frac{-20 \text{ k}\Omega}{300 \ \Omega} \approx -67$$

Notice that the voltage gain is stabilized through R_{mf}. The voltage gain for an actual circuit will depend on the type of circuit and in particular the source resistance. If the source resistance is comparable with the input resistance, then this will greatly reduce the voltage gain, because of the loss of signal across the source resistance. For the circuit shown in Figure 12.8 the overall voltage gain from the source to the output is:

$$A_{vtotal} = \frac{v_{out}}{v_s} = \frac{v_{out}}{v_i} \frac{v_i}{v_s} = A_{vf} \frac{R_{if}}{(R_s + R_{if})}$$

The source resistance and input resistance act as a potential divider. If R_s is large then there will be a large loss of signal. For voltage–shunt feedback a voltage source should have a small source resistance, or alternatively it should be a current source with a large parallel source resistance.

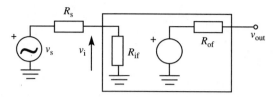

FIGURE 12.8 Effect of source resistance on the voltage gain of a feedback amplifier which has a low input resistance.

PSpice can be used to make a more detailed investigation for an actual circuit which involves voltage–shunt feedback, or any other form of feedback.

▶ **EXAMPLE 12.6** ─────────────────────────────────────

For the circuit in Example 12.5 the source resistance is 10 kΩ. Determine the overall voltage gain v_{out}/v_s with and without feedback.

▶ *Solution*

Without feedback the voltage gain is:

$$A_{vo} = \frac{-R_{mo}}{R_i} \frac{R_i}{(R_s + R_i)}$$

$$A_{vo} = \frac{-100 \text{ k}\Omega}{1.5 \text{ k}\Omega} \times \frac{1.5 \text{ k}\Omega}{(10 \text{ k}\Omega + 1.5 \text{ k}\Omega)} \approx -8.7$$

When feedback is applied then the gain is:

$$A_{vtotal} = A_{vf} \frac{R_{if}}{(R_s + R_{if})} = -67 \times 0.029 = -1.9$$

In practice an amplifier of this type would be used as a current to voltage amplifier, or transresistance amplifier, to convert a current from, say, a photodiode to a voltage.

12.6 ▶ **Other Factors Affected by Negative Feedback**

In the introduction it was noted that apart from stabilizing the gain and changing impedance levels, negative feedback also reduced distortion and noise.

Reduction of Distortion

Non-linear distortion usually arises when an amplifier is required to produce a large-amplitude signal, for example in a power amplifier. It is likely that under these conditions the signal will extend beyond the linear operating range of the active devices, and as a consequence the output signal is distorted. If the amount of distortion without feedback is D and with feedback D_f then when feedback is applied a fraction of this distortion BA_oD_f is returned to the input through the feedback network. The output now contains two terms, one due to the active device and one due to feedback. Thus:

$$D_f = D - BA_oD_f$$

or

$$D_f = \frac{D}{(1 + BA_o)} \tag{12.12}$$

That is, the distortion with feedback is reduced by the factor $1/(1 + BA_o)$.

EXAMPLE 12.7

Two amplifiers each with a gain of 15 are connected in cascade and negative feedback is applied to reduce the gain to 5. Determine

(a) the feedback ratio required
(b) the distortion with feedback if each amplifier contributes 5%.

Solution

(a) $A_f = \dfrac{15^2}{(1 + B \times 15^2)} = 5$

$\therefore B = 0.196$

(b) If each amplifier contributes distortion then the total is:

$D_{total} = D_1 + (D_1 \times D_2) + D_2$

$= 5\% + (5\% \times 5\%) + 5\%$

$= 10.25\%$

Then

$$D_f = \frac{D_{total}}{(1 + BA_o)} = \frac{10.25}{(1 + 0.196 \times 225)} = 0.23\%$$

The cancellation of the distortion suggested by equation 12.12 and the above example only occurs if the active devices are operating in their 'active' region. If the devices are driven into saturation or cut-off, then the output will be distorted irrespective of whether feedback is present or not. When the device is in saturation or cut-off the gain is very small or even reduced to zero and the loop gain in the denominator of equation 12.12 (BA_o) is also very small, or even zero. Under these conditions negative feedback has little effect. The 'active' region of an active device extends from saturation to cut-off, but may not be monotonically linear. Negative feedback will reduce any non-linearity which exists between the output and input signals in the 'linear' region.

▷ **EXAMPLE 12.8** ———————————————————————————————————————

In Figure 12.9 a non-inverting amplifier with a gain of 10 drives a push–pull output stage consisting of an npn transistor Q_1 and a pnp transistor Q_2. Negative feedback is applied to the operational amplifier by means of the potential divider R_2 and R_3. Because the transistors are operating as emitter followers the feedback connection may be moved to the junction of the emitters without changing the phase relationship within the feedback loop. With the aid of PSpice examine the output waveform for each feedback connection.

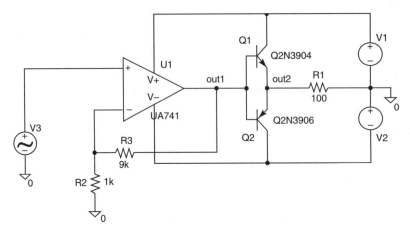

FIGURE 12.9   PSpice schematic of an amplifier output stage which exhibits distortion.

▷ *Solution*

The output voltage waveforms for the two positions of the feedback connection are shown in Figure 12.10. In 12.10(a) the feedback is only applied to the operational amplifier, while in 12.10(b) the feedback includes the output stage.

——

It can be seen from the two output waveforms that there is a reduction in the amount of distortion when the output stage is included in the negative feedback loop. However, the reduction is not as great as might be expected because the form of distortion known as *cross-over distortion* is caused by the transistors being cut off for a short period of time as the signal crosses the zero voltage axis. The loop gain during this period is very small so that the effect of negative feedback is greatly reduced.

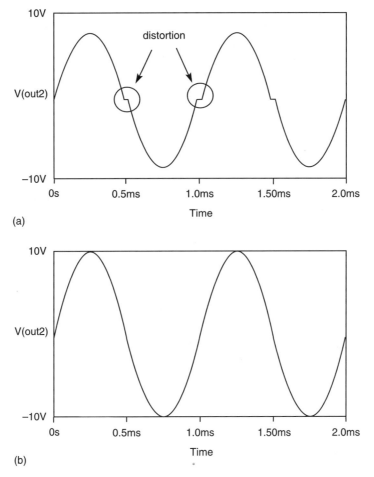

FIGURE 12.10   Output voltage across the load with (a) the feedback connection made to the common-base connection and (b) to the emitters.

Reduction of Electrical Noise

Negative feedback can be used to reduce electrical noise which is introduced into the amplifier signal path within the feedback loop, but note that feedback cannot reduce noise which enters with the signal from the source.

The reduction in the noise is given by:

$$N_f = \frac{N}{(1 + BA_o)} \tag{12.13}$$

One source of noise which may be represented by N in Figure 12.11 is the electrical signal which is introduced as a result of a poorly filtered dc power supply. This introduces 50 Hz (60 Hz) for half-wave rectification, or 100 Hz (120 Hz) for full-wave rectification, into the output of an amplifier. This form of noise is a particular problem for output stages, represented by the gain stage A_2, which operate at large current and voltage levels. It is difficult to provide adequate smoothing of the rectified output of a power supply if large currents are required. However, negative feedback can be used to reduce this form of noise. Provided that the pre-amplifier A_1 does not introduce any noise, then the noise represented by N is reduced when feedback is applied by a factor of $(1 + BA_0)$. For audio amplifier design A_1 represents the low-noise pre-amplifier stage and A_2 is the power-output stage.

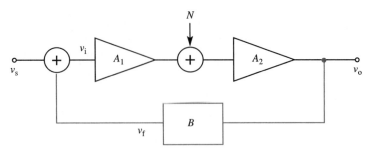

FIGURE 12.11   Reduction of noise by means of negative feedback.

Amplifier Bandwidth

The overall gain of an amplifier is given by equation 12.1 as:

$$A_f = \frac{v_o}{v_s} = \frac{A_o}{1 + BA_o}$$

and provided that the loop gain is large then the overall gain is approximated to:

$$A_f \approx \frac{1}{B}$$

There is no suggestion in either of these two equations that the gain varies with frequency. They are both based on the mid-band gain where the gain is constant and independent of frequency. In practice the gain of the amplifier is frequency dependent, and for a capacitively coupled transistor amplifier there will be a low-frequency 3 dB point caused by the coupling capacitors, and a high-frequency 3 dB point caused by the stray capacitance.

In Chapter 9 it was shown that the corner frequencies could be represented by simple RC networks. The voltage relationship for the low-frequency 3 dB point is given by equation 9.3 for a high-pass filter as:

$$v_{out} = \frac{v_{in}}{R + (1/j\omega RC)} R$$

If the network forms part of an amplifier then the frequency-dependent gain can be expressed as:

$$A(jf) = \frac{A_o}{1 + (1/j\omega RC)} = \frac{A_o}{1 - j(f_L/f)} \tag{12.14}$$

where A_o is the mid-band gain and $f_L = 1/2\pi RC$ is the 3 dB corner frequency for the high-pass filter. Notice the appearance of the negative sign in the denominator as a result of repositioning the j term.

Equation 12.14 describes the variation of gain with frequency for a high-pass filter, that is the low-frequency response of the amplifier. If this value of gain is substituted into the expression for the gain with feedback then:

$$A_f(jf) = \frac{A_o/[1 - j(f_L/f)]}{1 + \{BA_o/[1 - j(f_L/f)]\}}$$

and with some rearrangement:

$$A_f(jf) = \frac{A_o}{1 + BA_o - j(f_L/f)}$$

$$A_f(jf) = \frac{A_o/(1 + BA_o)}{1 - j[f_L/(1 + BA_o)f]}$$

and

$$A_f(jf) = \frac{A_{of}}{1 - j(f_{LF}/f)} \tag{12.15}$$

where A_{of} is the mid-band gain given by:

$$A_{of} = \frac{A_o}{1 + BA_o}$$

and the 3 dB corner frequency is:

$$f_{LF} = \frac{f_L}{1 + BA_o} \tag{12.16}$$

A similar expression may be obtained for the high-frequency 3 dB point by using equation 9.12 (see Problem 12.18). Then the 3 dB corner frequency for the low-pass filter is:

$$f_{HF} = f_H(1 + BA_o) \tag{12.17}$$

Examination of equations 12.16 and 12.17 reveals that the use of negative feedback has reduced the lower 3 dB point by $(1 + BA_o)$ and has increased the upper 3 dB point by the same factor. Thus the overall bandwidth $(f_H - f_L)$ has been increased. This is illustrated in Figure 12.12.

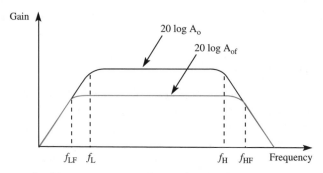

FIGURE 12.12   Bandwidth improvement with negative feedback.

It can be seen that the bandwidth has been improved but with a loss of gain. For most applications the improved bandwidth is of far greater benefit than the loss of gain. This is particularly true for operational amplifiers which have an excess of gain.

▶ **EXAMPLE 12.9** ——————————————————————————————————

An operational amplifier has an open-loop frequency-dependent gain given by:

$$A_f(jf) = \frac{10^5}{1 + j(f/10)}$$

and is connected as a non-inverting amplifier with a feedback factor $B = 0.01$. Determine the closed-loop gain expression, the mid-band gain and the 3 dB corner frequency.

▶ *Solution*

The frequency-dependent gain expression is:

$$A_f(jf) = \frac{10^5/[1 + (j(f/10))]}{1 + 0.01\{10^5/[1 + (j(f/10))]\}}$$

$$A_f(jf) = \frac{10^5}{1001 + j(f/10)}$$

This can be rewritten as:

$$A_f(jf) = \frac{10^5/1001}{1 + j(f/10010)} = \frac{A_{of}}{1 + j(f/f_H)}$$

where $A_{of} = 99.9$ and $f_H = 10.01$ kHz.

Practical Analysis of Feedback Amplifiers

The application of the basic feedback relationships described above depends on a knowledge of the feedback factor and the open-loop gain. To obtain these two parameters it is first necessary to identify within the complete circuit the basic amplifier, without feedback, and the feedback network.

The feedback forms a link between the output and input, and this link must be disabled to obtain the open-loop parameters. This process of disabling the feedback must be done for both the input and the output. At the input of the amplifier the feedback is disabled for voltage-derived feedback by setting the small-signal output voltage to zero, while for current-derived feedback the output current is set to zero.

At the output the feedback is disabled for a series-fed circuit by setting the small-signal input current to zero, while for shunt-fed feedback the input voltage is set to zero.

These actions ensure that the amplifier without feedback retains the same loading effects which are caused by the feedback network.

The unbypassed emitter resistor of a common-emitter amplifier is an example of current–series feedback. The basic amplifier is shown in Figure 12.13(a) which is redrawn in 12.13(b) to show the amplifier and feedback network as two separate parts of the circuit.

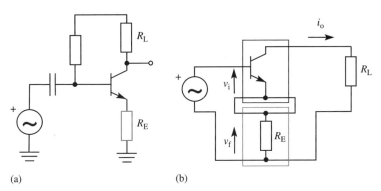

(a)   (b)

FIGURE 12.13   (a) Amplifier with unbypassed emitter resistor and (b) the amplifier and feedback network shown separately.

The configuration shown in Figure 12.13(b) corresponds to the topology for current–series (refer to Figure 12.4(c)) feedback. The feedback is disabled by open-circuiting the output, which places R_E in series with the input, and by open-circuiting the input which places R_E in the output circuit. The complete circuit with the feedback disabled, but including the effect of the feedback resistor, is shown in Figure 12.14.

The circuit in Figure 12.14 would be used to determine the open-loop gain, and the open-loop input and output resistances. In practice it is far simpler to analyze the circuit without reference to negative feedback. It is only presented here to illustrate how the analysis would proceed based on the application of negative feedback theory.

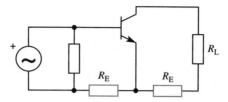

FIGURE 12.14   Amplifier without feedback but including the loading produced by the feedback network.

The emitter follower and source follower are both examples of voltage sampling with series feedback as illustrated in Figure 12.15(a) and with the amplifier without feedback shown in 12.15(b).

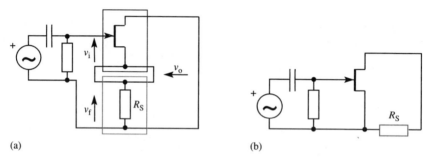

(a)   (b)

FIGURE 12.15   (a) Source follower with boxes around the amplifier and feedback network to show it to be voltage–series and (b) the amplifier without feedback.

Again it is far simpler to analyze the emitter follower by application of normal KVL rules to the equivalent circuit without recourse to negative feedback methods. However, for more complex circuits involving a number of active devices, then it may be easier to use feedback methods.

12.7 Some Examples of Negative Feedback

With four feedback configurations and many different transistor circuit configurations there are many examples of negative feedback, some of which are very specialized and apply to only one particular application. Only two examples are considered here to provide an indication of how feedback is applied and the circuits analyzed by application of feedback theory.

Voltage–Series Feedback

Voltage–series feedback offers a high input impedance and a low output impedance. This is a useful specification for many applications as it reduces the loading on the source and provides a low-resistance voltage source output. A two-transistor amplifier with voltage–series feedback is shown in Figure 12.16.

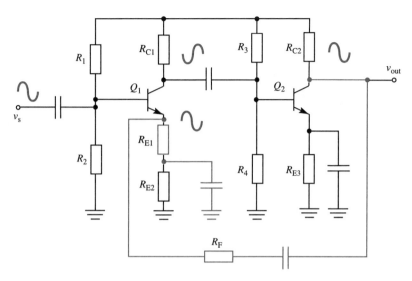

FIGURE 12.16   Two-stage amplifier with voltage–series feedback.

Notice the phase of the voltages at the different nodes, and in particular that the emitter of Q_1 generates a voltage across R_{E1} which opposes v_s. For negative feedback it is important for the phase of the signal at the output sampling point to generate a signal opposite to that of the input at the summing point, as is the case in Figure 12.16.

The output signal which is being sampled is a voltage, and the feedback resistor R_F together with the emitter resistor form a potential divider across the output to provide a voltage sample which is in series with the input signal as shown in Figure 12.17.

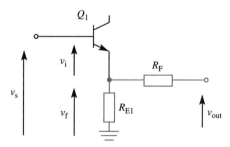

FIGURE 12.17   Simplified schematic showing the negative feedback.

The feedback corresponds to the feedback topology shown in Figure 12.4(a). The feedback factor is obtained by considering the relationship between v_f and v_{out}. From

Figure 12.17 it is:

$$v_f = \frac{v_{out} R_{E1}}{R_{E1} + R_F}$$

and the feedback factor is:

$$B = \frac{v_f}{v_{out}} = \frac{R_{E1}}{R_{E1} + R_f} \tag{12.18}$$

The transfer parameter for voltage–series feedback is voltage gain, and from Table 12.1 the gain with feedback is:

$$A_{vf} = \frac{A_{vo}}{(1 + BA_{vo})}$$

The input resistance is:

$$R_{if} = R_i(1 + BA_{vo})$$

This value of resistance only applies to the circuit contained within the feedback loop. This loop does not include the bias resistors for Q_1 and therefore the effective input resistance as seen by the source is:

$$R_{in(effective)} = R_1 \,\|\, R_2 \,\|\, R_{if} \tag{12.19}$$

With voltage sampling the output resistance is:

$$R_{of} = \frac{R_o}{(1 + BA_{vo})} \tag{12.20}$$

Open-Loop Gain

The equations for voltage gain, input and output resistance require a knowledge of the open-loop gain and the feedback factor. The feedback network has already been identified and the feedback factor determined in equation 12.18. The open-loop voltage gain is a little more difficult to calculate, because of the presence of the feedback network. For voltage sampling the input circuit is realized by shorting the small-signal output voltage to ground. This has the effect of placing the feedback resistor R_F in parallel with R_{E1}. Usually $R_F \gg R_{E1}$ and its effect on the input is very small. The gain of the first stage, including R_F, is:

$$A_1 = -\frac{R_{C1} \,\|\, R_{in2}}{r_{e1} + R_{E1} \,\|\, R_F}$$

where R_{in2} is the input resistance of the second stage.

For the output circuit to disable the feedback the input small-signal current is reduced to zero by making a break in the input at the emitter of Q_1. This action places R_{E1} and R_F in series as an additional load across the output. The gain of the second stage, including the effect of the feedback components, is:

$$A_2 = -\frac{R_{C2} \| (R_{E1} + R_F)}{r_{e2}}$$

The complete open-loop gain is:

$$A_{vo} = A_1 \times A_2$$

In practice it would be sufficiently accurate to ignore the effect of R_F by assuming that it is $\gg R_E$. The value of gain would then be slightly higher than the actual value, but the difference will not be great, and a circuit simulator would provide a more accurate check.

▷ **EXAMPLE 12.10** ————————————————————————————

An example of voltage–series feedback is shown in Figure 12.18 which is a schematic from PSpice. Determine the voltage gain, input resistance and output resistance assuming that β is 250 with a dc supply of 9 V.

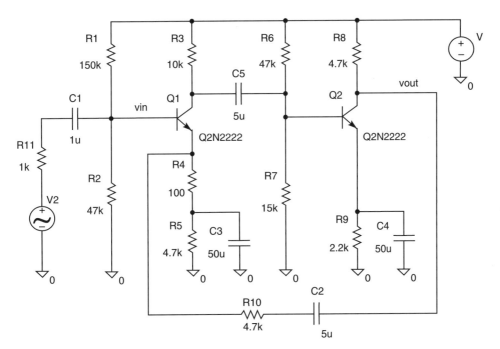

FIGURE 12.18   PSpice schematic of a two-transistor amplifier with voltage–series feedback.

▶ *Solution*

DC Conditions

For Q_1 the collector current is:

$$I_{C1} = \frac{V_T - 0.7}{4.8 \text{ k}\Omega + (R_T/250)} \approx 0.3 \text{ mA}$$

where $V_T = 2.15$ V and $R_T = 35.8$ kΩ.
  For Q_2 the collector current is:

$$I_{C2} = \frac{V_T - 0.7}{2.2 \text{ k}\Omega + (R_T/250)} \approx 0.7 \text{ mA}$$

where $V_T = 2.18$ V and $R_T = 11.4$ kΩ.

AC Conditions

The voltage gain for Q_1 taking account of R_F is:

$$A_{v1} = \frac{R_{C1} \| R_{i2}}{r_{e1} + R_{E1} \| R_F}$$

where

$$R_{i2} = 47 \text{ k}\Omega \| 15 \text{ k}\Omega \| \beta r_{e2}$$

and

$$r_{e1} = \frac{0.026}{0.3 \text{ mA}} = 86.6 \ \Omega$$

and

$$r_{e2} = \frac{0.026}{0.7 \text{ mA}} = 37.1 \ \Omega$$

$$\therefore \quad A_{v1} = -\frac{10 \text{ k}\Omega \| 5.1 \text{ k}\Omega}{86.6 \ \Omega + 98 \ \Omega} \approx -18$$

The voltage gain for Q_2 is:

$$A_{v2} = -\frac{R_{C2} \| (R_{E1} + R_F)}{r_{e2}}$$

$$A_{v2} = -\frac{4.7 \text{ k}\Omega \| 4.8 \text{ к}\Omega}{37.1 \ \Omega} = -64$$

The overall gain is:

$$A_{vo} = A_{v1} \times A_{v2} = 1170$$

The input resistance without feedback is:

$$R_{in} = \beta(r_{e1} + R_{E1} \| R_F) = 250(86.6 + 98) \approx 46 \text{ k}\Omega$$

Note that this calculation excludes the bias resistors for Q_1 because these are outside the feedback loop.

The output resistance is:

$$R_o = 4.7 \text{ k}\Omega \| 4.8 \text{ k}\Omega = 2.4 \text{ k}\Omega$$

The feedback factor is:

$$B = \frac{R_{E1}}{R_{E1} + R_F} = \frac{100}{100 + 4.7 \text{ k}\Omega} = 0.0208$$

The voltage gain with feedback is:

$$A_{vF} = \frac{A_{vo}}{1 + BA_{vo}} = \frac{1170}{1 + 0.0208 \times 1170} = 46 \approx 33 \text{ dB}$$

The input resistance with feedback, but excluding the bias resistors, is:

$$R_{if} = R_{in}(1 + BA_{vo}) = (46 \text{ k}\Omega)(1 + 0.0208 \times 1170) \approx 1 \text{ M}\Omega$$

The input resistance including the bias resistors is:

$$R_{if}|_{eff} = 150 \text{ k}\Omega \| 47 \text{ k}\Omega \| 1 \text{ M}\Omega \approx 35 \text{ k}\Omega$$

The output resistance with feedback is:

$$R_{of} = \frac{R_o}{(1 + BA_{vo})} = \frac{2.4 \text{ k}\Omega}{(1 + 0.0208 \times 1170)} \approx 95 \text{ }\Omega$$

PSpice Results

The overall gain with feedback is:

$$A_f = 32.9 \text{ dB}$$

The input resistance is:

$$R_{if} = 33 \text{ k}\Omega$$

The output resistance is obtained by connecting the signal source V_2 to the output by means of a capacitor (10 µF). From small-signal analysis the output resistance is the ratio of v_{out} to the current flowing in the 10 µF capacitor. These values are obtained from Probe in the mid-band and the output resistance is:

$$R_{of} = 116 \text{ }\Omega$$

It can be seen that there is good agreement between the calculated results and those obtained with PSpice.

Current–Shunt Feedback

The voltage–series feedback circuit provides a high input resistance and a low output resistance with a wide bandwidth which is an important requirement for many applications. As an alternative the current–shunt feedback circuit provides a low input resistance with a high output resistance. This is particularly useful as a current amplifier.

▶ **EXAMPLE 12.11** ────────────────────────

An example of current–shunt feedback is shown in Figure 12.19. The actual output is taken from the collector of Q_2, while the connection for the feedback network is to the emitter of Q_2. Since the output voltage is not being sampled it must be the output current, and because the feedback components comprising R_5 and R_2 are in parallel with the input, the feedback is current–shunt. Resistor R_6 is part of the dc bias for Q_1 and capacitors C_2 and C_3 are decoupling capacitors which act as short circuits over the operating frequency range.

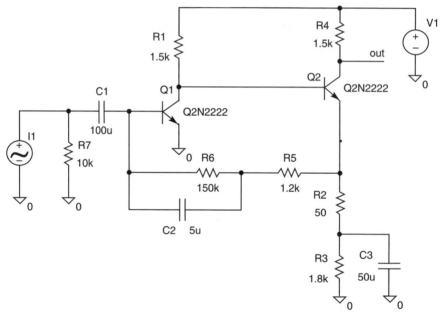

FIGURE 12.19 PSpice schematic of a two-transistor amplifier with current–shunt feedback.

The advantage of a current–shunt feedback circuit is the low input impedance and high output impedance. It is shown in Figure 12.19 with a current source which could represent a photodiode which generates a small-signal current proportional to the incident light intensity. The dc supply voltage is 9 V and $\beta = 250$ for each of the transistors. Determine the gain and input and output resistances.

▷ Solution

DC Conditions

Note that Q_1 derives its base current from the voltage generated across the emitter resistors for Q_2. This dc feedback stabilizes the current for the transistor pair.

A simplified bias circuit for Q_1 is shown in Figure 12.20(a), where the dc voltage which provides the base current is replaced by a voltage source (V_{RE2}). Applying KVL gives:

$$V_{RE2} = \frac{I_{C1}}{\beta}\, 151 \text{ k}\Omega + 0.7 \text{ V}$$

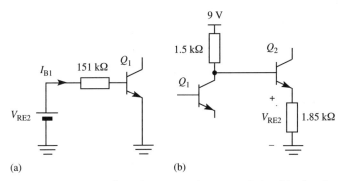

FIGURE 12.20   (a) Circuit used to determine the KVL relationship for Q_1 and (b) the circuit required for Q_2.

From the simplified circuit for Q_2 the application of KVL gives:

$$9 \text{ V} = I_{C1}(1.5 \text{ k}\Omega) + 0.7 \text{ V} + V_{RE2} \quad \text{ignoring } I_{B2}$$

Substituting for V_{RE2} and rearranging gives:

$$I_{C1} = \frac{9 \text{ V}}{1.5 \text{ k}\Omega + (151 \text{ k}\Omega/250)} = 4.3 \text{ mA}$$

and

$$r_{e1} = \frac{0.026}{4.3 \text{ mA}} = 6.0 \text{ }\Omega$$

Solving for V_{RE2} based on a knowledge of I_{C1} gives:

$$V_{RE2} = \frac{4.3 \text{ mA}}{250}\, 151 \text{ k}\Omega + 0.7 \text{ V} = 3.3 \text{ V}$$

and

$$I_{C2} = \frac{V_{RE2}}{R_{RE2}} = \frac{3.3 \text{ V}}{1.85 \text{ k}\Omega} = 1.78 \text{ mA}$$

and

$$r_{e2} = \frac{0.026}{1.78 \text{ mA}} = 14.5 \ \Omega$$

AC Conditions

The open-loop conditions may be determined by taking note of the type of feedback and then disabling the feedback with the appropriate open circuit or short circuit at the input and output. With current sampling the effect of the feedback network on the input may be obtained by setting the small-signal output current to zero. This may be achieved by disconnecting the emitter of Q_2 from the junction of R_2 and R_5. In the small-signal circuit this results in R_2 and R_5 appearing in series across the input of Q_1. For shunt comparison the output circuit is obtained by short-circuiting v_i which places R_5 in parallel with R_2. With these modifications the open-loop voltage gain of Q_1 is:

$$A_{vol} = -\frac{R_{C1} \| R_{i2}}{r_{e1}}$$

where R_{i2} is the input resistance of Q_2. It is given by:

$$R_{i2} = \beta(r_{e2} + R_2 \| R_5) = 250(14.5 \ \Omega + 50 \ \Omega \| 1.2 \text{ k}\Omega) \approx 16 \text{ k}\Omega$$

and

$$A_{vol} = -\frac{1.5 \text{ k}\Omega \| 16 \text{ k}\Omega}{6.0 \ \Omega} = -228$$

The open-loop voltage gain of Q_2 is:

$$A_{vo2} = -\frac{R_{C2}}{(r_{e2} + R_2 \| R_5)} = -\frac{1.5 \text{ k}\Omega}{(14.5 \ \Omega + 50 \ \Omega \| 1.2 \text{ k}\Omega)} = -24$$

The overall open-loop gain from the base of Q_1 to the collector of Q_2 is:

$$A_{voT} = A_{vol} \times A_{vo2} \approx 5477 \approx 75 \text{ dB}$$

The feedback network is shown in Figure 12.21(a) with the equivalent network representation showing current division in parallel resistors in 12.21(b).

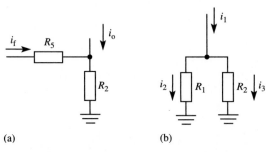

(a)   (b)

FIGURE 12.21 Simplified circuit of the feedback network with (a) the actual circuit and (b) parallel resistors showing current division.

From Figure 12.21(b) current division gives the current through R_1 as:

$$i_2 = \frac{i_1 R_2}{R_1 + R_2}$$

Thus based on the same relationship the current flowing in R_5 is:

$$i_f = -\frac{i_o R_2}{R_5 + R_2}$$

and the feedback factor is:

$$B = \frac{i_f}{i_o} = \frac{50\ \Omega}{1.2\ \text{k}\Omega + 50\ \Omega} = 0.04$$

The input resistance without feedback must take account of the loading produced by the feedback network, which for the modified circuit is obtained by open-circuiting the output to place R_5 and R_2 in series and in shunt with the input, that is:

$$R_i = (R_2 + R_5)\,\|\,\beta r_{e1} = (50\ \Omega + 1.2\ \text{k}\Omega)\,\|\,(250 \times 6.0\ \Omega) = 680\ \Omega$$

The open-loop output resistance is:

$$R_o = R_{C2} = 1500\ \Omega$$

The open-loop current gain is obtained from equation 12.9 as

$$A_{io} = A_{vo}\frac{R_i}{R_o} = 5472\ \frac{680\ \Omega}{1500\ \Omega} = 2480 \approx 68\ \text{dB}$$

The current gain with feedback is:

$$A_{if} = \frac{A_{io}}{1 + BA_{io}} = \frac{2165}{(1 + 0.04 \times 2480)} \approx 25 \approx 28\ \text{dB}$$

The input resistance with feedback is:

$$R_{if} = \frac{R_i}{1 + BA_{io}} = \frac{680\ \Omega}{(1 + 0.04 \times 2480)} \approx 7\ \Omega$$

The output resistance with feedback is:

$$R_{of} = R_o(1 + BA_{io}) = (1.5\ \text{k}\Omega)(1 + 0.04 \times 2165) \approx 1.3\ \text{M}\Omega$$

The effective output resistance, however, is:

$$R_{of}\big|_{\text{eff}} = R_{C2}\,\|\,R_{of} = 1.5\ \text{k}\Omega\,\|\,1.5\ \text{M}\Omega \approx 1.5\ \text{k}\Omega$$

Based on the PSpice simulation with and without feedback the following set of values are obtained. The feedback is disabled by either removing the capacitor C_2 or reducing it to a very small value for example 1 pF. The absence of the capacitor means that the 150 kΩ resistor becomes part of the small-signal feedback path, and with such a large value the feedback current i_f is very small.

$$A_{io} = \frac{I(R_4)}{I(C_1)} = 64 \text{ dB} \qquad A_{if} \approx 28 \text{ dB}$$

$$R_i = \frac{V(Q_1:b)}{I(C_1)} = 688 \ \Omega \qquad R_{if} = 10 \ \Omega$$

$$R_o = 1.5 \text{ k}\Omega \qquad\qquad R_{of} = 1.5 \text{ k}\Omega$$

There is good agreement between the hand calculations and PSpice simulated results, even though approximations have been used wherever possible for the hand-based calculations.

PROBLEMS

12.1 The nominal value of the open-loop gain of an amplifier is 1000. Determine the feedback factor (B) and the permissible range of open-loop gain if the gain with feedback is to be 100 ± 1.

[0.009, 908, 1110]

12.2 An amplifier has a nominal value of open-loop gain of 8000 with a minimum value of 4000 and a maximum value of 16000. If the feedback factor is 0.02, determine the closed-loop gain and the percentage variation.

[-0.6%, $+0.3\%$]

12.3 Identify the conditions, with respect to the loop gain, which determine whether the feedback is negative or positive. Under what circumstances may feedback change from being negative to becoming positive?

12.4 Identify the four feedback topologies and the units of the transfer parameter for both the amplifier and the feedback network for each topology.

12.5 State whether the input resistance is increased or decreased for each of the four topologies.

12.6 Repeat Problem 12.5 for the output resistance.

12.7 An amplifier without feedback has an open-loop gain of 1000 and an output resistance of 3 kΩ. Determine the type of feedback and the feedback factor which is required to reduce the output resistance to 30 Ω. Determine the closed-loop gain.

[$A_f = 10$]

12.8 Identify the type of feedback for the circuit shown and show that the input resistance with feedback is given by:

$$R_{if} = \frac{R_i}{(1 + BR_{mo})}$$

where B is the feedback factor $1/R_F$ and R_{mo} is the open-loop transfer function $-v_o/i_i$.

Hint: Start by summing the currents at the input node.

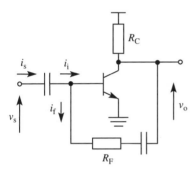

12.9   If for the circuit in Problem 12.8, $R_C = 1.5$ kΩ, $R_F = 4.7$ kΩ, $r_e = 25$ Ω and $\beta = 150$, determine the input resistance.

Hint: See Section 12.6 about the disabling of the feedback in order to determine the open-loop parameters.

[~100 Ω]

12.10 As an alternative approach to the feedback formula derived in Problem 12.8 for the input resistance, use Miller's theorem to determine the input resistance based on the values given in Problem 12.9.

Hint: The resistor R_F is in parallel with R_C for the gain calculation.

12.11 If the input resistance without feedback is 1.5 kΩ and the open-loop gain is 2000, determine the input resistance for series-applied feedback for which the closed-loop gain is 50.

[60 kΩ]

12.12 Identify the type of feedback for the circuit shown and show that the input resistance can be expressed as:

$$R_{if} = R_i(1 + BG_{mo})$$

where $B = -R_E$ and G_{mo} is the open-loop transfer function i_o/v_i.

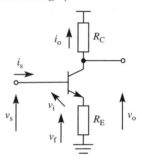

12.13 A feedback amplifier of the type illustrated in Problem 12.12 has an open-loop voltage gain of 100, an output resistance without feedback of 200 Ω and an input resistance of 1.5 kΩ. Determine the gain with feedback if the feedback factor is −100 Ω.

[~75 kΩ]

12.14 An audio amplifier consists of a pre-amplifier with a gain of 500 and distortion of 5%, and an output stage with unity gain and 20% distortion. Determine the feedback factor which is required to provide an overall gain of 10 and the resultant distortion.

[0.098, 0.52%]

12.15 A pre-amplifier has a distortion of 5% and a unity gain output stage has 12%. Determine the minimum value of gain for the pre-amplifier and the feedback factor to provide an overall distortion with feedback of 0.1% with an overall gain of 20.

[>350, 0.5]

12.16 An amplifier is designed to have a symmetrical undistorted sine wave output of ±10 V with a gain of 100. When an input signal of ±150 mV is applied severe distortion is noted caused by the amplifier saturating and going into cut-off. Will negative feedback improve the situation if the overall gain with feedback is 10 and the input signal is increased by a factor of 10 to compensate?

12.17 Identify a possible source of noise which can be reduced by application of negative feedback.

12.18 The frequency-dependent gain of an amplifier with a low-pass filter is given by:

$$A(jf) = \frac{A_o}{1 + j\omega RC}$$

Show that the 3 dB frequency with feedback is given by $f_{HF} = f_H(1 + BA_o)$ where $f_H = 1/2\pi RC$ and is the corner frequency without feedback.

12.19 An operational amplifier has an open-loop gain which is defined as:

$$A_{o(f)} = \frac{A_o}{1 + j(f/f_H)}$$

where A_o is 10^5 and f_H is the 3 dB corner frequency with a value of 10 Hz. The amplifier is connected as a non-inverting amplifier with a feedback factor (B) of 0.02. Determine the gain and the 3 dB frequency with feedback.

[50, 20 kHz]

12.20 An amplifier has an open-loop mid-band gain of 60 dB with lower and upper 3 dB corner frequencies each determined by single RC time constants of 100 Hz and 18 kHz. Determine the feedback factor for voltage–series feedback to give a mid-band gain of 40 dB, and determine the new corner frequencies.

[0.009, 10 Hz, 180 kHz]

12.21 For the amplifier in Problem 12.20 determine the changes which are permissible in the open-loop gain if the closed-loop gain is not to vary by more than ±0.5 dB.

[56–67 dB]

12.22   Verify that the feedback shown is voltage–series and that there is a correct phase relationship between the input and output nodes to which the feedback network is connected, and determine the closed-loop gain.

[40 dB]

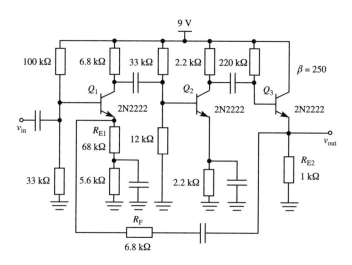

12.23 For the circuit shown in Problem 12.22 determine the open-loop gain if $r_{e1} = 98\ \Omega$, $r_{e2} = 34\ \Omega$, $r_{e3} = 6\ \Omega$ and $\beta = 250$.

[~1000]

12.24 For the circuit in Problem 12.22 determine the gain with feedback and the input and output resistances.

[40 dB, ~23 kΩ, ~1 Ω]

12.25   Use PSpice to verify the values of closed-loop gain, input resistance and output resistance for the circuit in Problem 12.22.

12.26   Identify the type of feedback for the circuit shown and show that the closed-loop gain is given approximately by:

$$A_f \approx \frac{R_1 + R_2}{R_2} = 21\ \text{dB}$$

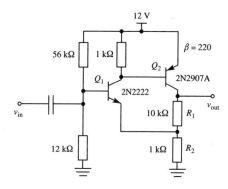

12.27 If for the circuit shown in Problem 12.26 $r_{e1} = 37 \ \Omega$, $r_{e2} = 31 \ \Omega$ and $\beta_1 = 220$, determine the open-loop voltage gain.

[~320]

12.28 For the circuit in Problem 12.26 determine the input and output resistances with feedback.

[~10 kΩ, ~360 Ω]

12.29 Use PSpice to verify the values of closed-loop gain and input and output resistance for the circuit shown in Problem 12.26.

CHAPTER 13

Operational Amplifier Applications

The differential amplifier described in Chapter 11 provides the basis for the first stage of the operational amplifier. Additional stages of gain are added to provide an overall gain which is generally in excess of 100000. This value of gain is far too large for normal use, and negative feedback, by means of external components, is used to obtain more realistic values of gain. It was shown in Chapter 12 that negative feedback can be used to stabilize the gain, modify impedance levels and generally improve circuit performance. For many of the expressions for gain, impedance, frequency response, distortion, it is usual to assume that the loop gain BA_o is large. For the operational amplifier this is easy to achieve with values of A_o in excess of 100000, and even though this value may vary by a factor of 2 or 3, this has little effect on the different parameters when feedback is applied.

A variety of simple circuits can be constructed from one or more operational amplifiers plus some external resistors to provide gain, to produce the sum or difference of two inputs, to act as comparators and to form simple switching circuits. There are many other applications, but there are also many specialist texts which describe such applications in detail, and it is not the purpose of this text to duplicate such material.

13.1 Non-Inverting Amplifier

In the circuit shown in Figure 13.1 the input signal is applied to the non-inverting terminal (+) and the output sine wave is shown to be in phase with the input sine wave.

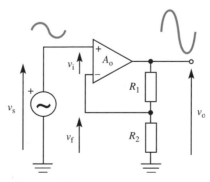

FIGURE 13.1 The non-inverting amplifier.

The resistors R_1 and R_2 act as a potential divider across the output and the signal from the centre-tap (v_f) is fed back to the inverting terminal.

The simple application of KVL to the input results in:

$$v_s = v_i + v_f$$

The output v_o is produced by the voltage which appears across the two input terminals, that is:

$$v_o = A_o v_i$$

The voltage which is returned to the inverting terminal from the potential divider at the output is:

$$v_f = \frac{v_o}{R_1 + R_2} R_2 = B v_o$$

where B is the feedback factor given by:

$$B = \frac{R_2}{R_1 + R_2} \tag{13.1}$$

and by substitution the input voltage can be expressed in terms of the output voltage as:

$$v_s = \frac{v_o}{A_o} + B v_o$$

The overall voltage gain of the amplifier can be expressed as the ratio of the output voltage to the source voltage, that is:

$$A_f = \frac{v_o}{v_s} = \frac{A_o}{1 + B A_o} \tag{13.2}$$

Equation 13.2 is of the same form as that developed for the general case of negative feedback in Chapter 12. The value of A_o is usually in excess of 10^5, and some texts describe the behaviour of an ideal operational amplifier as having an open-loop gain approaching infinity, that is $A_o \rightarrow \infty$. The resistor ratio B is typically 10^{-1}, so that $BA_o \gg 1$ and thus

$$A_f \approx \frac{1}{B} = \frac{R_1 + R_2}{R_2}$$

This is usually written as:

$$A_f = 1 + \frac{R_1}{R_2} \tag{13.3}$$

Equation 13.3 represents the gain expression for the non-inverting amplifier. It is independent of the actual gain (or open-loop gain) of the operational amplifier, and the absence of a negative sign indicates that there is no phase inversion of the output signal. The overall gain is very accurately predicted by means of the ratio of two resistors, which could be close tolerance and have a low temperature coefficient.

In terms of negative feedback topologies described in Chapter 12 the non-inverting amplifier may be redrawn as shown in Figure 13.2 to show the amplifier and feedback network separately.

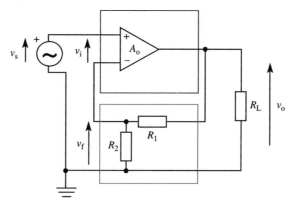

FIGURE 13.2 Non-inverting amplifier redrawn to show voltage–series feedback.

It can be seen that the network samples the output voltage and feeds a voltage (v_f) back to the input which is in series with v_s and v_i. It is an example of voltage–series feedback.

Input Resistance

For voltage–series feedback the input resistance is given by:

$$R_{if} = R_i(1 + BA_o) \tag{13.4}$$

This expression may also be obtained directly from the circuit for the non-inverting amplifier as shown in Figure 13.3. The input resistance of the basic amplifier is shown as R_i.

Applying KVL to the input gives:

$$v_s = v_i + v_f$$
$$= v_i + BA_o v_i$$
$$= v_i(1 + BA_o)$$

where v_f is replaced by the voltage developed across the potential divider, and v_o is replaced by $A_o v_i$. The voltage across the input terminals of the amplifier is developed

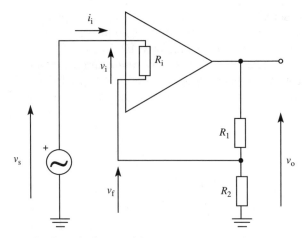

FIGURE 13.3   Circuit for the calculation of the input resistance of a non-inverting amplifier.

across the input resistance of the basic amplifier ($v_i = i_i R_i$) and thus:

$$v_s = i_i R_i (1 + BA_o)$$

The input resistance of the amplifier with feedback is described by v_s / i_i. Thus the input resistance with feedback is:

$$R_{if} = \frac{v_s}{i_i} = R_i(1 + BA_o)$$

and is the same as that predicted by equation 13.4. As with all series feedback circuits the input resistance is increased. The open-loop input resistance (R_i) is typically 1 MΩ for a bipolar 741 amplifier, but may be >100 MΩ for FET input amplifiers. With feedback this is increased still further.

Output Resistance

The output resistance is much more difficult to obtain directly from the circuit and it is simpler to take advantage of the relationship which exists for the output resistance for voltage sampling, which is:

$$R_{of} = \frac{R_o}{1 + BA_o} \tag{13.5}$$

where R_o is the output resistance of the basic amplifier as shown in Figure 13.4.

When feedback is applied the output resistance with feedback (R_{of}) is less than the output resistance without feedback. The output resistance without feedback may typically be less than 100 Ω so that with feedback it is reduced to a very small value.

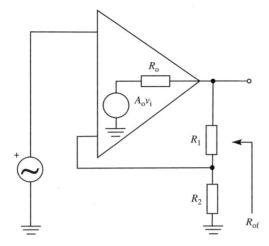

FIGURE 13.4   Output resistance of the non-inverting amplifier.

EXAMPLE 13.1

For the non-inverting amplifier shown in Figure 13.5 the data sheet for the amplifier specifies an input resistance of 2 MΩ, an output resistance of 100 Ω and an open-loop gain of 100000. Determine the input and output resistances with feedback and the closed-loop gain.

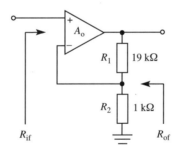

FIGURE 13.5   Non-inverting amplifier.

Solution

The feedback factor is obtained from the resistor network as:

$$B = \frac{R_2}{R_1 + R_2} = 0.05$$

The input resistance with feedback is:

$$R_{if} = R_i(1 + BA_o) = (2 \text{ M}\Omega)(1 + 0.05 \times 10^5) = 10000 \text{ M}\Omega$$

The output resistance with feedback is:

$$R_{of} = \frac{R_o}{1 + BA_o} = \frac{100 \ \Omega}{1 + 0.05 \times 10^5} = 0.02 \ \Omega$$

The closed-loop gain is:

$$A_f \approx \frac{1}{B} = 20$$

or using the complete equation:

$$A_f = \frac{A_o}{1 + BA_o} = 19.9996$$

It can be seen that the input resistance is extremely large and in practice it is unlikely that such a value would be achieved unless special precautions are taken to provide adequate spacing of the input conductor track with respect to any other tracks on the printed circuit board, and to protect the tracks from moisture which may be present in the atmosphere. This is achieved by coating the tracks with a protective varnish. The output resistance is very low and care would need to be taken to minimize any series resistance in the tracks by keeping track lengths as short as possible and as wide as possible, within the constraints of the pcb design rules. The closed-loop gain is given with sufficient accuracy by the reciprocal of the feedback factor because the open-loop gain is so large.

Unity Gain Buffer

The unity gain buffer amplifier is a special case of the non-inverting amplifier in which R_1 is reduced to zero ohms and R_2 is increased to infinity. The resulting circuit is shown in Figure 13.6.

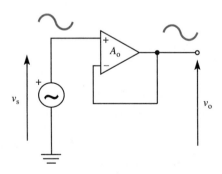

FIGURE 13.6   Unity gain buffer.

The closed-loop gain is now given by:

$$A_f = \frac{1}{B} = 1 + \frac{R_1}{R_2} = 1 + \frac{0}{\infty} = 1 \tag{13.6}$$

The input resistance is very large, because $BA_o = A_o$ since $B = 1$, and is:

$$R_{if} = R_i(1 + A_o) \tag{13.7}$$

and the output resistance is very small for the same reason, and is:

$$R_{of} = \frac{R_o}{1 + A_o} \tag{13.8}$$

The unity gain buffer provides a means for connecting a circuit which has a high output impedance to a low-impedance load without any loss of signal, and without any phase inversion.

▷ **EXAMPLE 13.2** ─────────────────────────────

For the same amplifier specified in Example 13.1, determine the input and output resistances when it is connected as a unity gain buffer.

▷ *Solution*

$$R_{if} = (2 \text{ M}\Omega)(1 + 10^5) = 200000 \text{ M}\Omega$$

and the output resistance is:

$$R_{of} = \frac{100 \text{ }\Omega}{1 + 10^5} = 0.0001 \text{ }\Omega$$

The input resistance is considerably larger than that for the non-inverting amplifier, and the output resistance is considerably smaller.

The importance of the unity gain buffer is that it provides a buffer between a circuit which may have a high output resistance and a load which may have a low resistance, without any loss of gain.

13.2 ▷ Inverting Amplifier

The circuit for an inverting amplifier is shown in Figure 13.7(a) together with a block diagram in 13.7(b) showing the feedback arrangement. The input is applied to the inverting terminal $(-)$.

From Figure 13.7(b) it can be seen that the output voltage is being sampled and a current i_f is created at the input which is determined by the feedback resistor and

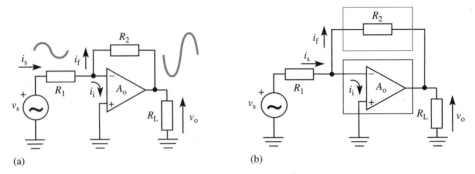

FIGURE 13.7   (a) The inverting amplifier and (b) amplifier redrawn to show voltage–shunt feedback.

the output voltage. The feedback configuration is voltage-shunt. The circuit may be analyzed by applying KVL to the input network and the feedback network as follows:

$$v_s = i_s R_1 + v_i$$
$$v_i = i_f R_2 + v_o$$
(13.9)

where v_i is the small-signal voltage across the input terminals of the amplifier. If the open-loop gain (A_o) is very large, then v_i is very small (v_o/A_o) and equations 13.9 may be simplified to:

$$v_s \approx i_s R_1$$
$$v_o \approx -i_f R_2$$
(13.10)

The input resistance of the differential input stage of the operational amplifier is very large, and consequently the input current i_i is very small. Thus the source voltage v_s drives a current i_s through R_1 and then most of this current flows through the feedback resistor. Thus:

$$i_s \approx i_f$$

and from equation 13.10:

$$\frac{v_s}{R_1} = -\frac{v_o}{R_2}$$

Defining the overall voltage gain as v_o/v_s gives:

$$A_f = \frac{v_o}{v_s} = -\frac{R_2}{R_1}$$
(13.11)

The presence of the negative sign confirms the fact that the output is 180° out of phase with respect to the input.

Virtual Ground (Virtual Earth)

In analyzing the circuit in Figure 13.7 it is assumed that the small-signal voltage between the two input terminals is very small ($v_i = v_o/A_o$). Consider, for example, an open-loop gain of 200000. If the dc supplies to the amplifier are ±15 V, then a sine wave output voltage cannot exceed 15 V peak. The input voltage required to produce 15 V peak is 75 µV. In practice the output is likely to be much smaller than 15 V, with a corresponding reduction in the amplitude of the input. With reference to Figure 13.7(a) this small-signal input voltage on the inverting terminal (−) is only a few microvolts above the level of ground (0 V) of the non-inverting terminal (+). It is virtually at ground – it is *a virtual ground (or virtual earth)*. The concept of the virtual earth simplifies the analysis of feedback circuits which employ shunt feedback.

Input Resistance

The input resistance may be obtained using the concept of the virtual earth, as shown in Figure 13.8. If the inverting terminal (−) is regarded as being the earth point of the circuit, then by simple inspection the input resistance as viewed from the source is:

$$R_{if} = R_1 \tag{13.12}$$

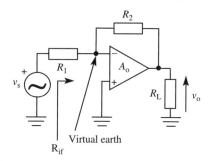

FIGURE 13.8 Inverting amplifier showing the position of the virtual earth.

An alternative to introducing the concept of the virtual earth is to use Miller's theorem. This is shown in Figure 13.9.

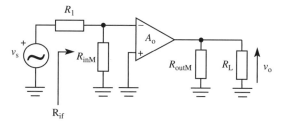

FIGURE 13.9 Miller equivalent of the inverting amplifier.

The feedback resistor R_2 is replaced by two resistors, R_{inM} at the input and R_{outM} at the output. The input Miller resistance is:

$$R_{inM} = \frac{R_2}{1 + A_{vo}} \tag{13.13}$$

where A_{vo} is the open-loop voltage gain. The input resistance is now obtained from Figure 13.9 by inspection as:

$$R_{if} = R_1 + R_{inM} \| R_i \approx R_1 \tag{13.14}$$

where R_i is the input resistance of the differential input stage of the amplifier, which is very large, and R_{inM} is very small and much less than R_1.

Output Resistance

From basic feedback theory the output resistance for voltage sampling is:

$$R_{of} = \frac{R_o}{1 + BR_{mo}} \tag{13.15}$$

where R_{mo} is the open-circuit transresistance (Note that this is not the same as R_{outM} mentioned above, which is the Miller output resistance.) The term in the denominator is large so that the output resistance with feedback is small.

▶ **EXAMPLE 13.3** _____

An operational amplifier has an open-loop gain of 2×10^5, an input resistance of 1 MΩ and an output resistance of 50 Ω. Determine the closed-loop voltage gain for the circuit shown in Figure 13.10 and the input resistance based on the concept of the virtual earth and Miller's theorem.

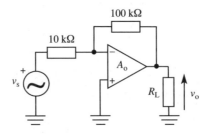

FIGURE 13.10   Circuit for Example 13.3.

▶ *Solution*

The closed-loop voltage gain is:

$$A_f = -\frac{100\,\text{k}\Omega}{10\,\text{k}\Omega} = -10$$

Using the concept of the virtual earth, the input resistance is:

$$R_{if} = 10 \text{ k}\Omega$$

Using Miller's theorem the feedback resistor is transposed to the input as:

$$R_{inM} = \frac{100 \text{ k}\Omega}{1 + 2 \times 10^5} \approx 0.5 \text{ }\Omega$$

and the total input resistance as seen by the source is:

$$R_{if} = 10 \text{ k}\Omega + 0.5 \text{ }\Omega \parallel 1 \text{ M}\Omega \approx 10 \text{ k}\Omega$$

▷ **EXAMPLE 13.4** ───────────────────

For the inverting amplifier shown in Example 13.3 determine the open-loop transresistance and use the feedback formula from Chapter 12 to verify the closed-loop voltage gain and the input resistance. Determine also the output resistance.

▷ *Solution*

The transresistance is defined as:

$$R_{mo} = -\frac{v_o}{i_i} = -\frac{v_o}{v_i}\frac{v_i}{i_i} = -A_{vo}R_i$$

and for the amplifier described in Example 13.3:

$$R_{mo} = 2 \times 10^5 \times 1 \times 10^6 = -2 \times 10^{11} \text{ }\Omega$$

The feedback factor is defined as:

$$B = \frac{i_f}{v_o} = -\frac{1}{R_2} \quad \text{assuming that } v_i \text{ is very small}$$

Thus the closed loop transresistance is:

$$R_{mf} = -\frac{R_{mo}}{1 + BR_{mo}} = -\frac{-2 \times 10^{11}}{1 - 10^{-5} \times 2 \times 10^{11}} = 100 \text{ k}\Omega$$

The input resistance is:

$$R_{if} = -\frac{R_i}{1 + BR_{mo}} = -\frac{10^6}{1 + 10^{-5} \times 2 \times 10^{11}} \approx 0.5 \text{ }\Omega$$

and the total resistance as seen from the source is:

$$R_{if}\big|_{source} = R_1 + R_{if} \approx 10 \text{ k}\Omega$$

The overall voltage gain is defined as:

$$A_{vf} = \frac{v_o}{v_s} = \frac{v_o}{i_i}\frac{i_i}{v_i}\frac{v_i}{v_s} = R_{mf}\frac{1}{R_{if}}\frac{R_{if}}{R_1 + R_{if}}$$

Using the values obtained above gives:

$$A_{vf} = -100 \text{ k}\Omega \times \frac{1}{0.5 \ \Omega} \times \frac{0.5 \ \Omega}{10 \text{ k}\Omega + 0.5\Omega} \approx -10$$

The output resistance is given as:

$$R_{of} = \frac{R_o}{1 + BR_{mo}} = \frac{50 \ \Omega}{1 + 10^{-5} \times 2 \times 10^{11}} = 2 \times 10^{-5} \ \Omega$$

From the above example it can be seen that feedback theory can be used to determine the different parameters for an inverting amplifier, but the more direct approach represented by the previous example is far simpler.

Summing Amplifier

The virtual earth associated with the inverting amplifier can be used as a summing junction to combine the currents from a number of sources, as shown in Figure 13.11.

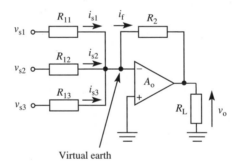

FIGURE 13.11   Summing amplifier.

With the virtual earth there is a simple relationship between the currents i_{s1}, i_{s2} and i_{s3} and their respective voltage sources, and as a result of the very high input resistance of the operational amplifier these currents all flow through R_f as i_f. Thus:

$$i_f = i_{s1} + i_{s2} + i_{s3}$$

With the inverting terminal being regarded as a virtual earth, the currents can be replaced by their respective voltages as:

$$-\frac{v_o}{R_f} = \frac{v_{s1}}{R_{11}} + \frac{v_{s2}}{R_{12}} + \frac{v_{s3}}{R_{13}}$$

With some rearrangement this becomes:

$$v_o = -\left(\frac{R_f}{R_{11}} v_{s1} + \frac{R_f}{R_{12}} v_{s2} + \frac{R_f}{R_{13}} v_{s3}\right)$$

(13.16)

From equation 13.16 it can be seen that the output is the sum of the input voltages, with each input voltage multiplied by a resistor ratio. If all the resistors are of equal value then the output voltage is:

$$v_o = -(v_{s1} + v_{s2} + v_{s3})$$

(13.17)

The summation does not have to be restricted to three inputs.

▶ **EXAMPLE 13.5** —————————————————————————

Determine the output voltage for the summing amplifier shown in Figure 13.12.

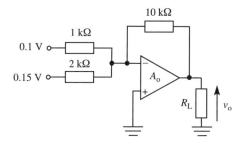

FIGURE 13.12   Summing amplifier.

▶ *Solution*

From equation 13.16 the output voltage is:

$$v_o = -\left(\frac{10\,k\Omega}{1\,k\Omega} 0.1\,V + \frac{10\,k\Omega}{2\,k\Omega} 0.15\,V\right) = -1.75\,V$$

A more practical application of the summing amplifier may be to introduce a dc offset to compensate for a dc level which may exist on a small-signal source, which is illustrated in the next example.

▶ **EXAMPLE 13.6** —————————————————————————

For the circuit shown in Figure 13.13 a small-signal source with a 100 mV amplitude has an associated dc offset of −0.5 V, represented by a dc source. Determine the value of the feedback resistor and the dc voltage which must be applied to the second input to provide a small-signal output with an amplitude of 1 V and zero offset.

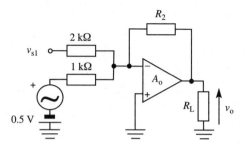

FIGURE 13.13   Summing amplifier used to compensate for a dc offset.

▶ *Solution*

The value of the feedback resistor is determined by the gain required to obtain the small-signal output of 1 V amplitude from an input of 100 mV amplitude, that is:

$$A_{vf} = -\frac{R_2}{R_{11}} = \frac{-1 \text{ V}}{100 \text{ mV}} = -10$$

$$\therefore R_2 = 10 \text{ k}\Omega$$

For the dc offset at the output to be 0 V:

$$v_o = 0 \text{ V} = -\left(\frac{10 \text{ k}\Omega}{1 \text{ k}\Omega} \times -0.5 \text{ V} + \frac{10 \text{ k}\Omega}{2 \text{ k}\Omega} \times v_{s1}\right)$$

$$\therefore v_{s1} = 1.0 \text{ V}$$

▶ **EXAMPLE 13.7** ————————————————————————————————

Use PSpice to examine the effect of changing the value of the dc voltage which is applied to the summing junction of the circuit in Example 13.6.

▶ *Solution*

The PSpice schematic is shown in Figure 13.14. For the small-signal source V_{S2} use a 100 mV sine wave with a frequency of 1 kHz and on offset (voff) of −0.5 V. For the other

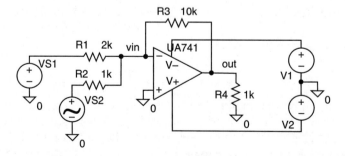

FIGURE 13.14   PSpice schematic of a summing amplifier.

summing input set V_{S1} to an initial value of 0 V. From **Analysis** select **Setup...** and **Transient...** and set the **Print step** to 10 µs and the **Final Time** to 2 ms. From **Parametric...** select **Voltage Source** and enter VS1 for **Name:**, select **Value List** and in **Value:** enter 0,1 to represent the 0 V and 1 V values for the dc voltage V_{S1}. Run the transient analysis and observe the output with Probe.

The two output waveforms for each value of V_{S1} are shown in Figure 13.15.

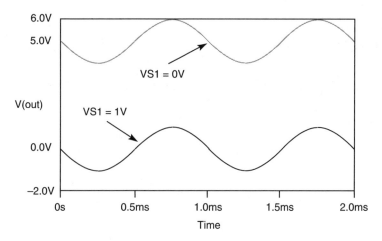

FIGURE 13.15 Probe output for the summing amplifier.

From the two waveforms it can be seen that the offset of the output can be changed by changing the value of V_{S1}. The amplitude of the output is 1 V and with V_{S1} set to 0 V the dc offset of the output is 5 V, while with V_{S1} set to 1 V the output is symmetrical about 0 V.

13.3 Difference Amplifier

An amplifier which produces an output which is proportional to the difference of two inputs is shown in Figure 13.16.

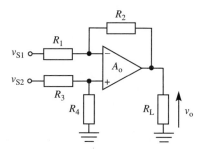

FIGURE 13.16 Difference amplifier.

The amplifier is analyzed by using superposition to determine the output from each source independently and then summing these outputs to obtain the overall result. First assume that v_{s1} is shorted to ground. The resultant circuit is shown in Figure 13.17(a) where it can be seen that the amplifier is a non-inverting amplifier with the input to the non-inverting terminal (+) reduced by the presence of the potential divider formed by R_3 and R_4.

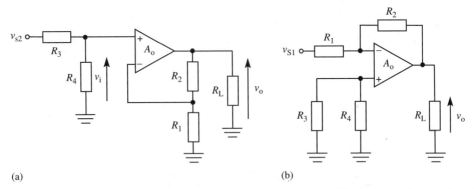

(a)   (b)

FIGURE 13.17   Application of superposition to the difference amplifier with (a) the circuit with v_{s1} shorted to ground and (b) v_{s2} shorted.

The voltage applied to the non-inverting terminal is:

$$v_i^+ = \frac{v_{s2}}{R_3 + R_4} R_4$$

and the output is:

$$v_o^+ = \left(1 + \frac{R_2}{R_1}\right)v_i^+ = \left(\frac{R_1 + R_2}{R_1}\right)\left(\frac{R_4}{R_3 + R_4}\right)v_{s2}$$

When v_{s2} is shorted to ground the circuit acts as an inverting amplifier with a resistor in series with the non-inverting terminal, as shown in Figure 13.17(b). Because of the high input resistance of the differential amplifier, the current which flows through the parallel combination of R_3 and R_4 is very small and can be ignored. The output voltage is:

$$v_o^- = -\frac{R_2}{R_1} v_{s1}$$

The effect of both signals is obtained by adding together the individual components to give:

$$v_o = v_o^+ + v_o^-$$

$$v_o = \left(\frac{R_4}{R_3 + R_4}\right)\left(\frac{R_1 + R_2}{R_1}\right)v_{s2} - \frac{R_2}{R_1} v_{s1} \tag{13.18}$$

This relationship is simplified by making $R_1 = R_3 = R_1$ and $R_2 = R_4 = R_2$. Then:

$$v_o = \frac{R_2}{R_1}(v_{s2} - v_{s1}) \tag{13.19}$$

From equation 13.19 it is seen that the output is proportional to the difference between the two inputs, where the inputs v_{s1} and v_{s2} are measured with respect to ground. An alternative interpretation is that the output is proportional to a voltage source which is connected directly between the two inputs, that is:

$$v_o = \frac{R_2}{R_1} v_d \tag{13.20}$$

where v_d is the difference voltage. This is a true differential amplifier where neither terminal of the input voltage source is connected to ground. An example of such an output is the output from a Wheatstone bridge, which could be a strain gauge transducer.

▶ **EXAMPLE 13.8** ————————————————————————————————

Design an operational amplifier circuit to produce an output of $0.5v_{s2} - 2v_{s1}$ if the two input resistors to the circuit are each 10 kΩ.

▶ *Solution*

From equation 13.18 the output is:

$$v_o = \left(\frac{R_4}{R_3 + R_4}\right)\left(\frac{R_1 + R_2}{R_1}\right)v_{s2} - \frac{R_2}{R_1} v_{s1}$$

For the coefficient of v_{s1} in the above equation to be 2 then:

$$\frac{R_2}{R_1} = \frac{R_2}{10\,\text{k}\Omega} = 2$$

$$\therefore R_2 = 20\,\text{k}\Omega$$

For the coefficient of v_{s2} to be 0.5 then:

$$\left(\frac{R_4}{R_3 + R_4}\right)\left(\frac{R_1 + R_2}{R_1}\right) = \left(\frac{R_4}{10\,\text{k}\Omega + R_4}\right)\left(\frac{10\,\text{k}\Omega + 20\,\text{k}\Omega}{10\,\text{k}\Omega}\right) = 0.5$$

$$\therefore R_4 = 2\,\text{k}\Omega$$

The circuit is shown in Figure 13.18.

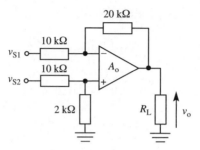

FIGURE 13.18   Circuit for Example 13.7.

Common-Mode Rejection

An important feature of difference amplifiers is the ability to reject common-mode signals. When the operational amplifier is used as a difference amplifier as in Figure 13.16, and when the signal is applied between the two inputs, then there is a good chance that there will be a large common-mode signal. This could be a fixed dc voltage which appears at both terminals, as may appear across a Wheatstone bridge, or it may be interference from the 50 Hz (60 Hz) line voltage. In either case an ideal difference amplifier should reject this signal. In practice the internal circuit components are not ideal and in particular the external resistor values are not precise. For the circuit shown in Figure 13.16 the imbalance which results from the resistors not having the precise values predicted by the ideal equations can be compensated by making R_4 adjustable. To establish the correct value of R_4 the two inputs are tied together and a sine wave signal applied, as shown in Figure 13.19.

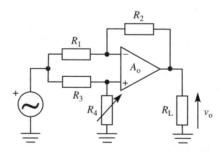

FIGURE 13.19   Difference amplifier connections for adjustment of common-mode rejection.

In a practical application the resistors R_1, R_2 and R_3 would have a tolerance of 1% or better. The signal from the ac source represents the common-mode signal and this would initially be set to a small value. The resistor R_4 is then adjusted to minimize the signal at the output. The amplitude of the input is increased as the optimum position of R_4 is reached.

▷ **EXAMPLE 13.9**

Use PSpice to examine the effect of adjusting the resistor R_4 in Figure 13.20. With
the resistor values shown the gain is 50. On the PSpice schematic the value of R_4
is defined as a parameter **{r4}** and within **Setup...** the **Parametric** box is used to
identify r4 as the global parameter which is to be varied from 49.9 kΩ to 50.2 kΩ.
(The curly brackets are a means for identifying a parameter value which is to be
varied.)

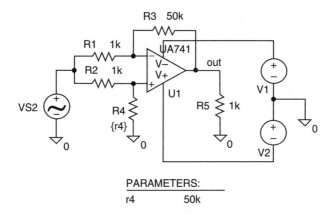

PARAMETERS:
r4   50k

FIGURE 13.20   PSpice schematic for investigating common-mode rejection.

▷ *Solution*

The output from Probe is shown in Figure 13.21.

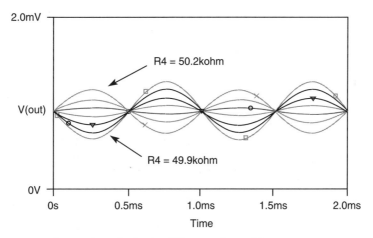

FIGURE 13.21   Output from Probe for different values of R_4.

A number of waveforms are superimposed in the one diagram shown in Figure 13.21 for different values of R_4. It can be seen that the amplitude of the output varies as R_4 is varied. For a value of 49.9 kΩ there is a large sine wave which starts with a negative-going peak. For a value of 50.2 kΩ the wave starts with a positive-going peak. At some intermediate value the output is reduced to zero, which represents the condition for zero common-mode output and maximum CMRR.

Instrumentation Amplifier

A disadvantage of the difference amplifier described above is the relatively low input resistance, equal to R_1 (or R_3) in Figure 13.16. A high input resistance is often desirable for accurate measurement of signals from transducers. This can be achieved with the addition of a non-inverting amplifier to each of the difference amplifier inputs. Non-inverting amplifiers have very high input resistance. The circuit is shown in Figure 13.22.

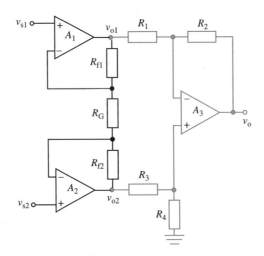

FIGURE 13.22 Instrumentation amplifier.

Amplifier A_3 and the resistors R_1, R_2, R_3 and R_4 form a difference amplifier, while amplifiers A_1 and A_2 are non-inverting amplifiers which share one of the feedback resistors R_G. The analysis involves the use of superposition and the use of the concept of the virtual earth. Consider the case when v_{s2} is reduced to zero. The resultant circuit is shown in Figure 13.23.

In Figure 13.23(a) the virtual earth, associated with A_2, which appears at the lower end of R_G means that A_1 acts as a non-inverting amplifier producing an output:

$$v_{o1} = v_{s1}\left(1 + \frac{R_{f1}}{R_G}\right)$$

(13.21)

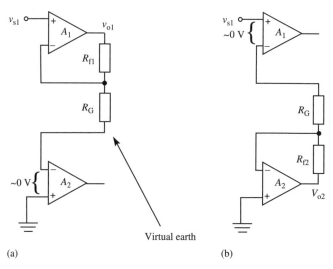

FIGURE 13.23   Input stages of the instrumentation amplifier when $v_{s2} = 0$.

There is also an output from A_2 also acting as an inverting amplifier as shown in Figure 13.23(b). The voltage across the terminals of A_1 is very small ($\sim$0 V) so that the voltage at the terminal of R_G is approximately equal to the input v_{s1}. The output from A_2 is:

$$v_{o2} = -v_{s1}\frac{R_{f2}}{R_G} \tag{13.22}$$

A similar set of relationships is obtained if v_{s1} is assumed to be zero and the effect of v_{s2} considered. The overall output from each amplifier is obtained by summing the individual outputs to give:

$$v_{o1} = v_{s1}\left(1 + \frac{R_{f1}}{R_G}\right) - v_{s2}\frac{R_{f2}}{R_G} \tag{13.23}$$

$$v_{o2} = v_{s2}\left(1 + \frac{R_{f2}}{R_G}\right) - v_{s1}\frac{R_{f1}}{R_G} \tag{13.24}$$

These two outputs are presented to the difference amplifier A_3 and if $R_1 = R_2 = R_3 = R_4$ then the output from A_3 is:

$$v_o = v_{o2} - v_{o1}$$

Substituting for v_{o2} and v_{o1} from equations 13.23 and 13.24 and setting $R_{f1} = R_{f2} = R_f$ gives:

$$v_o = (v_{s2} - v_{s1})\left(1 + \frac{2R_f}{R_G}\right) \tag{13.25}$$

or

$$v_o = v_d\left(1 + \frac{2R_f}{R_G}\right)$$ (13.26)

where v_d is the difference voltage.

In a practical instrumentation amplifier the three operational amplifiers are mounted in a single package with the resistor R_G being left as an external component to control the gain. The common-mode rejection is established by adjusting R_4 (Figure 13.22) during the manufacturing process. Values of 100 dB for the CMRR are not uncommon.

▷ **EXAMPLE 13.10** ——————————————————————

Determine the value of the external gain resistor for an instrumentation amplifier to provide a gain with feedback of 300 if the feedback resistor is 20 kΩ.

▷ *Solution*

$$A_f = \left(1 + \frac{2R_f}{R_G}\right) = 300$$

and

$$R_G = \frac{40\ \text{k}\Omega}{300 - 1} \approx 133\ \Omega$$

Instrumentation amplifiers are used when it is necessary to measure a very small differential voltage in the presence of a large common-mode signal as illustrated in Figure 13.24. This is often required when measuring the output from a transducer which may be mounted on pieces of equipment or structures which are remote from the test location. Long connecting leads may be required which are subject to interference from external signals such as the 50 Hz (or 60 Hz) line voltages.

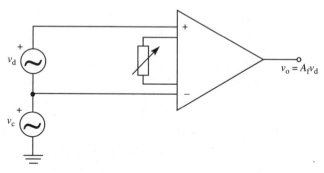

FIGURE 13.24   Instrumentation amplifier used to amplify a small signal v_d in the presence of a common-mode signal v_c.

NB The common-mode voltage v_c is shown as a single source located at the transducer, but in practice the interference is distributed along the length of the connecting leads.

Provided that the CMRR is large then the output is proportional to the signal produced by the transducer, and the common-mode signal is rejected.

13.4 Comparator

One of the simplest applications for the operational amplifier is one which uses the very large value of gain to detect when a signal is above or below a preset level. With a gain of 200000 an input of 65 µV is sufficient to cause the output of a typical amplifier operating from ±15 V power rails to saturate at +13 V or −13 V, depending on the polarity of the input. Thus this large output swing is used to determine when the input is within 65 µV of a preset input level as shown in Figure 13.25. The input is represented by a sine wave, but this could be a signal from a temperature or pressure sensor which is slowly varying, and the output changes when the input exceeds a preset level represented by V_{ref} produced by a potential divider.

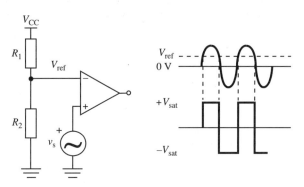

FIGURE 13.25 Operational amplifier used as a comparator.

EXAMPLE 13.11

A 1 kHz sine wave with an amplitude of 1 V is applied to the comparator shown in Figure 13.26. Sketch the output waveform as accurately as possible assuming that the amplifier saturates at ±13 V and estimate the width in seconds of the positive and negative halves of the output waveform.

Solution

$$V_{ref} = \frac{15 \text{ V}}{10 \text{ k}\Omega + 345 \text{ }\Omega} \, 345 \text{ }\Omega = 0.5 \text{ V}$$

The output switches each time the input exceeds 0.5 V and switches to +13 V. When the input falls below 0.5 V the output switches to −13 V, as shown in Figure 13.27.

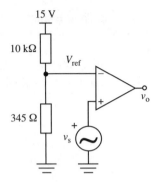

FIGURE 13.26   Comparator.

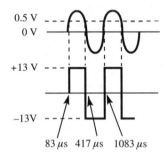

FIGURE 13.27   Output from the comparator.

Starting from 0 V the input reaches 0.5 V at:

$$t_1 = \frac{1}{2\pi \times 1 \times 10^3} \sin^{-1}\left(\frac{0.5}{1.0}\right) \approx 83 \ \mu s$$

It changes from $+13$ V to -13 V at:

$$t_2 = 500 \ \mu s - 83 \ \mu s = 417 \ \mu s$$

and it changes from -13 V to $+13$ V at:

$$t_3 = 1 \ ms + 83 \ \mu s = 1083 \ \mu s$$

The width of the positive excursion is:

$$t^+ = t_2 - t_1 = 334 \ \mu s$$

and the width of the negative excursion is:

$$t^- = t_3 - t_2 = 666 \ \mu s$$

One of the problems with the simple comparator shown in Figure 13.25 is the effect of electrical noise which may be combined with the input signal. This is illustrated in

Figure 13.28 where it is assumed that electrical noise is superimposed on the output from a transducer. The noise waveform is exaggerated for clarity.

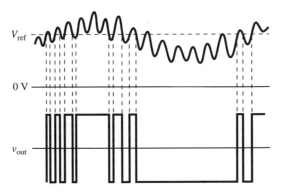

FIGURE 13.28   Effect of electrical noise.

When the signal is applied to a comparator which is set to respond to voltage levels greater than V_{ref} then as the signal approaches V_{ref} the electrical noise will momentarily take the signal above V_{ref} and the output of the comparator will change, only to change back again a short time later when the noise component takes the signal below V_{ref}. This is repeated several times before the signal is sufficiently above V_{ref} for the noise to have no effect. If the comparator is being used to control a relay, or computer circuitry, then this rapid switching may cause damage, or result in incorrect data being supplied to computer-based circuitry.

The Schmitt Trigger

The unwanted triggering of the output of the comparator can be prevented with a Schmitt trigger. The circuit is shown in Figure 13.29.

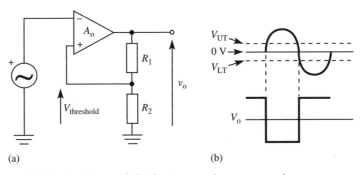

FIGURE 13.29   (a) Schmitt trigger and (b) the input and output waveforms.

The Schmitt trigger is very similar in appearance to the non-inverting amplifier of Figure 13.1, but note that the input is applied to the inverting input and the feedback is applied to the non-inverting input. The result is that any positive-going input signal causes the output to be inverted and to be negative going. The feedback from the potential divider comprising R_1 and R_2 is also negative going and is applied to the non-inverting terminal. A negative-going voltage applied to this terminal reinforces the negative-going output and forces the output to move further in the negative direction. The feedback is positive and once the output starts changing it very rapidly reaches the saturation value which for ±15 V supplies is about ±13 V. When power is first applied to a Schmitt trigger the output saturates to either $+V_{sat}$ or $-V_{sat}$. In Figure 13.29(b) it is assumed that the output is at $+V_{sat}$. A positive threshold voltage is applied to the non-inverting terminal by means of the potential divider with a value:

$$V_{UT} = \frac{R_2}{R_1 + R_2} (+V_{sat})$$ (13.27)

When the output switches to a negative value a negative threshold voltage is applied to the non-inverting terminal with a value:

$$V_{LT} = \frac{R_2}{R_1 + R_2} (-V_{sat})$$ (13.28)

The importance of the two threshold voltages may be realized by considering what happens when a positive-going voltage is applied to the input. Assume that the output is at $+V_{sat}$ so that the upper threshold exists at the non-inverting terminal. This is the situation which exists in Figure 13.29(b). When the input exceeds V_{UT} the output switches and the lower threshold is established on the non-inverting terminal. It is a simple matter to arrange for this voltage to be sufficiently below V_{UT} for it not to be affected by noise. The output will not change until the input has reduced to V_{LT} when the output will switch to a positive value and a positive threshold V_{UT} is reapplied to the non-inverting terminal. The switching of the Schmitt trigger occurs at different voltages, V_{UT} and V_{LT}. The difference between these two values is known as the *hysteresis* V_H. Thus:

$$V_H = V_{UT} - V_{LT}$$ (13.29)

The amount of hysteresis is controlled by the potential divider and can be selected to avoid false triggering caused by electrical noise.

▶ **EXAMPLE 13.12** ——————————————————————————

Determine the upper and lower trigger points and the hysteresis for the Schmitt trigger shown in Figure 13.30. Assume that the output saturates at ±13 V.

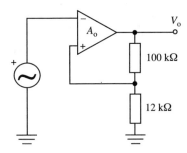

FIGURE 13.30   Schmitt trigger.

▷ *Solution*

The upper threshold is:

$$V_{UT} = \frac{12\ k\Omega}{100\ k\Omega + 12\ k\Omega}\ (13\ \text{V}) \approx 1.4\ \text{V}$$

and the lower threshold is:

$$V_{LT} = \frac{12\ k\Omega}{100\ k\Omega + 12\ k\Omega}\ (-13\ \text{V}) \approx -1.4\ \text{V}$$

The hysteresis is:

$$V_H = 1.4\ \text{V} - (-1.4\ \text{V}) = 2.8\ \text{V}$$

In the above example the noise fluctuations would have to exceed 2.8 V in order to trigger the circuit.

In the above examples the input is assumed to vary above and below 0 V. In some instances the input may be varying but be entirely positive or entirely negative. The Schmitt trigger may be required to establish threshold levels which are either both positive or both negative. This is achieved with the circuit shown in Figure 13.31.

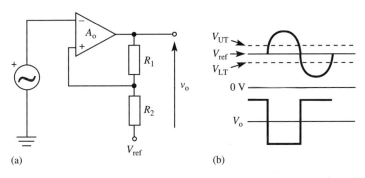

(a)   (b)

FIGURE 13.31   Schmitt trigger with a fixed dc offset.

A reference voltage (V_{ref}) is applied to one end of the potential divider. The voltage which is applied to the non-inverting terminal is now dependent on both the output (V_o) and reference voltage (V_{ref}) and the value is obtained by means of superposition. Assume first that V_o is zero. Then:

$$v_1^+ = \frac{R_1}{R_1 + R_2} V_{ref} \tag{13.30}$$

When V_{ref} is zero then:

$$v_2^+ = \frac{R_2}{R_1 + R_2} V_o \tag{13.31}$$

Thus when V_o is at its negative limit the lower threshold is:

$$V_{LT} = v_1^+ + v_2^+$$

$$= \frac{R_1}{R_1 + R_2} V_{ref} + \frac{R_2}{R_1 + R_2} (-V_{sat}) \tag{13.32}$$

and similarly when V_o is at its positive limit the upper threshold is:

$$V_{UT} = \frac{R_1}{R_1 + R_2} V_{ref} + \frac{R_2}{R_1 + R_2} (+V_{sat}) \tag{13.33}$$

The hysteresis is obtained as the difference between these two values as:

$$V_H = V_{UT} - V_{LT} \tag{13.34}$$

▶ **EXAMPLE 13.13** ───────────────────────────────

Determine the upper and lower threshold levels and the hysteresis for the Schmitt trigger shown in Figure 13.32. Assume that the output saturates at ±9 V.

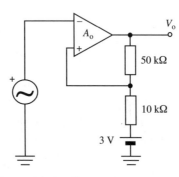

FIGURE 13.32   Schmitt trigger with dc offset.

▷ *Solution*

From equations 13.32 and 13.33:

$$V_{LT} = \left(\frac{50 \text{ k}\Omega}{10 \text{ k}\Omega + 50 \text{ k}\Omega}\right)(3 \text{ V}) + \left(\frac{10 \text{ k}\Omega}{10 \text{ k}\Omega + 50 \text{ k}\Omega}\right)(-9 \text{ V}) = 1 \text{ V}$$

$$V_{UT} = \left(\frac{50 \text{ k}\Omega}{10 \text{ k}\Omega + 50 \text{ k}\Omega}\right)(3 \text{ V}) + \left(\frac{10 \text{ k}\Omega}{10 \text{ k}\Omega + 50 \text{ k}\Omega}\right)(+9 \text{ V}) = 4 \text{ V}$$

and the hysteresis is:

$$V_H = V_{UT} - V_{LT} = 3 \text{ V}$$

▷ **EXAMPLE 13.14** ——————————————————————————

Design a Schmitt trigger to respond to a signal when it crosses thresholds at 2 V and 3 V as shown in Figure 13.33. Assume that the amplifier saturates at ±9 V and that R_2 (Figure 13.31(a)) is 1 kΩ. Verify your design by using PSpice to simulate a Schmitt trigger based on a 741 op-amp with ±10 V supplies and a 1 Hz sine wave input with an offset of 2.5 V and an amplitude of 1 V.

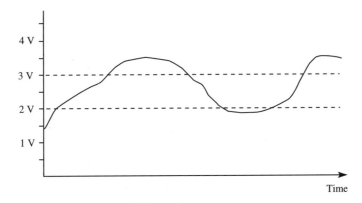

FIGURE 13.33   Signal to be applied to a Schmitt trigger.

▷ *Solution*

From equation 13.33 the upper threshold voltage level is:

$$3 \text{ V} = \frac{R_1}{R_1 + 1 \text{ k}\Omega} (V_{ref}) + \frac{1 \text{ k}\Omega}{R_1 + 1 \text{ k}\Omega} (9 \text{ V})$$

and the lower threshold is:

$$2 \text{ V} = \frac{R_1}{R_1 + 1 \text{ k}\Omega} (V_{ref}) + \frac{1 \text{ k}\Omega}{R_1 + 1 \text{ k}\Omega} (-9 \text{ V})$$

Solving for $R_1 V_{ref}$ from the two equations gives:

$$R_1 V_{ref} = 45k \frac{V^2}{I}$$

and solving for R_1 by substituting back into the equation for the upper threshold level gives:

$$R_1 = 17 \text{ k}\Omega$$

and finally V_{ref}

$$V_{ref} = 2.65 \text{ V}$$

The output obtained from Probe for a sine wave with an amplitude of 1 V and an offset of 2.5 V is shown in Figure 13.34.

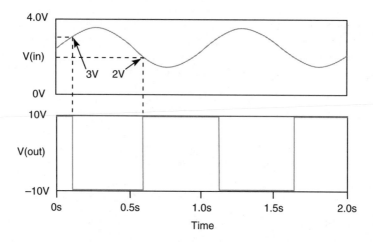

FIGURE 13.34   Output from Probe for a Schmitt trigger with a dc reference voltage.

13.5 Other Applications

The circuits described above are some of the more common applications for operational amplifiers. However, there are many more to be found in specialist texts on the subject. Some are described below.

Window Comparator

The circuit shown in Figure 13.35 detects when a signal exceeds preset upper and lower limits. It is known as a window comparator.

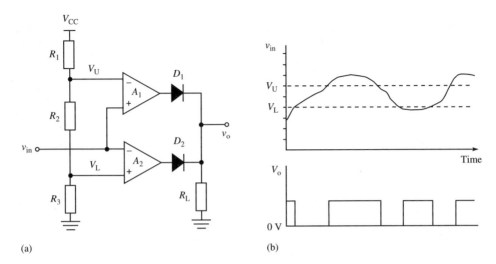

FIGURE 13.35 (a) Circuit of a window comparator and (b) the input and output waveforms.

The upper and lower limits are set by the potential divider, so that:

$$V_U = \frac{V_{CC}(R_2 + R_3)}{R_1 + R_2 + R_3} \tag{13.35}$$

and the lower limit is:

$$V_L = \frac{V_{CC}R_3}{R_1 + R_2 + R_3} \tag{13.36}$$

If v_{in} lies between the two limits then for comparator A_1 the non-inverting terminal is $<V_U$ and therefore the output is at its low saturation level, while the inverting terminal of comparator A_2 is $>V_L$ which results in its output also being at its low saturation level. With both outputs low the diodes are reverse biased and V_o is held at 0 V. When v_{in} exceeds V_U then the non-inverting terminal becomes positive with respect to the inverting terminal and the output is driven to its upper saturation level. This forward-biases diode D_1 which causes current to flow through the load resistor and V_o rises. When the input drops below V_U the output returns to 0 V. When the input falls below the lower threshold the inverting terminal of A_2 becomes negative with respect to the non-inverting terminal and the output is driven to its upper saturation level. Diode D_2 becomes forward biased and V_o again rises.

Current to Voltage Converter

Some transducers, for example photosensitive diodes, produce a current rather than a voltage output. The current may be converted to a voltage variation by passing it through a resistor. However, the junction capacitance associated with a photodiode, together with the resistance of the load, form a low-pass *RC* filter which may restrict the bandwidth. An alternative approach is to use a current to voltage converter which uses voltage–shunt feedback. The input resistance of these amplifiers is very small, which means that the frequency performance is improved over the use of a load resistor and a voltage amplifier. The circuit is shown in Figure 13.36.

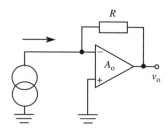

FIGURE 13.36   Current-to-voltage converter.

Because of the virtual earth at the inverting terminal the voltage across the resistor is v_o and because of the large input resistance between the input terminals of the amplifier the input current flows through the feedback resistor so that:

$$v_o = -iR \tag{13.37}$$

Constant Current Source

The same voltage–shunt feedback circuit can be used to produce a constant current source as shown in Figure 13.37.

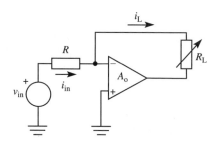

FIGURE 13.37   Constant current source.

It is assumed that the load resistance varies, but that a constant load current is required (e.g. a small electric motor may be required to run at constant speed even when the mechanical load varies). The input current is established by the input voltage and the input resistance. Since the load current is equal to the input current then:

$$i_L = i_{in} = \frac{v_{in}}{R} \tag{13.38}$$

The load current is independent of the load resistance (e.g. motor armature resistance) provided that the voltage across the load does not exceed the output saturation voltage of the operational amplifier. A disadvantage of the circuit is that both terminals of the load are live. There are variations of the circuit which allow one of the terminals to be at ground (see specialist text on operational amplifier applications).

Integrator

Consider the circuit shown in Figure 13.38.

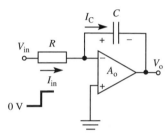

FIGURE 13.38   Circuit of an integrator.

For an ideal amplifier the output will be at 0 V if the input is also at 0 V. If now a positive voltage is applied to the input then because of the virtual earth the input current is:

$$I_{in} = \frac{V_{in}}{R} \tag{13.39}$$

Because the input resistance of the amplifier is very large this current flows in the capacitor. For a constant input voltage (v_{in}) the charging current for the capacitor remains constant. As the charge on the plates of the capacitor increases, the voltage across the capacitor also increases with the polarity shown in Figure 13.38, where the output of the amplifier is negative with respect to the capacitor plate attached to the inverting terminal. Since the inverting terminal is a virtual earth (0 V), then the output ramps down in a negative direction from 0 V.

The rate of change of the voltage may be obtained by considering the relationship between capacitance, charge and voltage:

$$V_C = V_o = \frac{Q}{C} \tag{13.40}$$

where V_C is the voltage across the capacitor, which is equal to the output voltage if account is taken of the virtual earth. The charge is the product of current and time:

$$Q = I \times t \tag{13.41}$$

so that the rate of change of voltage is:

$$\frac{V_o}{t} = \frac{I}{C} = \frac{-V_{in}}{RC} \qquad (13.42)$$

and the final voltage is:

$$V_o = \frac{-V_{in}}{RC} t \qquad (13.43)$$

Equation 13.43 shows that there is a simple *linear* relationship between the output voltage and time. This is in marked contrast to a capacitor which is charging through a resistor, which results in an *exponential* relationship. Therefore, for a step change in the input voltage there is a linear change in the output voltage with the polarity shown in Figure 13.39 with the negative sign in equation 13.43 resulting in a negative-going slope in the output, for a positive-going input.

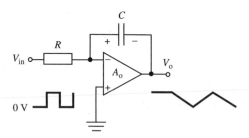

FIGURE 13.39   Relationship between input and output voltages for an integrator.

In Figure 13.39 the input voltage is assumed to be initially 0 V and the output is also 0 V. When the input goes positive then the output ramps down. If the input changes back to 0 V then provided that the output does not reach the negative saturation level, the direction of the charging current changes and the output starts to ramp back up towards 0 V. Thus a *square wave input* results in a *triangular wave output*. The relationship between the input waveform and the output waveform is more clearly shown in Figure 13.40.

The charging current for the capacitor (equation 13.39) is either $+V_{in}/R$ or $-V_{in}/R$ depending on the polarity of the input at any particular instant. For a positive value the output decreases, for example from A to B in Figure 13.40. When the input changes to a negative value the direction of the current through the capacitor changes direction and the output voltage moves in a positive direction B to C in Figure 13.40. The total change in the output voltage is:

$$\Delta V_o = V_A - V_B \quad (\text{or} \,|V_B - V_C|)$$

During the time that the voltage is moving from A to B the charging current is V_{in}/R. The time interval is $T/2$, as shown in Figure 13.40, where T is the period of the input

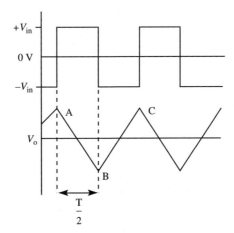

FIGURE 13.40   Input and output waveforms for an integrator.

waveform. Thus the change in the output voltage is:

$$\Delta V_\text{o} = \frac{V_\text{in} T}{2RC} \qquad (13.44)$$

▷ **EXAMPLE 13.15** ────────────────────────────────

The resistor in the integrator shown in Figure 13.39 is 1 kΩ and the capacitor is 100 nF. Sketch the output voltage waveform if the input is a square wave of 500 Hz with amplitudes of ±1 V.

What would happen if the frequency were to be reduced to 150 Hz? Assume that the amplifier is supplied from ±15 V supplies.

▷ *Solution*

For a positive-going step the output ramps down at a rate given by equation 13.42 as:

$$\frac{V_\text{in}}{RC} = \frac{1}{1\ \text{k}\Omega \times 100\ \text{nF}} = 10000\ \text{V s}^{-1}$$

The input voltage changes direction every half period or every 1 ms for a 500 Hz square wave. The change in the output voltage after 1 ms is:

$$\Delta V_\text{o} = 10000 \times 1\ \text{ms} = 10\ \text{V}$$

The waveforms for the input and output are shown in Figure 13.41.

Reducing the frequency to 150 Hz increases the time that charging current flows into the capacitor. After a half cycle (~3.35 ms) the output would change by 33 V. Since this exceeds the saturation voltage (~±13 V for total swing of 26 V) then the output would

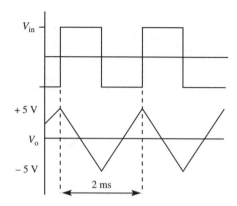

FIGURE 13.41   Input and output waveforms for an ideal integrator.

saturate after:

$$t = \frac{26 \text{ V}}{10000 \text{ V s}^{-1}} = 2.6 \text{ ms}$$

For input waveforms which are rectangular or square, the charging current is constant and the output wave is triangular, provided that the output voltage does not saturate or the rate of change does not exceed the slew rate. For continuously varying waves such as a sine wave then the charging current varies and the simple relationship shown in equation 13.42 or 13.44 no longer applies. For these waves the mathematical form of the relationship between output and input must be used, that is:

$$v_o = -\frac{1}{RC} \int_0^t v_{in} \, dt \tag{13.45}$$

Thus if the input is a sine wave $V_P \sin \omega t$ then the output is:

$$v_o = -\frac{1}{RC} \int_0^t V_P \sin(\omega t) dt$$

$$= -\frac{V_P}{\omega RC} (-\cos(\omega t)) \tag{13.46}$$

A cosine wave looks exactly the same as a sine wave except that there is a phase difference of 90° with respect to the sine wave. The most important property of equation 13.46 is the fact that the output is inversely proportional to the frequency. Thus if the output has an amplitude of 1 V for an input of 100 Hz, then it will be 0.5 V for a 200 Hz input, and for an input of 1 kHz it will be 0.1 V. For a change of one decade (100 Hz to 1 kHz) the output has changed from 1 V to 0.1 V, a factor of 10 (or 20 dB). This rate of change of voltage with frequency of −20 dB per decade is the same as that for a low-pass RC filter. The frequency response of the integrator is shown in Figure 13.42.

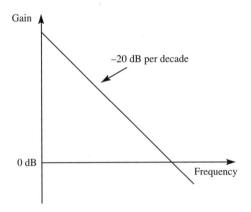

FIGURE 13.42 Frequency response of an integrator.

Practical Integrator

The graph shown in Figure 13.42 is for an ideal integrator with a frequency range which extends down to dc at which point the gain is in excess of 100000 or >100 dB. In a practical integrator the operational amplifier has offset voltages and input bias currents. The offset voltage results in a small dc voltage being present at the input terminals which causes a small charging current to flow to the capacitor. This appears as a linearly increasing voltage at the output. The bias current will also result in a charging current which adds to the output voltage. With dc gain values of 100 dB these two components cause the output to ramp up, or down, depending on the polarity of the offset voltage and/or bias current, until it reaches the saturation level. Thus the output of an ideal integrator in the absence of an input signal is likely to be offset towards the positive or negative saturation levels. This can be avoided by placing a resistor (R_F) in parallel with the capacitor. At dc the input resistor together with this shunt resistor establish the dc gain as $-R_F/R$ rather than the open-loop gain of 100 dB. With a resistor in shunt with the capacitor the preferred path for any current for the input bias, or that which is generated by the offset voltage, is through R_F rather than into the capacitor, and the output voltage is determined by the resistor ratio, which is typically about $\geqslant 10$ ($R_F \geqslant R$). True integration now only occurs at frequencies such that:

$$\frac{1}{2\pi fC} \ll R_F$$

When $X_C = R$ then:

$$f_C = \frac{1}{2\pi R_F C} \tag{13.47}$$

and this defines a corner frequency below which integration does not take place. Below f_C the gain approaches the dc value of $-R_F/R$. For true integration the frequency must be at least $10f_C$. The additional resistor has little effect on the output provided that the input frequency is greater than f_C.

▶ **EXAMPLE 13.16** ——

A 1 kHz square wave is applied to the integrator shown in Figure 13.43. The amplifier is provided with ±15 V supplies for which the output saturates at ±14 V (typical of a 741 op-amp). If the input square wave alternates between ±5 V:

(a) Determine the maximum output voltage and its shape.
(b) Determine the minimum slew rate for the amplifier.

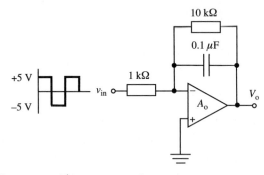

FIGURE 13.43   Integrator with square wave input.

▶ *Solution*

(a) For a square wave input the output is triangular provided that the output does not reach saturation. The period of the 1 kHz wave is 1 ms and from equation 13.44 the change in output voltage is:

$$\Delta V_o = \frac{V_{in}T}{2RC} = \frac{(5\text{ V})(1\text{ ms})}{(2)(1\text{ k}\Omega)(0.1\text{ }\mu\text{F})} = 25\text{ V}$$

This value appears to exceed the 14 V saturation level of the op-amp, but the 25 V represents the total change for the output. Each time the input changes from −5 V to +5 V the charging current to the capacitor changes direction, but its magnitude remains the same at 5 mA (5 V/1 kΩ). Provided that there are no dc offsets then the output alternates between +12.5 V and −12.5 V (25 V total), which is less than the saturation levels of the amplifier, and therefore the output is triangular.

(b) The rate of change is obtained from equation 13.42 as:

$$\frac{V_o}{t} = \frac{I}{C} = \frac{V_{in}}{RC} = \frac{(5\text{ V})}{(1\text{ k}\Omega)(0.1\text{ }\mu\text{F})} = 5 \times 10^4\text{ V s}^{-1}$$

The slew rate is usually expressed as V μs^{-1}. Therefore, the slew rate needs to be greater than 0.05 V μs^{-1} (typically 0.5 V μs^{-1} for a 741).

The waveforms may be examined by means of a PSpice simulation and are shown in Figure 13.44 where the square wave is applied to the integrator shown in Figure 13.43.

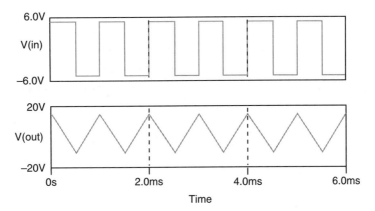

FIGURE 13.44   Integrator output for a square wave input.

It can be seen that the output voltage is triangular and varies between approximately +12 V and −12 V, resulting in a total swing of 24 V.

▷ **EXAMPLE 13.17** ——————————————————————————————————

For the circuit shown in Figure 13.43 use PSpice to examine the effect of a 10 mV dc offset applied to the input which is a square wave with an amplitude of ±0.1 V at a frequency of 100 Hz, without the shunt resistor and then with the shunt resistor.

▷ *Solution*

The 10 mV offset can be simulated by placing a dc source of this value in series with the square wave source. The 10 mV generates a current through the capacitor of 10 μA (10 mV/1 kΩ) when the shunt resistor is absent. Applied to the inverting terminal of the amplifier this current results in the output voltage moving in a negative direction as charge enters the capacitor. This movement is added to that produced by the square wave source when it is positive, and subtracts from it when the source is negative. The result is that the output drifts in a negative direction until it reaches the saturation level. The effect is seen in Figure 13.45 which shows the output from Probe.

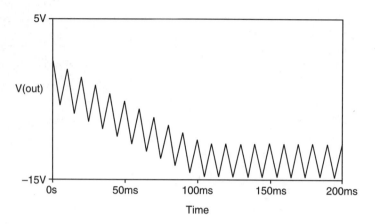

FIGURE 13.45 Output from Probe with R_F absent.

It can be seen that the output waveform reaches the negative saturation level after about 100 ms.

The output with the shunt resistor present is shown in Figure 13.46.

With the resistor present the output remains symmetrical about the zero-voltage level but with a small offset determined by the resistor ratio and the input dc offset.

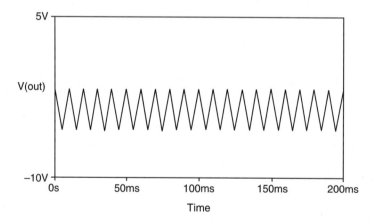

FIGURE 13.46 Output from Probe with the shunt resistor present.

▶**EXAMPLE 13.18** _____

For the integrator shown in Figure 13.47 determine the frequency above which integration takes place and determine the amplitude of the output if the input is a 5 V peak amplitude sine wave of 10 kHz.

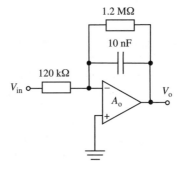

FIGURE 13.47   Practical integrator.

▷ *Solution*

The corner frequency is:

$$f_c = \frac{1}{2\pi R_F C} = \frac{1}{2\pi(1.2 \times 10^6)(10 \times 10^{-9})} \approx 13 \text{ Hz}$$

True integration will take place above frequencies of 130 Hz.

At 10 kHz the frequency is well above the corner frequency and the output is given by equation 13.46:

$$\left| \frac{V_o}{V_{in}} \right| = \frac{1}{2\pi f R C}$$

$$V_o = \frac{5 \text{ V}}{2\pi(120 \text{ k}\Omega)(10 \text{ nF})(10 \times 10^3)} = 66 \text{ mV}$$

The frequency response curve for the integrator is shown in Figure 13.48.

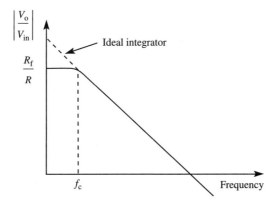

FIGURE 13.48   Frequency response for a practical integrator.

Single Supply Operation

Operational amplifiers normally have two dc supplies to provide positive and negative voltages with a common ground. These are clearly illustrated in the numerous PSpice examples. With this approach it is possible to design the internal circuitry of the amplifier to enable the inputs and output to be at, or very near, ground potential. In many cases a grounded signal source may be directly connected to an input, and the output signal is free to swing above and below ground to within about 2 V of each of the dc supplies. However, two supplies are sometimes inconvenient, and only one supply may be available. An operational amplifier may still be used with one supply, but some modification of the input circuitry is necessary to compensate for the fact that the inputs are no longer referenced to ground, but rather to a voltage mid-way between the value of the dc supply.

Connecting a grounded signal source directly to an input of an amplifier with only one dc supply would result in a very distorted output waveform. A dc bias of $0.5V_{CC}$ needs to be applied to the input. This produces an output of $0.5V_{CC}$ and the peak-to-peak output swing is a few volts less than V_{CC}. For a $+15$ V supply this means that the maximum peak-to-peak output is approximately $12-13$ V.

The input dc bias is easily achieved with an equal value resistor potential divider, as shown in Figure 13.49(a) for a non-inverting amplifier and 13.49(b) for an inverting amplifier. Note the need for capacitors because the inputs and output are no longer at ground potential and must be isolated from a grounded source and grounded load.

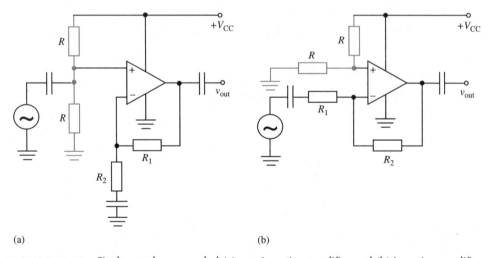

(a)   (b)

FIGURE 13.49   Single-supply ac-coupled (a) non-inverting amplifier and (b) inverting amplifier.

Other applications

There are many other applications for operational amplifiers, including differentiator circuits, precision rectifiers, peak detectors, clipping and clamping, to name but a few.

These are well covered in specialist texts. Operational amplifiers may also be used selectively to amplify or attenuate signals at different frequencies and to generate various types of waveform. These particular applications are covered in the following chapters.

▷ **PROBLEMS** ——————————————————————————————————

13.1 Determine the accurate and approximate values of gain for the circuit shown if $A_o = 150000$. Determine the value of A_o which would result in a 1% error between the approximate and accurate values.

  [22.99, 23, 2396]

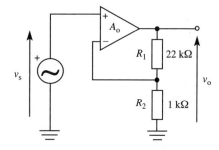

13.2 For the amplifier shown in Problem 13.1 determine the input and output resistances if R_i for the basic amplifier is 1 MΩ and R_o is 70 Ω.

  [6.5 GΩ, 0.01 Ω]

13.3 (a) If the input signal to the circuit shown is a low-frequency sine wave with a peak amplitude of 1 V, sketch the output waveform.
(b) Sketch the output waveform when the input amplitude is increased to ±16 V.

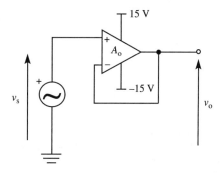

13.4 For the circuit shown sketch the output with the correct phase relationship with respect to the input, if the input is a 10 mV$_{\text{rms}}$ sine wave.

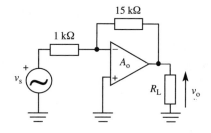

13.5 Determine the input resistance for the circuit in Problem 13.4.

13.6 Identify the type of feedback for the circuits shown in Problems 13.1 and 13.4 and write down expressions for the gain, input resistance and output resistance based on negative feedback models.

13.7 Determine the output voltage for the circuit shown.

[−6.5 V]

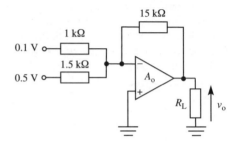

13.8 If v_{s1} is a 1 kHz sine wave with an rms value of 50 mV and v_{s2} is a 100 Hz square wave with amplitudes of ±1 V, sketch the output voltage for the circuit shown.

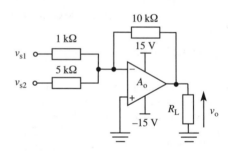

13.9 Use PSpice to verify the waveform obtained for Problem 13.8 by using the UA741 op-amp in the **eval.lib** library.

13.10 For the circuit shown determine the value of v_{s1} to ensure that the value of v_o is symmetrical about 0 V when v_{s2} is a sine wave of 100 mV sin ωt with a dc component of 2 mV.

[−4 mV]

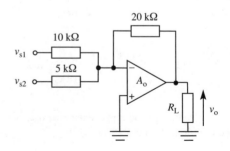

13.11 Determine v_o if $v_{s1} = 1$ mV + 2 mV sin ωt and $v_{s2} = -10$ mV.

[(400 mV)(1 + sin ωt)]

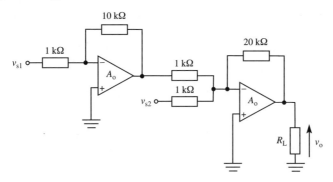

13.12 Determine the output voltage for the circuit shown.

[10 mV]

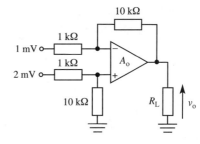

13.13 Determine the output voltage for the circuit shown.

[−5 mV]

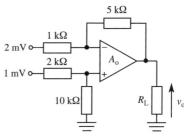

13.14 Determine v_o for the circuit shown.

[50 mV sin ωt]

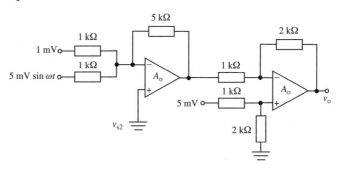

13.15   Determine the output voltage for the circuit shown.

[10 mV]

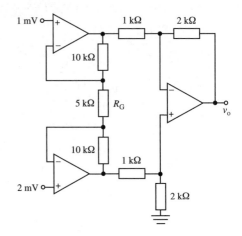

13.16   Select a suitable value for R_G in the circuit of Problem 13.15 to provide an overall gain of 1000 for the difference voltage.

[40.08 Ω]

13.17   Sketch the output waveform if v_s is a sine wave with an rms value of 6 V and if v_o saturates at ±9 V.

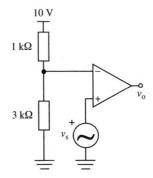

13.18   Use PSpice with the UA741 from the **eval.lib** library and ±10 V supplies to verify the result of Problem 13.17.

13.19   Sketch the output waveform for each of the circuits shown where v_s is a sine wave with an amplitude of 2 V.

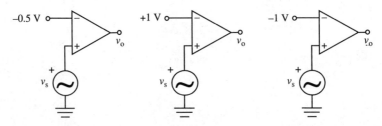

13.20 Sketch the output waveform if v_s is a sine wave with an amplitude of 10 V and if v_o saturates at ±13 V.

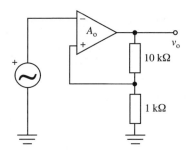

13.21 Determine the hysteresis for the circuit in Problem 13.20.

[2.36 V]

13.22 Determine the upper and lower threshold levels and the hysteresis for the circuit shown if the output saturates at ±13 V.

[−2.09 V, −0.27 V, 2.36 V]

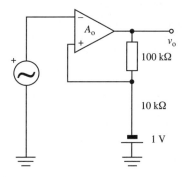

13.23 Determine the load current in the circuits shown.

[22 mA, 6.8 mA]

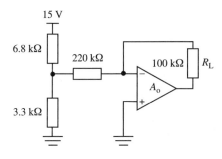

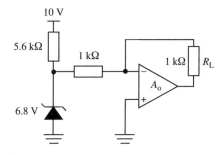

13.24 (a) For an ideal integrator the RC time constant is 200 ms. Determine the output voltage 2 seconds after applying a voltage to the input of 0.5 V.
(b) How long will it take to reach −15 V?

[(a) −5 V, (b) 6 s]

13.25   An ideal integrator has a 1 nF feedback capacitor. Select a suitable resistor to produce an integrator whose output will reach 8 V 100 ms after a −0.2 V dc voltage is applied to the input.

   [2.5 MΩ]

13.26   Design a circuit to convert a 100 Hz square wave with peak amplitudes of ±1 V to a triangular wave with peak amplitudes of ±10 V. Assume that the capacitor required for the circuit is 10 nF. Verify using PSpice and a UA741 amplifier with ±15 V supplies.

   [$R = 25$ kΩ]

CHAPTER 14

Active Filters

Analog circuits are used to process signals which have a wide range of different frequency components. An important application is to separate these different frequency components so that they can be processed either individually or as a selected group. This can be achieved with a filter. The frequency response of amplifier circuits has already been considered in Chapter 9, where the effect of coupling capacitors, decoupling capacitors and stray capacitance on the frequency response was considered in terms of RC networks. It was noted that RC networks could be divided into low-pass and high-pass filters depending on whether the network would pass low frequencies and attenuate high frequencies or vice versa. The filters and the corresponding frequency response curves are reproduced in Figure 14.1.

A combination of these two networks results in either a *band-pass filter* or a *band-stop filter*, as shown in Figure 14.2.

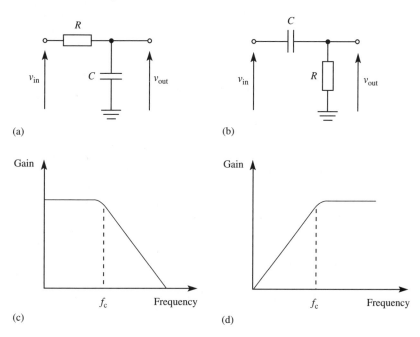

FIGURE 14.1 (a) Low-pass filter and (c) the frequency response; (b) high-pass filter with (d) the frequency response.

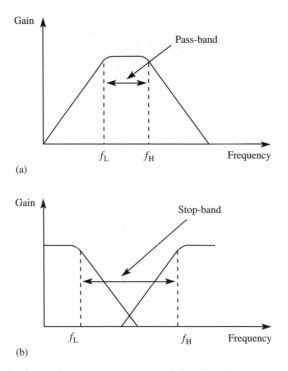

FIGURE 14.2   (a) Band-pass frequency response and (b) a band-stop frequency response.

Ideally the frequency response for each of the filters should be rectangular, as shown in Figure 14.3 where there is an abrupt change from the pass-band to the stop-band.

The ideal characteristics show a filter for which the gain is constant in the pass-band and at some frequency defined by f_c, f_L or f_H there is an abrupt change to a region where the signal is attenuated. A comparison of the ideal characteristics of Figure 14.3 with what is more likely to happen in practice, as shown in Figures 14.1 and 14.2, shows that the most immediate difference is in the rate of roll-off beyond the corner frequency. From Chapter 9 it was shown that for simple RC filters the rate of roll-off for a single RC network is 20 dB per decade. This figure can be increased by cascading RC sections, with 20 dB per decade being added for each section.

The different filters can be better achieved with the addition of an inductor. The simple parallel RLC network is an example of a band-pass filter. More complex networks with a number of inductors and capacitors can produce filters which closely approximate the ideal rectangular filters illustrated in Figure 14.3. These components are widely used in high-frequency communications circuits. However, for lower frequencies, and in particular for audio frequencies, filters which use inductors are not practical because of the large physical size of suitable inductors. Another type of filter has evolved which uses active gain stages, and in particular, operational amplifiers together with resistors and capacitors. These *active filters* can be used to create all of the

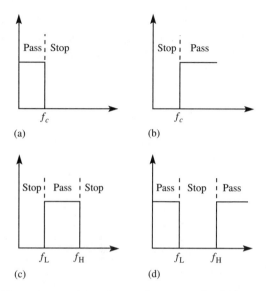

FIGURE 14.3 Ideal filter characteristics for (a) low-pass, (b) high-pass, (c) band-pass and (d) band-stop.

above filters without the use of inductors. The inclusion of gain greatly increases the versatility of the basic *RC* filter.

A practical example of the use of filters with a roll-off much greater than 20 dB per decade is the touch-tone dialling telephone. The digits 0 to 9 plus the * and # are represented by low-frequency tones in the frequency range 697 Hz to 1633 Hz. The different tones are generated from eight signals in two groups of four, which give a possible 16 combinations, as illustrated in Figure 14.4. Twelve tones are required for all touch-tone phones with an additional four tones for extra functions.

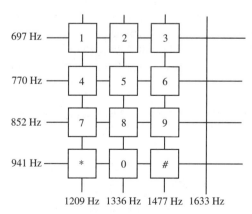

FIGURE 14.4 Signal arrangement for a touch-tone telephone dial.

Pressing a button generates a unique pair of signals which are transmitted to the exchange. Here they are decoded to provide a signal which operates the switching circuitry to make a connection with the telephone corresponding to the number dialled. The network which separates the signals is shown in Figure 14.5.

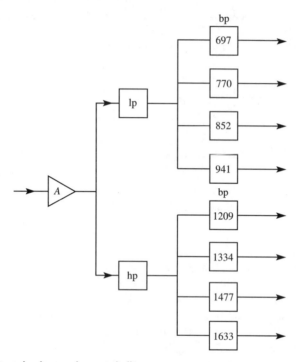

FIGURE 14.5   Decoder for touch-tone dialling.

After amplification to compensate for any loss of signal, the signal is split into two groups by means of a low-pass filter which passes all frequencies below (and including) 941 Hz, and a high-pass filter which passes all frequencies above (and including) 1209 Hz. The individual frequencies in each group are then separated by means of band-pass filters with their centre frequencies corresponding to the frequency of interest. The bandwidth of these filters $(f_H - f_L)$ is sufficiently narrow to reject all signals except the required signal.

The filters in this example are ideally suited to active filters based on operational amplifiers and RC networks.

14.1   Basic Filter Characteristics

There are many ways of constructing frequency-selective networks and they can be described mathematically by means of their transfer function. For example, in

Chapter 9 the relationship for the output voltage for a low-pass filter is given by:

$$v_{out} = \frac{v_{in}}{1 + j\omega RC} \tag{14.1}$$

or alternatively in terms of the transfer function (H) or 'gain' for the network as:

$$H(j\omega) = \frac{v_{out}}{v_{in}} = \frac{1}{1 + j\omega CR} \tag{14.2}$$

This is often written as:

$$H(j\omega) = \frac{1}{1 + (j\omega/\omega_c)} \tag{14.3}$$

where ω_c is $1/RC$. If only the magnitude is considered then the transfer function becomes:

$$|H(j\omega)|^2 = \frac{1}{1 + (\omega/\omega_c)^2} \tag{14.4}$$

Equation 14.4 describes the variation of the magnitude of the output voltage with respect to the input voltage for a low-pass RC filter with a corner frequency ω_c. It represents the response of a filter with the *order* of 1. In filter terminology it is a *one-pole filter*. The order of a filter is important as it affects the roll-off of the filter. As more sections are added to increase the order, the roll-off for this simple low-pass filter increases by 20 dB per decade for each section added. The higher the order the more closely the actual frequency response approaches the ideal rectangular response.

In addition to the number of filter sections, or order, a filter may also be classified by the response in the pass-band. The three most commonly used types of filter response are Butterworth, Chebyshev and Bessel. Each filter response can be described by a transfer function of the type shown in equation 14.4. Each response exhibits certain characteristics in the pass- and stop-bands which are unique to the particular type of filter and which offer particular advantages in different applications. The mathematical relationships for the transfer function for each type of response are obtained from consideration of general network theory and will not be considered in this text. Only the final relationships for each type of transfer function will be presented.

Butterworth Response

The Butterworth response for a low-pass filter is described by the transfer function:

$$|H(j\omega)|^2 = \frac{H_0}{1 + (\omega/\omega_c)^{2n}} \tag{14.5}$$

where n is the order and H_0 is a constant and represents the gain at $\omega = 0$. Notice that for $n = 1$, and $H_0 = 1$ equation 14.5 reverts to equation 14.4. Equation 14.5 is obtained by simply adding extra RC sections to the simple one-pole filter (however, there is the assumption that the sections do not interact with each other). The transfer function within the pass-band is flat and the roll-off between the pass-band and the stop-band increases by 20 dB per decade with each increase in order, as shown in Figure 14.6. The Butterworth response is often referred to as a *maximally flat response* because within the pass-band the gain remains constant.

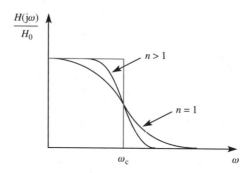

FIGURE 14.6   Low-pass Butterworth filter response for different orders.

The higher the order the closer the transition between the pass- and stop-bands approaches the ideal rectangular response, but of course the number of components increases. It would usually not be practical to exceed an order of 8.

Chebyshev Response

The Chebyshev response is described by the following relationship:

$$H(j\omega) = \frac{H_0}{1 + \varepsilon^2 C_n(\omega)^2 (\omega/\omega_c)} \tag{14.6}$$

where $C_n(\omega)$ is the nth-order Chebyshev polynomial (a mathematical relationship), ε is $\leqslant 1$ and H_0 is a constant. The Chebyshev polynomial is a mathematical relationship with properties which enable it to be simulated by means of an electrical circuit. It is described by:

$$C_n(\omega) = \cos(n \cos^{-1} \omega)$$

No attempt is being made in this text to describe this function in any more detail. Suffice it to say that there are specialist texts on filter design that show how this function is evaluated. The response may be plotted for different values of n and examples of the response for odd and even values are shown in Figure 14.7.

Because the polynomial is a mathematical expression, rather than an actual network, it is possible to have both negative and positive values of ω. The important point to note is that there are *ripples* in the pass-band and that the number of peaks corresponds to the order.

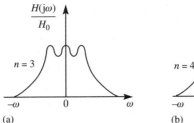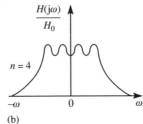

FIGURE 14.7   Plots of the Chebyshev polynomial for two different orders with (a) an order of 3 and (b) an order of 4.

The responses are symmetrical about $\omega = 0$ and a low-pass response has the appearance of that part of the response which extends from 0 to $+\omega$, while the high-pass response is effectively the mirror image, that is the part which starts from $-\omega$ and extends up to 0. In a real filter this would be from 0 Hz up to infinity. Notice that the order provides an indication of the number of troughs and peaks for the low-pass and high-pass filters. For a third-order low-pass or high-pass filter there are two peaks and one trough, while for the fourth-order filter there are two peaks and two troughs.

An important difference between this filter and the Butterworth is the rate of roll-off between the pass-band and the stop-band. For the Chebyshev it is much greater and the difference becomes more significant as the order increases and also as the amount of ripple increases. Thus for a given roll-off fewer sections are required for a Chebyshev filter than for a Butterworth filter, and therefore fewer components.

The amplitude of the ripple in the pass-band is a function of ε, with the ripple approaching 3 dB as ε approaches unity. The roll-off at the transition between the pass-band and stop-band is greatest for a large value of ε, but the ripple may be unacceptable for the transmission of the signal through the filter. There is a design compromise between the amount of ripple in the pass-band and the roll-off.

The concept of a 3 dB corner frequency does not apply to the Chebyshev filter as it does for the Butterworth filter, because of the presence of the ripples. For the Chebyshev the corner frequency refers to that frequency between the pass-band and the stop-band where the responses passes through the minimum within the ripple band. For 3 dB of ripple this would correspond to a difference of 3 dB between the peak in the pass band and the point at which the stop band starts, but for 0.5 dB of ripple, then this difference would only be 0.5 dB. Thus the corner frequency is determined by noting the frequency in the stop-band where the gain has dropped from a maximum value at the peak of one of the ripples to the value obtained at one of the troughs.

Bessel Response

The Butterworth and Chebyshev responses are both based on maintaining a constant magnitude of the signal with frequency within the pass-band (even though there are small variations for the Chebyshev filter), but there are considerable changes in the phase (or time delay) which vary non-linearly with frequency. These variations may not be important for analog signals, as in the majority of cases the information is contained in the amplitude of the signal, rather than the phase. However, for digital circuits where the signal consists of a train of pulses then the time domain properties of the filter are important. All components of the pulse must be delayed equally so that after passing through the filter, the pulse is accurately reproduced and the whole pulse is simply delayed in time, as shown in Figure 14.8. Here a single pulse is assumed to be applied to a filter, which has certain phase characteristics. For the ideal network the pulse appears at the output after a fixed delay τ_d.

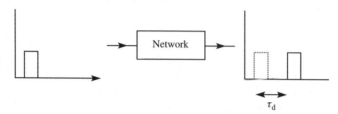

FIGURE 14.8   Pulse delay through an ideal network.

Time domain parameters which need to be considered in deciding whether the filter is suitable or not are *risetime, overshoot, ringing* and *settling time*. These parameters are illustrated in Figure 14.9 which show in more detail the leading edge of a pulse after it has passed through a network.

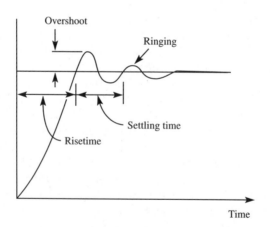

FIGURE 14.9   Detail of the leading edge of a pulse after passing through a filter.

The pulse has a finite risetime, and there may be some overshoot and oscillation which takes some time to settle before a flat pulse top is reached. The combination of each of these effects can affect the pulse emerging from the filter to such an extent that individual pulses will merge together, if they are not sufficiently spaced apart initially. These time domain effects are minimized with a Bessel filter which exhibits a linear phase characteristic where the phase shift varies linearly with frequency. This linear phase shift is equivalent to the constant time delay (τ_d) illustrated in Figure 14.8. The Bessel filter offers a *maximally flat time constant* within the pass-band. The disadvantage of the Bessel response is that the transition from pass- to stop-band is even worse than the Butterworth filter. However, where digital information, in the form of pulses, needs to be processed then the Bessel filter is far better than the Butterworth or the Chebyshev filters. More complex filter design can improve the Bessel roll-off.

Unlike the Butterworth and the Chebyshev there is no simple polynomial to solve for a Bessel filter, but there are relationships between $H(j\omega)$ and ωT which can approximate the linear phase relationship which is required.

The low-pass characteristics of the three filter types are shown in Figure 14.10.

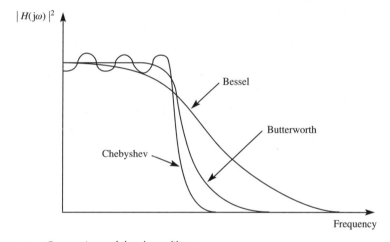

FIGURE 14.10   Comparison of the three filters.

Critical Frequency and Roll-Off

The corner frequency or critical frequency of an active filter is determined by the values of the resistors and capacitors and it is usual to simplify the design so that the corner frequency is given by:

$$f_c = \frac{1}{2\pi RC} \tag{14.7}$$

This formula applies for both low-pass and high-pass filters. The roll-off depends on the number of RC sections. A single RC section or first-order filter, or single-pole filter, has a roll-off of 20 dB per decade. A two-section filter has a roll-off of 40 dB per decade, and so on. Each section increases the roll-off by 20 dB per decade. The Chebyshev has a higher value of roll-off at the transition, but reverts to the above values further into the stop-band.

In designing a filter the critical frequency is obviously an important design parameter. Next it is necessary to decide on the amount of attenuation which is required just beyond the critical frequency in the stop-band. For the tone dialling filters this would be related to the separation of the different frequency components, about 80 Hz for the lower-frequency group and 150 Hz for the higher-frequency group. Thus for the lower frequencies the attenuation may need to be 60 dB down from the pass-band, 40 Hz from the critical frequency. This would require a roll-off in excess of 120 dB per decade, or six Butterworth sections.

14.2 Active Filters

In order to achieve the different filter characteristics it is necessary to *synthesize* the different responses with an appropriate selection of components. For modern electronic circuit design the most suitable components are resistors, capacitors and operational amplifiers. There are many different designs, but some of the most common are based on the *Sallen and Key filters* after their inventors. The original Sallen–Key filters were based on a unity gain amplifier, but the simple modification of allowing the amplifier to have gain increases the flexibility of the design. It is also known as a *VCVS* (voltage-controlled voltage source – see Chapter 12 for voltage–series feedback) filter.

The design of active filters is greatly simplified by the fact that all three types of filter (Butterworth, Chebyshev and Bessel) may be synthesized from the same basic circuit, but with different component values, and, more importantly, different values of gain for the gain block(s) for each filter type.

Low-Pass Filter

A second-order low-pass filter is shown in Figure 14.11.

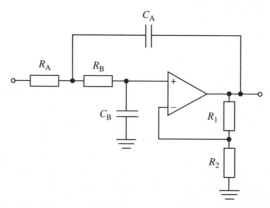

FIGURE 14.11   Low-pass Sallen–Key filter.

The operational amplifier is in a non-inverting configuration, or voltage-series connection, or voltage-controlled voltage source (VCVS). The input of the amplifier presents a very high impedance to the $R_B C_B$ section of the network and effectively isolates this section of the network from the load which is connected to the output of the amplifier. A second RC section comprises R_A and the feedback capacitor C_A. A detailed analysis (see Appendix 5) shows that the filter is a second-order filter with a corner frequency given by:

$$f_c = \frac{1}{2\pi\sqrt{R_A C_A R_B C_B}}$$

In practical designs it is usual to set $R_A = R_B = R$ and $C_A = C_B = C$ in which case the corner frequency is given by:

$$f_c = \frac{1}{2\pi RC}$$

The closed-loop gain of the amplifier, as determined by R_1 and R_2, is another factor in the design. Values for R, C and the gain are available in most standard texts on filter design and a selection of values for Butterworth, Chebyshev and Bessel filters are presented below.

High-Pass Filter

A second-order high-pass filter is shown in Figure 14.12.

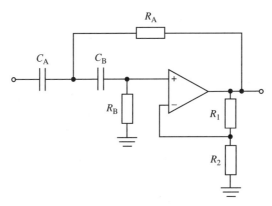

FIGURE 14.12 High-pass second-order filter.

As for the low-pass filter it is usual to let $R_A = R_B = R$ and $C_A = C_B = C$ so that the corner frequency is:

$$f_c = \frac{1}{2\pi RC}$$

The gain may be obtained from a table of values for each filter type.

Cascaded Filters for Greater Roll-Off

The transition between pass- and stop-bands is determined by the number of sections. The basic Sallen–Key filter is a two-pole or second-order filter. It is possible to construct a single-order filter, as shown in Figure 14.13. This filter has a roll-off of 20 dB per decade, but it still requires one operational amplifier which with the simple addition of an extra resistor and capacitor increases the order to 2. Since the ideal filter response is rectangular, any improvement in the roll-off which does not involve an excessive number of additional components is advantageous. Therefore, there is no particular merit in designing a single-pole filter when a second-pole filter can be created from the same operational amplifier. Thus when cascading filters to improve the roll-off it is usual to increase the order in steps of 2. A fourth-order (four-pole) filter is shown in Figure 14.14.

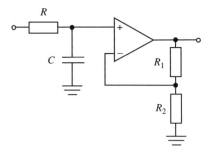

FIGURE 14.13   First-order low-pass filter.

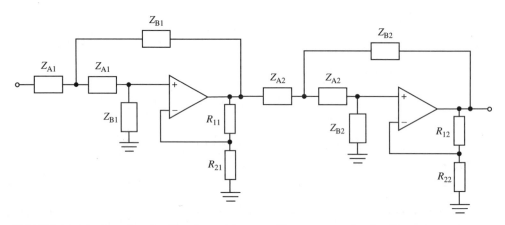

FIGURE 14.14   Fourth-order filter formed by cascading two second order filters.

For a low-pass filter the impedances Z_A are resistors and the impedances Z_B are capacitors, while for a high-pass filter Z_A represents capacitors and Z_B resistors. The RC time constants for each section are different, as are the gain factors for the amplifiers.

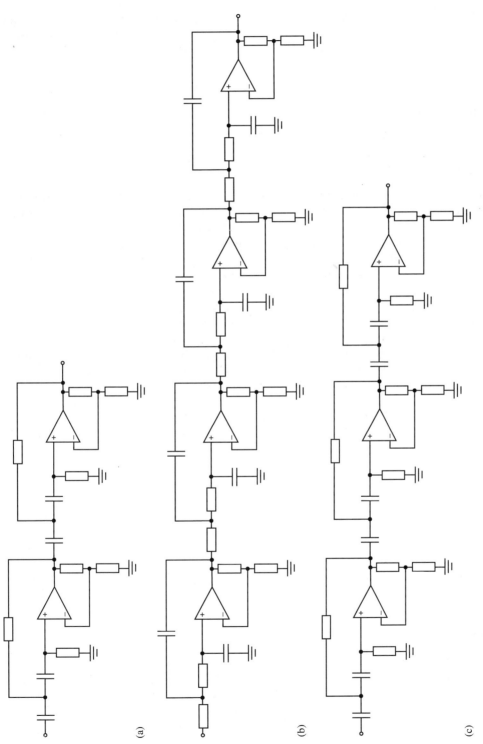

FIGURE 14.15 Sallen–Key cascaded filter section.

(a)

(b)

(c)

With the addition of further two-pole sections the roll-off is increased in steps of 40 dB per decade. Thus four sections would have a roll-of 80 dB per decade, six sections 120 dB per decade, and so on.

▶ **EXAMPLE 14.1** ——————————————————————————————

Identify the filter types, hp or lp, and roll-off for the filters shown in Figure 14.15.

────────────────────

Simplified Design Tables

All three types of filter (Butterworth, Chebyshev, Bessel) have the same basic format of resistors, capacitors and operational amplifier, but the component values and amplifier gains differ for each type. For multiple section filters the component and gain values may differ for each section. Low-pass and high-pass versions of the three filter types have been analyzed and the resultant equations simplified to provide sets of tables which are used to determine the component values for each type of filter and for each section. The key factors are the corner frequency and the gain of the amplifier. The Butterworth filter is the simplest to design in that the 3 dB corner frequency (f_c) is used to determine the same RC time constant for each section, but the gain for each section varies.

For the Bessel filter the gain varies for each section, but in addition the RC time constant for each section also varies. The RC time constant for each section is obtained from the 3 dB corner frequency multiplied by a normalizing factor (f_n).

The Chebyshev is the most complex because the amount of ripple presents an additional variable. The more specialist texts on filter design provide tables for Chebyshev designs in which the ripple is incremented in steps of 0.1 dB. However, for many filter designs it is adequate to consider a limited number of ripple values, say 0.5 dB and 2 dB. The Chebyshev design is based on values of gain for each section together with the normalizing factor necessary to determine the RC time constant for each section from the 3 dB corner frequency. Notice that for the Chebyshev filter the corner frequency corresponds to the point at which the response passes through the minimum of the ripple as it moves from the pass-band into the stop-band. This point is less than the 3 dB point associated with the Butterworth and Bessel filters.

A complete set of values for gain and multiplying factor f_n for the three filters is shown in Table 14.1.

Designing a Butterworth Low-Pass Filter

For the Butterworth filter all of the sections have the same values of R and C, given by the simple relationship $RC = 1/2\pi f_c$, where f_c is the required 3 dB corner frequency of the entire filter. The gain for each section of the filter is obtained from the table. For a two-pole filter the gain of the single amplifier is 1.586, for a four-pole filter the gains are 1.152 and 2.235, and so on for higher orders.

TABLE 14.1 Sallen–Key low-pass filter design parameters.

Order (poles)	Butter-worth	Bessel		Chebyshev (0.5 dB)		Chebyshev (2.0 dB)	
	Gain	Gain	f_n	Gain	f_n	Gain	f_n
2	1.586	1.268	1.274	1.842	1.231	2.114	0.907
4	1.152	1.084	1.432	1.582	0.597	1.924	0.471
	2.235	1.759	1.606	2.660	1.031	2.782	0.964
6	1.068	1.040	1.607	1.537	0.396	1.891	0.316
	1.586	1.364	1.692	2.448	0.768	2.648	0.730
	2.483	2.023	1.908	2.846	1.011	2.904	0.983
8	1.038	1.024	1.781	1.522	0.297	1.879	0.238
	1.337	1.213	1.835	2.379	0.599	2.605	0.572
	1.889	1.593	1.956	2.711	0.861	2.821	0.842
	2.610	2.184	2.192	2.913	1.006	2.946	0.990

▷ **EXAMPLE 14.2** ────────────────────────────────

Design a four-pole low-pass Butterworth filter which has a 3 dB corner frequency at 250 Hz and use PSpice to verify the design.

▷ *Solution*

The RC time constant for each section is:

$$RC = \frac{1}{2\pi(250 \text{ Hz})} \approx 0.637 \text{ ms}$$

Assume that the capacitor value is 220 nF; then the resistor is ~2.9 kΩ. The gain of the first stage is:

$$1.152 = 1 + \frac{R_1}{R_2}$$

If R_2 is 10 kΩ then R_1 is 1.52 kΩ. The gain of the second stage is:

$$2.235 = 1 + \frac{R_1}{10 \text{ k}\Omega}$$

$$R_1 = 12.4 \text{ k}\Omega$$

The PSpice schematic for this filter is shown in Figure 14.16. The UA741 amplifier is used with ±15 V supplies. The frequency scan is from 1 Hz to 10 kHz.

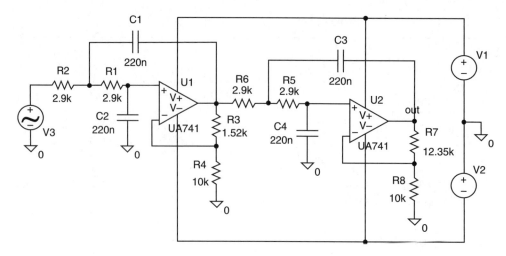

FIGURE 14.16   PSpice schematic for four-pole low-pass Butterworth filter.

The frequency response is shown in Figure 14.17.

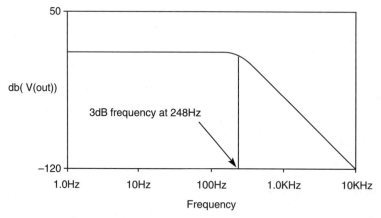

FIGURE 14.17   Probe output for low-pass filter.

From the Probe output the 3 dB frequency is measured as approximately 248 Hz and the slope is 80 dB per decade, which corresponds to the roll-off expected for a four pole filter (4 × 20).

Designing Bessel and Chebyshev Low-Pass Filters

For the Bessel and Chebyshev filters the *RC* time constants for the individual sections are different. The *RC* product for each section must be scaled by the normalizing factor

f_n (see Table 14.1). For each section the RC product is given by:

$$RC = \frac{1}{2\pi f_n f_c} \tag{14.8}$$

where f_c is the 3 dB corner frequency for the complete Bessel filter. For the Chebyshev filter it corresponds to the frequency at which the amplitude response passes through the minimum of the ripple band as it moves from the pass-band to the stop-band.

▶ **EXAMPLE 14.3** ───

Design a four-pole low-pass Bessel filter which has a 3 dB corner frequency at 300 Hz and verify the design using PSpice.

▶ *Solution*

A four-pole filter requires two two-pole sections. The RC time constant for the first section is:

$$RC = \frac{1}{2\pi(1.432)(300)} = 0.370 \text{ ms}$$

If R is 1 kΩ then C is 0.37 µF.
    For the second section the RC time constant is:

$$RC = \frac{1}{2\pi(1.606)(300)} = 0.330 \text{ ms}$$

If R is 1 kΩ then C is 0.33 µF.
    The gain of the first stage is:

$$1.084 = 1 + \frac{R_1}{R_2}$$

If R_2 is 10 kΩ then R_1 is 840 Ω.
    The gain for the second stage is:

$$1.759 = 1 + \frac{R_1}{R_2}$$

and if R_2 is 10 kΩ then R_1 is 7.6 kΩ
    The PSpice schematic for the filter is shown in Figure 14.18.
    The Probe output for the filter is shown in Figure 14.19.

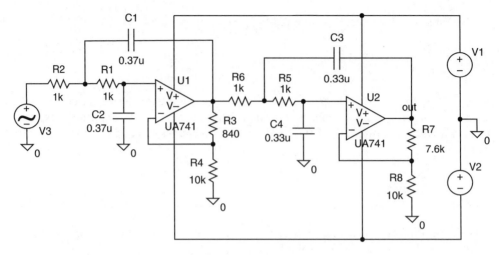

FIGURE 14.18   PSpice schematic for four-pole Bessel low-pass filter.

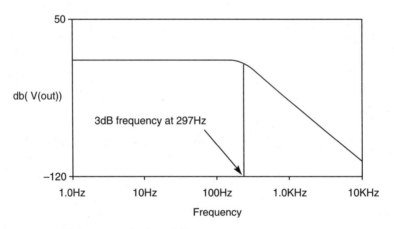

FIGURE 14.19   Probe output for Bessel filter.

Comparing the response to the Butterworth response shown in Figure 14.17 it may be seen that the roll-off for the Bessel is more gradual. There is a more rounded response at the corner frequency. Well beyond the corner frequency the roll-off is the same 80 dB per decade.

▶ **EXAMPLE 14.4** _____

Design a four-pole low-pass Chebyshev filter with a 2 dB ripple which has a corner frequency of 350 Hz and verify the design with PSpice.

▷ *Solution*

Two two-pole sections are required. The *RC* time constant for the first section is:

$$RC = \frac{1}{2\pi(0.471)(350)} = 0.965 \text{ ms}$$

If *R* is 1 kΩ then *C* is approximately 1 µF.
 The *RC* time constant for the second section is:

$$RC = \frac{1}{2\pi(0.964)(350)} = 0.472 \text{ ms}$$

If *R* is 1 kΩ then *C* is 0.47 µF
 The gain for the first stage is:

$$1.924 = 1 + \frac{R_1}{10 \text{ k}\Omega}$$

where R_2 is 10 kΩ which gives R_1 a value of 9.2 kΩ.
 The gain for the second stage is:

$$2.782 = 1 + \frac{R_1}{10 \text{ k}\Omega}$$

and R_1 is 17.8 kΩ.
 Simulating the filter with PSpice produces the output shown in Figure 14.20.

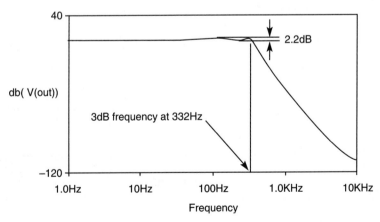

FIGURE 14.20 Low-pass Chebyshev output with 2 dB of ripple.

It can be seen that the ripple does not appear to be as symmetrical as suggested in Figure 14.7. However, the *x* axis in Figure 14.7 is linear rather than logarithmic, as in Figure 14.20. A detailed section of Figure 14.20 with a linear *x* axis is shown in Figure 14.21.

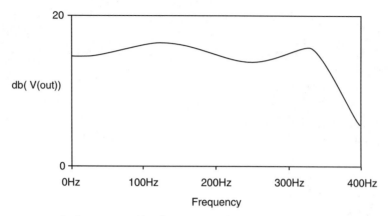

FIGURE 14.21   Probe output with a linear *x* axis.

It can be seen that there are two troughs and two peaks corresponding to the number of troughs and peaks for a low-pass four-pole Chebyshev filter.

A more important factor is that the roll-off approaches 100 dB per decade, and is in fact greater at the transition from pass-band to stop-band.

High-Pass Filters

A high-pass filter is formed by interchanging the R and the C. For the Butterworth the values remain the same for each section, as listed in Table 14.1, with the RC time constant determined by the 3 dB corner frequency, and the gain obtained from Table 14.1. For the Bessel and Chebyshev the gain values remain the same, as for the low-pass filters, but the normalizing factor f_n becomes $1/f_n$. Thus the RC time constant becomes:

$$RC = \frac{f_n}{2\pi f_c} \tag{14.9}$$

▷ **EXAMPLE 14.5** ——————————————————————————————

Design a Chebyshev high-pass filter with 0.5 dB of ripple, a corner frequency of 1200 Hz and a roll-off in excess of 120 dB per decade. Verify the design with PSpice.

▷ *Solution*

A roll-off in excess of 120 dB per decade implies an order of at least 6. From Table 14.1 the RC time constants are determined with the aid of equation 14.9. If all of the capacitors

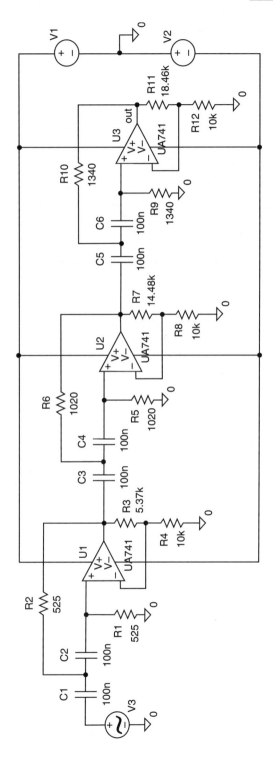

FIGURE 14.22   Schematic of six-pole 0.5 dB Chebyshev high-pass filter.

are assumed to be 100 nF then the resistors for the three two-pole sections are:

first section   525 Ω

second section   1020 Ω

third section   1340 Ω

If R_2 of the gain resistors is 10 kΩ then the values for R_1 for the different sections are:

first section   5.37 kΩ

second section   14.48 kΩ

third section   18.46 kΩ

The schematic is shown in Figure 14.22.
The frequency response is shown in Figure 14.23.

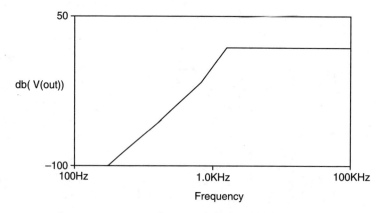

FIGURE 14.23   Frequency response for six-pole high-pass Chebyshev filter.

A more detailed view of the pass-band is shown in Figure 14.24.

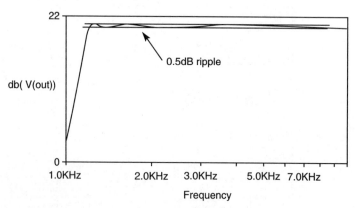

FIGURE 14.24   Detail of the pass-band with a linear x axis.

The 0.5 dB ripple is clearly seen in Figure 14.24. Only a small number of ripples are seen because the ideal frequency response for a high-pass filter extends to infinity.

Band-Pass Filters

A band-pass filter passes a limited band of frequencies and rejects all frequencies below and above the selected band. It can be made by cascading low-pass and high-pass filters as shown in Figure 14.25.

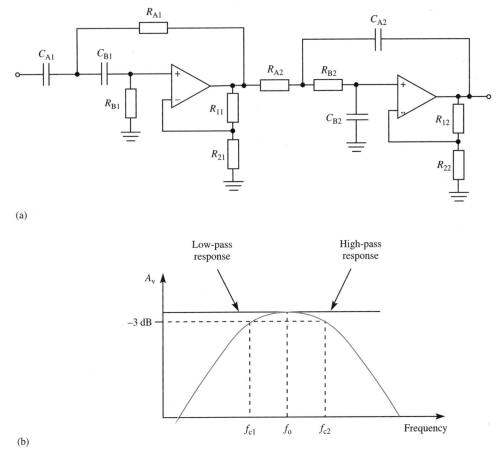

(a)

(b)

FIGURE 14.25 Band-pass filter formed by (a) cascading a high-pass and low-pass filters and (b) the frequency response.

Notice that the high-pass filter determines the lower corner frequency f_{c1}, while the low-pass filter determines the high corner frequency f_{c2}. The Q of the filter is defined as:

$$Q = \frac{f_0}{(f_{c2} - f_{c1})} \tag{14.10}$$

A problem with this approach is that it does not work very well if a high Q is required for which the bandwidth $(f_{c2} - f_{c1})$ is very small with respect to the value of the centre frequency f_0.

Band-stop filters can be obtained by summing the responses from low-pass and high-pass filters, but again this approach is not ideal if the rejected band is to be very narrow.

▶ **EXAMPLE 14.6** ─────────────────────────────────

Design a two-pole Butterworth band-pass filter with corner frequencies at 400 Hz and 4 kHz, and verify the design with PSpice.

▶ *Solution*

Low-Pass Section

The corner frequency is 4 kHz, f_{c2} in Figure 14.25. Assume $C = 10$ nF.

$R = 4.0$ kΩ $C = 10$ nF
$R_1 = 5.86$ kΩ $R_2 = 10$ kΩ

High-Pass Section

The corner frequency is 400 Hz, f_{c1} in Figure 14.25. Assume $C = 100$ nF.

$R = 4.0$ kΩ $C = 100$ nF
$R_1 = 5.86$ kΩ $R_2 = 10$ kΩ

The PSpice schematic is shown in Figure 14.26.

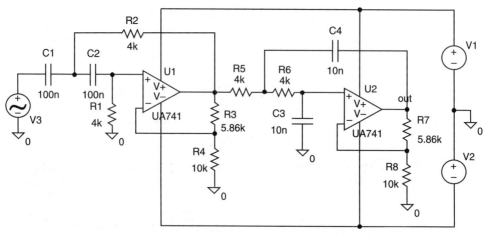

FIGURE 14.26   PSpice schematic of a Butterworth band-pass filter.

In the PSpice **Analysis...** and **Setup...** the frequency sweep is set to linear with 200 points between 100 Hz and 10 kHz. The Probe output is shown in Figure 14.27.

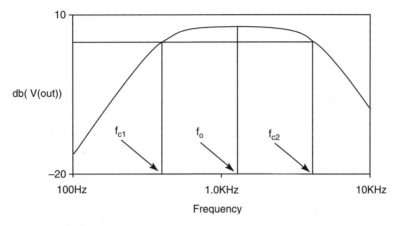

FIGURE 14.27   Probe output.

From the Probe output the corner frequencies are measured as $f_{c1} = 396$ Hz, $f_{c2} = 4$ kHz and $f_0 = 1.26$ kHz.

While the frequency response shown in Figure 14.27 is a band-pass response the Q is very small at:

$$Q = \frac{f_0}{f_{c2} - f_{c1}} = \frac{1.26 \text{ kHz}}{4 \text{ kHz} - 396 \text{ Hz}} = 0.35$$

For most applications involving a band-pass filter a Q of 10 or greater is required. The cascade approach does not lend itself to high Q.

▷ **EXAMPLE 14.7** ——————————————————————————————————

Repeat Example 14.6 but try to modify the design to have a Q of 10 about a centre frequency of 1.3 kHz.

▷ *Solution*

The corner frequencies may be obtained from equation 14.10 as:

$$Q = 10 = \frac{1.3 \text{ kHz}}{f_{c2} - f_{c1}}$$

That is, $f_{c1} = 1300 - 130/2 = 1235$ Hz and $f_2 = 1300 + 130/2 = 1365$ Hz.

Low-Pass Section

$R = 1.17$ kΩ $C = 100$ nF
$R_1 = 5.86$ kΩ $R_2 = 10$ kΩ

High-Pass Section

$R = 1.29$ kΩ $C = 100$ nF
$R_1 = 5.86$ kΩ $R_2 = 10$ kΩ

The frequency response obtained with Probe is shown in Figure 14.28.

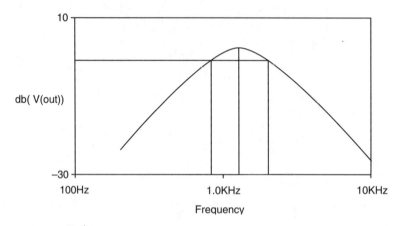

FIGURE 14.28   Probe output.

The centre frequency is correct at 1.28 kHz, but the corner frequencies are at 834 Hz and 2 kHz. The cascaded approach is only practical for very small values of Q, and when the corner frequencies are roughly a factor of 10 apart.

The basic Sallen–Key filter provides a filter of two poles per operational amplifier with the minimum number of components. Additional advantages are non-inverting gain and a low output resistance. However, it is very sensitive to component values and it is difficult to use in applications which require a tuneable filter. Other types of filter have been developed which are less sensitive to component values and one of these is the state-variable filter.

State-Variable Filters

A two-pole state-variable or biquad filter is shown in Figure 14.29. It is more complex than the Sallen–Key filter, but is more versatile, because its gain and corner frequencies are easily adjusted. There are output connections for low pass, high pass and band-pass, all within the one circuit. Because of its versatility it is widely available as a complete IC package from manufacturers such as Burr-Brown and National Semiconductors.

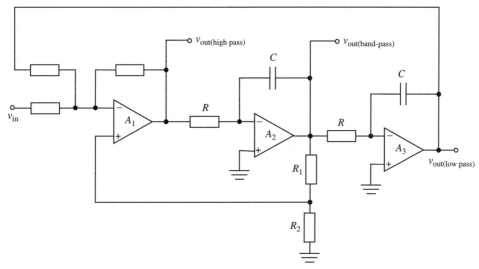

FIGURE 14.29   State-variable filter.

The circuit comprises a combined summing and difference amplifier (A_1) and two integrators $(A_2$ and $A_3)$. The integrators are single-pole low-pass filters each with a transfer function $-1/j\omega RC$ or $-\omega_c/s$ where $s = j\omega$. Acting in cascade produces a two-pole filter with the transfer characteristic ω_c^2/s. The corner frequency (f_c) is determined by the RC time constants of the integrators, and for the purposes of a band-pass filter they are usually equal and $f_c = f_0$.

There are two feedback paths, one from the output of A_2 back to the non-inverting terminal of A_1 and the other from A_3 to the inverting terminal of A_1. A block diagram of the filter is shown in Figure 14.30.

The basic frequency characteristics of the filter are shown in Figure 14.31.

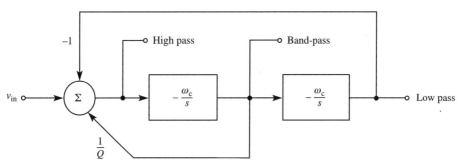

FIGURE 14.30   Block diagram of the two-integrator state-variable filter.

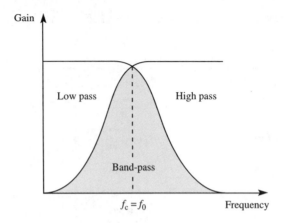

FIGURE 14.31   Basic frequency characteristics for state-variable filter.

For input frequencies less than f_c the input passes through the summing amplifier and the two integrators and arrives back at the second input of the summing amplifier $180°$ out of phase with v_{in}, which is shown as -1 in Figure 14.30. Thus at the summing node of A_1 the signals below f_c are cancelled. As the frequency approaches f_c this feedback signal is progressively attenuated by the two integrators and the fraction $(1/Q)$ fed back from A_2 becomes more significant. Above f_c the output from A_3 disappears because the signal is now attenuated by both integrators, and the signal from A_1 is now the combination of v_{in} and the fraction $(1/Q)$ from A_2.

The Q is established by the two feedback resistors R_1 and R_2 as:

$$Q = \frac{1}{3}\left(1 + \frac{R_1}{R_2}\right) \tag{14.11}$$

The centre frequency is given as:

$$f_0 = f_c = \frac{1}{2\pi RC} \tag{14.12}$$

▶ **EXAMPLE 14.8** ─────────────────────────────

Determine the centre frequency, Q and bandwidth of the state-variable filter shown in Figure 14.32 and confirm using PSpice.

▶ *Solution*

With both integrators having the same corner frequencies the centre frequency is:

$$f_0 = \frac{1}{2\pi(150 \text{ nF})(1.5 \text{ k}\Omega)} = 707 \text{ Hz}$$

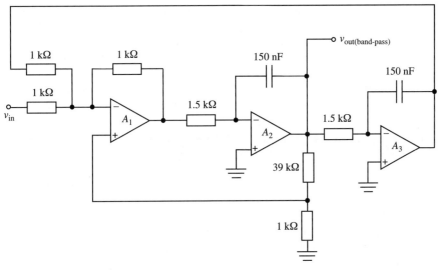

FIGURE 14.32   Band-pass state-variable filter.

From equation 14.11 the Q is:

$$Q = \frac{1}{3}\left(1 + \frac{39 \text{ k}\Omega}{1 \text{ k}\Omega}\right) = 13.3$$

The bandwidth is:

$$BW = \frac{f_0}{Q} = \frac{707}{13.3} = 53 \text{ Hz}$$

The output from Probe is shown in Figure 14.33.

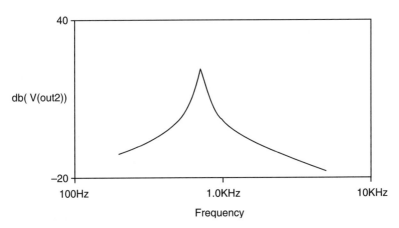

FIGURE 14.33   Band-pass response for state-variable filter.

The measured centre frequency is 705 Hz with a bandwidth of 55 Hz, which yields a Q of 13.5.

Manufacturers' literature for these filters provides information for selecting component values for Butterworth, Chebyshev and Bessel filters for low pass, high pass and band-pass. They can be cascaded to increase the order.

A band-stop filter may be obtained by summing the low-pass and high-pass outputs from the state-variable filter, as shown in Figure 14.34.

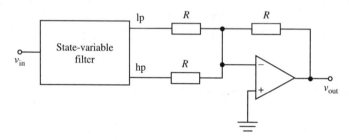

FIGURE 14.34   State-variable-based band-stop filter.

▶ **EXAMPLE 14.9** ——————————————

Verify that the PSpice schematic shown in Figure 14.35 exhibits a band-stop characteristic and has a centre frequency of 100 Hz. Determine the Q factor.

NB For version 6.0 of PSpice it is necessary to replace one of the op-amps, say u4 with an E-type source with a gain of 100 000, to avoid the 64 node limit of this version of the evaluation software.

▶ *Solution*

The output from Probe is shown in Figure 14.36.

The Q factor is determined by resistors R_5 (19 kΩ) and R_6 (1 kΩ), and from equation 14.11 the Q is:

$$Q = \frac{1}{3}\left(1 + \frac{19 \text{ k}\Omega}{1 \text{ k}\Omega}\right) \approx 7$$

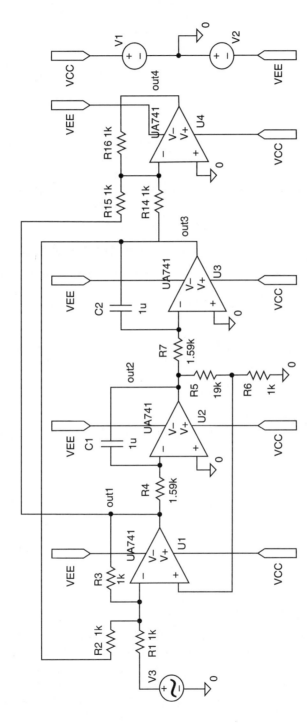

FIGURE 14.35   PSpice schematic of a state-variable band-stop filter.

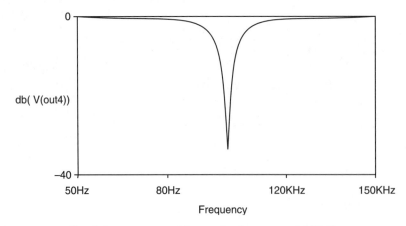

FIGURE 14.36   Band-stop response with a centre frequency at 100 Hz.

Twin-Tee Band-Stop Filter

An alternative band-stop filter may be obtained from the twin-tee configuration shown in Figure 14.37.

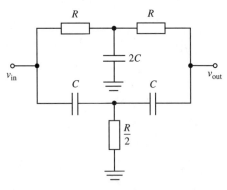

FIGURE 14.37   Twin-tee filter.

The network has infinite attenuation at $f_c = 1/2\pi RC$. This is achieved by adding together two signals which are 180° out of phase at the cut-off frequency. It requires very good component matching to achieve a good null at the cut-off frequency. However, like all RC filters the initial roll-off is very gradual. This can be improved by making the filter active with the addition of operational amplifiers, as shown in Figure 14.38.

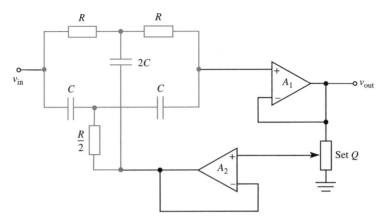

FIGURE 14.38   Active twin-tee notch filter.

The amplifier A_1 is a unity gain buffer to minimize the loading effects on the twin-tee, while A_2 is also a unity gain buffer which provides feedback from the output to the twin-tee. By including a potentiometer it is possible to adjust the Q. The Q is given by:

$$Q = \frac{1}{4(1-m)} \qquad (14.13)$$

where m is the potentiometer ratio. With the potentiometer wiper at ground, and $m = 0$, the Q is 0.25, and when the wiper is at the other extreme with $m = 1$, Q approaches 'infinity'.

▷ **EXAMPLE 14.10** ───────────────────────────────────────

Design a twin-tee active filter to reject 50 Hz line frequency and verify the design with PSpice.

▷ *Solution*

Assume a value of 1 µF for C in the twin-tee. Then for $f_c = 50$ Hz:

$$R = \frac{1}{2\pi 50 (1\ \mu\text{F})} = 3.18\ \text{k}\Omega$$

The PSpice schematic is shown in Figure 14.39.
The output from Probe is shown in Figure 14.40.
From the PSpice schematic the Q is:

$$Q = \frac{1}{4(1-m)} = \frac{1}{4}\left(\frac{1}{1 - 10\ \text{k}\Omega/(10\ \text{k}\Omega + 10\ \Omega)}\right) = 250$$

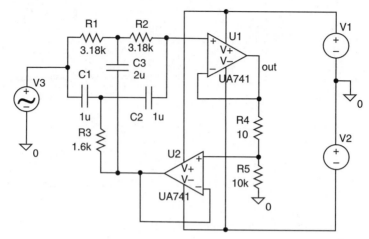

FIGURE 14.39   PSpice schematic of twin-tee filter.

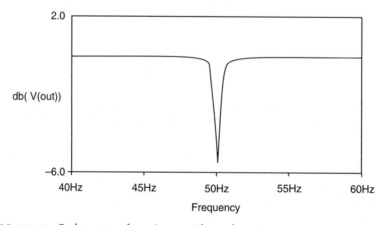

FIGURE 14.40   Probe output for twin-tee with notch at 50 Hz.

It can be seen that a very sharp notch occurs at 50 Hz. A difficulty with this type of filter is the need for accurate component values, which must also have stable temperature coefficients in order to achieve this sort of attenuation at the required frequency over a wide temperature range.

▷ **EXAMPLE 14.11** ─────────────────────────────────────

Repeat Example 14.10 for a frequency of 60 Hz.

▷ **PROBLEMS** ────────────

14.1 Draw the circuit diagram for non-active single-pole low-pass and high-pass RC filters.

14.2 What is the theoretical roll-off for an RC filter comprising three RC sections, assuming that there is no interaction between the different sections?

14.3 Show that the magnitude of a single-pole high-pass RC filter response can be expressed as:

$$|H(\text{j}\omega)|^2 = \frac{1}{1 + (\omega_c/\omega)^2}$$

14.4 Draw the comparative frequency responses for active high-pass multipole Butterworth, Chebyshev and Bessel filters.

14.5 Identify the important difference between the Bessel active filter and the Butterworth and Chebyshev filters.

14.6 Determine the corner frequency for the circuit shown, identify the filter type, estimate the roll-off and sketch the frequency response.

[1.06 kHz]

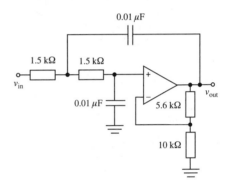

14.7 Determine the corner frequency, filter type, roll-off and sketch the frequency response for the circuit shown.

[133 Hz]

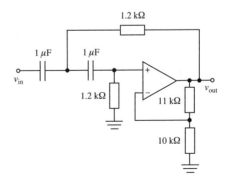

14.8   Design a maximally flat six-pole low-pass filter with a corner frequency of 3 kHz. Assume capacitor values of 22 nF and a value of 10 kΩ for the grounded gain resistor in each section. Verify the design using PSpice and UA741 amplifiers.

14.9   Modify the design of Problem 14.8 to a filter which has the same number of sections but a sharper roll-off, and which also has a 0.5 dB ripple in the pass band. Verify the design with PSpice.

14.10   Design an audio cross-over network with a critical frequency of 400 Hz to separate the low and high frequencies for the bass and tweeter loudspeakers. Assume a roll-off of 80 dB per decade with a maximally flat response. Use 47 nF capacitors and a 10 kΩ grounded gain resistor. Verify the design with PSpice.

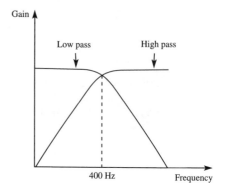

14.11   Design a high-pass filter which has a maximally flat time constant within the pass-band with a corner frequency of 5 kHz and a roll-off of greater than 60 dB per decade. Assume capacitor values of 47 nF and a grounded gain resistor of 10 kΩ. Verify the design with PSpice.

14.12   Determine the filter type, the corner frequency and sketch the frequency response for the circuit shown.

[800 Hz]

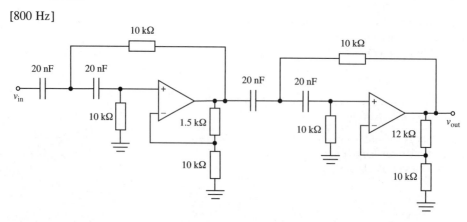

14.13   Determine the filter type, the corner frequency and sketch the frequency response for the circuit shown.

[298 Hz]

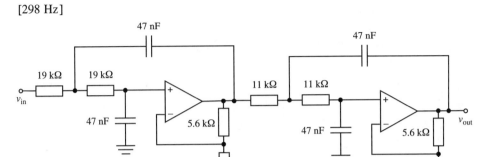

14.14   Identify the filter type for each stage, determine the corner frequencies and sketch the frequency response for the circuit shown.

[132 Hz, 1.6 kHz]

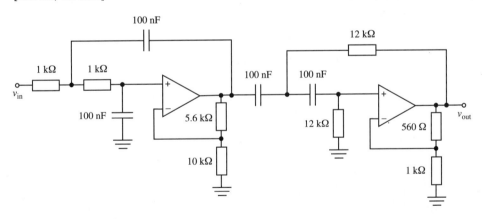

14.15   Design a Chebyshev band-pass filter with corner frequencies at 10 Hz and 100 Hz with not more than 0.5 dB of ripple in the pass-band and a roll-off in excess of 120 dB per decade. Assume capacitor values of 1 µF and a resistor next to the grounded end of the gain potential divider of 10 kΩ.

14.16   Determine the centre frequency, Q and bandwidth of the filter shown.

[994 Hz, 106 Hz]

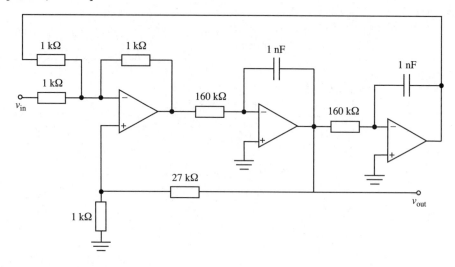

14.17   Design a state-variable band-pass filter with a centre frequency at 1209 Hz and a bandwidth of 40 Hz. Assume capacitor values of 10 nF. Verify the design using PSpice.

14.18   Identify the type of filter and determine the critical frequency and the bandwidth.

[48 Hz, 6 Hz]

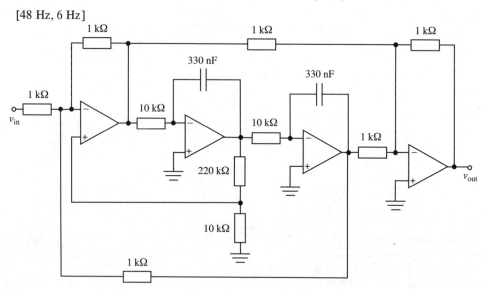

14.19 Design a state-variable band-reject filter for 1 kHz with a Q of 10. Use capacitors of 22 nF and a 1 kΩ grounded gain resistor. Verify the design with PSpice. (For the demonstration version it will be necessary to use an E-type source from the **analog.slb** library for the final summing amplifier to avoid exceeding the component count. Assume a gain for the source of 1000.)

14.20 Determine the critical frequency for the circuit shown and verify with PSpice.

[1 kHz]

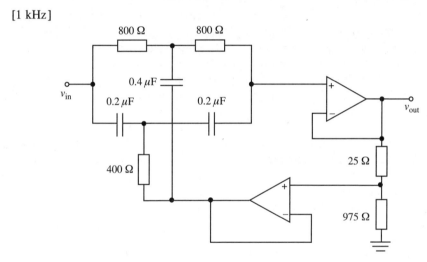

CHAPTER 15

Waveform Generation

An important requirement for many electronic circuits is the generation of a waveform of some sort. For analog circuits this may be a sine wave, while for digital circuits it is likely to be a square wave or rectangular pulses. A signal generator is an example of a test instrument which is used to produce waveforms to test electronic circuits, and where controls are provided to change the frequency, amplitude and shape of the waveform. Accuracy and stability are important features for the waveforms produced by a signal generator, where the waveform may be used to calibrate external circuits. Whether the waveform is generated internally within an electronic circuit, or as part of the circuitry of a signal generator, some means is required in each case to produce the waveform.

The source of square or rectangular waves is often the charging and discharging of a capacitor. This form of generation creates an exponential or triangular waveform which can be used to switch a comparator to generate a square wave. Sine waves can also be created from triangular waves by passing them through some form of shaping circuit which usually consists of diodes and resistors. Generators based on the charging action of a capacitor are referred to as non-linear oscillators or *relaxation oscillators*.

An alternative approach is to use positive feedback and a frequency-selective network. The circuit is designed to allow an active device to operate in its linear region by controlling the positive feedback by some means. Positive-feedback-based generators are usually referred to as *linear oscillators*, even though non-linear effects are used to limit the action of the positive feedback.

15.1 Relaxation Oscillators

A very simple form of relaxation oscillator can be produced by charging a capacitor from a resistor or current source, and then discharging it rapidly when it reaches some preset value, as shown in Figure 15.1.

Some form of electronic switch, contained within the box, closes the switch to discharge the capacitor at some preset voltage level. An approximately saw-tooth wave is created across the capacitor. In the past negative resistance devices, such as the unijunction transistor, were used to make relaxation oscillators based on this principle. These devices are difficult to manufacture and have been largely replaced by the operational amplifier which is far more versatile.

514

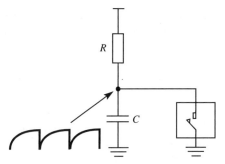

FIGURE 15.1   Simple relaxation oscillator.

Astable Multivibrator

Figure 15.2 shows a relaxation oscillator based on the same RC charging circuit, but with the addition of an operational amplifier, which is connected as a Schmitt trigger, to reverse the voltage applied to the capacitor at some preset threshold.

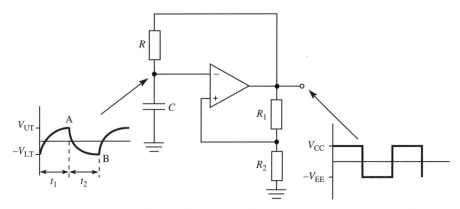

FIGURE 15.2   Schmitt-trigger-based relaxation oscillator.

The charging voltage for the capacitor is now provided by the output of the Schmitt trigger (see Section 13.4) rather than a fixed voltage source as depicted in Figure 15.1. The output from the Schmitt alternates between two saturation levels (for a ±15 V supply this will be approximately $V_{\text{sat}} = \pm13$ V). These two levels also establish, by means of R_1 and R_2, an upper threshold level (V_{UT}) and a lower threshold level (V_{LT}) at the non-inverting input. The operation is simple. Assume that when power is applied the output of the Schmitt trigger is positive (say $+13$ V). The capacitor now starts to charge exponentially towards the $+13$ V. The potential divider establishes an upper threshold level at the non-inverting terminal of V_{UT}.

When the voltage across the capacitor reaches and then exceeds V_{UT} the inverting terminal becomes positive with respect to the non-inverting terminal and the output of the operational amplifier moves in a negative direction. Positive feedback then causes the output of the Schmitt trigger to change polarity and to become -13 V. The charging current to the capacitor now changes direction and the voltage across the capacitor moves in a negative direction towards -13V with the same RC time constant. The potential divider applies a negative voltage to the non-inverting terminal of V_{LT}. When the voltage across the capacitor reaches V_{LT} the output of the Schmitt trigger switches back to its positive value. This process repeats with the voltage across the capacitor having an approximately triangular appearance, with maximum and minimum values of V_{UT} and V_{LT}, while the voltage at the output of the Schmitt trigger is a square wave, with maximum and minimum values of $\pm V_{sat}$.

In order to determine the frequency of the square wave it is necessary to consider the voltage across the capacitor with initial and final values (see Appendix 6). For a positive-going segment of the capacitor voltage versus time curve, the initial voltage is $-V_{LT}$ and the final voltage is $+V_{sat}$ and the equation for the capacitor voltage at any instant is:

$$v_C = V_{sat} - (V_{sat} + V_{LT})\exp\left(\frac{-t}{RC}\right) \tag{15.1}$$

For the next segment the voltage moves in a negative direction. The initial value is $+V_{UT}$ and the final value is $-V_{sat}$. The equation for the capacitor voltage is now:

$$v_C = -V_{sat} + (V_{sat} + V_{UT})\exp\left(\frac{-t}{RC}\right) \tag{15.2}$$

With reference to Figure 15.2 the positive-going segment of the voltage across the capacitor ends when $v_C = V_{UT}$ and from equation 15.1:

$$V_{UT} = V_{sat} - (V_{sat} + V_{LT})\exp\left(\frac{-t_1}{RC}\right)$$

and

$$\frac{t_1}{RC} = \ln\left(\frac{V_{sat} + V_{LT}}{V_{sat} - V_{UT}}\right) \tag{15.3}$$

and similarly for the negative segment:

$$\frac{t_2}{RC} = \ln\left(\frac{V_{sat} + V_{UT}}{V_{sat} - V_{LT}}\right) \tag{15.4}$$

If $V_{UT} = V_{LT} = V_{sat}/2$ then the period, which is the sum of t_1 and t_2, is:

$$T = 2RC \ln\left(\frac{1.5V_{sat}}{0.5V_{sat}}\right) = 2.2RC \tag{15.5}$$

The frequency of oscillation is the reciprocal of the period:

$$f = \frac{1}{T} = \frac{0.455}{RC} \text{ Hz} \tag{15.6}$$

This equation represents the frequency of the square wave output when the potential divider is selected to make the threshold voltages applied to the non-inverting terminals of the operational amplifier equal to half the output saturation voltage, that is $R_1 = R_2$.

An asymmetric waveform may be obtained with the circuit shown in Figure 15.3.

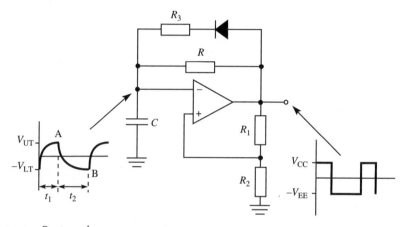

FIGURE 15.3   Rectangular wave generator.

The addition of the diode permits two different time constants. When the output voltage of the operational amplifier is positive the diode is forward biased and R and R_3 are in parallel. When the output of the operational amplifier is negative the diode is reverse biased and the capacitor charges through R alone. Thus the two time constants are different and the output is a rectangular wave.

▷ **EXAMPLE 15.1** ─────────────────────────────────

Determine the output waveform in Figure 15.3 if $R = 10$ kΩ, $C = 1$ μF, $R_3 = 1$ kΩ and $R_1 = R_2$. Use PSpice to verify the waveform based on the UA741 operational amplifier from the **eval.lib**.

▷ *Solution*

It is first necessary to determine the two time constants for the circuit. For the positive-going voltage across the capacitor:

$$\tau_1 = (R \| R_3)C = 0.91 \text{ ms}$$

and for the negative-going voltage:

$$\tau_2 = RC = 10 \text{ ms}$$

From equation 15.1:

$$v_{C1} = V_{sat} + (-V_{LT} - V_{sat})\exp\left(\frac{-t_1}{0.91 \text{ ms}}\right)$$

If $R_1 = R_2$ then the output of the Schmitt trigger changes when this voltage reaches $V_{sat}/2$. Thus:

$$\frac{V_{sat}}{2} = V_{sat} + \left(-\frac{V_{sat}}{2} - V_{sat}\right)\exp\left(\frac{-t_1}{0.91 \text{ ms}}\right)$$

and

$$\exp\left(\frac{t_1}{0.91 \text{ ms}}\right) = 3$$

and $t_1 = 0.1$ ms.

For the negative-going voltage across C:

$$v_{C2} = -V_{sat} + (V_{UT} + V_{sat})\exp\left(\frac{-t}{10 \text{ ms}}\right)$$

The transition occurs when $v_{C2} = V_{UT} = V_{sat}/2$ and:

$$\exp\left(\frac{t_2}{10 \text{ ms}}\right) = 3$$

and $t_2 = 11$ ms.

The period is:

$$T = t_1 + t_2 = 12 \text{ ms}$$

The schematic from PSpice is shown in Figure 15.4.

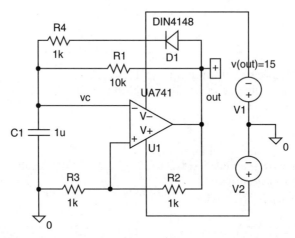

FIGURE 15.4   PSpice schematic for astable multivibrator.

Notice the use of the **initial-condition** box attached to the output. This is necessary for circuits for which there is no input since in the ideal world of simulation there is no electrical noise to initiate positive feedback which is necessary for the operation of the circuit. The initial condition chosen in the schematic is for v_{out} to be 15 V.

The output from Probe is shown in Figure 15.5. Notice that the periodic nature of the waveform is established after the first transition, which is based on the initial conditions set in the schematic.

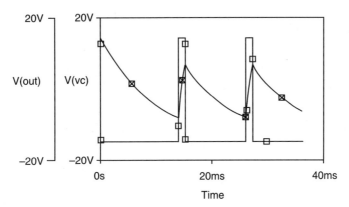

FIGURE 15.5   Voltage waveforms across the capacitor and from the output.

The 555 Timer

The astable is so widely used in electronic circuits as a waveform generator that a special purpose IC has been designed – the 555 timer. A simplified schematic of the internal design of the IC is shown in Figure 15.6.

The circuit is more complex than the simple astable shown in Figure 15.2, but it is also more versatile. It is a mix of analog and digital circuitry with two comparators which monitor the voltage across an external capacitor and an RS flip-flop which produces a unidirectional (0 V to V_{CC}) pulsed output voltage. The transistor acts as a switch to discharge the external capacitor. The potential divider establishes the voltage levels for the comparators. With three equal value resistors the supply voltage is divided into three with $V_{CC}/3$ applied to comparator C_2 and $2V_{CC}/3$ applied to comparator C_1.

An example of a complete circuit for an astable using the 555 timer is shown in Figure 15.7.

Notice that the threshold and trigger terminals are connected together, that is the non-inverting terminal of C_1 and the inverting terminal of C_2 are both connected to the external capacitor and are, therefore, monitoring the capacitor voltage. When power is initially applied the capacitor is discharged which causes the output of C_1 to be low and the output of C_2 to be high, forcing the output of the flip-flop and the base of the transistor low, thus turning the transistor OFF. The capacitor now charges through R_1 and R_2 towards V_{CC}. When the voltage across C reaches $V_{CC}/3$ the output of the

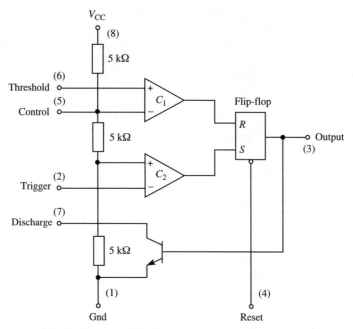

FIGURE 15.6   Simplified schematic of the 555 timer (the numbers in parentheses are the IC pin numbers).

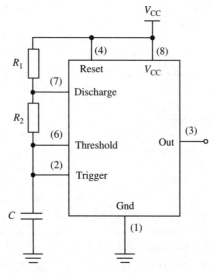

FIGURE 15.7   The 555 timer connected as an astable multivibrator.

comparator C_2 switches to a low state, and when the voltage across C reaches $2V_{CC}/3$ the output of comparator C_1 switches to a high state. This resets the flip-flop and causes the output to go high, which turns the transistor ON. The transistor now provides a discharge path for the capacitor through R_2. As the voltage across C falls to a value of $V_{CC}/3$ the comparator C_2 switches high causing the output of the flip-flop to go low, which turns the transistor OFF, and the cycle repeats. The output is a rectangular waveform with a duty cycle which depends on the values of R_1 and R_2.

After the initial cycle the voltage across the capacitor is varying between $V_{CC}/3$ and $2V_{CC}/3$. During the charging cycle the time constant is:

$$\tau_c = (R_1 + R_2)C$$

and during the discharge the time constant is:

$$\tau_d = R_2 C$$

During the charging cycle the capacitor charges from $V_{CC}/3$ towards V_{CC} and the equation for the voltage across C is:

$$v_C = V_{CC} + \left(\frac{V_{CC}}{3} - V_{CC}\right)\exp\left(-\frac{t}{\tau_c}\right)$$

and switching occurs when v_C is $2V_{CC}/3$. Substituting this value in the equation gives:

$$\exp\left(\frac{t}{\tau_c}\right) = 0.5$$

or

$$t = 0.696\tau_c \tag{15.7}$$

A similar argument can be made for the discharge of C from $2V_{CC}/3$ to $V_{CC}/3$ to yield a value for τ_d.

The charging time, which corresponds to a high at the output, is given by:

$$t_1 = 0.693(R_1 + R_2)C$$

and the discharge time is given by:

$$t_2 = 0.693R_2 C$$

The total period for the output waveform is the sum of these two values:

$$T = 0.693(R_1 + 2R_2)C$$

The frequency is the reciprocal of this time:

$$f = \frac{1.44}{(R_1 + 2R_2)C} \tag{15.8}$$

The duty cycle is defined as the ratio of the time when the output is high to the period, that is:

$$\text{duty cycle} = \frac{t_1}{t_1 + t_2} \times 100\%$$

$$\text{duty cycle} = \frac{R_1 + R_2}{R_1 + 2R_2} \times 100\% \tag{15.9}$$

If $R_1 \ll R_2$ then the duty cycle approaches 50%, while if $R_1 \gg R_2$ it approaches 100%. To achieve a duty cycle of less than 50% it is a simple matter to add an inverter to the output of the 555 timer to invert the waveform, in which case the duty cycle is:

$$\text{duty cycle} = \frac{R_2}{R_1 + 2R_2} \times 100\% \tag{15.10}$$

Now if $R_1 \ll R_2$ the duty cycle approaches 50%, while if $R_1 \gg R_2$ it approaches 0%.

▶ **EXAMPLE 15.2** ──────────────────────────────

Determine the frequency and duty cycle of the 555 waveform generator shown in Figure 15.8.

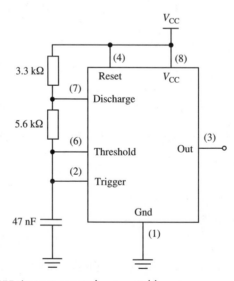

FIGURE 15.8   The 555 timer connected as an astable.

▶ *Solution*

From equation 15.8:

$$f = \frac{1.44}{(3.3\ \text{k}\Omega + 11.2\ \text{k}\Omega)47\ \text{nF}} = 2.1\ \text{kHz}$$

From equation 15.9 the duty cycle is:

$$\text{duty cycle} = \frac{3.3 \text{ k}\Omega + 5.6 \text{ k}\Omega}{3.3 \text{ k}\Omega + 11.2 \text{ k}\Omega} \times 100\% \approx 61\%$$

▷ **EXAMPLE 15.3** _____

Determine the duty cycle and sketch the waveform if an inverter is placed at the output of the 555 timer.

Voltage-Controlled Oscillator

Pin 5 on the 555 timer provides access to the point on the potential divider where the voltage is normally $2V_{CC}/3$. By connecting an external voltage (V_{CONT}) to this node it is possible to override the internal voltage. This is shown in Figure 15.9. The inverting terminal of comparator C_1 is now held at V_{CONT}, while the potential divider, comprising the two 5 kΩ resistors, holds the non-inverting terminal of comparator C_2 at $V_{CONT}/2$. The external capacitor now charges and discharges between V_{CONT} and $V_{CONT}/2$. By varying V_{CONT} it is possible to vary the charge and discharge times and hence the frequency of the output waveform.

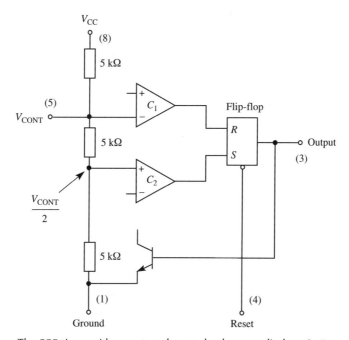

FIGURE 15.9   The 555 timer with an external control voltage applied to pin 5.

An example of a voltage-controlled oscillator is shown in Figure 15.10.

The diagram shows a potentiometer being used to supply the control voltage but it could also be obtained from another part of an electronic circuit, in which case the frequency may be controlled by some property of the circuit, an external sensor or an external signal so that the output frequency would be a function of the sensor output.

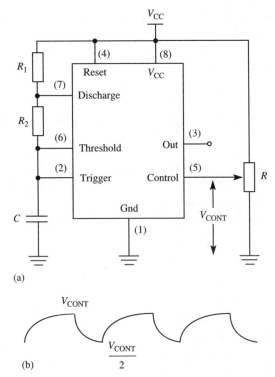

(a)

(b)

FIGURE 15.10   (a) Voltage-controlled oscillator using a 555 timer and (b) the waveform across the capacitor.

Triangular Wave Generator

A triangular wave is useful because it can be used to create other waveforms, for example a sine wave, by using shaping networks. In the simple square wave generator shown in Figure 15.2, the voltage waveform across the capacitor is approximately triangular. A true triangular wave could be generated if the capacitor were charged and discharged through a constant current source. This is precisely what happens in the integrator (see Section 13.5). One approach to producing a triangular wave is to use the circuit shown in Figure 15.2 to generate a square wave, and to apply the square wave to an integrator. A more elegant approach is to combine the integrator with a Schmitt trigger as shown in Figure 15.11.

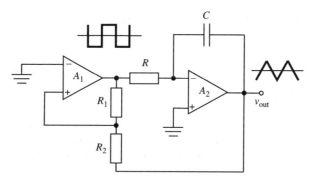

FIGURE 15.11   Feedback circuit with Schmitt trigger and integrator to produce triangular waves.

The amplifier A_1 is connected as a Schmitt trigger with the potential divider R_1 and R_2 providing the positive feedback. The amplifier A_2 acts as an integrator with the output connected back to the base of the potential divider. (Section 13.4 describes a Schmitt trigger with a reference voltage applied to the potential divider.) In this circuit the reference voltage is the output from the integrator, while the input to the inverting terminal of the Schmitt trigger is fixed at 0 V. The waveform at the output and input of the Schmitt trigger are shown in Figure 15.12.

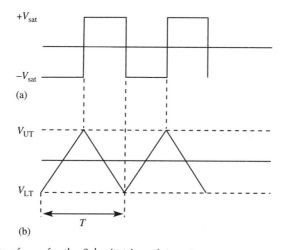

FIGURE 15.12   Waveforms for the Schmitt trigger integrator.

The output from the Schmitt trigger is either $+V_{sat}$ or $-V_{sat}$. If it is initially negative as shown in Figure 15.12(a) then the integrator produces a positive-going ramp, as shown in 15.12(b). This voltage is applied to the non-inverting terminal of the Schmitt trigger. An expression was obtained for the voltage on the non-inverting terminal of the

Schmitt trigger in Section 13.4 and is reproduced here for convenience:

$$v^+ = \frac{R_1}{R_1 + R_2} v_{\text{out}} + \frac{R_2}{R_1 + R_2} (-V_{\text{sat}}) \tag{15.11}$$

where v_{out} is the voltage at the output of the integrator (and represents the reference voltage in equation 13.32). This equation assumes that the output of the Schmitt is at $-V_{\text{sat}}$; thus the output of the integrator is ramping up. With the inverting terminal of A_1 at ground the comparator will trigger, and the output change state, when v^+ reaches 0 V. Substituting 0 V for v^+ in equation 15.11 gives the upper threshold value for v_{out} as:

$$V_{\text{UT}} = \frac{R_2}{R_1} V_{\text{sat}} \tag{15.12}$$

Similarly the lower threshold may be obtained as:

$$V_{\text{LT}} = -\frac{R_2}{R_1} V_{\text{sat}} \tag{15.13}$$

The period may be determined by considering the time taken for the positive- (or negative-) going ramps, and noting that for a symmetrical waveform this time is $T/2$. The charging current in the capacitor is:

$$i = -C \frac{dv_c}{dt} \tag{15.14}$$

where v_c is the capacitor voltage. For the positive-going ramp and an ideal integrator for which the current is constant, the current is:

$$i = I = \frac{-V_{\text{sat}}}{R} \tag{15.15}$$

The rate of change of voltage is:

$$\frac{dv_c}{dt} = \frac{V_{\text{UT}} - V_{\text{LT}}}{T/2} = \frac{4V_{\text{sat}}R_2}{TR_1} \tag{15.16}$$

Substituting equations 15.15 and 15.16 into equation 15.14 gives:

$$T = \frac{4RR_2C}{R_1} \tag{15.17}$$

The frequency is the reciprocal of the period:

$$f = \frac{R_1}{4R_2RC} \tag{15.18}$$

Note that the frequency is independent of the saturation voltage (V_{sat}) of the comparator and is indirectly proportional to R so that the frequency may be controlled by varying R. The maximum frequency will be limited by the slew rate of the integrator.

▷ **EXAMPLE 15.4** ⎯⎯⎯⎯⎯⎯⎯⎯⎯⎯⎯⎯⎯⎯⎯⎯⎯⎯⎯⎯⎯⎯⎯⎯⎯⎯⎯

Determine the frequency of the triangular wave generator shown in Figure 15.13. What must the value of R be to obtain a frequency of 3 kHz? Verify using PSpice.

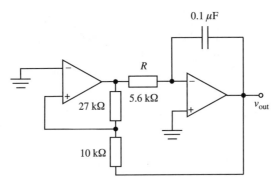

FIGURE 15.13 Triangular wave generator.

▷ *Solution*

From equation 15.18 the frequency is:

$$f = \frac{27 \text{ k}\Omega}{4(5.6 \text{ k}\Omega)(10 \text{ k}\Omega)(0.1 \text{ μF})} = 1.2 \text{ kHz}$$

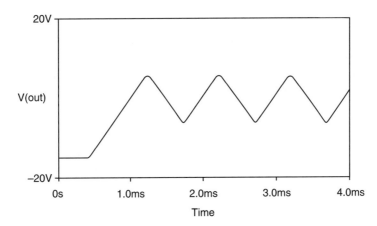

FIGURE 15.14 Triangular wave generator output.

In order to increase the frequency to 3 kHz the value of R must be:

$$R = \frac{27 \text{ k}\Omega}{4(3 \text{ kHz})(10 \text{ k}\Omega)(0.1 \text{ } \mu\text{F})} = 2.25 \text{ k}\Omega$$

The output from Probe for the component values shown in Figure 15.13 is shown in Figure 15.14.

As for other oscillator circuits it is necessary to include an initial condition in the PSpice simulation in order to start the sequence and this influences the initial part of the waveform, as shown by the initial, large positive-going ramp in Figure 15.14. The measured frequency is 1 kHz, which is not as close to the theoretical value as one would expect.

Wave-Shaping Network

The circuit shown in Figure 15.11 generates a square wave and a triangular wave. With a wave-shaping circuit it is possible also to generate a sine wave. This type of circuit forms the basis for many low-cost function generators. It is also a convenient way of generating sine waves at very low frequencies.

The sine wave is produced by applying the triangular wave to a diode–resistor-shaping network, as shown in Figure 15.15.

The network is symmetrical with diodes D_1 and D_2 responding to positive-going input voltages and diodes D_3 and D_4 responding to negative-going voltages. Resistors R_3, R_4 and R_5 act as a potential divider for the positive supply $+V$ and a similar potential divider acts for the symmetrical negative supply $-V$. The potential dividers establish reference voltages $+V_1$, $+V_2$, $-V_1$ and $-V_2$. With a triangular wave input the output consists of a waveform composed of three straight line segments for each quarter cycle.

Consider v_{in} at zero and increasing in a positive direction. Initially the two diodes D_1 and D_2 are reverse biased as a result of the voltages $+V_1$ and $+V_2$ and no current flows through R_1, and the output voltage equals the input voltage. When the input reaches the first threshold $+V_1$ the diode D_2 becomes forward biased and begins to conduct. Assuming an ideal diode with no forward voltage drop or series resistance then the output voltage is:

$$v_{out} = IR_2 + V_1$$

where the current is:

$$I = \frac{v_{in} - V_1}{(R_1 + R_2)}$$

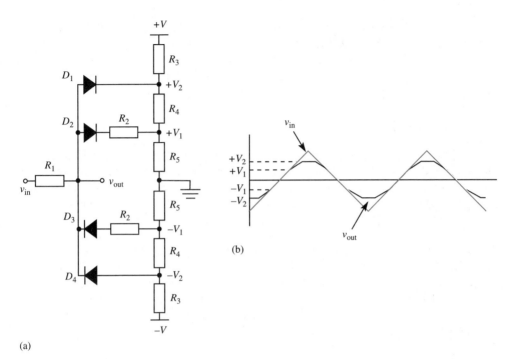

(a)

FIGURE 15.15   (a) Diode shaping network and (b) the triangular input and the piece-wise approximation of a sine wave output.

which gives the output as:

$$v_{out} = V_1 + (v_{in} - V_1) \frac{R_2}{(R_1 + R_2)}$$

This shows that the output continues to rise above V_1, as v_{in} continues to rise, but at a reduced rate determined by the resistor ratio $R_2/(R_1 + R_2)$. This straight line segment continues until the second threshold is reached.

When the input reaches V_2 diode D_1 conducts and clamps v_{out} at V_2 (less the forward voltage drop across the diode in the case of a real diode). This results in a straight line segment at a constant voltage of V_2. Once the input reaches its peak value the process reverses when segments with opposite slopes are produced until the input voltage reaches 0 V. During the negative-going cycle the diodes D_3 and D_4 operate.

By increasing the number of diodes and resistors more straight line segments can be included in the output waveform and a closer approximation made to a sine wave. The transition between the straight line segments is smoothed by the non-ideal $I–V$ characteristics of practical diodes, with a gradual transition occurring as each diode begins to conduct. A measure of the quality of the sine wave is obtained by measuring the harmonic distortion of the sine wave. This is the ratio of the rms voltage of all the harmonics above the fundamental frequency to the rms voltage of the fundamental. An example of the above wave-shaping network is presented as a problem at the end of the chapter (Problem 15.8). The applied voltage needs to be several volts, say ±10 V, in order to swamp the effect of the diode turn-on voltages. Function generator IC chips are available which contain operational amplifiers to generate the triangular and square waves, plus the diodes and resistors for the diode-shaping network.

15.2   Sine Wave Oscillators

Many of the general purpose function generators which produce square, triangular and sine waves use wave shaping to generate the sine wave as described above. A single IC containing the majority of the circuit elements may be purchased to build such generators, for example the ICL8038. However, for many applications in telecommunication systems, the frequency range of the function generator is inadequate, and the sine waves are not sufficiently free from distortion and harmonics. For these applications it is necessary to generate the waves directly.

For the audio frequency range sine waves can be generated with *RC* selective networks, such as a Wien bridge circuit, or a phase-shift oscillator, but for frequencies beyond 1 MHz it is usual to use *LC* circuits, such as the Colpitts, Hartley and Clapp. For very accurate and stable frequency generation a piezoelectric crystal may be used in the feedback loop of a suitable amplifier.

Positive Feedback

To generate sine waves directly rather than by using wave shaping, it is necessary to use an amplifier, a frequency-selective network and positive feedback. However, unlike the relaxation oscillator, the amplifier must remain in its linear range so that the output wave is not distorted. This is difficult to achieve and a figure of merit for a sine wave oscillator is the amount of distortion which is present in the output sine wave.

The basic idea of positive feedback is illustrated in Figure 15.16.

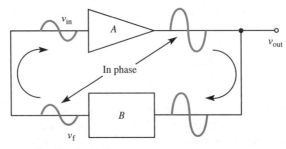

FIGURE 15.16   Positive feedback circuit.

The output from the amplifier is fed back to the input through a feedback network, and unlike negative feedback, the phase of the returned signal is the same as the original signal at the input. Thus the signal is reinforced by the feedback and it increases. If the feedback network is frequency selective then this phase condition only exists for one frequency and a self-sustained generation of this frequency takes place and the circuit oscillates.

If the gain of the amplifier is A then the output is Av_{in} where v_{in} is the input voltage. The amplifier output is applied to the input of the feedback network which has a 'gain' B so that the output v_f from the feedback network is Bv_{out}. Since v_{out} is equal to Av_{in} then the voltage being returned to the input of the amplifier is ABv_{in}. If this voltage is less than v_{in} the signal is attenuated and no oscillation takes place. If this signal is greater than v_{in} then the input is reinforced and the voltage around the loop increases. If AB is unity then the voltage fed back is equal to v_{in} and a steady-state condition is reached where the oscillation is sustained.

Thus the conditions for oscillation are:

1   The phase shift around the feedback loop must be zero, that is the phase of the signal at the input of the amplifier must be the same as that of the returned signal from the feedback network.
2   The gain around the loop must be unity.

In terms of the basic feedback relationship given in Chapter 12 and represented by equation 12.1:

$$A_f = \frac{A_o}{1 + BA_o}$$

positive feedback occurs when BA_o approaches -1, which results in the closed-loop gain A_f becoming infinite. Under these conditions an infinitesimal signal anywhere within the feedback loop can produce a significant output from the circuit without an external input signal, but such conditions imply that the signal goes on increasing. There must be some way to limit the signal so that the output remains within the linear operating range of the active device, and there must also be some means of starting the process.

Starting Conditions

How do the oscillations start? When power is applied to the circuit what is it about the circuit or components which causes a signal to be generated when there is no external input, apart from dc power? At room temperature resistors generate electrical 'noise' as a result of the random movement of the electrons which are present in all conductors. The resistors act like voltage sources each generating a noise voltage of a few microvolts. Because the motion of the electrons is random the fluctuations are also random and as a result there are voltage variations within this electrical noise which extend from dc to very high frequencies.

When power is applied the noise from those resistors at the input of the amplifier is amplified and appears at the output. This amplified signal is applied to the frequency-selective feedback network and returned to the input. If there is no phase shift then this signal arrives back at the input to reinforce the input signal and the signal increases. Because the feedback network is frequency selective only those components of the original noise signal which match the frequency of the feedback circuit are amplified. For signal components above and below this frequency the phase shift is not zero and the condition for oscillation does not apply and they are attenuated.

Sustaining Conditions

Initially the loop gain (AB) must be greater than unity so that the signal increases, but this process cannot be allowed to continue indefinitely because the signal will become too large for the active components in the amplifier to handle. There must be some means of reducing the loop gain to unity in order to sustain the signal at a level appropriate for the dc voltage levels in the circuit, so that there is no distortion. This is illustrated in Figure 15.17.

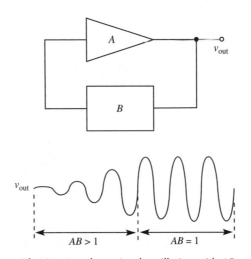

FIGURE 15.17   Start-up with $AB > 1$ and sustained oscillation with $AB = 1$.

The process by which AB is reduced to unity usually involves some non-linear action either in the amplifier or in the feedback loop. In the amplifier as the signal increases the active devices eventually reach saturation or cut-off, at which point the gain is effectively reduced to zero and the oscillation starts to decay. This would tend to produce a clipped waveform where the positive and negative peaks are clipped, and therefore distorted. For the complete cycle the average value of AB is unity. Alternatively a non-linear device may be included in the feedback loop to limit the waveform at some predetermined level, which restricts the gain of the feedback circuit to a level at which $AB = 1$.

15.3   Wien Bridge Oscillator

The Wien bridge oscillator is generally used for frequencies of less than 1 MHz. The frequency-selective network is formed from resistor–capacitor networks based on the low-pass and high-pass filters described in Chapter 9. The two networks together form a simple band-pass filter, as shown in Figure 15.18.

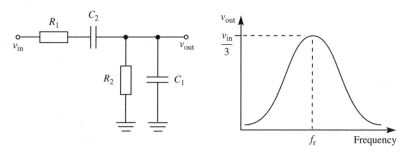

FIGURE 15.18   Frequency-selective network.

The low-pass network consists of R_1 and C_1 while the high-pass section consists of R_2 and C_2. At low frequencies C_2 of the high-pass section presents a very high impedance to the signal and v_{out} is small, while at high frequencies C_1 of the low-pass section presents a short circuit to the signal and again v_{out} is small. In between the output reaches a maximum at f_r. Assuming that the associated amplifier does not introduce any phase shifts then the operation of an oscillator based on this network is determined by the transfer characteristics of this frequency-selective network.

The output of the network in Figure 15.18 is:

$$v_{out} = \frac{R_2 \| 1/j\omega C_1}{R_1 + (1/j\omega C_2) + R_2 \| 1/j\omega C_1} \, v_{in}$$

With some rearrangement this can be written as:

$$\frac{v_{out}}{v_{in}} = \frac{\omega C_2 R_2}{\omega(C_2 R_1 + C_1 R_2 + C_2 R_2) + j(\omega^2 C_1 C_2 R_1 R_2 - 1)} \tag{15.19}$$

The phase angle of this expression is $\theta = \tan^{-1}(\mathscr{IP}/\mathscr{RP})$ and for oscillation to occur this must be zero. This can be achieved by ensuring that the imaginary part $(\mathscr{IP})$ is zero, that is:

$$\omega^2 C_1 C_2 R_1 R_2 = 1$$

This is simplified if $R_1 = R_2 = R$ and $C_1 = C_2 = C$. Then:

$$f_r = \frac{1}{2\pi RC} \tag{15.20}$$

where f_r is the resonant frequency of the network corresponding to the peak in the gain versus frequency graph in Figure 15.18. Applying the same conditions to equation 15.19 and setting the imaginary part to zero to gives:

$$\frac{v_{out}}{v_{in}} = \frac{1}{3} \tag{15.21}$$

Thus at the resonant frequency the frequency-selective network attenuates the signal by a factor of 3. For the loop gain to be unity it is necessary for the amplifier in the circuit to have a gain of 3 to compensate for the loss. The basic circuit for the Wien bridge oscillator is shown in Figure 15.19.

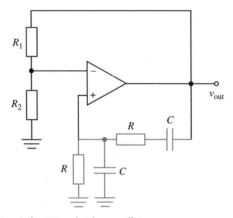

FIGURE 15.19   Basic circuit for Wien bridge oscillator.

The amplifier is connected as a non-inverting amplifier so that the gain is determined by the resistors R_1 and R_2 as:

$$\text{gain} = 1 + \frac{R_1}{R_2}$$

Connecting the feedback network between the output and the non-inverting terminal ensures that no phase change is introduced by the amplifier and because the phase angle for the selective network is zero, then the condition for oscillation is established. For oscillation to be sustained the gain of the amplifier must compensate for the attenuation of $1/3$ in the feedback circuit. Thus the gain of the amplifier must be 3 and:

$$R_1 = 2R_2 \tag{15.22}$$

This condition for the gain will sustain the oscillation, but is not sufficient to start the process. There are a number of different approaches to ensuring that the gain is greater than 3 initially and then reduces to 3 when the oscillation has started, and one simple approach involves a tungsten lamp used as a variable resistance, as shown in Figure 15.20.

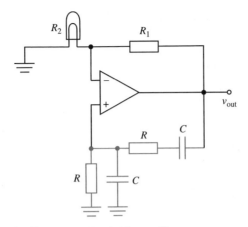

FIGURE 15.20 Practical self-starting Wien bridge oscillator.

The value of R_1 is chosen so that when the lamp is cold the gain is greater than 3. As the oscillation starts the output level rises and an alternating current, at the oscillation frequency, flows in the lamp. As current flows its temperature rises, which causes its resistance to rise and to reduce the gain. When the gain is less than 3 the amplitude of the oscillation begins to fall and the ac current through the lamp falls. This causes the resistance to fall and increases the gain. A balance is reached for which $AB = 1$.

In a second circuit an FET is used as a variable resistance. The circuit is shown in Figure 15.21. Provided that the source to drain (V_{ds}) of the FET is less than the pinch-off voltage then the FET acts like a variable resistor (R_{DS}) where the magnitude of the resistance is determined by the gate voltage. In the circuit the output voltage is rectified

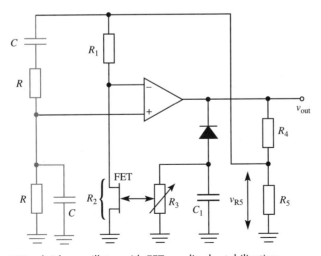

FIGURE 15.21 Wien bridge oscillator with FET amplitude stabilization.

by the diode to produce a dc voltage across the parallel combination of R_3 and C_1 which act as a low-pass filter. A proportion of this voltage necessary to maintain the oscillation is applied by means of a potentiometer R_3 to the gate of the FET. The amplitude of the voltage applied to the FET must be small enough to ensure that it is working in its linear region. This is achieved with the potential divider R_4 and R_5 so that the peak value of V_{R5} is less than the pinch-off voltage of the FET.

When the power is first applied to the circuit the output voltage is zero and $V_{GS} = 0$ V, R_{DS} is low and the gain is >3. As the amplitude increases, a dc voltage is produced and applied to the gate of the FET, R_{DS} increases and the gain decreases until a stable situation is reached.

An important feature of both circuits is that the time constant of the stabilizing circuit is greater than the period of oscillation to avoid distortion. For the tungsten lamp it is a thermal time constant and for the FET circuit it is the time constant of the low-pass filter $R_3 C_1$.

▶ EXAMPLE 15.5

Determine the frequency of the output of the Wien bridge oscillator shown in Figure 15.22. Determine the minimum value of R_{DS} for the FET at $V_{GS} = 0$ V required to ensure start-up.

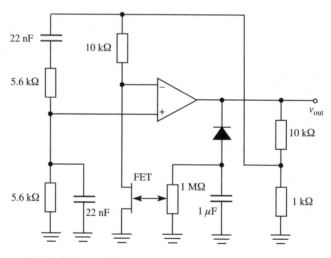

FIGURE 15.22   Wien bridge oscillator.

▶ Solution

The frequency is:

$$f_r = \frac{1}{2\pi RC} = \frac{1}{2\pi (5.6 \text{ k}\Omega)(22 \text{ nF})} \approx 1.3 \text{ kHz}$$

Initially the amplifier gain must be greater than 3:

$$\text{gain} = 1 + \frac{10\ \text{k}\Omega}{R_{\text{DS}}} > 3$$

Therefore $R_{\text{DS}} < 5\ \text{k}\Omega$.

After start-up the output voltage increases, a dc bias is applied to the gate of the FET and R_{DS} increases to a value of 5 kΩ at which value the oscillation is sustainable.

A PSpice schematic of the above circuit is shown in Figure 15.23.

Some trial and error was necessary to obtain the correct value for the gain resistor R_1 but the value of 2 kΩ was found to produce a sine wave output, as shown in the Probe output in Figure 15.24.

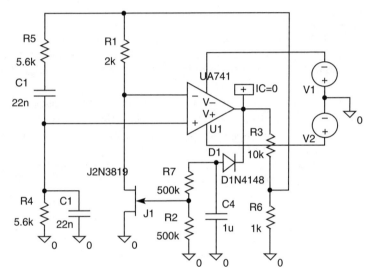

FIGURE 15.23   PSpice schematic of a Wien bridge oscillator.

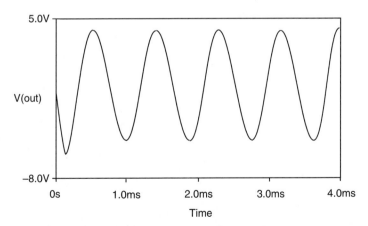

FIGURE 15.24   Probe output for Wien bridge oscillator.

Practical Considerations

In a practical Wien bridge oscillator, used as a signal generator, the frequency-selective capacitors would be attached to a switch to provide different frequency ranges from about 10 Hz to 1 MHz, say in decade steps – 10 Hz, 100 Hz, 1 kHz, 10 kHz, 100 kHz and 1 MHz. The two frequency-selective resistors would be coupled potentiometers to provide variation within each frequency range. The output from the oscillator would be taken to a buffer amplifier and attenuator to provide adequate and variable power output.

15.4 LC Oscillators

The *RC*-based Wien bridge oscillator is not suited to frequencies above 1 MHz because of the phase shift through the amplifier. This phase shift adds to the phase shift through the frequency-selective *RC* network and causes the resonant frequency to depart from the value determined by the resonant circuit. A more suitable frequency-selective circuit is one based on inductance and capacitance. *LC* oscillators are suitable for frequencies from 1 MHz to 500 MHz. Because these frequencies are far in excess of the operating frequency of operational amplifiers, *LC* oscillators use discrete bipolar transistors and FETs.

The design of *LC* oscillators is difficult because of the presence of stray capacitance associated with the components and the electrical wiring. It is usual to use approxima-

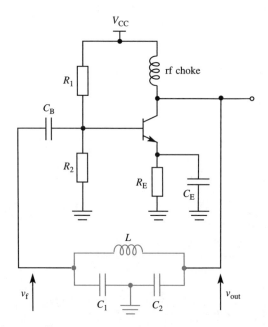

FIGURE 15.25   Colpitts oscillator.

tions for the initial design and to provide sufficient adjustment within the tuned circuit to optimize the performance after it has been constructed.

Colpitts Oscillator

A Colpitts oscillator which uses a bipolar transistor is shown in Figure 15.25.

The bipolar transistor is connected as a common-emitter amplifier with the base potential divider and the emitter resistor providing the dc bias. The emitter resistor is decoupled to provide maximum gain. The collector load is a radio frequency choke. This is simply an inductor which has a large inductive impedance over the frequency range of interest.

Positive feedback is provided by the tuned circuit comprising L, C_1 and C_2. Notice that the output voltage appears across C_2 and that the voltage fed back to the base appears across C_1. There are many variations of the Colpitts oscillator but this arrangement of the two capacitors which act as a potential divider is a common feature. The equivalent circuit is shown in Figure 15.26.

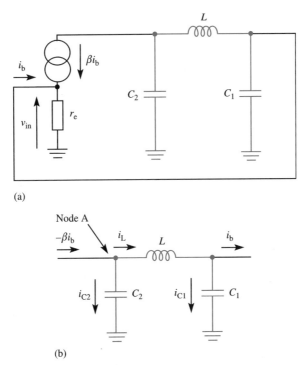

FIGURE 15.26 (a) Equivalent circuit for Colpitts oscillator and (b) simplified network for the analysis.

The transistor is represented by its low-frequency equivalent circuit, and this is adequate to obtain a first-order relationship for the frequency of oscillation and gain. The circuit is analyzed by equating the currents at node A in the simplified circuit in

Figure 15.26(b). The current at this node may be determined as follows:

$$i_b = \frac{v_{in}}{(1 + \beta)r_e} \tag{15.23}$$

$$i_L = i_b + i_{C1}$$

$$i_L = \frac{v_{in}}{(1 + \beta)r_e} + \frac{v_{in}}{X_{C1}} \tag{15.24}$$

$$V_A = v_{in} + i_L X_L$$

$$i_{C2} = \frac{V_A}{X_{C2}} = \frac{v_{in}}{X_{C2}} \left[1 + X_L \left(\frac{v_{in}}{(1 + \beta)r_e} + \frac{v_{in}}{X_{C1}} \right) \right] \tag{15.25}$$

Equating the currents gives:

$$\beta i_b + i_L + i_{C2} = 0$$

Substituting from equations 15.23, 15.24 and 15.25 and using the relationship $\beta \gg 1$ gives

$$\frac{1}{r_e} + \frac{1}{\beta r_e} + \frac{1}{X_{C1}} + \frac{1}{X_{C2}} + \frac{X_L}{\beta r_e X_{C2}} + \frac{X_L}{X_{C1} X_{C2}} = 0$$

Replacing X_{C1}, X_{C2} and X_L and rearranging the terms into real and imaginary parts gives:

$$\left(\frac{1}{r_e} + \frac{1}{\beta r_e} - \frac{\omega^2 L C_2}{\beta r_e} \right) + j\omega(C_1 + C_2 - \omega^2 L C_1 C_2) = 0$$

This is a general equation for the network and the condition for oscillation is that the imaginary part is zero (corresponding to zero phase shift). Equating the imaginary part to zero gives the frequency of oscillation as:

$$f_o = \frac{1}{2\pi \sqrt{L[C_1 C_2 / (C_1 + C_2)]}} \tag{15.26}$$

Equating the real part to zero together with the condition in equation 15.26 gives:

$$\beta = \frac{C_2}{C_1} \tag{15.27}$$

This describes the value of current gain required for oscillation to be sustained. In practice the capacitor ratio should be less than the small-signal value of current gain to ensure that the oscillator is self-starting. As the oscillations build up the non-linearity of the transistor causes the current gain to decrease and a balance is obtained where the loop gain is unity, thus sustaining the oscillations.

Loading Effects

The resonant frequency given by equation 15.26 assumes a high-Q circuit with no external loading. The resonant frequency of a parallel tuned circuit is given by

$$f_o = \frac{1}{2\pi\sqrt{LC_{eq}}}\left(\frac{Q^2}{Q^2+1}\right) \qquad (15.28)$$

where C_{eq} is the parallel combination of C_1 and C_2. For values of Q greater than 10 the equation reduces to equation 15.26.

The input resistance of the transistor appears across the tuned circuit, as shown in Figure 15.27(b), and lowers the Q. From Figure 15.26(a) the input resistance at the base is βr_e, and since r_e is inversely proportional to the emitter current it is preferable to operate the transistor at low rather than high bias currents. An FET may be used in place of a bipolar transistor as shown in Figure 15.27(b). The very high input resistance of the FET reduces the loading effect on the tuned circuit. The expression for the resonant frequency is the same as for the bipolar transistor, and it can be shown (see Problem 15.22) that the condition for sustaining the oscillation is given by:

$$g_m = \frac{1}{r_{ds}}\frac{C_1}{C_2} \qquad (15.29)$$

where r_{ds} is the drain to source resistance of the FET.

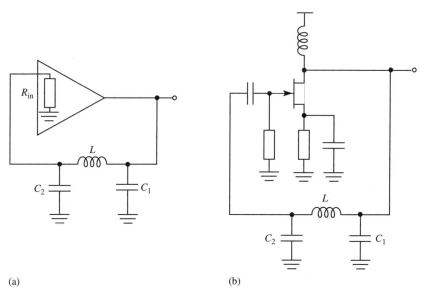

(a)   (b)

FIGURE 15.27   (a) Input resistance loading reduces the resonant frequency and can be minimized by (b) using an FET.

▷ **EXAMPLE 15.6** ───────────────────────────────

An FET has a minimum g_m of 3000 μS and r_{ds} of 50 kΩ. Design a Colpitts oscillator to operate at 1 MHz with an inductor of 5 μH. Determine the value of C_1 and C_2 so that the oscillations are sustainable.

▷ *Solution*

From equation 15.26:

$$f_o = 1\text{ MHz} = \frac{1}{2\pi\sqrt{(5\text{ μH})C_{eq}}}$$

$$C_{eq} = \frac{C_1C_2}{C_1 + C_2} = 5.07\text{ nF}$$

From equation 15.29:

$$g_m = 3000\text{ μS} = \frac{1}{50\text{ kΩ}}\frac{C_1}{C_2}$$

and

$$\frac{C_1}{C_2} = 150$$

Therefore $C_2 = 33.6$ pF and $C_1 = 5$ nF.

In a practical circuit there must be an external load in the form of the circuit which is driven by the oscillator. If the input impedance to this external circuitry is low then the resonant frequency will be affected (see equation 15.28). Two possible solutions for attaching an external load are shown in Figure 15.28, with capacitive coupling shown in 15.28(a) and transformer coupling in 15.28(b).

For the capacitive coupling the reactance of the coupling capacitor is large compared with the load resistor and prevents excessive loading, but there is a loss of signal across the capacitor. In the transformer coupling a few turns of wire form a loosely coupled secondary to couple the low-impedance load to the high-impedance tuned circuit. The selection of the capacitor or the number of turns is often based on trial and error, particularly at higher frequencies.

At high frequencies the transistor interelectrode capacitors must be included in the design. For the base circuit C_{be} (C_{gs} for the FET) must be added to C_2 and for the collector circuit C_{bc} (C_{ds} for the FET) must be added to C_1. Since both of these capacitors are voltage dependent the design becomes very much more complicated. Computer simulation can be used provided an accurate model is available for the transistor.

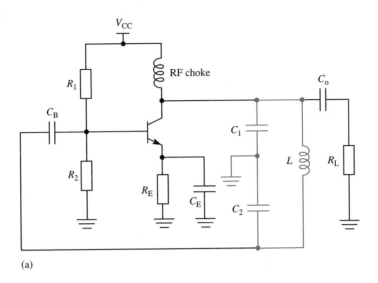

(a)

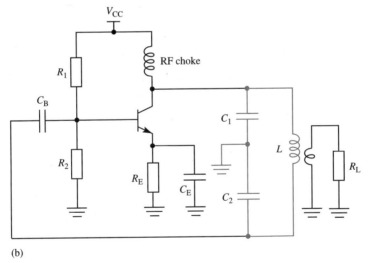

(b)

FIGURE 15.28   Output coupling with (a) capacitive coupling and (b) transformer coupling.

▶ **EXAMPLE 15.7** ─────────────────────────────────

Determine the frequency of oscillation of the circuit shown in Figure 15.29 and use PSpice to verify the design calculation.

▶ *Solution*

The frequency of oscillation from equation 15.26 is:

$$f_o = \frac{1}{2\pi\sqrt{(5\ \mu H)(1\ nF \| 10\ nF)}} = 2.36\ \text{MHz}$$

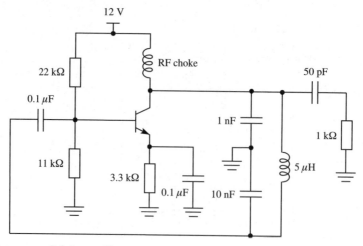

FIGURE 15.29   Colpitts oscillator.

The output from Probe is shown in Figure 15.30.

After some trial and error an **initial-condition** of 3.8 V for the base voltage was used to obtain the waveform shown. The build-up of the oscillation can be seen, and the limit is reached when the transistor is being driven into saturation. The measured frequency is 2.11 MHz. From equation 15.28 the Q may be calculated as:

$$Q = \sqrt{\frac{f_m}{f_o - f_m}} = \sqrt{\frac{2.11}{2.36 - 2.11}} \approx 3$$

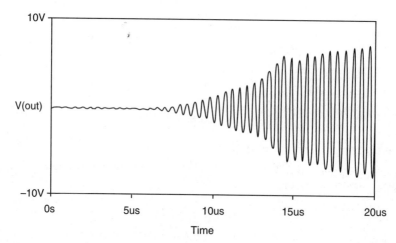

FIGURE 15.30   Probe output from the 1 kΩ load resistor.

Clapp Oscillator

The addition of an extra capacitor in series with the inductor, as shown in Figure 15.31, results in the Clapp oscillator.

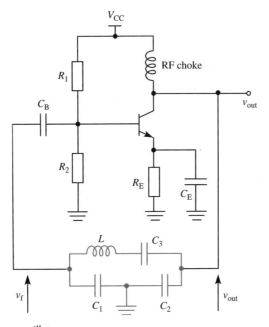

FIGURE 15.31   Clapp oscillator.

The equivalent capacitance is now given by:

$$C_{eq} = \frac{1}{1/C_1 + 1/C_2 + 1/C_3} \tag{15.30}$$

If C_3 is much smaller than C_1 and C_2 then $C_{eq} \approx C_3$ and the resonant frequency is not dependent on C_1 and C_2. Since the interelectrode capacitances of the transistor contribute to C_1 and C_2 but not to C_3 the Clapp oscillator provides a more accurate and stable frequency of oscillation, which is independent of the transistor interelectrode capacitances.

Hartley Oscillator

In the Hartley oscillator the reactive components in the tuned circuit are interchanged, that is there are two inductors and one capacitor, as shown in Figure 15.32.

It can be shown (see Problem 15.20) that the resonant frequency is:

$$f_o = \frac{1}{2\pi\sqrt{(L_1 + L_2)C}} \tag{15.31}$$

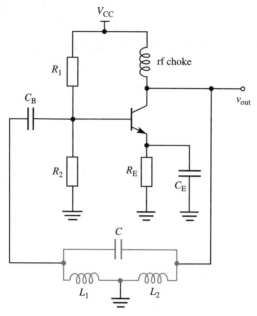

FIGURE 15.32 Hartley oscillator.

and to ensure start-up the ratio of the inductors is typically equal to or less than 10, that is:

$$\frac{L_1}{L_2} \le 10 \tag{15.32}$$

Crystal-Controlled Oscillators

The most accurate and stable oscillators use piezoelectric crystals in place of an LC circuit. The piezoelectric effect is found in a number of materials, including quartz and certain manufactured ceramic materials. Quartz is the material most commonly used for very stable oscillators.

The piezoelectric effect is an electromechanical process where the application of an alternating voltage creates a mechanical stress in the material, which causes the material to vibrate. The crystal has a natural mechanical resonant frequency and the vibrations are greatest when the applied electrical signal corresponds to this mechanical resonant frequency.

The frequency of vibration of the crystal is inversely proportional to its thickness, and there are mechanical limits to how thin a crystal can be cut and polished. Crystals can be manufactured to have *fundamental* frequencies from a few kHz up to about 10 MHz. Higher frequencies can be obtained by operating the crystal in *overtone* mode, which produces multiples of the fundamental. The overtones are usually odd multiples of the fundamental (3, 5, 7, ...).

Equivalent Circuit

A crystal consists of a thin plate of quartz, a few millimetres in diameter, with metal electrodes on opposite faces. Wires are attached to the metal electrodes and it is sealed into a metal case. At low frequencies the crystal is equivalent to a capacitor C_m.

When the crystal is vibrating there is a large increase in the amplitude of the electric signal when the frequency of the applied signal corresponds to the mechanical resonant frequency of the crystal. This is similar to the increase in the electric signal across a parallel *LC* circuit, and in fact the equivalent circuit for the crystal is based on a parallel–series *RLC* circuit as shown in Figure 15.33.

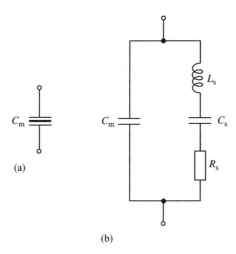

(a)

(b)

FIGURE 15.33 (a) Symbol with low-frequency capacitance C_m and (b) the ac equivalent circuit.

The crystal can operate in either series resonance when the resonant frequency is determined by L_s and C_s, or parallel resonance when the frequency is determined by L_s and C_m. The series resonant frequency is:

$$f_s = \frac{1}{2\pi\sqrt{L_s C_s}} \tag{15.33}$$

The parallel resonant frequency is slightly higher than the series resonant frequency and is given by:

$$f_s = \frac{1}{2\pi\sqrt{L_s C_{\text{effective}}}} \tag{15.34}$$

where

$$C_{\text{effective}} = \frac{C_{\text{m}}C_{\text{s}}}{C_{\text{m}} + C_{\text{s}}}$$

To use the crystal it must be connected into a circuit so that one or other mode is selected.

Series Resonance

To excite the crystal in series resonant mode the crystal can be connected in the feedback path so that at resonance the impedance is low and maximum positive feedback occurs. An example of a series resonant circuit using a bipolar transistor is shown in Figure 15.34.

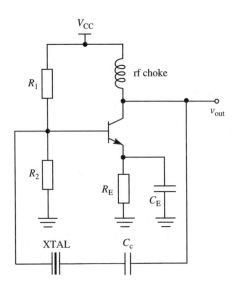

FIGURE 15.34   Crystal-controlled oscillator using series resonance.

This form is often referred to as a Pierce oscillator. The capacitor C_{c} can be used to provide fine tuning of the resonant frequency. The bipolar transistor can be replaced by an FET.

Parallel Resonance

In parallel resonance the crystal acts as a high-impedance circuit with a large inductive element. The crystal can replace the inductor in a Colpitts oscillator, as shown in Figure 15.35.

The frequency will lie somewhere between f_{s} and f_{p}.

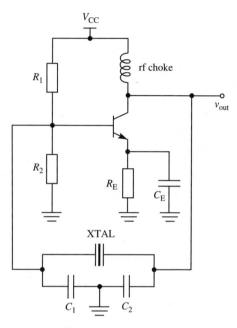

FIGURE 15.35   Colpitts crystal oscillator.

▶ **EXAMPLE 15.8** ——————————————————————

A quartz crystal has the following parameters: $L = 0.5$ H, $C_s = 0.022$ pF, $C_m = 5.5$ pF and $R_s = 150$ Ω. Determine the series and parallel resonant frequencies.

▶ *Solution*

From equation 15.32:

$$f_s = \frac{1}{2\pi\sqrt{(0.5 \text{ H})(0.022 \text{ pF})}} = 1.517 \text{ MHz}$$

From equation 15.33:

$$C_{\text{effective}} = \frac{(5.5 \text{ pF})(0.022 \text{ pF})}{(5.5 \text{ pF}) + 0.022 \text{ pF}} = 0.0219 \text{ pF}$$

and

$$f_p = \frac{1}{2\pi\sqrt{(0.5 \text{ H})(0.0219 \text{ pF})}} = 1.521 \text{ MHz}$$

Summary

Relaxation oscillators provide a simple means for generating square or rectangular waves, and in particular the 555 timer is widely used for this purpose.

Wave shaping with diode networks provides a convenient means for producing sine waves from triangular waves which are produced with a relaxation-based oscillator. This approach can be used for very low-frequency sine waves, and for multifrequency function generators. For the Wien bridge oscillator, buffer amplifiers and attenuators can be added to make a complete laboratory instrument.

The relatively low cost of crystals means that for single-frequency applications it is generally simpler to use a crystal rather than discrete components. For many digital applications involving a fixed frequency clock, it is usual to use a crystal, and if necessary to divide down from a high frequency to obtain the necessary clock frequencies. The advantages of the crystal are accuracy and stability over a wide range of temperatures, and circuit simplicity without the need for tuning components.

For variable frequency oscillators the Wien bridge can provide sine wave output from about 10 Hz up to about 1 MHz. The frequency can be varied in steps by means of ganged switched capacitors and a ganged potentiometer for continuous variation for each switch position. The output from the basic oscillator can be further amplified and a buffer amplifier provided to drive a variety of external loads. An attenuator would also be included to provide calibrated and variable output levels.

For higher frequencies, associated with telecommunication circuits, the Colpitts or Hartley oscillators can be used. For the Colpitts oscillator the frequency can be adjusted by means of a tuned inductor. Alternatively the series capacitor in the Clapp oscillator provides a means for greater variation, or the single capacitor in the Hartley can be used for adjustment. An important consideration for any oscillator designed for frequencies greater than 1 MHz is to include the effects of stray capacitance. Simulation can be used but accurate component models are required, particularly for the transistors. Stray capacitance can be modelled as lumped capacitors (1–5 pF) attached to appropriate nodes to represent the effect of wiring capacitance.

► PROBLEMS

15.1 Determine the minimum and maximum frequency of the waveform from the circuit shown for the minimum and maximum position of the potentiometer. Sketch the waveforms which appear across the capacitor and the output.

[94 Hz, 17 Hz]

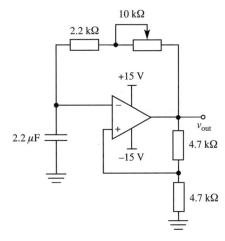

15.2 Determine the frequency of the output waveform assuming that the output saturates at ±14 V. Verify using PSpice.

[135 Hz]

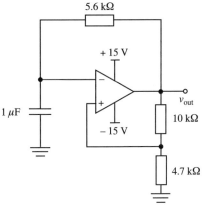

15.3 Determine the frequency and sketch the output waveform for the circuit shown. Verify using PSpice.

[381 Hz]

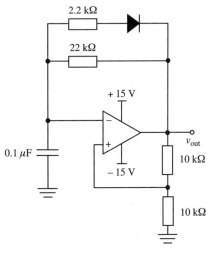

15.4 Design an astable multivibrator to generate a rectangular wave as shown. Assume that the capacitor has a value of 1 µF and that the Schmitt trigger feedback resistors are of equal value. Verify the design using PSpice.

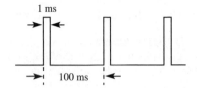

15.5 Determine the frequency and duty cycle of the output waveform.

[35 Hz, 90%]

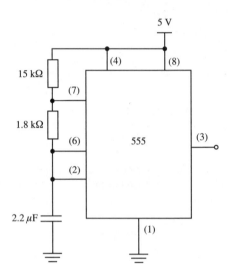

15.6 Modify the design of the circuit shown in Problem 15.5 to provide a duty cycle of 10% with the same frequency.

15.7 Sketch the waveforms at v_{out1} and v_{out2}, determine the frequency and amplitude of each waveform assuming that the output of each operational amplifier saturates at ±10 V. Verify using PSpice with a UA741 op-amp and ±11 V supplies.

[161 Hz]

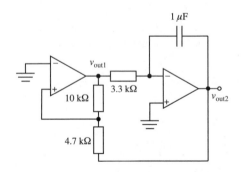

15.8 For the waveform generator in Problem 15.7 determine the effect of increasing the saturation voltage to ±14 V. Sketch the output waveforms.

15.9 Using the PSpice schematic for the circuit in Problem 15.7 add the wave shaping circuit shown and verify that the output is approximately sinusoidal with a distortion of 4% for the first five harmonics. (When using PSpice set **Fourier Analysis** in **Setup...** and observe the distortion in the text output file, that is select **Examine Output**.)

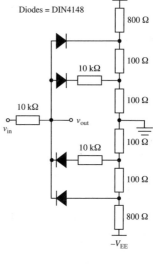

15.10 The circuit shown is for a Wien bridge oscillator with switched capacitors and variable ganged potentiometer. Determine the minimum and maximum frequency for the switch position shown.

[0.8 Hz, 106 Hz]

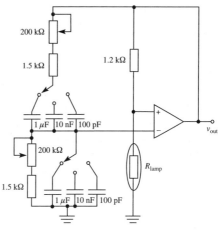

15.11 Determine the minimum and maximum for the other two switch positions.

15.12 Determine the maximum value of R_{lamp} required to ensure that the oscillator starts.

[400 Ω]

15.13 Determine the rms lamp current for sustained oscillation if the rms output voltage is 5 V.

[2.8 mA]

15.14   For the circuit shown determine the small-signal input resistance at the base of the transistor and the dc voltage between the collector and the emitter.

[11.6 kΩ, 12.8 V]

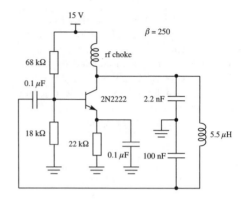

15.15   For the circuit shown in Problem 15.14 determine the frequency of oscillation.

[1.46 MHz]

15.16   For the circuit shown in Problem 15.14 determine the minimum value of β for sustained oscillation.

[45]

15.17   If the unloaded Q of the tuned $Q = R_p/\omega_o L$ is 200, where R_p is the effective parallel resistance of the tuned circuit at resonance and ω_o is the resonant frequency, determine the Q in the circuit shown in Problem 15.14.

[106]

15.18   For the same circuit determine the value of the coupling capacitor required to couple a 500 Ω load to ensure that the Q does not drop below 50.

[25 pF]

15.19   Determine the frequency if the 500 Ω load is coupled to the collector of the oscillator by means of a 0.1 µF capacitor instead of the capacitor determined in Problem 15.18.

[1.44 MHz]

15.20   Based on the approach used to analyze the Colpitts oscillator derive the following equation for the Hartley oscillator:

$$\frac{1}{r_e} + \frac{1}{\beta r_e} - \frac{1}{\beta\omega^2 L_2 C r_e}$$

$$+ j\left(\frac{1}{\omega^3 L_1 L_2 C} - \frac{1}{\omega L_1} - \frac{1}{\omega L_2}\right) = 0$$

Hence show that the resonant frequency is given by:

$$f_o = \frac{1}{2\pi\sqrt{C(L_1 + L_2)}}$$

and the the minimum value of β is L_2/L_1.

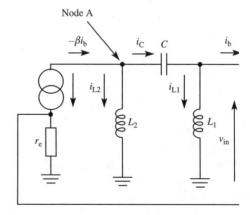

15.21 The simplified equivalent circuit for an FET-based Colpitts oscillator shown in Figure 15.27 is shown here. Derive the following network equation:

$$g_m + \frac{1}{r_{ds}} - \frac{\omega^2 L C_1}{r_{ds}}$$
$$+ j(\omega C_1 + \omega C_2 - \omega^3 L C_1 C_2) = 0$$

Hint: Assume that because the input impedance of the FET is very large that $i_L = i_{C1}$.

Hence show that the resonant frequency is:

$$f_o = \frac{1}{2\pi\sqrt{L[C_1 C_2/(C_1 + C_2)]}}$$

and the minimum value of g_m for sustaining the oscillation is:

$$g_m = \frac{1}{r_{ds}} \frac{C_1}{C_2}$$

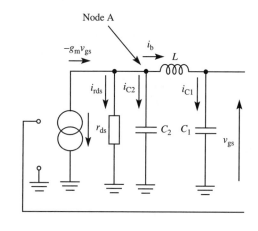

Node A

15.22 Determine the frequency of oscillation for the circuit shown.

[10 MHz]

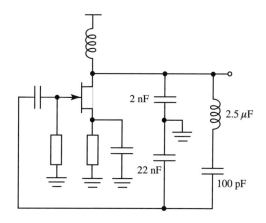

15.23 For the circuit shown in Problem 15.22 determine the minimum value of g_m for sustained oscillation if r_{ds} is 40 kΩ.

[275 µS]

15.24 For the circuit shown determine the frequency of oscillation and the minimum value of β for sustained oscillation.

Hint: Compare the circuit with the more conventional layout of the Colpitts or Hartley oscillators.

[1.5 MHz, 10]

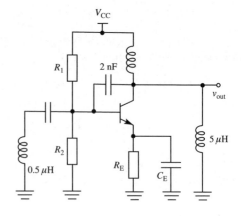

15.25 A quartz crystal has values of $L = 5$ mH, $C_s = 0.05$ pF, $R_s = 100 \ \Omega$ and $C_m = 15$ pF. Determine the series and parallel resonant frequencies.

[10.06 MHz, 10.08 MHz]

CHAPTER 16

Power Amplifiers

Amplifiers are used to increase the level of a signal and depending on the increase required, stages are cascaded to increase the gain. The last stage of the cascade may be required to drive some form of load, for example a loudspeaker, a servo mechanism or a coaxial cable for rf applications. In each case the load resistance is likely to be much lower than the output resistance of a typical amplifier stage and this can result in a considerable loss of signal, unless the stage is designed to operate with low-impedance loads. In addition to providing a match to the load, the last stage is also usually required to generate a large signal in order to provide an adequate volume of sound in large loudspeakers, or mechanical torque in a servo mechanism. The last stage must increase the voltage and current levels of the signal without distortion. These requirements form the basis of power amplifier design.

The analysis of the output stage is complicated by the fact that the signal levels are not small and the equivalent circuits for the transistors based on small-signal levels no longer apply. Large-signal circuit models are available for CAD simulation (see Chapter 4 for details of the Ebers–Moll and Gummel–Poon models), but they are too complicated for hand-based calculations, and a much simpler form of analysis is used which is based on load-lines and maximum symmetrical swings. Maximum power output may be obtained by calculating the maximum output voltage and maximum output current based on optimum biasing to provide maximum symmetrical swing. An important parameter for an output stage is the amount of distortion resulting from non-linearity in the output transistor current–voltage characteristics. A figure of merit for the audio power amplifier is the amount of distortion which it produces measured in terms of harmonics of the fundamental. The *total harmonic distortion* is the ratio of the rms value of the harmonics in the output signal to the rms of the fundamental. For a good-quality audio amplifier the figure is typically 0.01%. Distortion figures are best obtained from CAD simulators by using the Fourier analysis feature which is present in most SPICE-type simulators.

An important design requirement for an output stage is the efficient transfer of power to the load, which may be hundreds of watts. This power is provided by the dc supply, which for low-voltage transistor circuits must be capable of supplying the large currents required for outputs of hundreds of watts. It is also necessary to minimize the power dissipation in the output transistors. Apart from temperature limitations for transistors of about 150°C, there are also requirements to prolong the life of batteries in portable equipment by ensuring that the power is used as efficiently as possible and not dissipated as heat.

Bipolar transistors are widely used for output stages because of their excellent current-handling capability, but a special form of MOSFET known as a VMOS is also used. Power amplifiers are classified by the nature of the collector current waveform into class A, class B and class C. In addition the VMOS transistors are used in a mode of operation known as class D in which the transistors are used as switches.

16.1 Classification of Output Stages

Output stages involving bipolar transistors may be classified according to the waveform of the collector current when a sine wave is applied to the input. The different conditions are illustrated in Figure 16.1.

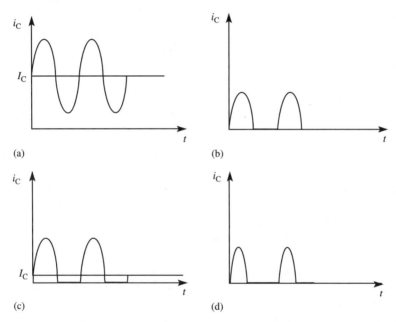

FIGURE 16.1   Collector current waveforms for (a) class A, (b) class B, (c) class AB and (d) class C.

In the simplest class A output stage the transistor conducts for the complete cycle, as shown in Figure 16.1(a). The transistor is biased with a collector current I_C which is larger than the peak value of the output current, and the transistor conducts for the complete cycle of the input signal. When the input signal is removed a steady collector current (I_C) flows in the load. If the transistor is biased so that no collector current flows when the input signal is absent then the condition shown in Figure 16.1(b) is obtained. The transistor only conducts when the input signal is present and is positive. When the input signal is absent no collector current flows. This form of operation is known as class B. Two transistors are required in order to reproduce the complete input signal for a complete cycle. A more detailed examination of the waveform produced by the basic class B stage would show that there is distortion at the instant when the transistor starts to conduct. This is due to the fact that a finite voltage is required at the base–emitter junction (0.5 V to 0.7 V) before the transistor conducts fully. This source of distortion can be reduced by allowing a small dc collector current to flow so that the transistors are not fully OFF when the input signal is reduced to zero. This current is much smaller than the peak ac current. This class AB operation is illustrated in Figure 16.1(c). Again a second transistor is required to complete the negative half cycle, and at the instant when the input sine wave is crossing zero, both transistors are conducting.

It will be shown that by reducing the dc component of current in the output stage, the average power dissipated in the transistor is reduced and efficiency is improved. In the class C amplifier this approach is taken to the extreme by only allowing the current to flow for a small fraction of each cycle, as shown in Figure 16.1(d). This pulsating collector current is used to generate a sine wave in a high-Q tuned LC circuit. Provided that the timing of the pulsating current corresponds to the resonant frequency of the tuned circuit, then a continuous sine wave output is obtained. Tuned LC loads exist in rf transmitter circuits and class C amplifiers are used to drive these circuits. The design of class C amplifiers is very specialized and usually forms part of a course on communications and will not be covered in any detail in this text.

In the class D mode of operation the transistor is used as a switch with a separate higher-frequency signal being used to switch the transistor ON and OFF many times during a cycle of the input signal. The resultant output is a train of pulses having widths which are proportional to the amplitude of the input signal. Although a filter is required to recover the original signal in the load, the advantage of the class D stage is that the efficiency is very high, approaching 100%. The switching properties of the MOSFET make it the preferred device for class D operation.

16.2 Class A Output Stage

For a class A output stage the transistor must conduct continuously and the quiescent collector current must equal or exceed the peak collector signal current. The common-emitter and emitter follower are examples of class A amplifiers. In both cases the output waveform is a faithful reproduction of the input wave, but both stages are very inefficient in the conversion of dc power from the supply to signal power in the load. However, the simplicity of the class A means that it is useful for low-power-output stages of a few hundred milliwatts to drive a small loudspeaker or earphones. The design involves the correct choice of dc biasing to ensure maximum symmetrical output swing.

Common-Emitter Output Stage

The circuit for consideration together with a diagram showing the dc and ac load-lines is shown in Figure 16.2.

The diagram shows a basic common-emitter stage, but the design objectives are to maximize the voltage swing across the load (R_C) with maximum transfer of ac power from the dc source to the load. These conditions may be attained by referring to the

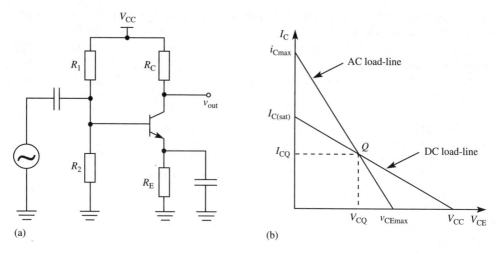

FIGURE 16.2   (a) Common-emitter output stage and (b) the ac and dc load-lines.

load-lines for both dc and ac conditions. These were considered in some detail in Chapter 6, but are summarized again here for convenience.

For the dc load-line the intercept with the current axis represents the maximum current ($I_{C(sat)}$) which would flow when V_{CE} is zero, that is $I_{C(sat)} = V_{CC}/(R_E + R_C)$. The intercept with the voltage axis corresponds to zero current flowing in the transistor, that is $V_{CE(cut-off)} = V_{CC}$. The slope of the dc load-line is $-1/(R_E + R_C)$.

The load is the collector resistance and the ac load as seen by the collector is $R_{ac} = R_C$. Therefore, the ac load-line has a slope $-1/R_{ac}$ and passes through the Q point. The maximum symmetrical swing of the voltage across the load is determined by the position of the Q point, and following the procedure used in Chapter 6, Section 6.2, the optimum Q point for maximum symmetrical swing is obtained by allowing i_{Cmax} to have a value of $2I_{CQ}$ which gives:

$$I_{CQ} = \frac{V_{CC}}{2R_{ac} + R_E} \tag{16.1}$$

and

$$V_{CEQ} = \frac{V_{CC}}{2 + R_E/R_{ac}} \tag{16.2}$$

Common-Emitter Stage Efficiency

An important factor in the design of a power amplifier is the efficient transfer of ac power from the dc source to the load. Efficiency is defined as:

$$\eta = \frac{\text{average signal power delivered to load}}{\text{average power drawn from dc source}} \tag{16.3}$$

The basic definitions of sinusoidal average power are:

$$P = V_{rms}I_{rms} = V_pI_p/2$$
$$P = V_{rms}^2/R = V_p^2/2R$$
$$P = I_{rms}^2R = I_p^2R/2$$

where the subscript 'p' refers to peak value.

The dc power taken from the supply is:

$$P_{CC} = I_{CQ}V_{CC} \tag{16.4}$$

Note that this ignores the power lost in the base-bias potential divider. This loss is usually not significant because the potential divider current is usually small compared with the collector current. Substituting from equation 16.1 gives:

$$P_{CC} = \frac{V_{CC}^2}{(2R_{ac} + R_E)} \approx \frac{V_{CC}^2}{2R_{ac}} \quad \text{if } R_{ac} \gg R_E \tag{16.5}$$

The power in the load is:

$$P_{ac} = \frac{i_p^2}{2}R_{ac} = \frac{1}{2}I_{CQ}^2R_{ac}$$

and substituting from equation 16.1 gives:

$$P_{ac} = \frac{1}{2}\frac{V_{CC}^2}{(2R_{ac} + R_E)^2}R_{ac} \approx \frac{V_{CC}^2}{8R_{ac}} \quad \text{if } R_{ac} \gg R_E \tag{16.6}$$

The efficiency is:

$$\eta = \frac{P_{ac}}{P_{CC}} = \frac{V_{CC}^2/8R_{ac}}{V_{CC}^2/2R_{ac}} = 0.25 \quad R_{ac} \gg R_E \tag{16.7}$$

With an efficiency of only 0.25, or 25%, this type of amplifier is not ideal for large power outputs. Much of the lost power is dissipated in the transistor. The collector current (I_{CQ}) is flowing continuously through the collector, and because most of the voltage drop is across the collector–base junction the power dissipation in the transistor may be assumed to occur at the collector, that is:

$$P_C = I_{CQ}V_{CEQ} \tag{16.8}$$

This represents the maximum power dissipation at the collector when the signal is zero. Substituting from equations 16.1 and 16.2 and using the approximation $R_{ac} \gg R_E$ gives:

$$P_C = \frac{V_{CC}^2}{4R_{ac}} \tag{16.9}$$

The ratio of collector power to the maximum ac load power is:

$$\frac{P_C}{P_{ac}} = \frac{V_{CC}^2/4R_{ac}}{V_{CC}^2/8R_{ac}} = 2 \qquad\qquad (16.10)$$

Thus for every watt dissipated in the load, 2 W are dissipated in the transistor. This is very wasteful and the class A amplifier is generally only suitable for outputs of a few hundred milliwatts, if only to limit the heat dissipated in the transistor, and is adequate for driving a small loudspeaker or earphones.

▶ **EXAMPLE 16.1** ————————————————————————————

Determine the Q point, the maximum symmetrical swing, the power dissipated in the collector and the efficiency for the circuit shown in Figure 16.3. Assume that $\beta = 100$ and $V_{BE} = 0.8$ V. Use PSpice to examine the output voltage.

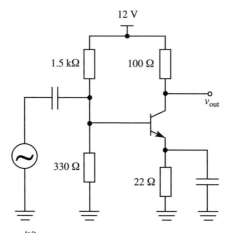

FIGURE 16.3 Class A amplifier.

▶ *Solution*

The collector current is:

$$I_{CQ} = \frac{(2.16\ \text{V} - 0.8\ \text{V})}{22\ \Omega + (270\ \Omega/100)} = 55\ \text{mA}$$

and V_{CEQ} is:

$$V_{CEQ} = 12\ \text{V} - 55\ \text{mA}(100\ \Omega + 22\ \Omega) = 5.29\ \text{V}$$

The maximum ac collector current is:

$$i_{Cmax} = I_{CQ} + \frac{V_{CEQ}}{R_{ac}} = 108\ \text{mA}$$

Since i_{Cmax} is less than $2I_{CQ}$ the peak ac current for symmetrical swing is limited to V_{VEQ}/R_{ac} rather than I_{CQ}, that is 53 mA.

The maximum small-signal collector to emitter voltage is:

$$v_{CEmax} = V_{CEQ} + I_{CQ}R_{ac} = 5.29 \text{ V} + 5.5 \text{ V} = 10.79 \text{ V}$$

The peak ac collector to emitter voltage is 5.29 V.

These values are shown in Figure 16.4.

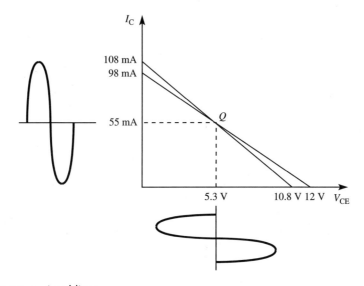

FIGURE 16.4   Load-lines.

The power provided by the supply is:

$$P_{CC} = V_{CC}I_{CQ} = (12 \text{ V})(55 \text{ mA}) = 660 \text{ mW}$$

The power dissipated in the load resistor is:

$$P_{ac} = \frac{i_p^2}{2}R_{ac} = \frac{(53 \text{ mA})^2}{2} \; 100 \; \Omega = 140 \text{ mW}$$

The power dissipated in the collector is:

$$P_C = V_{CEQ}I_{CQ} = (5.3 \text{ V})(55 \text{ mA}) = 292 \text{ mW}$$

The efficiency is:

$$\eta = \frac{P_{ac}}{P_{CC}} = \frac{140 \text{ mW}}{660 \text{ mW}} \times 100\% = 21\%$$

This is close to the theoretical value, but assumes that the maximum symmetrical swings can be obtained. In practice this is generally not possible without serious distortion. The output from Probe for a sinusoidal input of 30 mV peak at 1 kHz, which is necessary to approach the required peak-to-peak output of 10.6 V, is shown in Figure 16.5.

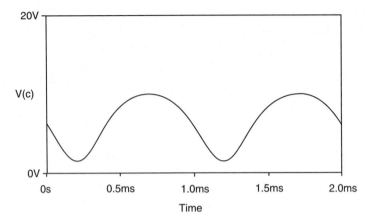

FIGURE 16.5   Probe output for 30 mV peak input signal.

It can be seen that there is serious distortion, particularly as the transistor approaches cut-off.

It is not possible to approach the maximum theoretical current and voltage swings in a class A stage because the transistor characteristics are very non-linear as the transistor approaches cut-off or saturation. The efficiency decreases considerably as the input is reduced in order to reduce the distortion, because even though the peak values of current and voltage are reduced, the dc quiescent value remains the same. The ac power is reduced, but the dc power remains the same. Much of the distortion can be removed with the application of negative feedback, but only so long as the input does not drive the transistor into saturation or cut-off.

Class A with Negative Feedback

The distortion is greatly reduced if negative feedback is applied. The simplest solution is to remove the decoupling capacitor from the circuit shown in Figure 16.2. This has the effect of making the dc and ac load-lines identical, as shown in Figure 16.6.

The maximum collector current is $I_{C(sat)} = V_{CC}/(R_C + R_E)$, and since the dc and ac load-lines are identical, the quiescent current for maximum symmetrical swing is:

$$I_{CQ} = \frac{V_{CC}/2}{R_C + R_E} \qquad\qquad (16.11)$$

and

$$V_{CEQ} = \frac{V_{CC}}{2} \qquad\qquad (16.12)$$

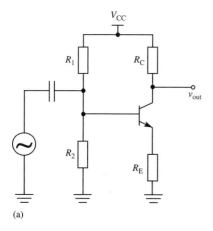
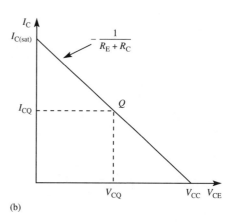

(a)   (b)

FIGURE 16.6 Common-emitter amplifier with negative feedback.

The efficiency is the same as before at 25%, with again twice as much power dissipated in the collector as in the load, but the input signal required to obtain maximum swing is much larger because the gain is greatly reduced.

▶ **EXAMPLE 16.2** ──────────────────────────────

Modify the base potential divider for the circuit shown in Figure 16.3 to provide maximum symmetrical swing when the decoupling capacitor is removed. Assume $\beta = 100$ and $V_{BE} = 0.8$ V.

▶ *Solution*

$$I_{C(sat)} = \frac{V_{CC}}{R_C + R_E} = \frac{12\ \text{V}}{122\ \Omega} \approx 98\ \text{mA}$$

$$\therefore \quad I_{CQ} = 49\ \text{mA}$$

The voltage drop across R_E is:

$$V_{RE} = (49\ \text{mA})(22\ \Omega) = 1.1\ \text{V}$$

and the base voltage is:

$$V_B = 1.1\ \text{V} + 0.8\ \text{V} = 1.9\ \text{V}$$

The Thévenin resistance of the base potential divider is:

$$R_T = 0.1\,(\beta_{min})R_E = 220\ \Omega$$

The base resistances are:

$$R_1 = \frac{V_{CC}}{V_B} R_T = \frac{12\ V}{1.9\ V}\ 220\ \Omega \approx 1.4\ k\Omega$$

$$R_2 = \frac{V_{CC}}{(V_{CC} - V_B)} R_T = \frac{12\ V}{(12\ V - 1.9\ V)}\ 220\ \Omega \approx 260\ \Omega$$

The Probe output for maximum output without serious distortion is shown in Figure 16.7.

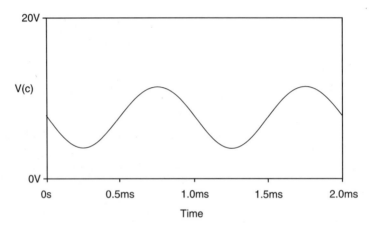

FIGURE 16.7   Probe output for class A amplifier with negative feedback.

By trial and error the input sine wave required to produce the waveform shown in Figure 16.7 has an amplitude of 0.9 V and the peak signal across the load is 3.8 V with 1.3% distortion. This results in an average power in the load of $(3.8^2/2)/100\ \Omega = 72$ mW. The collector current is 41 mA which results in a dc power consumption of $(41\ mA)(12\ V) = 492$ mW. This results in an efficiency of:

$$\eta = \frac{72\ mW}{492\ mW} \times 100\% = 14.6\%$$

The collector to emitter voltage is 6.95 V resulting in a collector dissipation of $(41\ mA)(6.95\ V) \approx 285$ mW. Note that this is considerably greater than that dissipated in the collector load.

The efficiency observed in Example 16.2 is less than the theoretical efficiency because some of the ac power is now lost across the emitter resistor which is no longer decoupled.

Emitter Follower Output Stage

The low output resistance of the emitter follower means that it is ideally suited to driving low-resistance loads. A typical emitter follower stage is shown in Figure 16.8(a) with the dc and ac load-lines in 16.8(b). The load is capacitively coupled to the emitter.

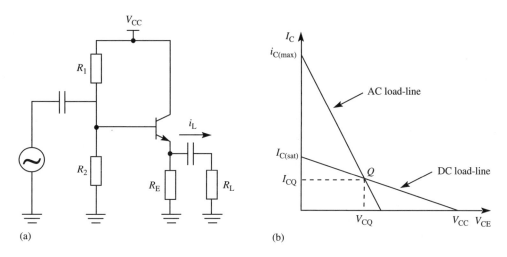

(a)   (b)

FIGURE 16.8   (a) Emitter follower with (b) the dc and ac load-lines.

For the dc load-line the intercept with the current axis represents the maximum current ($I_{C(sat)}$) and it occurs when V_{CE} is zero, that is $I_{C(sat)} = V_{CC}/R_E$. The intersection with the voltage axis corresponds to zero current flowing in the transistor, that is $V_{CE(cut-off)} = V_{CC}$. The slope of the dc load-line is $-1/R_E$.

The ac load as seen by the emitter is $R_{ac} = R_E \| R_L$. The ac load-line has a slope $-1/R_{ac}$ and passes through the Q point. The maximum symmetrical swing of the voltage across the load is determined by the ac load-line and the position of the Q point. Following the procedure used in Chapter 6, Section 6.2, for the common-emitter amplifier the optimum Q point for maximum symmetrical swing can be obtained by letting $i_{C(max)} = 2I_{CQ}$ which gives:

$$I_{CQ} = \frac{V_{CC}}{R_E + R_{ac}} \qquad R_{ac} = R_E \| R_L \qquad (16.13)$$

and

$$V_{CEQ} = \frac{V_{CC}}{2 + (R_E/R_L)} \qquad (16.14)$$

In practice the quiescent current would need to be larger than $0.5i_{Cmax}$ to avoid non-linearity of the transistor as it approaches saturation and cut-off.

▷ **EXAMPLE 16.3** ———————————————————————

Design an emitter follower output stage, with a base potential divider for dc bias, to provide 5 V peak across a 50 Ω load and use PSpice to verify the design. The supply voltage is 24 V and the minimum value of β is assumed to be 100.

▷ *Solution*

To allow for non-linearity near saturation and cut-off let $V_{CEQ} = 6$ V. Then from equation 16.14:

$$V_{CEQ} = 6 \text{ V} = \frac{24 \text{ V}}{2 + (R_E/50 \text{ }\Omega)}$$

and

$$R_E = 100 \text{ }\Omega$$

From equation 16.13:

$$I_{CQ} = \frac{24 \text{ V}}{100 \text{ }\Omega + 33.3 \text{ }\Omega} = 180 \text{ mA}$$

For dc biasing $V_{RE} = I_{CQ}R_E = 18$ V. The base voltage is:

$$V_T = 18 \text{ V} + 0.8 \text{ V} = 18.8 \text{ V}$$

The Thévenin resistance for the base circuit is:

$$R_T = (0.1)\beta R_E = 1 \text{ k}\Omega$$

The two base-bias resistors are:

$$R_1 = \frac{24 \text{ V}}{18.8 \text{ V}} \; 1 \text{ k}\Omega \approx 1.3 \text{ k}\Omega$$

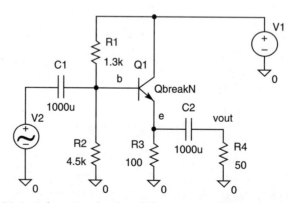

FIGURE 16.9   PSpice schematic of emitter follower output stage.

and
$$R_2 = \frac{24\ \text{V}}{(24\ \text{V} - 18.8\ \text{V})}\ 1\ \text{k}\Omega = 4.6\ \text{k}\Omega$$

The PSpice schematic is shown in Figure 16.9.

An input sine wave of 5 V amplitude at 1 kHz is applied and the output waveform is shown in Figure 16.10.

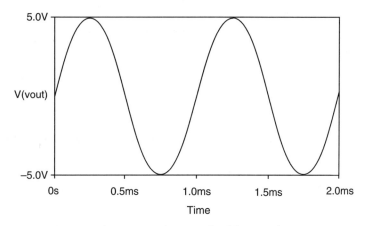

FIGURE 16.10   Output voltage across the 50 Ω load from Probe.

It can be seen that a 5 V peak amplitude sine wave is obtained and from the Fourier analysis the total harmonic distortion is 0.24%.

▷ **EXAMPLE 16.4** ————————————————————————————

Repeat the above example with $V_{\text{CEQ}} = 5$ V instead of 6 V and note the distortion which results from changing the operating point.

Class A Emitter Follower Efficiency

For maximum symmetrical swing at the output the Q point is established by equations 16.13 and 16.14. These equations represent the peak values for collector current and collector to emitter voltage. The peak current in the load is obtained by current division between R_E and R_L (see Figure 16.8(a)) as:

$$i_{\text{pL}} = \left(\frac{V_{\text{CC}}}{R_E + R_{\text{ac}}}\right)\left(\frac{R_E}{R_E + R_L}\right) \tag{16.15}$$

The average ac power in the load is:

$$P_{\text{ac}} = \frac{i_{\text{pL}}^2}{2}\,R_L = \left[\left(\frac{V_{\text{CC}}}{R_E + R_{\text{ac}}}\right)\left(\frac{R_E}{R_E + R_L}\right)\right]^2 \frac{R_L}{2} \tag{16.16}$$

The average power provided by the dc supply is:

$$P_{CC} = V_{CC}I_{CQ} = \frac{V_{CC}^2}{R_E + R_{ac}} \tag{16.17}$$

Thus the efficiency for maximum symmetrical swing is:

$$\eta = \frac{P_{ac}}{P_{CC}} = \left[\left(\frac{V_{CC}}{R_E + R_{ac}}\right)\left(\frac{R_E}{R_E + R_L}\right)\right]^2 R_L \left/ 2\left(\frac{V_{CC}^2}{R_E + R_e}\right)\right.$$

which simplifies to:

$$\eta = \frac{R_{ac}}{2(R_E + 2R_L)} \tag{16.18}$$

If $R_E \gg R_L$ then all of the ac power would be dissipated in the load and the efficiency would be:

$$\eta = \frac{R_L}{2(2R_L)} = 0.25$$

This value of 0.25, or 25%, corresponds to the theoretical maximum for class A operation, but is unrealistic, because the inequality is impractical. An alternative condition for maximum power transfer is that the source impedance is equal to the load impedance. Under these conditions if $R_E = R_L = R$ then the efficiency is:

$$\eta = \frac{R/2}{2(R + 2R)} = 0.083$$

This value of 8.3% is considerably lower than the theoretical maximum of 25%, but even this value assumes that the collector current swings from 0 mA to $i_{C(max)}$ and that the collector–emitter voltage varies from 0 V to $v_{CE(max)}$. In practice this would result in considerable distortion. To avoid this distortion it is usual to increase the quiescent values of current and voltage so that for maximum swing the values do not reach the extremes of 0 mA, $v_{CE(max)}$ or $i_{C(max)}$, 0 V. While these design compromises reduce the distortion they also result in a reduction in the efficiency, because the quiescent current drawn from the supply is greater than that predicted by equation 16.13. As a result the efficiency will be even less than the value predicted by equation 16.18.

▶ **EXAMPLE 16.5** ————————————————————————————————

Determine the quiescent conditions for the circuit shown in Figure 16.11, the power dissipated in the load, in the transistor and the efficiency. Use PSpice to examine the output and to measure the peak current and hence the power dissipated in the load for maximum symmetrical swing. Assume that $\beta = 100$ and $V_{BE} = 0.8$ V.

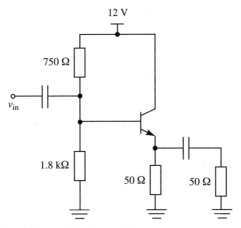

FIGURE 16.11   Emitter follower class A amplifier.

▷ *Solution*

The dc collector current is:

$$I_{CQ} = \frac{8.47 \text{ V} - 0.8 \text{ V}}{50 \ \Omega + (529 \ \Omega/100)} \approx 139 \text{ mA}$$

The quiescent collector–emitter voltage is:

$$V_{CEQ} = 12 \text{ V} - (139 \text{ mA})(50 \ \Omega) \approx 5 \text{ V}$$

The maximum current is:

$$I_{C(sat)} = \frac{12}{50 \ \Omega} = 240 \text{ mA}$$

The peak ac collector current is:

$$i_{C(max)} = I_{CQ} + \frac{V_{CEQ}}{R_{ac}} = 139 \text{ mA} + \frac{5 \text{ V}}{25 \ \Omega} = 339 \text{ mA}$$

The load line diagram is shown in Figure 16.12.

From the load-line it can be seen that the maximum symmetrical collector current swing cannot exceed 139 mA, and will be less than this if distortion is to be avoided, that is:

$$i_{pC} = 139 \text{ mA}$$

The maximum voltage swing will be:

$$v_{L(max)} = I_{CQ}R_{ac} \approx 3.5 \text{ V}$$

Using equation 16.15 the maximum load current is:

$$i_{pL} = i_{pC}\left(\frac{R_E}{R_E + R_L}\right) = (139 \text{ mA})(0.5) \approx 70 \text{ mA}$$

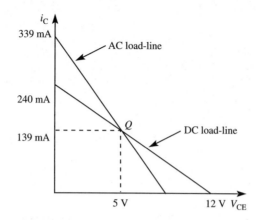

FIGURE 16.12   Load-line diagram.

The maximum power dissipated in the load is:

$$P_{ac} = \frac{i_{pL}^2}{2} R_L = \left(\frac{70 \text{ mA}^2}{2}\right) 50 \ \Omega = 122 \text{ mW}$$

The power provided by the dc supply is:

$$P_{CC} = I_{CQ} V_{CC} = (139 \text{ mA})(12 \text{ V}) \approx 1.7 \text{ W}$$

and the efficiency is:

$$\eta = \frac{P_{ac}}{P_{CC}} \times 100\% = \frac{122 \text{ mW}}{1.7 \text{ W}} \times 100\% \approx 7\%$$

The power dissipated in the transistor is:

$$P_C = I_{CQ} V_{CEQ} = (139 \text{ mA})(5 \text{ V}) \approx 0.7 \text{ W}$$

The PSpice schematic for the stage is shown in Figure 16.13.

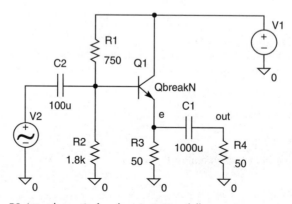

FIGURE 16.13   PSpice schematic for class A emitter follower stage.

The waveform at the emitter for an input sinusoid of 3.5 V peak is shown in Figure 16.14.

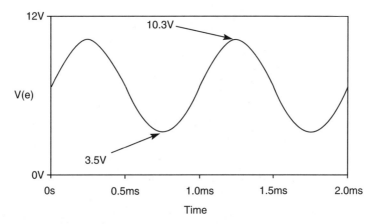

FIGURE 16.14   Probe output at the emitter.

The peak-to-peak amplitude of the voltage across the load is 6.8 V which results in a power dissipation in the load of:

$$\frac{1}{2}\left(\frac{6.8\ \text{V}}{2}\right)^2 \frac{1}{50\ \Omega} = 115\ \text{mW}$$

The dc collector current for the simulation is 136 mA resulting in a dc power dissipation of $(136\ \text{mA})(12\ \text{V}) = 1.6\ \text{W}$. This results in an efficiency of:

$$\eta = \frac{115\ \text{mW}}{1.6\ \text{W}} \times 100\% \approx 7\%$$

Emitter Follower Output Stage – Practical Considerations

A class A emitter follower output stage capable of producing many watts in a low-impedance load such as a loudspeaker is shown in Figure 16.15.

The first step to improving the basic circuit is to use two power supplies. This enables the output at the emitter of Q_2 to be biased to zero voltage, which allows a grounded load, such as a loudspeaker, to be connected directly to the emitter without the need for a coupling capacitor. An additional transistor Q_1 is added to reduce the base drive requirements for the output transistor Q_2. This is the familiar Darlington connection where the effective current gain is $\beta_1 \times \beta_2$, and the base current drawn through R_1 is greatly reduced in relation to the potential divider current.

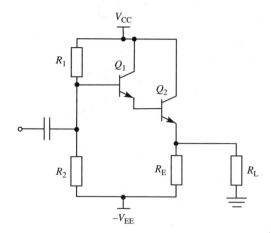

FIGURE 16.15   Class A emitter follower power output stage.

The PSpice schematic for a class A power-output stage capable of producing 10 W in an 8 ohm loudspeaker is shown in Figure 16.16.

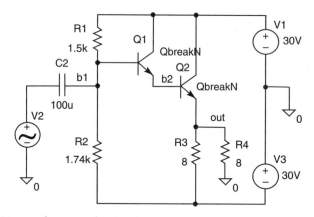

FIGURE 16.16   PSpice schematic of 10 W class amplifier.

The output waveform is shown in Figure 16.17 for a 13 V peak amplitude sine wave input. The distortion is 0.1%.

The dissipation in the output transistor, Q_2, is 110 W and a further 115 W are dissipated in the 8 ohm emitter resistor. This is a characteristic feature of class A amplifiers. They are capable of producing low-distortion output, but at the expense of excessive power dissipation in the output transistor and bias components. The class A stage is not suitable for battery-operated circuits beyond a few hundred milliwatts.

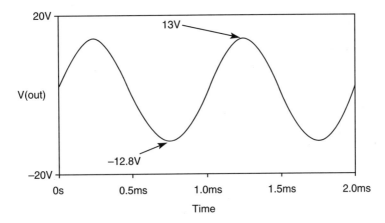

FIGURE 16.17   Probe output across 8 Ω for 13 V peak input.

▶ **EXAMPLE 16.6** _____

Verify the quiescent conditions for the circuit shown in Figure 16.6 and estimate
the maximum efficiency. Assume that $\beta_1 = \beta_2 = 100$ and $V_{BE1} = V_{BE2} = 0.8$ V.

▶ *Solution*

The Thévenin resistance for the base potential divider is:

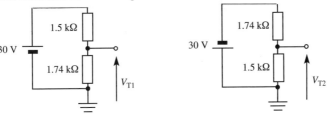

$R_T = 1.5 \text{ k}\Omega \,\|\, 1.74 \text{ k}\Omega = 805 \ \Omega$

With two power supplies the Thévenin voltage can be obtained by superposition as:

$V_T = V_{T1} + V_{T2} = 16.1 \text{ V} - 13.9 \text{ V} = 2.2 \text{ V}$

The collector current in Q_2 may be obtained by replacing the base-bias circuit with the

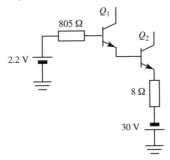

equivalent Thévenin circuit as shown. Then:

$$I_{C2Q} = \frac{32.2 \text{ V} - 1.6 \text{ V}}{8 \ \Omega + (805 \ \Omega/100 \times 100)} = 3.8 \text{ A}$$

$$V_{CE2Q} = 60 \text{ V} - (3.8 \text{ A})(8 \ \Omega) = 29{,}6 \text{ V}$$

The maximum collector current in Q_2 is:

$$I_{C2(sat)} = \frac{V_{CC} + V_{EE}}{R_E} = \frac{60 \text{ V}}{8 \ \Omega} = 7.5 \text{ A}$$

The maximum ac current in Q_2 is:

$$i_{C2(max)} = I_{C2Q} + \frac{V_{CE2Q}}{R_{ac}} = 3.8 \text{ A} + \frac{29.6 \text{ V}}{4 \ \Omega} = 11.2 \text{ A}$$

The load line is shown in Figure 16.18.

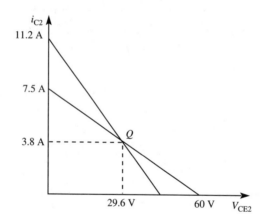

FIGURE 16.18   Load-line for class A power amplifier.

Examination of the load-line shows that the maximum current swing in the ac load is equal to I_{C2Q}, that is 3.8 A.

The peak current in the load is:

$$i_{pL} = 3.8 \text{ A} \ \frac{8 \ \Omega}{8 \ \Omega + 8 \ \Omega} = 1.9 \text{ A}$$

and the maximum power dissipated in the load is:

$$P_{ac} = \frac{i_{pL}^2}{2} R_L = \frac{(1.9 \text{ A})^2}{2} \ 8 \ \Omega = 14.4 \text{ W}$$

The power provided by the supply is:

$$P_{CC} = I_{CQ2}(V_{CC} + V_{EE}) = (3.8 \text{ A})(60 \text{ V}) = 228 \text{ W}$$

which gives an efficiency of:

$$\eta = \frac{P_{ac}}{P_{CC}} = \frac{14.4 \text{ W}}{228 \text{ W}} \times 100\% = 6.3\%$$

16.3 Class B Output Stage

When the transistor conducts for only half of a sinusoidal cycle and is cut off for the remaining half, it is said to be a class B amplifier. The Q point represents zero collector current, and the applied signal drives the transistor into conduction, but only for half a cycle, as shown in Figure 16.19.

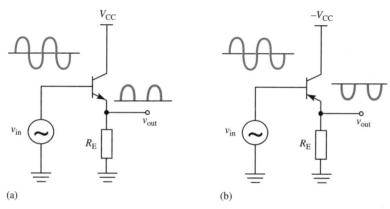

FIGURE 16.19 Emitter follower with a zero-current Q point for (a) an npn transistor and (b) a pnp transistor.

An npn transistor conducts during the positive half cycles of the input waveform, while the pnp conducts during the negative half cycle. In order to obtain a complete sinusoidal output it is necessary to use both types of transistor in a *push–pull* amplifier, as shown in Figure 16.20.

When the input is positive the npn transistor conducts, but a positive input voltage turns the pnp transistor OFF. Current flows from V_{CC}, through the npn transistor and through the load resistor to ground. For the negative half cycle the npn transistor is turned OFF and the pnp transistor is turned ON and current flow is from the load through the pnp transistor and to V_{EE}. Thus a complete sinusoidal cycle is created across the load.

Ignoring the small saturation voltage across the transistors, the maximum peak current through the transistor and the load is:

$$I_{C(sat)} = i_{pL(max)} = \frac{V_{CC}}{R_L} \quad \left(\text{or} \ \frac{-V_{EE}}{R_L} \right) \tag{16.19}$$

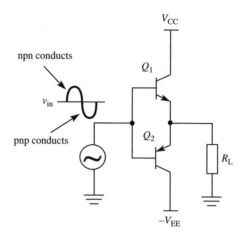

FIGURE 16.20   Class B push–pull amplifier.

The maximum peak ac voltage across the load is:

$$v_{\mathrm{pL(max)}} = V_{\mathrm{CC}} \quad (\text{or} -V_{\mathrm{EE}}) \tag{16.20}$$

Class B Efficiency

In the ideal class B amplifier no current flows when the input signal is absent. When the signal is present, first one transistor conducts and then the other for alternate halves of the input sinusoid. The waveforms are shown in Figure 16.21.

The power dissipated in the load is:

$$P_{\mathrm{ac}} = \frac{v_{\mathrm{pL}}^2}{2R_{\mathrm{L}}} \tag{16.21}$$

Current is only drawn from the supply when one or other of the transistors conducts and it has a peak value of $i_{\mathrm{pC}} = v_{\mathrm{pL}}/R_{\mathrm{L}}$. For a complete input cycle the combined waveform for the supply current from both supplies resembles the output of a full-wave rectifier, as shown in Figure 16.21(d). The average value of a full-wave rectified sine wave is $2/\pi$ times its peak value (see Appendix 6). Thus:

$$P_{\mathrm{CC}} = \frac{2}{\pi} i_{\mathrm{pC}} V_{\mathrm{CC}} = \frac{2}{\pi} \frac{v_{\mathrm{pL}}}{R_{\mathrm{L}}} V_{\mathrm{CC}} \tag{16.22}$$

The efficiency is:

$$\eta = \frac{P_{\mathrm{ac}}}{P_{\mathrm{CC}}} = \frac{v_{\mathrm{pL}}^2/2R_{\mathrm{L}}}{2v_{\mathrm{pL}}V_{\mathrm{CC}}/\pi R_{\mathrm{L}}} = \frac{v_{\mathrm{pL}}}{V_{\mathrm{CC}}} \frac{\pi}{4} \tag{16.23}$$

The maximum efficiency is obtained when the peak voltage across the load has its

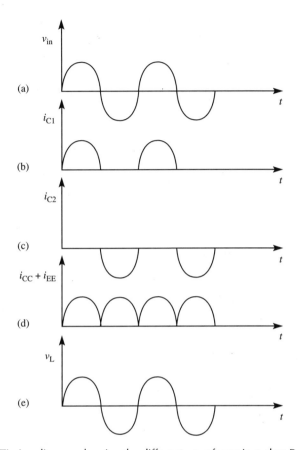

FIGURE 16.21   Timing diagram showing the different waveforms in a class B amplifier.

maximum value, that is $v_{pL(max)} = V_{CC}$. Then:

$$\eta_{(max)} = \frac{\pi}{4} = 78.5\% \tag{16.24}$$

This value is considerably greater than that obtained for the class A amplifier (25%).

The class A amplifier dissipates maximum power in the transistor when the input signal is zero, whereas the quiescent power dissipation in the class B amplifier is zero. When an input signal is applied the average power dissipated in the transistor under class B operation is:

$$P_C = P_{CC} - p_{ac}$$

Substituting from equations 16.21 and 16.22 for P_{ac} and P_{CC} gives:

$$P_C = \frac{2v_{pL}V_{CC}}{\pi R_L} - \frac{v_{pL}^2}{2R_L} \tag{16.25}$$

The power dissipated in the transistors is a function of the peak voltage across the load. The power is zero when the signal is absent ($v_{pL} = 0$) and increases as v_{pL} increases. However, the negative term increases as the square of v_{pL} and at some point the power must start to decrease as v_{pL} continues to increase. The position of the maximum value of P_C may be obtained by differentiating the expression for P_C with respect to v_{pL} and setting the result equal to zero. Thus:

$$\frac{d}{dv_{pL}}(P_C) = \frac{2V_{CC}}{\pi R_L} - \frac{2v_{pL}}{2R_L} = 0$$

and

$$v_{pL}\big|_{max} = \frac{2V_{CC}}{\pi} \tag{16.26}$$

Substituting this value into equation 16.25 gives:

$$P_{C(max)} = \frac{2V_{CC}^2}{\pi R_L} \tag{16.27}$$

This is the total power dissipated in the two transistors. Thus each transistor must be capable of dissipating half this amount, that is:

$$P_{CN(max)} = P_{CP(max)} = \frac{V_{CC}^2}{\pi R_L} \tag{16.28}$$

The ratio of collector power to maximum ac load power is:

$$\frac{P_C}{P_{ac(max)}} = \frac{V_{CC}^2/\pi^2 R_L}{V_{CC}^2/2R_L} = \frac{2}{\pi^2} \approx 0.2 \tag{16.29}$$

This represents an improvement by a factor of 10 over the same relationship for the class A stage (see equation 16.10). Thus for every watt dissipated in the load only 0.2 W is dissipated in each transistor.

Since the power dissipated in the transistors depends on the value of the peak voltage across the load, and since the efficiency is also dependent on the load voltage (equation 16.23), it is of interest to determine the efficiency for maximum dissipation in the transistors. Substituting $2V_{CC}/\pi$ for v_{pL} in equation 16.23 for the overall efficiency gives:

$$\eta = \frac{2V_{CC}}{\pi} \frac{\pi}{4V_{CC}} = 50\% \tag{16.30}$$

A plot of the variation of power dissipated in the transistors versus the value of the peak ac voltage across the load is shown in Figure 16.22 (see Problem 16.12). Note that the power dissipated in the transistors decreases, and the power dissipated in the load increases, when the amplitude of the load voltage is increased beyond $2V_{CC}/\pi$

($\approx 0.64 V_{CC}$). However, non-linear distortion would also increase as the saturation condition of the two transistors is approached.

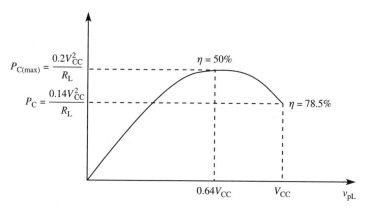

FIGURE 16.22 Transistor dissipation versus amplitude of load voltage for class B amplifier.

An important consideration for the class B transistor is its voltage rating. The maximum peak-to-peak swing across the load is $2V_{CC}$ for maximum output. Referring to Figure 16.20, when the input is positive Q_2 is cut off, Q_1 conducts and when I_{C1} is maximum $V_{CE1(sat)} \approx 0$ V. The voltage across the load is $+V_{CC}$ and the voltage across Q_2 is $V_{CC} + V_{EE}$. If $V_{CC} = V_{EE}$ as is usual then the maximum collector to emitter voltage which each transistor is subjected to is $2V_{CC}$, that is each transistor must be rated at least to this value. (A transistor in a class A stage need only be rated to V_{CC}.)

▷ **EXAMPLE 16.7** ────────────────────────────────

For the class B amplifier shown in Figure 16.20 the supply voltages are ±15 V and the load resistance is 8 Ω. Determine the power dissipated in the load, the power drawn from the supply, the power dissipated in the transistors and the conversion efficiency for a peak input of 10 V.

▷ **Solution**

The power in the load is:

$$P_{ac} = \frac{v_{pL}^2}{2R_L} = \frac{10 \text{ V}^2}{2 \times 8 \text{ Ω}} = 6.25 \text{ W}$$

The power provided by the supplies is:

$$P_{CC} = \frac{2}{\pi} \frac{v_{pL}}{R_L} V_{CC} = \frac{2 \times 10 \text{ V} \times 15 \text{ V}}{\pi \times 8 \text{ Ω}} \approx 11.9 \text{ W}$$

The power dissipated in the transistors is:

$$P_{CN} = P_{CP} = \frac{2v_{pL}V_{CC}}{\pi R_L} - \frac{v_{pL}^2}{2R_L} = 11.937 \text{ W} - 6.25 \text{ W} = 5.7 \text{ W}$$

The conversion efficiency is:

$$\eta = \frac{P_{ac}}{P_{CC}} = \frac{6.25 \text{ W}}{11.9 \text{ W}} \times 100\% = 52\%$$

▶ **EXAMPLE 16.8** ────────────────────────────

Repeat the calculations if the amplitude of the input voltage is 5 V.

Cross-Over Distortion

The circuit shown in Figure 16.20 is very simple, and it produces a large amount of distortion. With no signal applied neither transistor conducts and no current flows from either supply, as expected for a class B amplifier. However, in order that the transistors conduct when a signal is applied the input signal must exceed approximately 0.5 V. Thus there is a delay before the npn transistor starts to conduct when the signal is increasing from zero voltage in a positive direction. Similarly as the input approaches zero voltage at the end of a positive half cycle, the npn transistor stops conducting when the voltage drops below 0.5 V. A similar sequence of events takes place for the pnp transistor for the negative half cycle. This results in the output waveform across the load having a discontinuity as the signal crosses the 0 V axis, as shown in Figure 16.23. This discontinuity in the waveform is known as *cross-over distortion*.

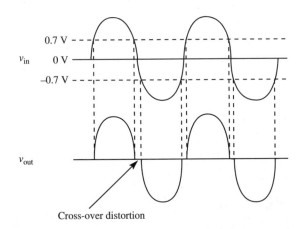

FIGURE 16.23   Input waveform and the corresponding output waveform showing the cross-over distortion.

Class AB Amplifier

The cross-over distortion can be reduced by the application of negative feedback, but not entirely removed, because at the point of cross-over neither transistor is conducting, the gain is effectively zero, and negative feedback has little effect under these conditions. A more satisfactory solution is to provide a small forward bias so that the transistors conduct when the input signal is reduced to zero, as shown in Figure 16.24.

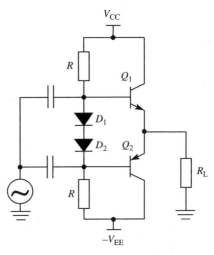

FIGURE 16.24   Biasing a push–pull amplifier to prevent cross-over distortion.

The potential divider comprising two resistors and two diodes provide a bias to the bases of the two transistors. The value of the resistors is selected to enable the diodes to conduct so that there is a voltage of between 1.2 V and 1.4 V, or $2V_{BE}$, between the bases of the transistors. This allows the transistors to conduct when the input signal is zero.

This mode of operation is known as class AB, because it is still a push–pull amplifier where first one and then the other transistor conducts, but in addition there is also a finite current present for zero input signal, as in a class A amplifier.

When a positive-going input signal is applied the forward bias applied to Q_1 is increased and the collector current increases. At the same time the V_{BE} applied to Q_2 decreases by an equal amount so that the collector current in Q_2 decreases. For negative-going input signal the opposite process occurs with the current in Q_1 decreasing and the current in Q_2 increasing. For small values of input (less than 0.6 V amplitude) both transistors conduct, with one conducting more than the other, depending on the polarity of the input. When the input exceeds approximately 0.8 V, most of the current flow will be confined to one transistor as in the ideal class B operation. The important feature of the class AB mode is the smooth transition as the input signal crosses the zero-voltage axis.

Thermal Stability

The use of diodes rather than an additional resistor in the potential divider is important to improve the temperature stability of the circuit. Output stages are required to produce large-output signals in low-resistance loads and this inevitably results in large currents and significant power dissipation at the collector of the transistor. Power dissipation results in a temperature rise. For every 1°C rise in temperature, V_{BE} decreases by approximately 2 mV for a silicon transistor. Thus a temperature rise of 40°C results in a fall in V_{BE} of 80 mV. For a fixed base bias, which would result if resistors were used in place of the diodes, the voltage at the base of each transistor would be 80 mV larger than required for the room temperature collector current. Thus a temperature rise of 40°C would result in the collector current increasing by a factor of ~10 for fixed bias. This would increase the power dissipation causing a further increase in the temperature, a further reduction in V_{BE} and a further rise in the collector current. This is a form of positive feedback which can lead to *thermal runaway*.

The use of silicon diodes, with a forward voltage across the diode which has the same temperature characteristics as the V_{BE} of the transistors, compensates for the temperature changes in the transistors. The diodes need to be in close thermal contact with the transistors so that they experience the same temperature changes. If the transistors require a V_{BE} of 0.7 V to establish a current of 5 mA then a 40°C rise in temperature would result in V_{BE} dropping to 0.62 V. If the diodes experience the same temperature rise then their forward voltage will also fall by 80 mV, and the voltage applied to the base of each transistor is correct for the new temperature. In practice the match between V_{BE} and the diode forward voltage drop is not perfect, but it is sufficient to prevent thermal runaway.

The bias resistors (R in Figure 16.24) are chosen to provide sufficient base current for the transistors to reach the peak output voltage across the load. For example, if 5 W are required across an 8 ohm load from a ±15 V supply, then the peak voltage across the load is 8.9 V, and the base voltage for Q_1 is approximately 9.7 V (assuming $V_{BE} = 0.8$ V). Thus with $V_{CC} = 15$ V there will be 5.3 V (15 V − 9.7 V) available across the bias resistor to provide the current for the diodes and also the base current for Q_1. Assuming a value of 50 for the common-emitter current gain for the transistors, then the peak base current is 22 mA, and the maximum value for the bias resistor is 5.3 V/22 mA = 240 Ω, say 220 Ω from the preferred standard range of resistors.

▶ **EXAMPLE 16.9** ─────────────────────────────────

Design a class AB stage to provide 16 W into an 8 ohm load from ±24 V supplies. Assume that $\beta = 50$ and $V_{BE} = 0.8$ V. Use PSpice to verify the design.

▶ *Solution*

The peak values of voltage and current across the load are:

$$v_{pL} = \sqrt{2 \times 8\ \Omega \times 16\ W} = 16\ V$$

$$i_{pL} = \sqrt{\frac{2 \times 16\ W}{8\ \Omega}} = 2\ A$$

The peak base voltage for Q_1 is 16.8 V and the peak base current is $2\,\text{A}/50 = 40$ mA. The bias resistor must be able to provide this current to the base, that is:

$$R = \frac{24\,\text{V} - 16.8\,\text{V}}{40\,\text{mA}} = 180\,\Omega$$

The quiescent diode current will be:

$$I_{DQ} = \frac{48\,\text{V} - 1.6\,\text{V}}{2 \times 180\,\Omega} \approx 130\,\text{mA}$$

The circuit for the class B stage is shown in Figure 16.25.

For the PSpice simulation use transistors and diodes from the breakout library and set β to 50 for the transistor models (bf = 50 using **Edit** and selecting **model...**). Set the amplitude of the input to 16 V at a frequency of 1 kHz and run the transient simulation for 2 ms. The output from Probe for the load voltage and current is shown in Figure 16.26.

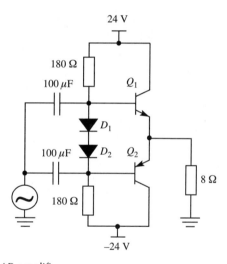

FIGURE 16.25 Class AB amplifier.

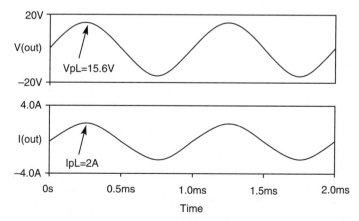

FIGURE 16.26 Probe plots of v_{out} and i_{out} for the class B amplifier.

An examination of the PSpice output file shows that the quiescent diode current is 129 mA and that the distortion is 0.4%.

A further improvement in thermal stability can be achieved by employing negative feedback, as shown in Figure 16.27, with the inclusion of two emitter resistors (R_E). These are typically less than 5 ohms. A voltage drop of a few hundred millivolts across each resistor minimizes the effect of changes in V_{BE} caused by changes in temperature. However, this improvement is achieved at the expense of reduced efficiency and reduced output voltage swing across the load.

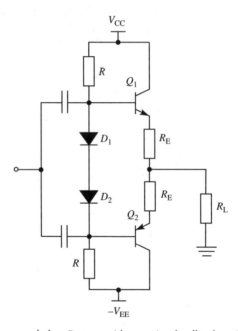

FIGURE 16.27 An improved class B stage with negative feedback resistors.

Darlington Class B Output Stage

From Example 16.9 it can be seen that in order to provide adequate base drive for the output transistors, the bias resistor must have a low value. The consequence of this is a large current drain through the base potential divider – 130 mA in the above example. The base drive is greatly reduced if the current gain can be increased. This is achieved with the Darlington connection, as shown in Figure 16.28.

The base current of Q_2 ($I_{B2} = I_{C2}/\beta_2$) is the collector current of Q_1, and the base current of Q_1 is $I_{B1} = I_{C2}/\beta_1\beta_2$. Thus the effective current gain for the composite transistor is equal to the product of the individual current gains of the two transistors. With a gain of 50 for the output power transistor (Q_2) and 100 for Q_1 the combined

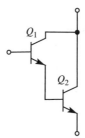

FIGURE 16.28   Darlington configuration.

gain is 5000. The effective base to emitter voltage drop is $2V_{BE}$ plus the dc voltage drop across the two emitter feedback resistors (R_E). Additional diodes and/or resistors are required in the base-bias potential divider, as shown in Figure 16.29. An additional diode is included plus a variable resistor which can be used to optimize the bias to minimize cross-over distortion (R_2 is typically less than 100 ohms).

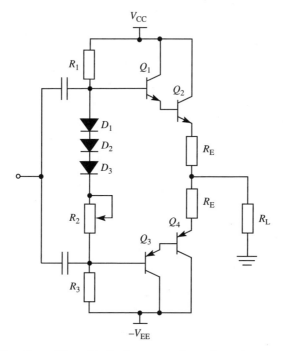

FIGURE 16.29   Class B amplifier with Darlington-configured transistors.

Single-Supply Operation

A disadvantage of the class B circuits shown above is the need for two power supplies. In the circuit shown in Figure 16.30 only one power supply is used.

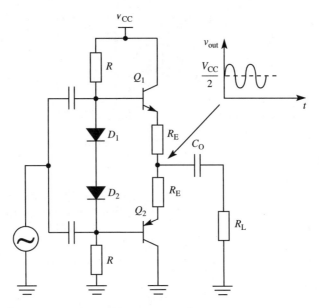

FIGURE 16.30   Class B stage with a single power supply.

With matched components in the base potential divider the base of Q_1 is one V_{BE} above $V_{CC}/2$, while the base of Q_2 is one V_{BE} below $V_{CC}/2$. The junction of the two emitter resistors will be at half the supply voltage. A coupling capacitor C_O is now required to couple the load resistor to the emitters. This capacitor must be large to ensure that its reactance at the lowest frequency to be used is small compared with the load resistance.

When a signal is applied Q_1 conducts when the input is positive and current is drawn from the supply and flows through Q_1 and the load. When the input is negative Q_1 is cut off and no current can flow from the supply. For the negative half cycle current is provided by the coupling capacitor C_O as it discharges through Q_2. Thus for a positive half cycle Q_1 conducts and current flows from the supply and charges C_O, while during the negative half cycle Q_2 conducts and the load current is provided by C_O. Since C_O is providing the current for one half cycle the time constant $R_L C_O$ must be much larger than the period of the lowest input frequency.

The lowest frequency as a result of C_O is:

$$f_{c(C_O)} = \frac{1}{2\pi(R_L + R_E)C_O} \tag{16.31}$$

To determine the efficiency it is necessary to consider the peak values of current and voltage in the load. The peak current is:

$$i_{pL} = \frac{v_{pL}}{R_L + R_E}$$

where v_{pL} is the peak voltage across the load. The ac power dissipated in the load is:

$$P_{\mathrm{ac}} = \frac{i_{\mathrm{pL}}^2}{2} R_{\mathrm{L}} = \frac{v_{\mathrm{pL}}^2 R_{\mathrm{L}}}{2(R_{\mathrm{L}} + R_{\mathrm{E}})} \tag{16.32}$$

Current is only drawn from the supply for half a cycle and, therefore, the average value is $1/\pi$ times the peak value. Thus the power provided by the supply is:

$$P_{\mathrm{CC}} = \frac{i_{\mathrm{pL}}}{2} V_{\mathrm{CC}} = \frac{1}{\pi} \frac{v_{\mathrm{pL}}}{(R_{\mathrm{L}} + R_{\mathrm{E}})} V_{\mathrm{CC}} \tag{16.33}$$

The efficiency is:

$$\eta = \frac{P_{\mathrm{ac}}}{P_{\mathrm{CC}}} = \left(\frac{\pi}{2}\right)\left(\frac{v_{\mathrm{pL}}}{V_{\mathrm{CC}}}\right)\left(\frac{R_{\mathrm{L}}}{R_{\mathrm{L}} + R_{\mathrm{E}}}\right) \tag{16.34}$$

The efficiency reaches a maximum when the peak voltage has its maximum value of $V_{\mathrm{CC}}/2$, that is:

$$\eta = \left(\frac{\pi}{2}\right)\left(\frac{V_{\mathrm{CC}}/2}{V_{\mathrm{CC}}}\right)\left(\frac{R_{\mathrm{L}}}{R_{\mathrm{L}} + R_{\mathrm{E}}}\right) = \left(\frac{\pi}{4}\right)\left(\frac{R_{\mathrm{L}}}{R_{\mathrm{L}} + R_{\mathrm{E}}}\right) \tag{16.35}$$

The efficiency reaches a value of 78.5% if $R_{\mathrm{E}} = 0$ which is the theoretical maximum value for the class B amplifier.

▶ **EXAMPLE 16.10** ―――――――――――――――――――――――――――――

Determine the current through the diodes and the voltage at the bases and emitters of the transistors for the circuit shown in Figure 16.31. Assume a forward voltage drop for the diodes of 0.7 V.

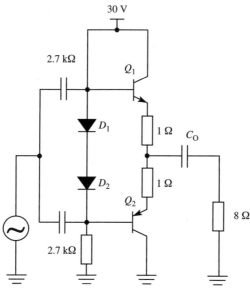

FIGURE 16.31   Class AB amplifier with single power supply.

▶ *Solution*

The current through the base-bias circuit is:

$$I_{\text{Bias}} = \frac{V_{\text{CC}} - V_{\text{D1}} - V_{\text{D2}}}{R_1 + R_2} = \frac{28.6 \text{ V}}{5.4 \text{ k}\Omega} \approx 5.3 \text{ mA}$$

The voltage at the base of Q_1 is:

$$V_{\text{B1}} = V_{\text{CC}} - I_{\text{Bias}} R_1 = 30 \text{ V} - (5.3 \text{ mA})(2.7 \text{ k}\Omega) \approx 15.7 \text{ V}$$

and for Q_2 the base voltage is:

$$V_{\text{B2}} = I_{\text{Bias}} R_2 = (5.3 \text{ mA})(2.7 \text{ k}\Omega) = 14.3 \text{ V}$$

The emitter voltage for both transistors is:

$$V_{\text{E}} = V_{\text{B2}} + V_{\text{D1}} = 14.3 \text{ V} + 0.7 \text{ V} = 15 \text{ V}$$

or

$$V_{\text{E}} = V_{\text{B2}} - V_{\text{D2}} = 15.7 \text{ V} - 0.7 \text{ V} = 15 \text{ V}$$

The collector to emitter voltages are:

$$V_{\text{CE1}} = V_{\text{CE2}} = \frac{V_{\text{CC}}}{2} = 15 \text{ V}$$

▶ **EXAMPLE 16.11** ───────────────────────────────

Determine the peak values for the voltage and current in the load resistor for the circuit in Figure 16.31 if the input signal has a peak value of 10 V and thus determine the efficiency.

Determine the value of the load coupling capacitor if the lowest frequency is 20 Hz.

▶ *Solution*

The peak output voltage is approximately equal to the peak input voltage since each transistor is acting as an emitter follower. Thus:

$$v_{\text{pL}} = v_{\text{pin}} = 10 \text{ V}$$

The peak current in the load is:

$$i_{\text{pL}} = \frac{v_{\text{pL}}}{R_{\text{L}} + R_{\text{E}}} = \frac{10 \text{ V}}{9 \Omega} \approx 1.1 \text{ A}$$

The power dissipated in the load is:

$$P_{\text{ac}} = \frac{i_{\text{pL}}^2}{2} R_{\text{L}} = \frac{(1.1 \text{ A})^2}{2} 8 \Omega = 4.84 \text{ W}$$

The average current taken from the supply is:

$$i_{CC(average)} = \frac{1}{\pi} i_{pL} = \frac{1.1 \text{ A}}{\pi} = 0.35 \text{ A}$$

and the average power provided by the supply is:

$$P_{CC} = (0.35 \text{ A})(30 \text{ V}) = 10.5 \text{ W}$$

The efficiency is:

$$\eta = \frac{P_{ac}}{P_{CC}} \times 100\% = \frac{4.84 \text{ W}}{10.5 \text{ W}} \times 100\% = 46\%$$

The value of the coupling capacitor is obtained from equation 16.31 as:

$$C_O = \frac{1}{2\pi(8 \text{ }\Omega + 1 \text{ }\Omega)(20 \text{ Hz})} = 884 \text{ }\mu\text{F}$$

The value of capacitor obtained for the above example illustrates a major disadvantage of using a single power supply for a class B amplifier, namely the very large value of the coupling capacitor required to connect the load in order to achieve good low-frequency response. A low-frequency response of 20 Hz is typical for an audio amplifier.

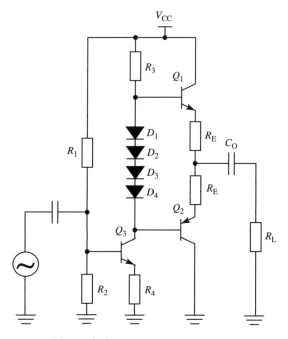

FIGURE 16.32   Class AB amplifier with driver.

Class AB Driver

In the examples for the class AB amplifiers above capacitors have been used to connect the signal source to the output stage. This is not ideal, particularly for low frequencies when large values of capacitor are necessary, and a more suitable method is direct coupling, as shown in Figure 16.32.

The transistor Q_3 is a common-emitter amplifier with R_3 acting as the collector load. Resistors R_1, R_2 and R_4 provide the dc bias for Q_3 which also establishes the correct operating point for the complementary output stage. Voltage gain is usually not important for an output stage, as this is best achieved at lower signal levels in preceding stages. Thus the emitter resistor R_4 is not provided with a decoupling capacitor and the gain of the driver is approximately $-R_3/R_4$.

▶ **EXAMPLE 16.12** ───

Determine the quiescent currents and voltages and the overall gain for the circuit shown in Figure 16.33. Assume that $\beta_{1-3} = 100$ and $\beta_{4-5} = 50$ and all the $V_{BE} = 0.7$ V. Estimate the peak output voltage across the load and hence the maximum power output. Verify the values with PSpice using transistors and diodes from the breakout library and an input signal of 1 kHz with an amplitude of 1 V.

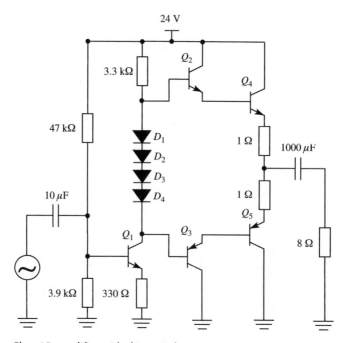

24 V

Q_2

3.3 kΩ

Q_4

47 kΩ

D_1
D_2
D_3
D_4

1 Ω

1000 μF

10 μF

1 Ω

Q_5

Q_1

Q_3

8 Ω

3.9 kΩ   330 Ω

FIGURE 16.33   Class AB amplifier with driver and gain stage.

▷ *Solution*

The collector current for Q_1 is:

$$I_{C1Q} = \frac{1.84 \text{ V} - 0.7 \text{ V}}{330 \text{ }\Omega + (3.6 \text{ k}\Omega/100)} = 3.1 \text{ mA}$$

and

$$r_{e1} = \frac{0.026}{3.1 \text{ mA}} \approx 8.4 \text{ }\Omega$$

The voltage at the base of Q_2 (ignoring the effect if I_{B2}) is:

$$V_{B2} = 24 \text{ V} - (3.1 \text{ mA})(3.3 \text{ k}\Omega) \approx 13.8 \text{ V}$$

while the voltage at the base of Q_3 is:

$$V_{B3} = 13.8 \text{ V} - (4 \times 0.7 \text{ V}) = 11 \text{ V}$$

The gain of Q_1 is given approximately by $R_{C1}/(r_{e1} + R_{E1})$ and is:

$$A_{v2} = \frac{-3.3 \text{ k}\Omega}{8.4 \text{ }\Omega + 330 \text{ }\Omega} \approx -9.8$$

The gain of the class AB output stage is approximately unity so that the overall gain is -9.8.

The maximum power output is determined by the maximum voltage swing across R_{C2}. For a positive-going swing it is:

$$v_{RC1\,(max\,+)} = 24 \text{ V} - 13.8 \text{ V} = 10.2 \text{ V}$$

and for a negative-going swing it is:

$$v_{CE1\,(max\,-)} = 11 \text{ V} - (3.1 \text{ mA})(330 \text{ }\Omega) \approx 10 \text{ V}$$

Thus the output voltage swing is almost symmetrical at 10 V. The maximum power across the 8 ohm load is:

$$P_{ac(max)} = \frac{10 \text{ V}^2}{2} \frac{1}{8 \text{ }\Omega} = 6.25 \text{ W}$$

From the PSpice simulation with a peak input signal of 1 V the following information is obtained:

$$I_{C1} = 2.7 \text{ mA}$$
$$V_{B2} = 14.7 \text{ V}$$
$$V_{B3} = 12 \text{ V}$$
$$A_{vtotal} = -7.4$$
$$P_{ac} = 3.4 \text{ W}$$

Short-Circuit Protection

A problem with any output stage, and particularly the class B and class AB stages, is a short-circuited load. In Figure 16.24, for instance, if the load is short-circuited, then the full supply voltage is applied to each of the transistors. Depending on the amount of base current which can be drawn, the collector current may exceed the maximum rating and the transistor is destroyed. The presence of emitter feedback resistors may limit the current, but typically these resistors are only a few ohms in value. For the single-supply output stage shown in Figure 16.30, the coupling capacitor C_O provides protection, but for the dual-supply output there is a direct dc connection to ground in the event of a short. Protection is provided with the addition of another transistor, as shown in Figure 16.34.

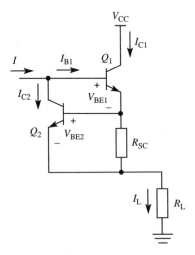

FIGURE 16.34   Short-circuit protection for one-half of a class B output stage.

Under normal operation the voltage across R_{SC} is less than the turn-on voltage (~ 0.6 V) for Q_2, and this transistor does not conduct. The load current I_L is simply:

$$I_L = I_{C1} = \beta_1 I_{B1} = \beta_1 I$$

which represents normal operation. If the current through the load increases, then the voltage drop across R_{SC} increases. When this voltage reaches the turn-on voltage for Q_2, then Q_2 starts to conduct. The current I must now supply both I_{B1} and I_{C2}, and because it is limited in value, then the base current to Q_1 is limited as more current flows through Q_2. The maximum load current is:

$$I_{Lmax} = \frac{V_{BE2(on)}}{R_{SC}} \approx \frac{0.6 \text{ V}}{R_{SC}} \tag{16.36}$$

▶ **EXAMPLE 16.13** ————————————————————————————————

Estimate the value of R_{SC} to provide protection for a 50 W audio amplifier driving an 8 ohm loudspeaker.

▶ *Solution*

The peak load current is:

$$I_{Lp} = \sqrt{\frac{2P_{ac}}{R_L}} = 3.5 \text{ A}$$

If 3.5 A is to be the maximum permitted current then the value of R_{SC} for short-circuit protection is:

$$R_{SC} = \frac{0.6 \text{ V}}{3.5 \text{ A}} = 0.17 \ \Omega$$

A similar circuit is required to protect the pnp transistor when two power supplies are used.

16.4   Class C Amplifier

A class C amplifier is one in which the load current flows for less than half a cycle. For an amplifier with a resistive load this would result in severe distortion, but if the load is a tuned *LC* resonant circuit, then an undistorted sine wave can be obtained. The mechanical analogy of the class C amplifier is a pendulum of a clock. The pendulum will continue to oscillate to and fro if a small impulse is applied during each swing. The impulse is only applied for a very small part of the complete mechanical swing. In the same way the current in a high-*Q* tuned *LC* network will continue to flow back and forth between the capacitor and the inductor if a small impulse is applied at regular intervals which are synchronized with the natural frequency of the network.

A tuned class C amplifier is shown in Figure 16.35. The input is a sine wave which has a frequency equal to the natural frequency of the tuned circuit which acts as a collector load. A negative base bias V_{BB} is applied by means of a radio frequency choke (rfc) which has a very high impedance at the frequency of the input signal. The bias is adjusted so that the transistor only conducts for a small fraction of the positive input cycle.

The efficiency of the class C amplifier can approach 90%, because power is only drawn from the dc supply for a small fraction of the ac cycle. The class C amplifier is used primarily for high-power, high-frequency applications such as in radio frequency transmitters. The design of class C amplifiers is complex and is more appropriately covered in a course on communication theory.

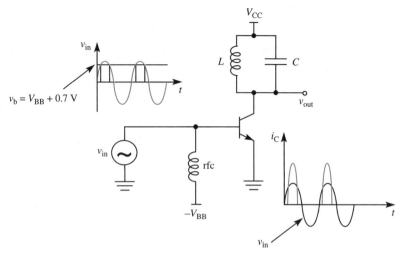

FIGURE 16.35   Class C amplifier with an *LC* tuned circuit as a load.

16.5   Class D Amplifier

The class D amplifier is even more non-linear than the class C in its mode of operation. It is an amplifier in which the output is switched ON and OFF at a variable rate determined by the input signal. The transistor acts as a switch and consequently does not operate in its linear region, except when it is changing from one state to another. Consequently the quiescent power dissipation is zero.

In the class D amplifier the input signal is converted into a series of pulses having widths which correspond to the amplitude of the input. For low-level signals the pulses are narrow, while for high-level signals the pulses are wide. The basic components of a *pulse-width modulator* are shown in Figure 16.36. A saw-tooth wave is applied to one terminal of a comparator and the input signal to the other terminal. The comparator output switches between high and low values depending on whether the saw-tooth is greater or less than the input signal. As the level of the input signal rises the width of the output pulses increases. As the level of the input signal falls the width of the output pulses decreases, as shown in Figure 16.36(b). The peak-to-peak amplitude of the saw-tooth must exceed the maximum amplitude of the input wave.

The average power of the output wave changes as the input signal varies. Thus the average is high for maximum amplitude and low for low input signal levels. The output transducer must only respond to the average levels and not the instantaneous changes caused by the switching of the comparator. This can be achieved by ensuring that the frequency of the saw-tooth wave is at least 10 times greater than the maximum frequency of the input signal.

The output from the pulse-width modulator is used to drive the class D output stage. The most suitable output device is a MOSFET. The device must be able to switch large currents ON and OFF at high frequencies. Bipolar transistors suffer from charge-storage

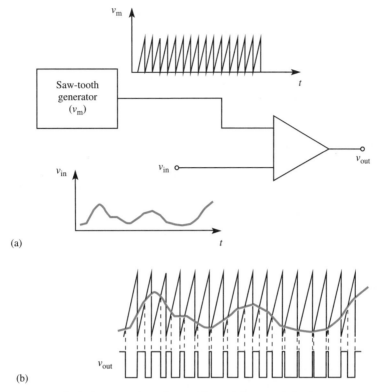

FIGURE 16.36   (a) Block diagram of a pulse-width modulator and (b) the output wave.

effects and are not ideally suited to switching operations when large currents are involved. The FET, in which only one carrier type is involved in conducting current, does not suffer from charge storage, and the MOSFET with its very high input impedance is ideally suited to this mode of operation. A development of the MOSFET manufacturing process has resulted in the VMOS. This device is ideally suited to high-speed switching of large currents. An attractive feature of the VMOS is that it has a negative temperature coefficient. Thus if the temperature rises the current is reduced, which prevents thermal runaway. This allows VMOS transistors to be placed in parallel to increase power output. If one of the transistors takes more current its temperature rises, and the current is reduced. (For bipolar transistors which have a positive temperature coefficient, the transistor that takes more current is likely to be destroyed by thermal runaway.)

An example of a MOSFET output stage which is capable of switching large currents through a load is shown in Figure 16.37.

When the input is high Q_1 and Q_2 are turned ON, while the gate of Q_3 is low as a result of the voltage drop across the drain resistor of Q_1 and is OFF. The output terminal is effectively at ground and there is no current through the load. When the input goes low, Q_1 and Q_2 are turned OFF. The voltage at the gate of Q_3 rises turning this

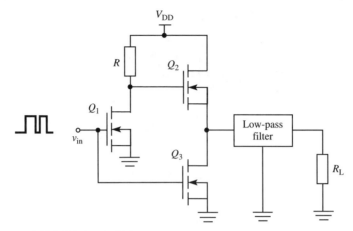

FIGURE 16.37   MOSFET class D high-current output stage.

transistor ON. Current now flows through the load. The advantage of this form of output, which is known as a *totem-pole output*, is that the load is driven from the transistor for both ON and OFF cycles.

In order to recover the original input signal the class D amplifier requires a low-pass filter to remove the switching waveform, as shown in Figure 16.37. The design of the filter is complicated by the fact that it must not attenuate the large load currents. The design is usually based on an *LC* low-pass filter configuration.

The main advantage of the class D amplifier is improved efficiency which can approach 100%. The main disadvantages are the need for very good low-pass filters and the fact that high-current switches can generate electromagnetic interference (EMI). All modern electronic circuits must comply with EMI regulations which limit the amount of interference produced by the operation of any piece of electrical equipment.

16.6   Transistor Power Dissipation

Transistors which are used in the output stages of power amplifiers are required to carry large currents and to dissipate many watts at the collector–base junction. This power dissipation heats the transistor and causes the temperature in the vicinity of the collector–base junction to rise. For silicon transistors the maximum permitted temperature is between 150°C to 200°C. Power transistors must be capable of dissipating this heat and are mounted in metal or plastic cases which are much larger than the cases used for small-signal transistors in order to improve heat dissipation. The process is further improved by attaching the case of the transistor to a heat sink. The heat sink may simply be the metal chassis of the instrument case or it may be a separate aluminium finned heat sink. The main purpose of the heat sink is to increase the surface area in order to improve the transfer of heat away from the transistor collector–base junction and into the ambient.

The analysis of heat flow makes use of a simple analogy of Ohm's law in which current is replaced by power, voltage by temperature, and electrical resistance by thermal resistance (θ). Thus for an electrical circuit:

$$V = IR$$

and for the thermal equivalent:

$$\Delta T = P_j \theta$$

where ΔT is the temperature difference between the junction and the surroundings, P_j is the power dissipated at the junction and θ is the thermal resistance between the junction and the case of the transistor in °C W^{-1}. It is a measure of the ease with which heat flows through a material, or across a boundary: the lower the value, the easier it is for the heat to flow. The thermal resistance depends on the method of construction of the transistor and the materials used, and is specified by the manufacturer.

The electrical analogy can be used to investigate heat flow across several boundaries, each with a different thermal resistance. For example, consider a transistor chip with a thermal resistance between the junction and the case θ_{jc}, attached to a case which has a thermal resistance to the air surrounding it of θ_{ca}. If the temperature of the junction is T_j, the temperature of the case T_c, and the temperature of the surrounding air is T_a, and if the power dissipated at the junction is P_j, then for the junction to the case the temperature difference is:

$$T_j - T_c = \theta_{jc} P_j \qquad (16.37)$$

and for the case to the ambient (surrounding air) the temperature difference is:

$$T_c - T_a = \theta_{ca} P_j \qquad (16.38)$$

The electrical analogy is shown schematically in Figure 16.38(b).

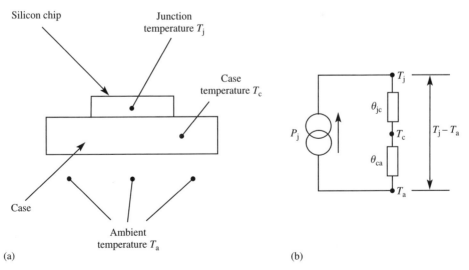

(a)

(b)

FIGURE 16.38 Thermal analog of (a) the mounted transistor chip and (b) the electrical analog.

From the electrical analogy and equations 16.37 and 16.38 the junction temperature can be expressed as:

$$T_j = T_a + (\theta_{jc} + \theta_{ca})P_j \tag{16.39}$$

The analogy can be extended to include a heat sink by including a thermal resistance for the flow of heat from the case of the transistor to the heat sink (θ_{cs}) and a thermal resistance for the flow of heat from the heat sink to the ambient (θ_{sa}). Then the temperature of the junction is:

$$T_j = T_a + (\theta_{jc} + \theta_{cs} + \theta_{sa})P_j \tag{16.40}$$

▷ **EXAMPLE 16.14** ────────────────────────────────

A transistor has a power rating of 2 W. The thermal resistance from the junction to the case is 5°C W^{-1}, for the case to a heat sink it is 2°C W^{-1} and for the heat sink to the ambient it is 10°C W^{-1}. Determine the junction temperature, case temperature and the temperature of the heat sink if the ambient temperature is 25°C.

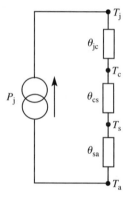

▷ *Solution*

The electrical analog is as shown. The power flow is constant through each of the thermal regions, that is the chip, the case and the heat sink. Therefore, by equation 16.40 the junction temperature is:

$$T_j = 25 + (10 \times 2) + (2 \times 2) + (5 \times 2) = 59°C$$

The temperature of the case is:

$$T_c = 25 + (10 \times 2) + (2 \times 2) = 49°C$$

and the temperature of the heat sink is:

$$T_s = 25 + (10 \times 2) = 45°C$$

To improve the flow of heat away from the transistor it is usual to use a silicone grease between the contact surface of the transistor case and the heat sink. The manufacturers of this material and of heat sinks quote the *thermal conductivity k* (W m^{-1}°C^{-1}) and the thermal resistance is obtained as:

$$\theta = \frac{1}{k} \frac{t}{area} \text{ °C W}^{-1} \tag{16.41}$$

where t is the thickness. For example, mica is often used to provide electrical insulation between the transistor case, which is the collector for power transistors, and the heat sink, which may be at ground potential. The thermal conductivity of mica is 0.6 W m^{-1}°C^{-1}. The thermal resistance of a mica washer 50 μm thick with an area of 2 cm^2 is:

$$\theta = \frac{1}{0.6} \frac{50 \times 10^{-6}}{2 \times 10^{-4}} \approx 0.4 \text{°C W}^{-1}$$

The need for a heat sink is determined by estimating the junction temperature for the particular operating conditions. The temperature of the junction depends on the rate at which heat is removed, and this in turn depends on the thermal resistance of the attached heat sink.

▷ **EXAMPLE 16.15** ─────────────────────────

A power transistor is to operate with a power dissipation of 20 W in an ambient temperature of 50°C and the junction temperature must not exceed 150°C. The thermal resistance of the junction to case (θ_{jc}) is 1°C W^{-1} and for the case to ambient (θ_{ca}) 5°C W^{-1}. Determine the junction temperature without a heat sink and determine the thermal resistance necessary for a heat sink to avoid exceeding the maximum junction temperature. Assume that a mica washer is used with a thermal resistance of 0.5°C W^{-1}.

▷ *Solution*

The junction temperature without a heat sink is:

$T_j = 50 + (1 \times 20) + (5 \times 20) = 170$°C

As the temperature exceeds 150°C a heat sink is necessary. The thermal equation, including the mica washer, is:

$T_{j(max)} = 150 = 50 + (1 \times 20) + (0.5 \times 20) + (\theta_H \times 20)$

where the thermal resistance of the case has been replaced by the thermal resistance of the mica washer and θ_H is the thermal resistance for the heat sink. Solving for θ_H gives:

$\theta_H = (150 - 50 - 20 - 10)/20 = 3.5$°C W^{-1}

A heat sink is required which has a thermal resistance of 3.5°C W^{-1}. The manufacturer's data sheets would be examined to find one which closely matches this figure.

Derating Curve

Manufacturers specify a maximum permissible power for their transistors at certain ambient temperatures or case temperatures; for example, a transistor may be rated at 15 W for an ambient temperature of 25°C. Provided that the temperature of the ambient does not exceed 25°C then the transistor may dissipate 15 W. However, if the ambient temperature increases, then the power dissipation must be reduced. The manufacturer provides a derating curve of power versus temperature. There is a linear decrease of power above the normal ambient temperature, as shown in Figure 16.39.

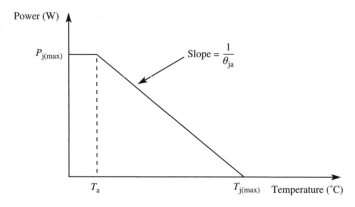

FIGURE 16.39   Derating curve.

The slope of the curve beyond the critical ambient temperature is the reciprocal of the thermal resistance for the junction to ambient. That is:

$$\text{slope} = \frac{P_{j(max)}}{T_{j(max)} - T_a} = \frac{1}{\theta_{ja}}$$

or

$$P_{j(max)} = \frac{T_{j(max)} - T_a}{\theta_{ja}} \tag{16.42}$$

In the absence of a derating curve then this equation can be used to determine the maximum permitted power when the ambient temperature exceeds 25°C.

▶ **EXAMPLE 16.16** ─────────────────────────────────────

A transistor is rated at 12 W for an ambient temperature of 25°C. The thermal resistance of the junction to case is 2.5°C W^{-1} and for the case to ambient 8°C W^{-1}. Determine the power dissipation if the ambient temperature rises to 70°C.

▷ *Solution*

At 25°C the junction temperature is:

$T_j = 25 + (2.5 + 8)12 = 151°C$

At an ambient temperature of 70°C the permitted power dissipation for a junction temperature of 151°C, from equation 16.42, is:

$$P_j = \frac{151 - 70}{2.5 + 8} = 7.7 \text{ W}$$

▷ **PROBLEMS** —————————————————————————————————

16.1 For the class A amplifier shown $\beta = 50$ and $V_{BE} = 0.8$ V. Determine the quiescent collector current and collector–emitter voltage and the peak load current for maximum symmetrical swing.

[44 mA, 13 V, 44 mA]

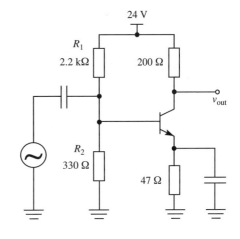

16.2 If R_2 is changed to 470 Ω determine the new values of quiescent current and voltages and the peak load current for maximum symmetrical swing.

[62 mA, 8.7 V, 43 mA]

16.3 Determine the power dissipated in the load and the transistor for each value of R_2 (330 Ω, 470 Ω) and the power conversion efficiency.

[194 mW, 570 mW, 18.3%]
[185 mW, 540 mW, 12.4%]

16.4   Determine the value of R_2 to maximize the symmetrical peak-to-peak swing of the load current. Assume that $\beta = 50$ and $V_{BE} = 0.8$ V.

[147 Ω]

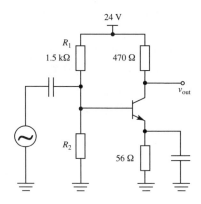

16.5   For the circuit shown assume that $\beta = 100$ and $V_{BE} = 0.7$ V. Determine the quiescent collector current and collector–emitter voltage and the peak collector current for symmetrical swing.

[7 mA, 8.5 V, —]

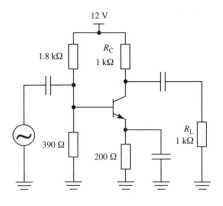

16.6   For the circuit in Problem 16.5 determine the peak load current for maximum symmetrical swing and hence determine the power dissipated in the load and the power conversion efficiency.

[3.5 mA, 7%]

16.7   Determine the peak current in the load for maximum symmetrical swing and the efficiency. Assume a value of 50 for β and $V_{BE} = 0.8$ V.

[53 mA, 21%]

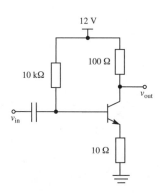

16.8 Select suitable values for R_B and R_E to enable a 2 V peak-to-peak sine wave to be produced across the 10 Ω load. To avoid distortion let $V_{CEQ} = 2$ V. Assume $\beta = 100$ and $V_{BE} = 0.8$ V. Use PSpice and a generic transistor to verify the design.

[$R_B = 270$ Ω, $R_E = 100$ Ω]

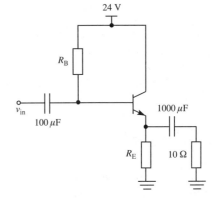

16.9 Determine the efficiency for the amplifier in Problem 16.8 and suggest why it is so much smaller than the value predicted by equation 16.18.

[0.9%]

16.10 Determine the power dissipated in the load, the transistor and also determine the efficiency for the circuit shown. Assume a value for β of 50 and V_{BE} of 0.8 V.

[0.64 W, 10 W, 4%]

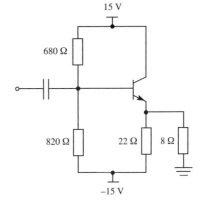

16.11 For the class B amplifier determine the output power, the supply power and the power dissipated in each transistor if the output is a sine wave with a peak amplitude of 10 V. (Ignore the presence of cross-over distortion.)

[5 W, 6.37 W, 1.37 W]

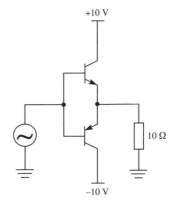

16.12 Repeat the power calculations for Problem 16.11 if the output has peak amplitudes of 8 V, 6 V, 4 V and 2 V. Plot the variation of transistor power dissipation versus output peak amplitude and compare the result with Figure 16.22.

16.13 Determine the maximum peak value of the input voltage for the circuit shown, the power delivered to the load, power dissipated in the transistors and the maximum breakdown voltage rating for each transistor.

[24 W, 6.6 W, —]

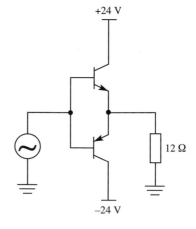

16.14 If the power output for the circuit in Problem 16.13 is to be increased by 40%, determine the new value of the supply voltages and the minimum value of the breakdown voltages for the transistors.

[28.4 V, —]

16.15 Determine the voltage between the bases of Q_1 and Q_2 in the circuit shown and the peak base current which can be supplied to each transistor.

[1.43 V, 30.4 mA]

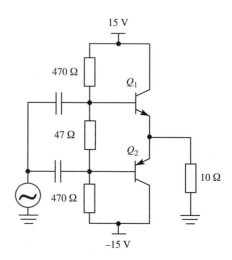

16.16 If for the circuit in Problem 16.15 $\beta = 50$ determine the maximum power in the load for an undistorted output.

[11.5 W]

16.17 Design a class AB output stage based on the circuit in Figure 16.24 to produce 20 W into an 8 ohm loudspeaker with ± 30 V supplies. Assume $\beta = 50$ and $V_{BE} = 0.8$ V. Use PSpice to verify the design.

[The required bias resistor $R \approx 250\ \Omega$

16.18 For the circuit shown $\beta_1 = \beta_3 = 100$, $\beta_2 = \beta_4 = 50$, $V_{BE1} = V_{BE3} = 0.7$ V, and $V_{BE2} = V_{BE4} = 0.8$ V. Determine the quiescent current in the diodes and the peak collector current in the transistor pairs Q_1, Q_3 and Q_2, Q_4 when the input is a sinusoid with a peak value of 22 V. What is the peak voltage change across the 1 kΩ bias resistors? Use PSpice to verify the results.

[22 mA, 44 mA, 2.2 A, 0.4 V]

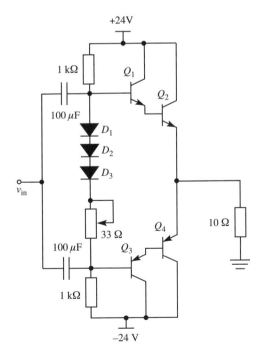

16.19 If, for the circuit shown, the peak input voltage is 11 V, determine the power dissipated in the transistors and the overall efficiency. Determine the minimum corner frequency.

[3.4 W, 63%, 18 Hz]

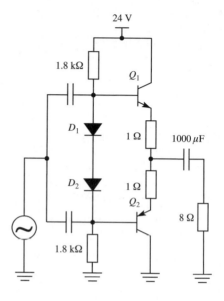

16.20   For the circuit shown assume that $V_{BE1} = 0.7$ V, $V_{BE2} = V_{BE3} = 0.8$ V and $\beta_1 = 100$. Determine the collector current for Q_1 and the dc voltage at node A.

[3 mA, 11.5 V]

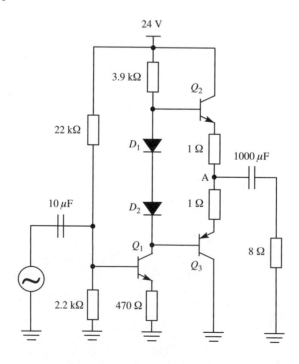

16.21 The dc voltage value of 11.5 V at node A in the circuit shown in Problem 16.20 would suggest a maximum ac voltage swing across the load of 8–10 V, but if $\beta_2 = \beta_3 = 50$ use PSpice to show that the maximum output is only about 800 mV for an input of about 1.2 V before the output is seriously distorted. Why is this?

16.22 For the circuit shown in Problem 16.20 increase β_2 and β_3 to 5000 (equivalent to a Darlington connection) and repeat the PSpice exercise of Problem 16.21 to show that the output voltage can be increased to about 8 V for the same input voltage of 1.2 V peak. Does this modification provide the answer for Problem 16.21?

16.23 Draw the circuit diagram of a class C rf output stage and sketch the collector current waveform with respect to a sine wave input.

16.24 Sketch a block diagram of a class D amplifier and identify the practical problems associated with such an amplifier.

16.25 A transistor dissipates 15 W. If the thermal resistance of the junction to case is $1°C\ W^{-1}$ and case to ambient is $5°C\ W^{-1}$, determine the junction temperature and case temperature when operating in an ambient of 70°C.

[$T_j = 160°C$, $T_c = 145°C$]

16.26 A transistor dissipates 50 W in an ambient of 60°C. The thermal resistances are $\theta_{jc} = 0.5°C\ W^{-1}$, $\theta_{ca} = 4°C\ W^{-1}$. Determine the junction temperature without a heat sink. Determine the thermal resistance of a heat sink to avoid the junction temperature exceeding 170°C.

[285°C, $1.7°C\ W^{-1}$]

16.27 A power transistor is rated at 20 W with a case temperature of 50°C and a maximum junction temperature of 160°C. Determine the power dissipation if the case temperature is 100°C.

[11 W]

16.28 If the transistor of Problem 16.27 is mounted on a heat sink for which the thermal resistance, θ_{sa}, is $10°C\ W^{-1}$ and which results in a thermal resistance from case to heat sink, θ_{cs}, of $0.4°C\ W^{-1}$, determine the maximum allowable power dissipation if the ambient is 25°C.

[8.5 W]

16.29 If a mica washer is placed between the transistor and heat sink described in Problem 16.28 and has a thermal resistance of $1°C\ W^{-1}$, determine the maximum allowable power dissipation for the same ambient.

[8 W]

16.30 The manufacturer's data for heat sinks quote thermal resistance values of $13°C\ W^{-1}$, $5.4°C\ W^{-1}$ and $1.3°C\ W^{-1}$. Determine the power dissipation for the transistor of Problem 16.27 if it is mounted on each of the heat sinks with a mica washer and in an ambient of 40°C.

[6 W, 9.7 W, 14.6 W]

CHAPTER 17

Analog/Digital Interface

Many analog electronic circuits are used to process low-level signals which may be derived from transducers (microphone, temperature sensor, strain gauge, etc.) or from within a communication system (telephone, radio, television, radar) or within an instrumentation system (voltmeter, oscilloscope, signal generator). Analog circuits are used to provide gain, frequency selection and to generate waveforms. The output generated can provide an end result in the form of signals to drive a loudspeaker, signals to drive a meter, an oscilloscope trace, a relay or an electric motor. However, an alternative for many electronic designs is to convert the analog signal into a digital signal at a fairly early stage and to complete the processing of the signal by means of digital circuitry. Many more complex operations may be achieved with digital signals than with analog signals and the results can be presented in a more easily understood manner; for example, the liquid crystal display of the modern digital voltmeter has now completely replaced the moving-coil meter. Digital signals are more easily transmitted over long distances without any deterioration, unlike analog signals which become corrupted with electrical noise, and thus digital communication systems are replacing analog communication systems. Modern instruments such as oscilloscopes are now able to provide digital outputs of voltage levels and time intervals on liquid crystal displays.

It may appear that analog circuitry is no longer required and that all circuits should be digital, but digital circuits cannot handle very small signal levels and analog circuits are required to provide gain. For many operations a simple analog circuit is often more efficient than the equivalent digital circuit; for example, the operations of multiplication and frequency selection are more easily achieved with analog circuits than with their digital equivalents. Very high-frequency signals are more easily handled with analog circuits rather than digital, and in fact to appreciate fully the operation of digital circuits at frequencies in excess of 100 MHz it is necessary to resort to analog circuit analysis. There will always be a need for analog circuits, but there is also a need to provide an interface between analog and digital.

Analog-to-digital (A/D) converters are now available as integrated circuits and it is not the intention of this chapter to consider the design of such converters in any detail, but rather to consider the different methods of conversion and relative merits of each method. While A/D converters provide outputs which are suitable for processing with a microcomputer, there are other forms of conversion which still produce discrete signals but not necessarily in the form of a binary code; for example, voltage-to-frequency and pulse-width modulation are two examples of a continuously varying signal being converted into a fixed amplitude signal but with time-varying intervals between pulses.

A more complex circuit which combines analog and discrete signal methods is *the phase-locked loop* or PLL. The PLL is used extensively to extract signals which may be buried in other signals, for frequency synthesis and for frequency multiplication. Finally it is also necessary to consider how digital signals are converted into analog signals.

17.1   Analog-to-Digital Conversion

Analog signals vary continuously, as illustrated in Figure 17.1(a) which shows the output of an analog circuit for some arbitrary analog signal. The signal varies both in amplitude and in rate of change of amplitude with time. Some means is required to convert the continuously varying amplitude into a digital signal which has a fixed amplitude, but which is coded into groups of pulses, as shown in Figure 17.1(b) where the input is assumed to be an analog signal and where the output is digital with 4 bits. The four output lines D_0, D_1, D_2 and D_3 are coded at discrete time intervals t_0, t_1, t_2, etc., as shown in Figure 17.1(b). In Figure 17.1(c) the coded digital data on these lines is shown in tabular form as 4 bit binary words.

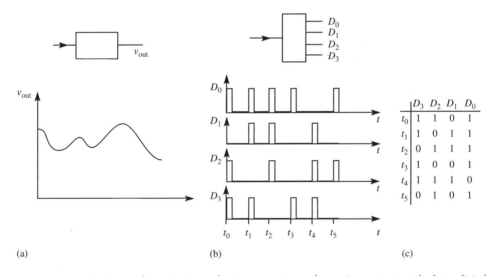

	D_3	D_2	D_1	D_0
t_0	1	1	0	1
t_1	1	0	1	1
t_2	0	1	1	1
t_3	1	0	0	1
t_4	1	1	1	0
t_5	0	1	0	1

(a)   (b)   (c)

FIGURE 17.1   (a) An analog circuit producing a continuously varying output with (b) a digital circuit having four outputs, and (c) the tabular form of the digital output.

In order to change the analog signal into a digital one, some means is required for sampling the analog signal at discrete time intervals corresponding to t_0, t_1, t_2, etc., shown in Figure 17.1(b). A possible scheme is illustrated in Figure 17.2.

The analog signal is sampled at discrete time intervals by a sampling circuit which effectively takes a 'snap-shot' of the analog signal each time a clock pulse is applied. The output from the sampling circuit increases and decreases in discrete steps, as shown in the diagram. Thus for a short period of time, corresponding to the period between the

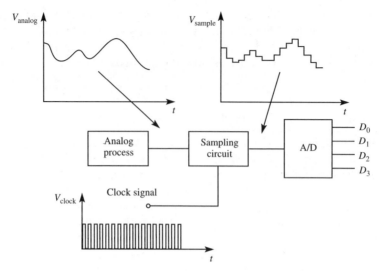

FIGURE 17.2   Basic requirements for an A/D converter.

clock pulses, the signal applied to the A/D converter is of a constant amplitude and the conversion process can take place and a coded output – $D_3D_2D_1D_0$ – produced for each discrete signal level obtained from the sampling circuit. The frequency of the clock is determined by the maximum rate of change of the analog signal, or its maximum frequency component. The clock frequency needs to be greater than the maximum frequency of the analog signal to ensure that no information is lost in the conversion process.

The sampling circuit is known as a *sample-and-hold* circuit and will be considered below. There are a number of A/D converter methods and four will be considered here: counting A/D, successive approximation, dual-slope and parallel-comparator or flash converter.

Sample-and-Hold

An analog circuit produces a signal which varies continuously, but the very nature of digital circuits is such that they operate in discrete time steps determined by an oscillator which generates a clock frequency. The digital signal only changes in time with the clock signal, either for each clock pulse or for some multiple. In order to provide an accurate recording of the analog signal some means is required to 'freeze' the analog signal while the digital circuitry converts the analog signal into a digital code. The sample-and-hold performs this task.

The sample-and-hold circuit in its simplest form is a switch and a capacitor, as shown in Figure 17.3(a).

The switch is operated by a clock signal which closes the switch briefly at discrete intervals. When the switch is closed the analog voltage is applied to the capacitor and the voltage across C takes a value equal to V_{analog}. The switch only remains closed for a

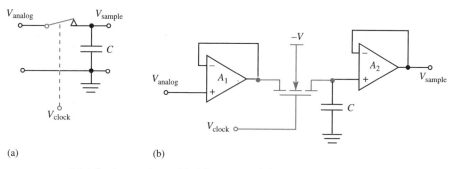

FIGURE 17.3   (a) A basic sample-and-hold circuit and (b) a practical realization.

short time and then opens. When the switch opens the voltage across C remains essentially constant and allows the A/D conversion process to take place with a constant input voltage applied to the A/D converter. The time during which the switch is closed is called the *sample interval*, which is very much shorter than the reciprocal of the highest frequency of the analog signal so that during the sampling interval the amplitude of the sample voltage may be considered to be essentially constant. The time when the switch is open is called the *hold interval*, which is also less than the reciprocal of the highest signal frequency, but only by a factor of between 2 and 10. There must be sufficient samples to ensure that the sampled waveform V_{sample} which is applied to the A/D converter bears a close resemblance to the original V_{analog}, as shown in Figure 17.2, and a communications sampling theorem (the Nyquist theorem) states that the sampling frequency should be at least twice the frequency of the signal being sampled to ensure that the content of the original signal is not lost.

A practical realization of a sample-and-hold circuit is shown in Figure 17.3(b). The switch is an enhancement-mode MOSFET with the clock waveform applied to the gate. These devices are capable of rapid switching and ensure that the sampling period can be short. They also have a low ON resistance. It is important that the capacitor is fully charged during the sampling period, and for this to happen, the output resistance for the analog source, and the ON resistance of the switch, must be low. A low output resistance is achieved with the unity gain buffer amplifier A_1. The buffer must also be able to supply adequate charging current $I = C\, dV/dt$ and must have sufficient slew rate to follow the input signal. During the hold period the capacitor must retain its voltage and the leakage current must be small, that is the rate of change of voltage is small ($dV/dt = I_{\text{leakage}}/C$). This is again achieved with a unity gain buffer which has a very high input resistance. The OFF resistance of the MOSFET switch is also very large, which minimizes the leakage current from the capacitor. Note that the operational amplifiers have a finite switching speed determined by their frequency response and slew rate.

Counting A/D Converter

The digital output from the A/D converter must represent the different voltage levels of the analog input signal. The most convenient binary code for the output is binary-coded decimal (BCD) and for a 4 bit code there are 16 possible binary codes. Thus a 4 bit A/D converter should be capable of distinguishing 16 different input voltage levels. One approach to designing an A/D converter is to generate BCD code by means of a digital counter, to convert the code back into an analog form with a digital-to-analog (D/A) converter, and to make a comparison, using a comparator, with the actual analog input. When the two voltages (the actual analog voltage, and the simulated voltage from the D/A converter) are equal, then that particular BCD code is transferred to the output of the A/D converter. The basic circuit is shown in Figure 17.4.

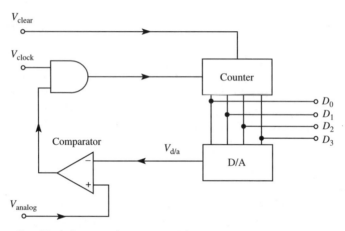

FIGURE 17.4   Simplified diagram of a counting A/D converter.

Clock pulses are applied to a counter through an AND gate. At the start of each sampling period a CLEAR pulse is applied to the counter so that the count starts from zero each time. The output from the counter is applied to a D/A converter so that its output is proportional to the count. At the start of each count the voltage from the D/A converter ($V_{d/a}$) is less than the analog input to the comparator and the output from the comparator is $+V$ and represents a '1' which is applied to the AND gate. As the count proceeds the $V_{d/a}$ increases until it exceeds V_{analog}, at which point the output from the comparator drops to 0 V which, when applied to the AND gate, stops the clock pulses reaching the counter. The count represented by $D_3D_2D_1D_0$ is the digital output for this particular value of analog voltage. For a 4 bit counter 1111 represents 15 increments of the input voltage. It starts at 0000 and 15 clock pulses are required to reach 1111. Thus for an input of 0 V to 5 V each count represents $(5 V)/15 = 0.333$ V. A smaller increment is obtained by using a greater number of bits. Thus an 8 bit counter would provide 256 increments.

A disadvantage of the counting A/D converter shown in Figure 17.4 is that the counter and $V_{d/a}$ are reset to zero for each measurement. Because the analog input is being sampled, the time required to measure each sample places a limit on the speed of the A/D converter. Hence, if the analog input is close to its maximum value, then many clock pulses are required to bring $V_{d/a}$ up to the level of the analog input before a reading is obtained. This represents the worst-case *conversion time* which is the product of the maximum number of counts and the period of the clock. Thus for the 4 bit counter and a 20 kHz clock the worst-case conversion time is 75 µs ($1/20$ kHz × 15).

For many applications the analog input varies in a continuous manner, and may be increasing or decreasing, but from one sample to the next the difference between the two samples is not very great. It is a waste of time to have to reset $V_{d/a}$ to zero each time. The process can be speeded up by modifying the circuit, as shown in Figure 17.5.

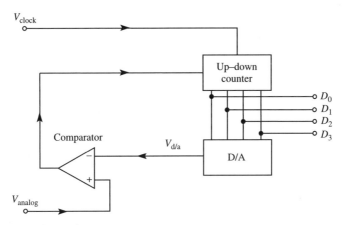

FIGURE 17.5   A tracking A/D converter.

The simple binary counter is now replaced by an up–down counter and the output from the comparator controls the direction of the count. If initially the count is zero, then for the first measurement the count increases until $V_{d/a} > V_{analog}$ and the comparator output changes and reverses the direction of the counter. When $V_{d/a} < V_{analog}$ the direction of the count is again changed. Thus the counter will track the analog input and this converter is known as a *tracking converter*. A tracking converter must be able to follow the variations of the input voltage and a worst-case conversion time might be defined as $dt = dV/S$, where dV is one voltage increment from the D/A and S is the maximum slew rate of the input voltage. The reciprocal of dt would provide the minimum value of the clock frequency. Thus for the 4 bit counter and an input voltage range of 4 V, a single increment is 267 mV. If the maximum slew rate is 0.01 V µs^{-1}, then dt is 26.7 µs or the clock frequency is 37 kHz.

Successive Approximation A/D Converter

In the successive approximation converter the counter is replaced by some logic circuitry which generates a binary code which is processed by the D/A converter and compared with the analog input. Depending on whether $V_{d/a}$ is greater or less than V_{analog} the logic circuitry makes changes to the code to bring about a match. The changes are made in a controlled manner so that by a successive number of approximations a match is obtained. A simplified schematic is shown in Figure 17.6.

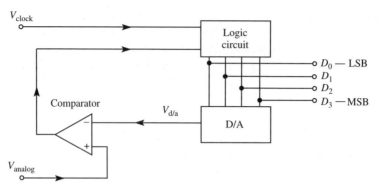

FIGURE 17.6   A simplified diagram of a successive approximation A/D converter.

At the start of a measurement the most significant bit (MSB) is set to a 1 and the other bits are set to 0. An output is generated by the D/A converter and compared with the analog input. If $V_{d/a} > V_{analog}$ then the number 1000 is too large and the MSB should be 0. If $V_{d/a} < V_{analog}$ then the MSB is a 1. Hence after the first trial the MSB has been selected. The process is repeated with D_2 and then D_1 and D_0.

The process of successive approximation is illustrated in the table below:

Trial	Code	$V_{d/a} > V_i$	$V_{d/a} < V_i$
First	1000	$0D_2D_1D_0$	$1D_2D_1D_0$
Second	#100	$00D_1D_0$	$11D_1D_0$
Third	##10	$000D_0$	$111D_0$

At the first trial the code is:

1000

This number is input to the D/A converter and the output $V_{d/a}$ is compared with the analog input. If $V_{d/a} > V_i$ then the number is too large and the MSB is 0. If $V_{d/a} < V_i$ then the MSB is left at 1. After the first trial the MSB is known and is shown in the table for the second trial as # (either a 1 or 0).

For the second trial the number applied to the D/A is:

#100

The comparison is repeated and the first two MSB bits can now be determined so that for the third trial the code is:

##10

The process is repeated until $V_{d/a} \approx V_i$. The degree of equality depends on the number of bits. For an 8 bit D/A the degree of inaccuracy between $V_{d/a}$ and V_i is $V_i/256$.

In the schematic the logic circuit is such that at each clock pulse a proper binary code is generated for the D/A converter. If the output from the comparator is a 1 then the bit being tested is unchanged. If it is a 0 then the bit under test is changed to a 0. When all bits are selected the correct code is passed to the output terminals of the A/D.

The conversion time with N bits is N clock pulses as opposed to a worst-case conversion time for the counting-type converter of 2^N.

Dual-Slope A/D Converter

In an earlier chapter it was shown that linear voltage ramps could be generated with an operational amplifier connected as an integrator where a capacitor is placed in the feedback path. The output voltage increases or decreases in a linear fashion at a controlled rate, depending on the polarity of the input dc source. The integrator provides a variable voltage which changes at a controlled rate of so many volts per second. If this voltage is applied to a comparator, the other input of which is the analog voltage to be measured, then the comparator output will change when the voltage from the integrator exceeds the analog voltage. If at the start of the measurement a binary counter is started and is stopped when the comparator output changes, the output from the counter is proportional to the analog input voltage. This *single-slope* converter is very simple but not very accurate, because of the demands placed on the stability of the integration process. The *dual-slope integrator* is a development of this simple conversion process.

A block diagram of the dual-slope A/D converter is shown in Figure 17.7(a).

The dual-slope converter comprises an integrator, a comparator and a counter, plus two electronic switches Sw_1 and Sw_2 which are controlled by logic circuitry. The logic circuits also control the counter, clearing it at the start of a measurement. At the start of a sequence the logic circuits clear the counter, set Sw_1 to read the input voltage (V_{analog}) and open switch Sw_2. Assuming that V_{analog} is positive and constant (it is obtained from the sample-and-hold circuit) then the output of the integrator starts to ramp down at a constant rate and after a time t is given by:

$$V_{int} = -\frac{1}{RC} \int_0^t V_{analog} \, dt = -\frac{V_{analog} t}{RC} \tag{17.1}$$

This voltage is applied to the comparator and with a reference voltage of 0 V the output voltage is positive. This voltage represents a 1 at the input of the AND gate and the clock pulses are transferred to the counter. The counter now counts clock pulses. The count continues until the MSB bit (D_3 for the 4 bit counter) changes to a 1. The next clock pulse will set all the bits to zero. (For the 4 bit counter this would represent 2^4 or 16 clock pulses.) At this instant, represented by a time T_1, the voltage at the output of the integrator is:

$$V_{int}(T_1) = -\frac{V_{analog} T_1}{RC} \tag{17.2}$$

The switch Sw_1 is now moved to connect a reference voltage $-V_{ref}$ to the input of the integrator. The integrator now integrates a known negative voltage and the output ramps up, as shown in Figure 17.7(b).

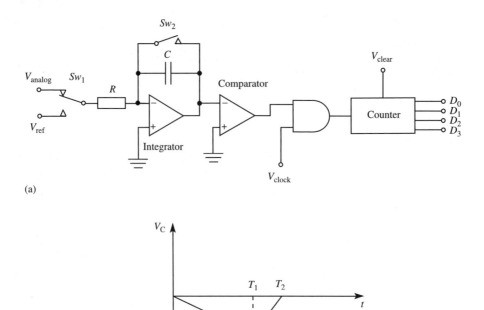

(a)

(b)

FIGURE 17.7 Dual-slope A/D converter with (a) the block diagram and (b) the output voltage from the integrator.

At the instant the switch Sw_1 is activated the initial voltage is given by equation 17.2. For $t > T_1$ the output voltage from the integrator is:

$$V_{int} = -\frac{V_{analog}T_1}{RC} - \frac{1}{RC}\int_{T_1}^{t} -V_{ref}\,dt = -\frac{V_{analog}T_1}{RC} - \frac{V_{ref}(t-T_1)}{RC}$$

When the output from the integrator becomes slightly positive the output from the comparator becomes negative, or zero, which deactivates the AND gate and prevents the clock pulses reaching the counter. At $t = T_1$ all the bits of the counter were zero. Thus the new count at T_2 represents the time $T_2 - T_1$. At T_2 the value of V_{int} is zero (or slightly positive) and from the above equation:

$$0 = -\frac{V_{analog}T_1}{RC} + \frac{V_{ref}(T_2-T_1)}{RC}$$

and with some manipulation:

$$T_2 - T_1 = \frac{V_{analog}}{V_{ref}}T_1 \tag{17.3}$$

The time T_1 is known because it represents the time required for the counter to cycle through all of its codes, that is for an N bit counter it represents 2^N counts. If the clock period is T_{clk} then T_1 is:

$$T_1 = 2^N T_{clk} \qquad (17.4)$$

and substituting into equation 17.3 gives:

$$V_{analog} = V_{ref} \frac{(T_2 - T_1)}{2^N T_{clk}} \qquad (17.5)$$

If during the interval $T_2 - T_1$ there are n clock pulses (nT_{clk}) then:

$$V_{analog} = V_{ref} \frac{n}{2^N} \qquad (17.6)$$

Thus the unknown voltage (V_{analog}) is proportional to the number of counts n. Note that the unknown voltage is independent of the RC product of the integrator and also the clock period (T_{clk}). Because the conversion is based on the measurement of number of clock pulses it is a simple matter to increase the accuracy by increasing the number of counts (use a higher clock frequency) and thus 16 bit accuracy is easily achieved.

A longer integration time is a distinct advantage for digital voltmeters because it cancels the 50 Hz or 60 Hz line voltage interference. In measuring a dc voltage it is very likely that there will also be a small component of the 50 Hz or 60 Hz line voltage superimposed on the dc voltage, as shown in Figure 17.8.

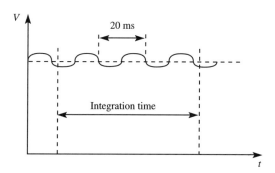

FIGURE 17.8   A dc voltage level with superimposed mains interference.

For a 50 Hz line voltage the period is 20 ms and if the integration time is 60 ms, then there will be three positive excursions and three negative excursions, and over the integration time there is a good probability that these positives and negatives will cancel out. If the integration time is not a multiple of the interference frequency, then the interference will not cancel and there will be a small error in the reading of the dc voltage.

The dual-slope A/D converter is widely used in digital voltmeters and often forms part of a single integrated circuit which contains the sample-and-hold, the dual-slope A/D and the logic circuitry for an LCD display. For example, the ICL7106 is a CMOS chip which only requires the resistors and capacitors for the integrator and for the internal clock, plus a display.

Parallel-Comparator or Flash A/D Converter

In the parallel-comparator converter the input is applied simultaneously to the inputs of a number of comparators, each of which has a different value of reference voltage. Thus depending on the value of the input voltage a certain number of the comparator outputs will change, while the remaining comparators will be unaffected. Logic is used to process the outputs to provide a binary-coded reading of the input voltage. A simplified schematic is shown in Figure 17.9.

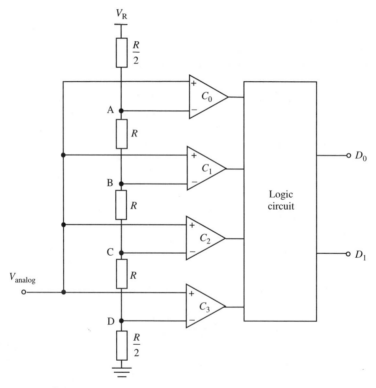

FIGURE 17.9 Parallel-comparator A/D converter.

The resistors form a potential divider which establishes known voltages at the inputs of each comparator of $7V_R/8$, $5V_R/8$, $3V_R/8$ and $V_R/8$ at A, B, C and D respectively. For the arrangement shown the comparator output will be a 1 if the input voltage is larger than the reference voltage and 0 otherwise. Thus if the outputs of C_0, C_1 and C_2 are 0 and the output of C_3 is 1 then the input voltage lies between $3V_R/8$ and $V_R/8$. The output from the comparators changes at discrete intervals of the input voltage, as determined by the voltages at different points on the potential divider. For the converter shown in Figure 17.8 with four comparators, the resolution is $V_R/4$. For eight comparators it is $V_R/8$.

The state of the four comparators shown in Figure 17.9 can be represented by a 2 bit binary code, as shown in Table 17.1.

TABLE 17.1   Truth table for the 2 bit A/D converter

V	C_3	C_2	C_1	C_0	D_1	D_0
$0-V_R/8$	0	0	0	1	0	0
$V_R/8-3V_R/8$	0	0	1	1	0	1
$5V_R/8-3V_R/8$	0	1	1	1	1	0
$7V_R/8-5V_R/8$	1	1	1	1	1	1

For eight comparators it would require 3 bits, 16 comparators would require 4 bits and for an 8 bit output there would have to be 256 comparators.

One of the advantages of the parallel-comparator A/D converter is speed of conversion as there are no counters. The unknown voltage is simply applied to the input and the binary-coded output is obtained. The speed is determined by the rate of change of the comparator outputs, say $10-20$ ns. These converters are sometimes known as *flash converters*. The disadvantage is the large number of comparators – 256 for 8 bits – and the need for accurate resistor values in the potential divider. They tend to be very expensive and are assembled from individual integrated circuits on a pcb. However, they are the only converters which are capable of handling analog signals such as TV or radar.

Comparison of A/D Converters

The parallel-comparator or flash converter is the fastest, but also the most complicated, in that it requires a large number of comparators and accurate resistors. It is the only method for converting video signals into digital, but it is expensive because of the complexity. The successive approximation converter is slower but much simpler to implement. The converters which are based on counting are the slowest, and this includes the dual-slope converter. Dual-slope converters are very accurate with 12 or more bits being easily achieved, and they are widely used in digital multimeters where accuracy is more important than speed.

17.2   Voltage-to-Frequency Conversion

It is not always necessary to convert an analog signal into a digitally coded signal. An alternative is to generate a pulsed signal with a frequency which is proportional to the analog control signal. Voltage-to-frequency converters are useful when it is necessary to transmit a signal over a cable when the 'digital' signal is less likely to be corrupted by interference.

Voltage-to-frequency (V/F) conversion can be achieved with discrete components and operational amplifiers, 555 timers and special purpose integrated circuits.

Op-Amp V/F Converter

In this method the analog signal is converted into a pulse train whose frequency varies with the input analog voltage. A very simple approach is to use the triangular waveform generator described in Chapter 15, and which is shown in its modified form in Figure 17.10.

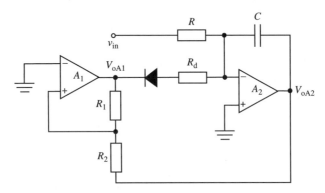

FIGURE 17.10   Simple voltage-to-frequency converter.

In comparison with the triangular wave generator, the feedback path is broken between the output of A_1 and the input of A_2, and a diode inserted. The effect of this modification is to change the charging conditions of the integrating capacitor, C. When the output of the comparator, V_{oA1}, is negative the diode is forward biased and the charging current for C is $-V_{oA1}/R_d$ and provided that $R_d \ll R$ then the capacitor charges very rapidly. The output of A_2 ramps up until it reaches the threshold voltage of the comparator, when the output of A_1 changes and becomes positive. The diode is now reverse biased and the feedback loop is open-circuited. The input voltage v_{in} now provides the charging current for the integrator, and the output of A_2 ramps down at a rate determined by v_{in}. The waveforms at the outputs of A_1 and A_2 are shown in Figure 17.11.

If the time required for the integrator to reset, t_1, is very small compared with the time for the integrator to run down under the control of v_{in} then the frequency of oscillation is proportional to v_{in}. An expression for t_2 may be obtained following the

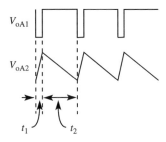

FIGURE 17.11   Output waveforms for A_1 and A_2.

procedure used to obtain equation 15.18. Ignoring the time interval t_1 then:

$$\frac{dV_C}{dt} = \frac{V_{UT} - V_{LT}}{t_2} \tag{17.7}$$

The charging current during this time interval is:

$$i = \frac{v_{in}}{R} \tag{17.8}$$

Since $i = C\, dV/dt$, substituting for I and dV/dt from equations 17.7 and 17.8 gives:

$$t_2 = \frac{RC(V_{UT} - V_{LT})}{v_{in}}$$

Now if $V_{UT} = -V_{LT} = (R_2/R_1)V_{sat}$, then:

$$f \approx \frac{1}{t_2} = \frac{R_1}{2R_2 RCV_{sat}}\, v_{in} \tag{17.9}$$

where V_{sat} is the output saturation voltage of the operational amplifiers. In practice the reset time (t_1) and the finite switching times of the comparator cause the frequency to be slightly less than that predicted by equation 17.9. The lower frequency limit is set by integrator drift caused by bias current and offset voltages. The upper limit is set by the reset time and comparator switching times. Typically a frequency range of two to three decades may be obtained.

▶ **EXAMPLE 17.1** ————————————————————————————————

Use PSpice to evaluate the performance of the V/F converter shown in Figure 17.12.

▶ *Solution*

An example of the output waveforms for v_{in} of 1 V is shown in Figure 17.13.

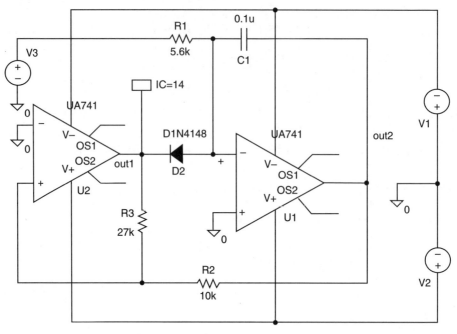

FIGURE 17.12 PSpice schematic of voltage-to-frequency converter.

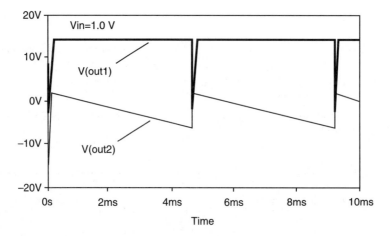

FIGURE 17.13 Example of voltage-to-frequency converter waveforms for $v_{in} = 1$ V.

Assuming that V_{sat} is 13 V then a comparison of both the calculated and measured values of the frequency of the pulse waveform is shown below:

v_{in} (V)	Calculated (Hz)	Measured (Hz)
0.1	18.5	21.7
1.0	185	220
10	1850	2087

555 Timer as V/F Converter

The 555 timer may be configured to act as a V/F converter by applying the analog signal to the control terminal. This mode of operation is described in Chapter 15, Section 15.1, and the circuit is reproduced here in Figure 17.14 for convenience.

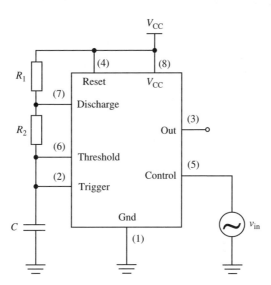

FIGURE 17.14   The 555 timer as a voltage-to-frequency converter.

From Chapter 15 it was noted that with a control voltage the capacitor charges and discharges between $V_{control}$ and $V_{control}/2$. When the capacitor is charging the starting voltage is $V_{control}/2$ and the end voltage is $V_{control}$. The equation for the charging time is (see equation A6.4):

$$t_H = C(R_1 + R_2) \ln\left(\frac{V_{CC} - V_{control}/2}{V_{CC} - V_{control}}\right) \tag{17.10}$$

For the discharge the starting voltage is $V_{control}$ and the end voltage is $V_{control}/2$. This yields a discharge time of:

$$t_L = CR_2 \ln\left(\frac{V_{control}}{V_{control}/2}\right) \tag{17.11}$$

From equation 17.11 it is seen that the time constant t_L is independent of the control voltage and is simply $0.69R_2C$. In the case of the charging time, however, the time constant increases with increasing values of control voltage, but not in a simple fashion. The frequency, which is equal to the reciprocal of the sum of the two time constants, decreases with increasing value of $V_{control}$.

▷ **EXAMPLE 17.2** ————————————————————————

Use PSpice to evaluate the effect of changing the control voltage from 1 V to 10 V for the schematic shown in Figure 17.15.

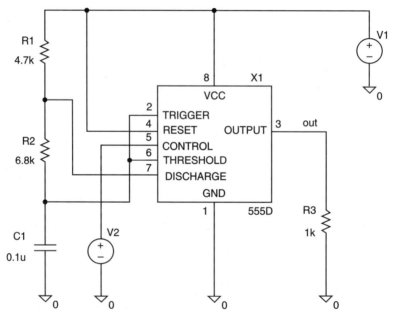

FIGURE 17.15   PSpice schematic for a voltage-to-frequency converter based on the 555 timer.

▷ *Solution*

From equation 17.10 the charging time when the control voltage is 1 V is:

$$t_H = (4.7k + 6.8k)(0.1\mu) \ln\left(\frac{15\,V - 1\,V}{15\,V - 0.5\,V}\right) = 40\,\mu s$$

From equation 17.11 the discharge time is:

$t_L = 0.69(6.8k)(0.1\mu) = 470$ μs

An example of the waveform with a 1 V control voltage is shown in Figure 17.16.

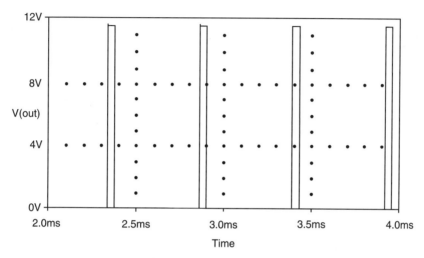

FIGURE 17.16   555 output with $V_{control} = 1$ V.

The measured times for 1 V and 10 V are as follows:

$V_{control}$	t_H (μs)	t_L (μs)	Frequency
1 V	39	487	1.9 kHz
10 V	39	258	3.4 kHz

In addition to the 555 timer there are a number of special purpose V/F converters in integrated circuit form, e.g. the National Semiconductor LM331 and the Texas Instruments 74LS5124.

17.3   Digital-to-Analog Conversion

As well as being able to convert an analog signal into a digital signal, the reverse may also be required, that is to convert a digital signal into an analog one. A *digital-to-analog converter or D/A converter* is required. The simplest approach uses scaled resistor values, for example the summing amplifier with each resistor following a binary progression of 2, 4, 8, 16, etc. This can lead to very large resistor ranges which require high accuracy, and which may be impossible to implement in IC form. To overcome this problem the $R-2R$ network has been developed where only two resistor values are required. An alternative is to use current sources which are easily implemented in ICs and by scaling transistor areas it is a simple matter to scale currents.

Summing Amplifier Binary-Weighted D/A Converter

One simple approach is to use a summing amplifier (see Section 13.2) and scaled resistors, as shown in Figure 17.17. The input switches are electronically controlled and are used to apply either 0 V or V_{ref} to each of the input resistors. These switches are driven by the digital input signal. The values of the input resistors follow a binary sequence.

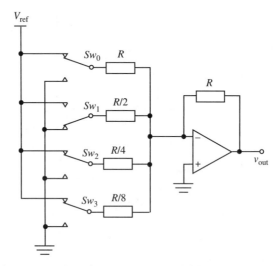

FIGURE 17.17　D/A converter based on a summing amplifier.

From equation 13.16 the output voltage is:

$$V_{out} = -R\left(\frac{Sw_0}{R} + \frac{Sw_1}{R/2} + \frac{Sw_2}{R/4} + \frac{Sw_3}{R/8}\right) \tag{17.12}$$

where the value of each switch is either 0 V or V_{ref}, or in general:

$$v_{out} = -V_{ref}(2^0 b_0 + 2^1 b_1 + \cdots + 2^n b_n) \tag{17.13}$$

where b_0, b_1, b_n are either 0 or 1 depending on the position of the switches. For the switch positions shown in Figure 17.17 the output is:

$$v_{out} = -V_{ref}(1 + 0 + 4 + 8) = -13V_{ref}$$

With four switches the output can have 16 different levels from 0 V, when all the switches are 0 (0000), to $15V_{ref}$ when the input code is 1111. With eight switches there are 256 levels.

The output of the D/A converter is not monotonic but rather a series of steps, as shown in Figure 17.18 for a 3 bit D/A converter.

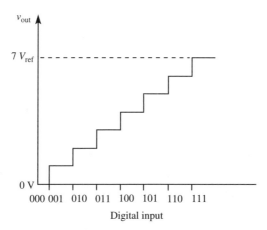

FIGURE 17.18   Graph of output voltage versus input digital code for a D/A converter.

In order that the variation of the output analog voltage approaches one which is continuous it is necessary to increase the number of bits in the input code. For an 8 bit D/A converter there would be 256 steps from 0 V to a maximum value of, say, 10 V while for 12 bits there would be 2048 steps. The accuracy and stability of the D/A converter is primarily determined by the accuracy and temperature tracking of the resistors. For a 12 bit D/A converter with a minimum resistor value of 1 kΩ, the maximum value is 2.048 MΩ. It is very difficult to obtain accurate and stable resistors over such a large range, and is impossible to reproduce such values in an integrated circuit. If the maximum resistor were reduced to 20.48 kΩ, then the minimum would be 10 Ω and at this level it would be affected by stray series resistance associated with the electronic switches (some form of FET). This type of D/A converter is rarely suitable for more than about 4 bits. A ladder-type converter with only two resistor values can overcome the problems of large resistor ranges.

Ladder-Type D/A Converter

A ladder-type or $R-2R$ converter with only two resistor values of R and $2R$ is shown in Figure 17.19.

The circuit contains more resistors than the summing amplifier type of converter, but they perform the same function as the resistors in the summing amplifier shown in Figure 17.17. Consider the situation when Sw_0 is connected to V_{ref} and Sw_1 and Sw_2 are connected to ground; then the equivalent circuit is shown in Figure 17.20. The diagram shows the step-by-step reduction of the network with the aid of Thévenin until the circuit is reduced to an inverting amplifier in 17.20(d) having an input of $V_{ref}/8$ and the output is $v_{out} = -V_{ref}/8$, or $v_{out} = -V_{ref}/2^3$. With Sw_1 connected to V_{ref} the output is $v_{out} = -V_{ref}/2^2$ and with Sw_2 connected to V_{ref} the output is $v_{out} = -V_{ref}/2^1$. If all three

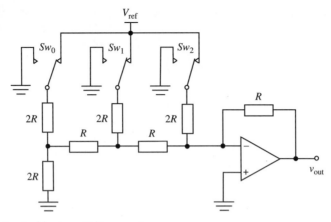

FIGURE 17.19   Ladder-type D/A converter.

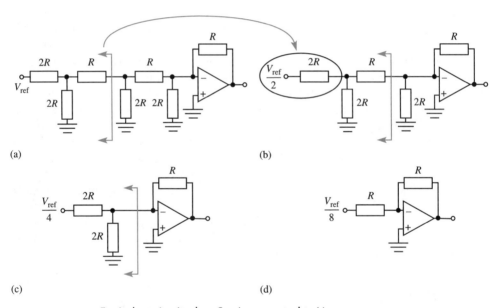

FIGURE 17.20   Equivalent circuit when Sw_0 is connected to V_{ref}.

switches are connected to V_{ref} then, by superposition, the output is:

$$v_{out} = -V_{ref}\left(\frac{1}{2^3} + \frac{1}{2^2} + \frac{1}{2^1}\right) = -\frac{V_{ref}}{2^3}(2^0 + 2^1 + 2^2)$$ (17.14)

The format of equation 17.14 is the same as equation 17.13. The advantage of this circuit, however, is that only two resistor values are required which are more easily implemented in IC form.

Scaled Current Sources D/A Converter

For ICs current sources (Section 11.4) are often easier to fabricate than resistors, and an example of a multiple current source is shown in Figure 17.21. The emitter areas of the current mirrors (Q_0, Q_1, Q_2) are scaled as shown so that the currents I_0, I_1, I_2 are binary multiples of the reference current (I_{ref}). Electronic switches connect the current sources to the output so that the output current is:

$$I_{out} = kI_{ref}(2^0 b_0 + 2^1 b_1 + 2^2 b_2) \tag{17.15}$$

where b_0, b_1 and b_2 represent the switch positions of 0 or 1 depending on whether the current source is connected to the output or not. The current can be converted to a voltage with the simple addition of a resistor connected between the output pin and V_{CC}.

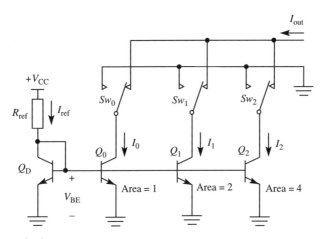

FIGURE 17.21   Multiple current source.

D/A Converter Parameters

Resolution, linearity and accuracy are some of the more important parameters which need to be considered in selecting a D/A converter. An additional parameter is the time taken to respond to a sudden change in the input code, the transient response.

The greater the number of bits the smaller the step size in the output voltage. This is usually referred to as the resolution: 8 bit D/A converters are very common and provide 256 levels in the output voltage which may range from 0 V to +10 V; 12 bit D/A converters with 2048 levels are also readily available.

Ideally the output voltage should vary linearly with the input binary code, but in practice all D/A converters will depart from this ideal. Resistor values may not be precise, transistor areas for current sources may not be precise multiples; the electronic switches are not perfect and introduce series resistance in the ON condition. Any of these factors can result in the discrete steps in the output voltage varying from one input code to the next, as illustrated in Figure 17.22.

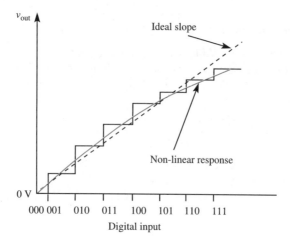

FIGURE 17.22   Non-linear output voltage from a D/A converter.

Accuracy and linearity are related, but changes to the reference voltage, or the feedback resistor in Figure 17.17 or 17.19, could affect the accuracy without affecting the linearity. Accuracy could also be affected by operating conditions, for example temperature changes could alter resistor values and/or the reference voltage. Typically a D/A converter will have a temperature sensitivity of ±50 ppm per °C, which represents a change of ±0.005% in the output voltage for each degree Celsius.

The output voltage does not change instantaneously when the input code changes. Some time is required for the output to establish its final value after a change to the input. This is known as the *settling time*. Typically it is of the order of 500 ns.

17.4   Phase-Locked Loop

The phase-locked loop (PLL) is a very useful circuit for signal processing, and is now available from a number of manufacturers as an integrated circuit. The PLL contains a phase detector, a low-pass filter, an amplifier and a voltage-controlled oscillator (VCO). A simplified block diagram of the PLL is shown in Figure 17.23.

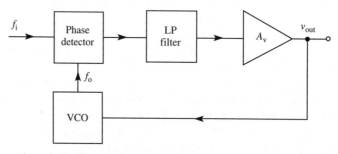

FIGURE 17.23   Phase-locked loop system.

The phase detector is a circuit with two inputs which produces an output which is proportional to the phase difference between the two input signals. To avoid the problem caused by variations in the amplitude of either of the two input signals, it is usual to design the circuit to accept pulse waveforms. A very simple example of such a circuit is the exclusive-OR shown in Figure 17.24.

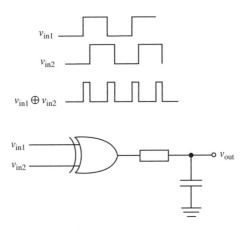

FIGURE 17.24   Exclusive-OR phase detector.

If the two input signals are in phase there is no output, but when a phase difference exists, as shown in Figure 17.24, there is an output which can be filtered with the RC low-pass filter to produce a dc voltage which is proportional to the phase difference. In practice the circuit is more complex than the simple exclusive-OR, but the principle remains the same. Note that a difference in frequency can also be described as a phase difference, and in Figure 17.23 the output from the phase detector is proportional to the frequency difference between f_i and f_o. The dc output from the low-pass filter is amplified and applied to the VCO with a polarity which causes the frequency f_o to change in the direction of f_i.

When the PLL is operating correctly the control signal to the VCO is a measure of variations in the input frequency. The most direct application of the PLL is the demodulation of an fm signal, either for radio/TV, or for decoding digital fm telephony signals.

Capture and Lock

When the frequency of the input signal and the VCO are the same the output from the phase detector is zero and the PLL is said to be *in lock*. While the system remains in lock the VCO control signal is a faithful reproduction of the frequency variation of the input signal, assuming that the variations in the input frequency are neither too large nor too rapid as to throw the system out of lock.

When power is initially applied to the circuit the VCO operates at its free-running frequency and is unlocked. The free-running frequency of the VCO corresponds to zero input to its control terminal. The circuit is capable of applying either a positive or negative voltage to the VCO in order to change the frequency. The free-running frequency must be within the *capture* range of the circuit in order that it acquires lock.

The process of capture is very dependent on the frequency response of the low-pass filter, as shown in Figure 17.25.

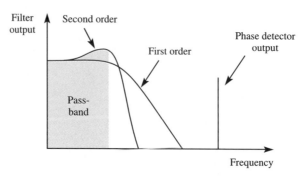

FIGURE 17.25   PLL frequency response.

The diagram shows a first-order frequency response, which is typical of a free-standing low-pass RC network, and a second-order response, which is more typical of the closed-loop response of the PLL, which incorporates an RC network as a low-pass filter. The shaded area represents the pass-band within which the input signal with a frequency f_i may vary, and an output will be produced. If at the instant of switch-on there is a large difference between the input frequency and the free-running frequency of the VCO, then the output from the phase detector will be a sine wave corresponding to the difference, and this is the input to the low-pass filter, and is represented by the vertical line in the diagram. For capture to occur this line (or frequency) must lie within the pass-band, or at least within the roll-off region of the low-pass characteristic. For the position shown in Figure 17.25 it does not lie within either the roll-off region or the pass-band, and consequently there is no output from the filter and no control signal to apply to the VCO. The VCO would continue to run at its free-running frequency, but there would be no output from the PLL. Provided that the signal from the phase detector is within the pass-band or roll-off region then an output will be generated from the filter and there will be a feedback signal to the VCO which will move the VCO frequency in the direction of the input frequency. Depending on the time constant of the system, the VCO frequency may overshoot and a reverse polarity control voltage will be generated to move the VCO frequency back in the opposite direction. This process may be repeated until the system settles to a steady state.

Capture and Lock Range

When the system first starts the VCO operates at its free-running frequency (f_o) and capture will take place provided that the difference between f_o and the input frequency is less that the *capture range* ($\pm f_C$).

When the system is in lock, the VCO and signal are in synchronism and will remain in lock over the *lock-in range* (±f$_L$). As illustrated in Figure 17.26 the lock-in range is greater than the capture range.

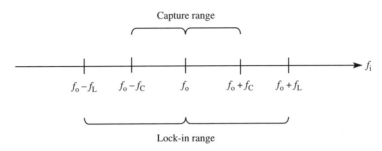

FIGURE 17.26 Lock-in range and capture range.

Two examples of integrated circuit PLLs are the 4046 produced by Philips and SGS-Thomson and the 565 made by National Semiconductor and Signetics Corporation. For both devices the manufacturers' data sheets specify a capture frequency of:

$$f_C = \pm \frac{1}{2\pi} \sqrt{\frac{2\pi f_L}{RC}}$$

where f_L is the lock-in range and RC represents the resistor and capacitor values for the low-pass filter.

PLL Applications

FM demodulation of fm signals is one of the simplest applications. In an fm signal information is encoded onto a carrier by varying the frequency in response to the information waveform, for example speech or music. The received signal, either the carrier for a cable-based signal, or an intermediate-frequency signal for radio or television, is applied to the PLL. The bandwidth of the low-pass filter within the PLL is adjusted to be wide enough to pass the modulating signal. A high degree of linearity in the VCO is necessary to avoid distortion of the audio signal.

Frequency multiplication is used to generate a multiple of an input frequency in digital frequency synthesizers. A typical requirement is to generate multiples of the line frequency (60 Hz or 50 Hz) for dual-slope A/D converters. Consider a 12 bit dual-slope A/D converter to operate from a 50 Hz supply and where four readings are to be made each second. The integration time is 80 ms and the required clock frequency is $2^{12}/80$ ms, or 51200 Hz. This frequency is to be synchronized with the line frequency. The frequency-multiplying PLL is shown in Figure 17.27.

The input frequency (f_i) is 50 Hz; thus the frequency from the output of the divider must also be 50 Hz. The clock frequency is to be 51200 Hz, and thus the division

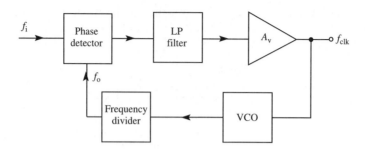

FIGURE 17.27   Frequency multiplier block diagram.

factor is 1024. The free-running frequency of the VCO is 51.2 kHz and this will lock to the line frequency and will take account of any small variations in the line frequency.

The same circuit may be used in a frequency synthesizer. Different output frequencies (f_{clk} in Figure 17.27) may be obtained by replacing the fixed frequency divider by a variable divider with switches to control the division factor. The input frequency would be a crystal oscillator rather than the line supply.

Another use of a PLL is in the recovery of digital signals in a communications channel or from a magnetic tape head. The signal may be corrupted with noise, or the pulses distorted. The low-pass filter of the PLL can be adjusted to eliminate the noise, while the inherent gain in the loop regenerates the signal to produce an output which is free from noise, and with the pulses restored.

Summary

A/D converters and D/A converters are readily available in IC form from many manufacturers, all of whom provide detailed data sheets for their products: 8 bit converters are widely available but 12 bit and 16 bit converters are also available, particularly the dual-slope A/D converters for digital instrumentation. High-speed flash converters are usually built from discrete components and are very expensive, but do allow conversion at very high speeds.

In addition to A/D and D/A there are a wide range of voltage-to-frequency and frequency-to-voltage circuits which are available in integrated circuit form and PLL chips. The PLL chips may be complete but often are available as an integrated VCO and a separate phase detector or analog multiplying circuit.

▶ **PROBLEMS** ──────────────────────────

17.1   If the IC A_1 in the sample-and-hold circuit of Figure 17.3 supplies 5 mA of output current and if the hold capacitor is 10 nF, determine the maximum rate of change which the circuit can accurately follow.

[0.5 V µs^{-1}]

17.2   If the combined leakage current for the FET and the input IC A_2 in Figure 17.3 is 100 pA, determine the change in voltage during the hold cycle if the hold period is 1 ms.

[10 µV]

17.3   The clock frequency in an 8 bit counting A/D converter is 10 kHz. Determine the maximum conversion time.

[25.5 ms]

17.4   If the input voltage range for a 10 bit counting A/D converter is 0 V to 10 V, determine the conversion time for an input voltage of 4 V if the clock frequency is 25 kHz.

[16 ms]

17.5   Describe the operation of the tracking A/D converter, noting the differences between it and the counting converter.

17.6   A 6 bit tracking A/D converter is required to track a voltage source which can vary from 0 V to 5 V with a maximum rate of change of 0.004 V µs^{-1}. Determine a suitable clock frequency.

[50 kHz]

17.7   Describe the operation of the successive approximation A/D converter.

17.8   What increase in speed can be achieved by using a 10 bit successive approximation converter in place of a 10 bit counting converter assuming full-scale input?

[100 times]

17.9   Describe the operation of the dual-slope A/D converter.

17.10 Very accurate resistor and capacitor values are required for the dual-slope A/D converter when it is used for a digital voltmeter which is to have a 0.01% reading accuracy. Is this statement true?

17.11 What is the shortest integration time for dual-slope A/D converters used in a digital volt–ammeter and when operating from 50 Hz and 60 Hz line frequencies?

[60 ms, 50 ms]

17.12 Design a 6 bit flash converter and determine the minimum input voltage change which can be detected if the input voltage varies from 0 V to 10 V.

[156 mV]

17.13 What is the input voltage for the 6 bit flash converter if the output code is 100111?

[6.08 V]

17.14 Draw the circuit diagram for a D/A converter based on a summing amplifier with 64 levels, and select suitable resistor values based on a minimum value of 1.5 kΩ.

[48 kΩ, 24 kΩ, 12 kΩ, 6 kΩ, 13 kΩ, 1.5 kΩ]

17.15 Determine the output voltage for the D/A converter described in Problem 17.14 if the input code is 110011, the reference voltage is 1.2 V and the feedback resistor is 150 Ω.

[6.12 V]

17.16 For the D/A converter described in Problem 17.15 determine the largest error in the output voltage when the input code changes by 1 bit.

[±60 mV]

17.17 For the ladder D/A converter shown in Figure 17.19 show that the output voltage is $-0.75V_{ref}$ if the switches Sw_1 and Sw_2 are both connected to the reference voltage, and Sw_0 is connected to ground.

CHAPTER 18

Voltage Regulators

Voltage regulators provide a constant dc output voltage which is almost completely unaffected by changes in the load current, the input voltage or the temperature. It forms the basis of most power supplies which are required by all electronic equipment which derives its power from the ac line voltages. The input voltage to a regulator is the filtered output from a diode rectifier. Half-wave and full-wave diode rectifier circuits with a capacitive filter were described in Chapter 3. The diode rectifier converts the ac supply to dc but with a large ripple voltage which is very dependent on the load current – the larger the current the larger the ripple. The voltage regulator removes the ripple and provides a dc voltage which is largely independent of variations in the load resistance.

There are two main categories of regulator, *linear regulators* and *switching regulators*. The simplest linear regulator uses a resistor and a Zener diode, as described in Chapter 3, but the performance is limited and most practical linear regulators use a transistor as a control element and an amplifier to provide an error signal within a negative feedback loop. The switching regulator incorporates a high-speed switch which is controlled by a feedback loop which monitors the output voltage. Variations of the output voltage affect the switching speed in such a way as to compensate for the variation, and thereby maintain a constant output voltage.

Because all mains-powered electronic equipment requires a voltage regulator, there are many linear and switching-type regulators which are manufactured as integrated circuits, and it is unlikely that a designer will be required to custom-design a voltage regulator. Manufacturers' application notes provide detailed design rules for each type and the purpose of this chapter is to provide an understanding of the operating principles to enable a sensible choice to be made and to ensure that the design rules are understood.

18.1 Linear Regulators

The linear voltage regulator uses a transistor to regulate the flow of current and the voltage supplied to the load from an unregulated source. The unregulated source is usually a full-wave rectifier with capacitive smoothing. The purpose of the regulator is to monitor the output voltage and, by means of feedback, to maintain it as constant as possible for all load currents. Transistors can be incorporated into a regulator either as a series control element or as a shunt element.

639

Series Regulator

A simplified block diagram of series regulators is shown in Figure 18.1.

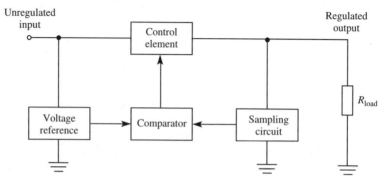

FIGURE 18.1 Block diagram of a series voltage regulator.

The unregulated input is the filtered output from a rectifier; it is unipolar but will contain a large ripple voltage. The control element is a linear device whose input is the unregulated voltage plus a control signal from a comparator, and whose output is a regulated voltage. Control is achieved by comparing the required output with a fixed voltage reference. If the output voltage increases, for example as a result of a variation in the load current, the input to the comparator from the sampling circuit increases and a voltage difference will exist between the reference voltage and the sampled voltage. This results in an output being generated by the comparator which causes the control element to decrease the output voltage. The reduction continues until the comparator detects no difference between the reference voltage and the sampled voltage. Similarly if the output voltage decreases the comparator causes the control element to increase the output.

In Figure 18.2 the control element is an npn bipolar transistor with a Zener diode acting as the voltage reference.

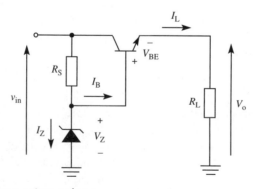

FIGURE 18.2 Transistor series regulator.

The comparator function, as illustrated in Figure 18.1, is performed by the base to emitter voltage which controls current flow through the transistor. Application of Kirchhoff's voltage laws around the output circuit gives:

$$V_{BE} = V_Z - V_O \qquad (18.1)$$

If V_Z is regarded as a constant then from equation 18.1 any change in V_O will result in a change in V_{BE}. If V_O increases then V_{BE} decreases to maintain the balance described by the equation. A reduction in V_{BE} causes the transistor to conduct less heavily and the load current is reduced. The reduction in the load current flowing through the load resistance will cause the voltage across the load to fall, which will restore the balance described by equation 18.1.

A suitable value for R_S is given by (see equation 3.25):

$$R_S = \frac{v_{in(min)} - V_Z}{I_{Z(min)} + I_B}$$

and since $I_B \approx I_L / \beta$ then:

$$R_S = \frac{v_{in(min)} - V_Z}{I_{Z(min)} + I_L / \beta} \qquad (18.2)$$

The minimum value of v_{in} is determined by the rectifier and the associated capacitive filter which determines the amount of ripple. The minimum value of the Zener diode current is determined by the manufacturer's data for the diode and is usually sufficient to avoid the knee of the current–voltage curve at breakdown. With reference to Chapter 3 and equation 3.25, the important difference in the regulator shown in Figure 18.2 and the simple Zener diode regulator described in Chapter 3 is the removal of the load current from the Zener diode circuit, and in particular R_S. The fact that the load current does not have to flow through R_S means that it can have a much larger value. Thus the stability given by $\Delta V_O / \Delta v_{in} = r_Z / R_S$ (see equation 3.23) is greatly improved.

▷ **EXAMPLE 18.1** ────────────────────────────────

For the regulator shown in Figure 18.2 v_{in} has an average value of 20 V with a ripple voltage of 2 V. The minimum Zener current is 5 mA, $V_Z = 12$ V, $r_Z = 5\ \Omega$, $V_{BE} = 0.8$ V and $\beta = 75$. Determine the output voltage if the load is 37 Ω, the value of R_S and the variation of the output voltage (ΔV_O).

▷ *Solution*

From equation 18.1 the output voltage is:

$$V_O = V_Z - V_{BE} = 12\ \text{V} - 0.8\ \text{V} = 11.2\ \text{V}$$

The load current is:

$$I_L = \frac{V_O}{R_L} = \frac{11.2\ \text{V}}{37\ \Omega} \approx 300\ \text{mA}$$

A suitable value for R_S is obtained with the aid of equation 18.2. With a ripple voltage of 2 V (peak to peak) the minimum value of v_{in} is 20 V − 1 V = 19 V, and R_S is:

$$R_S = \frac{19\,\text{V} - 12\,\text{V}}{(5\,\text{mA} + 300\,\text{mA})/75} \approx 778\,\Omega$$

The maximum variation of the input voltage is ±1 V and the variation of the output is:

$$\Delta V_O = \Delta v_{in}\frac{r_Z}{R_S} = 1\frac{5\,\Omega}{778\,\Omega} = 6\,\text{mV}$$

The transistor regulator is a considerable improvement over the simple Zener diode regulator described in Chapter 3 but does not offer the performance required for modern electronic circuits. The output voltage determines the value of the Zener voltage. However, Zener diodes with very low temperature cocfficients usually have voltage values of between 3 V and 5 V, and this severely restricts the controlled voltage range of such a regulator, assuming that a low temperature coefficient is required. Some means is required to provide a greater range of output voltage which is independent of the Zener voltage.

Also for greatest stability of the output voltage the value of the series resistor R_S should be as large as possible, and ideally the only current flowing through it should be that required by the Zener diode. Finally the feedback shown in Figure 18.1 from the output and back to the control element should have a large loop gain to ensure that very small changes are detected and rapidly corrected. Figure 18.3 shows an improved series regulator which incorporates a gain stage in the feedback loop.

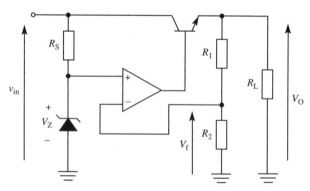

FIGURE 18.3   Series regulator with an op-amp gain stage.

Resistors R_1 and R_2 act as a voltage divider across the output which feeds back a sample V_f to the inverting terminal of an operational amplifier. The Zener voltage is connected to the non-inverting terminal so that the amplifier output is proportional to $V_Z - V_f$. If V_O increases then V_f increases and the output from the amplifier decreases. The reduction in the voltage applied to the base of the npn transistor causes it to conduct less heavily, which compensates for the rise in the output voltage. Thus the feedback stabilizes the output. A similar action occurs when the output tries to decrease, with the transistor being made to pass more current to compensate for the fall in output voltage.

The operational amplifier in Figure 18.3 may be regarded as a non-inverting amplifier where V_Z is the input and V_f is the feedback. The closed-loop gain is:

$$A_{cf} = \frac{V_O'}{V_Z} = 1 + \frac{R_1}{R_2} \tag{18.3}$$

Thus if the base–emitter voltage is neglected, then the output voltage is:

$$V_O \approx V_O' = \left(1 + \frac{R_1}{R_2}\right) V_Z \tag{18.4}$$

The output voltage can now be varied from a minimum value of V_Z to a maximum value determined by the resistor ratio and the value of v_{in}.

Because of the very high input resistance of the operational amplifier the only current which need be considered in R_S is that required by the Zener diode and consequently a larger value may be selected for R_S, which improves the regulation.

▶ **EXAMPLE 18.2** ─────────────────────────

Design a series voltage regulator based on the circuit shown in Figure 18.3 using a 5.6 V Zener diode to provide a regulated output of 20 V. Assume an unregulated input of 28 V with a ripple voltage of 8 V and a minimum Zener diode current of 12 mA.

▶ *Solution*

From equation 18.4:

$$20\,V = \left(1 + \frac{R_1}{R_2}\right) \times 5.6\,V$$

Let $R_1 = 10$ kΩ; then $R_2 \approx 3.9$ kΩ.
The current through R_S is 12 mA and is given by:

$$12\,mA = \frac{v_{in(min)} - V_Z}{R_S}$$

The minimum value of v_{in} is $28\,V - V_{ripple}/2 = 24\,V$ and $R_S \approx 1.5$ kΩ.

───

The use of an operational amplifier greatly improves the stabilization of the output with respect to variations of the input. Since the main component of the input variation is ripple voltage, the ripple rejection is improved by a factor $1 + A_oB$, where A_o is the open-loop gain of the amplifier and B is the feedback resistor ratio $R_2/(R_1 + R_2)$.

Negative feedback also reduces the output resistance of the voltage regulator. The basic regulator output impedance is that of the emitter follower formed by the control transistor, which is low. Negative feedback reduces this further so that it becomes:

$$R_{out(CL)} = \frac{R_{out}}{1 + A_o B} \tag{18.5}$$

Typically the output resistance of an integrated circuit voltage regulator is a few milliohms and may be regarded as an almost perfect voltage source.

Short-Circuit Protection

A serious limitation of the series regulator shown in Figure 18.3 is the strong probability that the transistor will be damaged if the output is short-circuited. This is a particular problem for laboratory power supplies where accidental shorting of the output may occur. Under short-circuit conditions the full input voltage appears across the transistor and with V_f reduced to zero V_Z appears at the non-inverting terminal of the amplifier, which drives the output high thus forcing the transistor to conduct heavily. As well as damaging the transistor, the short-circuit current may also damage the rectifier diodes. Some means is required to limit the current.

Current Limiting

A simple form of current limiter is shown in Figure 18.4.

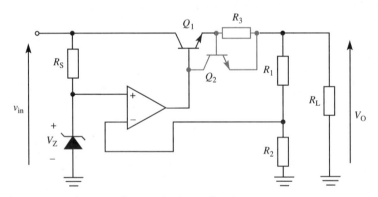

FIGURE 18.4   Series voltage regulator with current limiting.

This form of limiting is the same as that used in class B power amplifiers where the voltage across R_3 is the base–emitter voltage for Q_2. When V_{BE2} reaches 0.6–0.7 V Q_2 begins to conduct heavily and consequently current which should be flowing into the base of Q_1 is diverted to the collector of Q_2 and thus Q_1 is protected. The current through Q_2 is limited by the current sourcing capability of the operational amplifier, which may be a few tens of mA. The current–voltage characteristic of the output is shown in Figure 18.5.

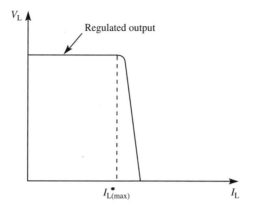

FIGURE 18.5   Typical load current versus load voltage with current limiting.

Beyond $I_{L(max)}$ the output voltage reduces to zero with only a relatively small additional increase in current. The maximum short-circuit current is that value required to produce 0.7 V across R_3, that is:

$$I_{L(max)} = \frac{0.7}{R_3} \tag{18.6}$$

▶ **EXAMPLE 18.3** _____

If R_3 in Figure 18.4 is 0.5 Ω determine the maximum current and if the output voltage is 24 V determine the output voltage if $R_L = 100\ \Omega$ and 10 Ω.

▷ *Solution*

From equation 18.6:

$$I_{L(max)} = \frac{0.7\,\text{V}}{0.5\,\Omega} = 1.4\,\text{A}$$

The load current when $R_L = 100\ \Omega$ and the output voltage is 24 V is:

$$I_L = \frac{24\,\text{V}}{100\,\Omega} = 0.24\,\text{A}$$

Since this value is less than $I_{L(max)}$ then the output voltage is 24 V.
    When the load resistance is reduced to 10 Ω the load current is:

$$I_L = \frac{24\,\text{V}}{10\,\Omega} = 2.4\,\text{A}$$

This is greater than the $I_{L(max)}$ (1.4 A) and current limiting takes place which reduces the output voltage to:

$$V_O = (1.4\ \text{A})(10\ \Omega) = 14\ \text{V}$$

▶ **EXAMPLE 18.4** ————————————————————————————————

Use PSpice to simulate the current-limiting action of the circuit shown in Figure 18.4 with $R_3 = 0.5\ \Omega$.

▶ *Solution*

To simplify the simulation a behavioural model (ETABLE) is used for the operational amplifier. Generic devices are used for the npn transistors and Zener diode from the **breakout.slb**. The PSpice schematic is shown in Figure 18.6.

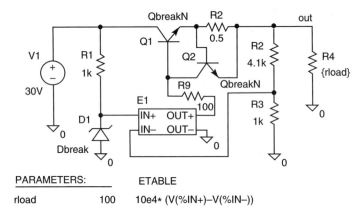

FIGURE 18.6   PSpice schematic of series regulator with current limiting.

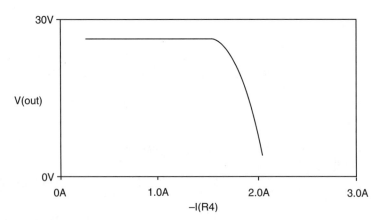

FIGURE 18.7   Probe output showing current limiting.

The **Expression** (**Expr**) for the ETABLE behavioural model is:

Expr = 10e4 * (V(%IN+)-V(%IN−))

where 10e4 is the gain and V(%IN+) and V(%IN−) are the voltages for the two input terminals. The TABLE values are set to [0, 0][30, 30]. The load resistor is described as a parameter (r_{load}). Notice the curly brackets for the value of R_4. The initial value is set as 100 Ω. Within the **Setup...** a **DC sweep...** is selected with a **Global Parameter**. The name of the parameter is r_{load} with a sweep range of 100 Ω to 1 Ω in steps of 1 Ω. In Probe the X-axis variable is changed to be the current through R_4 (-I(R4)). The plot of v_{out} versus I_{load} is shown in Figure 18.7.

A limitation of the simple current limiting shown in Figure 18.4 is that the transistor and rectifier diodes must be capable of conducting $I_{L(max)}$ for some time, depending on how quickly the short circuit is detected and removed. A further development of short-circuit protection is to reduce the current as well as the voltage when a short circuit occurs.

Fold-Back Current Limiting

In Example 18.3 it is seen that the output voltage decreases when the current exceeds $I_{L(max)}$. In fold-back limiting this reduction in the output voltage is detected and used to decrease the current supplied to the load. As the load resistance approaches 0 Ω both current and voltage in the load approach 0. The basic circuit is shown in Figure 18.8.

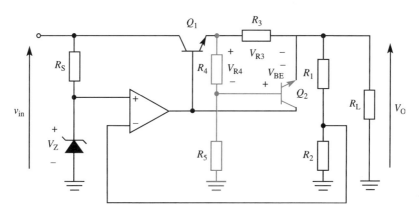

FIGURE 18.8 Series regulator with fold-back current limiting.

The circuit is similar to the basic current-limiting circuit except that the base of Q_2 is taken to a potential divider comprising R_4 and R_5. Applying Kirchhoff's law to the base−emitter loop of Q_2 gives:

$$V_{BE} = V_{R3} - V_{R4} \tag{18.7}$$

Under normal operating conditions V_{R4} is a constant value determined by the values of R_4 and R_5 and the voltage applied to the base of Q_2 is negative with respect to the emitter and it is turned OFF. As the maximum load current is approached the voltage drop across R_3 increases until it exceeds that across R_4. When the difference reaches 0.7 V, Q_2 begins to conduct and limits the current through Q_1. However, as the output voltage decreases the voltage across R_4 decreases and as can be seen from equation 18.7 if V_{R4} decreases then a smaller value of V_{R3} is required to maintain V_{BE} at 0.7 V. The reduction in V_{R3} is achieved by a reduction in the load current. A further reduction in the load resistance further reduces the output voltage, which results in a further reduction of the load current. The graph of load current versus load voltage is shown in Figure 18.9.

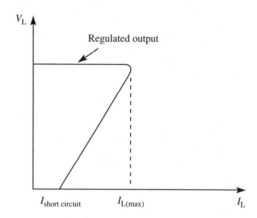

FIGURE 18.9   Fold-back current limiting.

▶ **EXAMPLE 18.5** ────────────────────────────────

For the circuit shown in Figure 18.8 the unregulated input is 24 V, $R_4 = 1.5$ kΩ, $R_5 = 8.5$ kΩ and the output voltage is 12 V. Determine the value of R_3 to provide current limiting at 1 A. If $V_Z = 4.7$ V and $R_1 = 2$ kΩ determine the value of R_2.

▶ *Solution*

The voltage across R_4 for normal operation and, therefore, at the onset of current limiting is:

$$V_{R4} = \frac{12\,\text{V}}{1.5\,\text{k}\Omega + 8.5\,\text{k}\Omega} \times 1.5\,\text{k}\Omega = 1.8\,\text{V}$$

From equation 18.7:

$$V_{BE} = 0.7\,\text{V} = V_{R3} - 1.8\,\text{V}$$

and

$$V_{R3} = 2.5 \text{ V}$$

Thus:

$$R_3 = \frac{V_{R3}}{I_{L(max)}} = \frac{2.5 \text{ V}}{1 \text{ A}} = 2.5 \, \Omega$$

For an output of 12 V and with $V_Z = 4.7$ V then the voltage across R_2 must be approximately 4.7 V so that:

$$R_2 = \frac{(4.7 \text{ V})(2 \text{ k}\Omega)}{12 \text{ V} - 4.7 \text{ V}} \approx 1.3 \text{ k}\Omega$$

▷ **EXAMPLE 18.6** ────────────────────────────────────

Use PSpice to simulate the current fold-back shown in Figure 18.8.

▷ *Solution*

The PSpice schematic is shown in Figure 18.10.

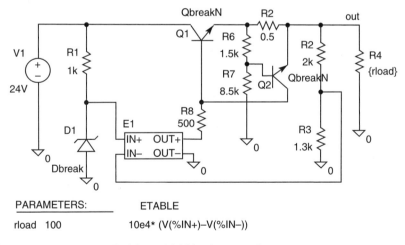

FIGURE 18.10   PSpice schematic of fold-back current limiting.

The settings for the behavioural model for the operational amplifier are the same as those in Example 18.4 except for the TABLE values which are [0,0][24,24]. The value of the load resistor is varied in the **DC Sweep...** as a **Global Parameter** from 100 Ω to 1 Ω in 1 Ω steps. The Probe output is shown in Figure 18.11.

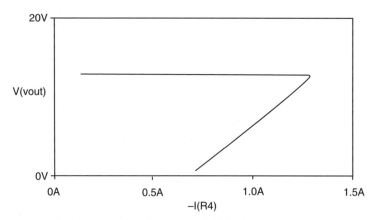

FIGURE 18.11   Probe output showing current fold-back.

Shunt Regulator

A block diagram of a shunt regulator is shown in Figure 18.12. The same basic components are used as for the series regulator, with the addition of a series resistor R_{series}. Notice that the control element is now in parallel with the load.

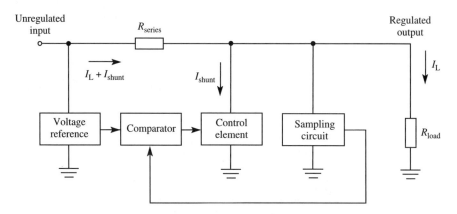

FIGURE 18.12   Block diagram of a shunt regulator.

As for the series regulator the current through the control element is determined by the output voltage sample which is applied to the comparator. However, instead of controlling the load current directly, the shunt regulator bypasses current, I_{shunt}, to ground. Because this current flows, together with the load current, through the series resistor R_{series} there is a variable voltage drop V_{series} which affects the output.

Consider an increase in the output voltage. The output from the comparator increases which causes the control element to conduct more current. Thus the voltage drop across the series resistor increases and the available voltage to the output falls. If the output voltage falls then the control element conducts less current which results in a smaller voltage drop across the series resistor, and a rise in the output voltage.

A circuit diagram of a shunt regulator with an operational amplifier is shown in Figure 18.13.

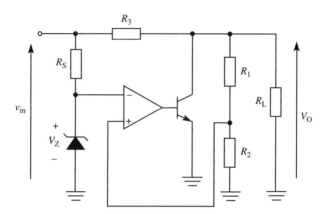

FIGURE 18.13   Shunt regulator incorporating an operational amplifier.

Notice the polarity of the operational amplifier input connections. For the series regulator an increase in the output is counteracted with a *decrease* in the current flowing through the control element, whereas for the shunt regulator an increase in the output requires an *increase* in the current through the control element.

One advantage of the shunt regulator is that it is not damaged by a short-circuited output. The short-circuit current is limited to v_{in}/R_3 and provided the power dissipation of R_3 is sufficient to withstand the short-circuit current, then there is no damage.

▷ **EXAMPLE 18.7** _____

For the circuit shown in Figure 18.13 determine the power rating of the series resistor, R_3, if it has a value of 47 Ω and if the maximum value of the input voltage is 30 V.

▷ *Solution*

The maximum power is dissipated in the series resistor when the output is short-circuited and when v_{in} is at its maximum value, that is:

$$P_{R3} = \frac{V_{R3}^2}{R_3} = \frac{(30\,\text{V})^2}{47\,\Omega} = 19\,\text{W}$$

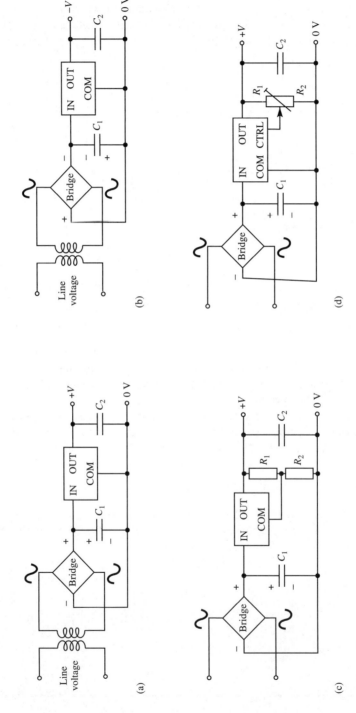

FIGURE 18.14   Examples of regulated power supplies constructed from integrated circuit three- and four-pin regulators.

Integrated Circuit Regulators

For most applications it is not necessary to custom-design a voltage regulator for a piece of electronic equipment, but rather to use one of the many 'off-the-shelf' integrated circuit regulators. The IC chip contains the transistor control element, the operational amplifier and the voltage reference. Because of the integration of the different parts of the circuit only three pins are required for the most basic regulator: an unregulated input, the regulated output and a common connection which may be taken to ground or to a potential divider. Some have four or more pins to provide additional control functions. Some typical configurations for positive and negative fixed voltage and variable voltage supplies are shown in Figure 18.14.

Fixed voltage positive and negative supplies are shown in Figures 18.14(a) and 18.14(b). Commonly available voltages are 5 V, 12 V and 15 V for both positive and negative polarities. Capacitor C_1 is the capacitive filter of the bridge rectifier. Notice the polarity for this large-value electrolytic capacitor (typically 1000 μF). Capacitor C_2 is typically a non-polarized 100 nF and improves the transient response when the current drawn from the supply is subject to rapid changes. Figure 18.14(c) shows how some three-terminal regulators are connected to provide variable voltage outputs. Alternatively Figure 18.14(d) shows a four-terminal regulator which has an additional control terminal which can be used to provide a variable output. Some of the characteristics of commonly used regulators are listed in Table 18.1.

Table 18.1   IC voltage regulators

Type	Output current	Output voltage	Load regulation	Ripple rejection
LM78L05	100 mA	+5 V	0.4%	62 dB
LM78L12	100 mA	+12 V	0.25%	54 dB
LM78L15	100 mA	+15 V	0.25%	51 dB
LM79L05	100 mA	−5 V	0.2%	60 dB
LM79L12	100 mA	−12 V	0.2%	55 dB
LM79L15	100 mA	−15 V	0.3%	52 dB
LM317LZ	100 mA	+1.2 V–37 V	0.1%	80 dB
LM317MP	500 mA	+1.2 V–37 V	0.1%	80 dB
LM317T	1.5 A	+1.2 V–37 V	0.1%	80 dB
LM78MGCP	500 mA	+5 V–30 V	1%	62 dB
LM78GCP	1 A	+5 V–30 V	1%	62 dB

The table provides an indication of the output current ranges and voltages which may be found for integrated circuit voltage regulators. The load regulation is defined as:

$$\text{load regulation} = \frac{V_{\text{no load}} - V_{\text{full load}}}{V_{\text{no load}}} \times 100\% \qquad (18.8)$$

For example, if the no-load voltage is 12 V and the full-load voltage is 11.9 V, then the load regulation is:

$$\frac{12\,\text{V} - 11.9\,\text{V}}{12\,\text{V}} \times 100\% = 0.8\%$$

The ripple rejection is defined as:

$$\text{ripple rejection} = 20\log \frac{V_{\text{ripple(out)}}}{V_{\text{ripple(in)}}} \tag{18.9}$$

For example, a ripple rejection of 60 dB means that the output ripple is 1000 times less than the ripple at the input to the regulator.

▷ **EXAMPLE 18.8** ──────────────────────────────

Determine the load current and output voltage ripple for the power supply shown in Figure 18.15.

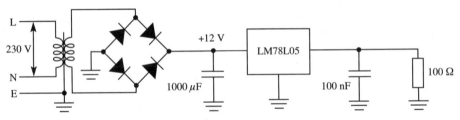

FIGURE 18.15   Regulated power supply.

▷ *Solution*

The LM78L05 has a fixed voltage output of 5 V with a maximum output current of 100 mA. With a 100 Ω load the output current is:

$$I_{\text{L}} = \frac{5\,\text{V}}{100\,\Omega} = 50\,\text{mA}$$

The unregulated input is 12 V. The voltage across the 1000 μF capacitor varies in a triangular fashion with a ripple voltage approximately given by:

$$V_{\text{ripple}} = \frac{V_{\text{dc(unregulated)}}}{2 f R_{\text{L(unregulated)}} C} \quad \text{if } V_{\text{dc}} \approx V_{\text{peak}}$$

where $R_{\text{L(unregulated)}}$ is the load seen by the 1000 μF capacitor. This can be written as:

$$V_{\text{ripple}} = \frac{I_{\text{L}}}{2 f C} = \frac{50\,\text{mA}}{(2)(50)(1000\,\mu\text{F})} = 0.5\,\text{V}$$

From Table 18.1 the ripple rejection ratio for the LM78L05 is 62 dB or 1260:1. Thus the output voltage ripple is:

$$V_{\text{ripple(out)}} = \frac{0.5\,\text{V}}{1260} \approx 0.4\,\text{mV}$$

18.2 Switched-Mode Regulator

A major disadvantage of the series and shunt regulators discussed in Section 18.1 is the dissipation in the series element in each type, the pass transistor for the series regulator and the series resistor for the shunt regulator. In both cases the input voltage is

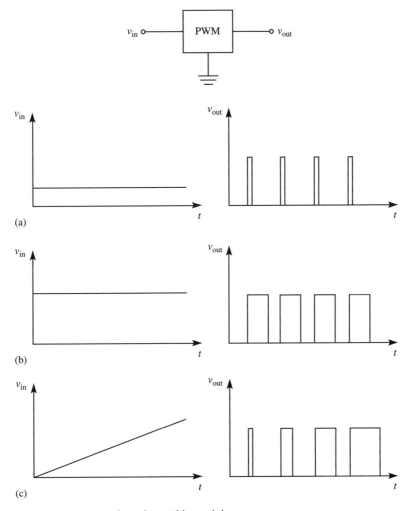

FIGURE 18.16 Operation of a pulse-width modulator.

greater than the output voltage and the difference is dropped across the series element. The product of this voltage and the load current represents a power loss. When efficiency is important, then a switching-mode regulator, which rapidly turns the control element ON and OFF, can greatly reduce static power dissipation.

The main component of a switching regulator is a *pulse-width modulator*. This is illustrated in Figure 18.16.

The pulse-width modulator generates a train of pulses whose width is a function of the dc input voltage level as shown in Figures 18.16(a), (b) and (c). In Figure 18.16(a) the dc input voltage is small and the pulses are narrow; in 18.16(b) the input voltage is large and the width of the pulses is greater. In Figure 18.16(c) the input is increasing linearly and the width of the pulses increases linearly with time.

The pulses are unidirectional, and therefore the output has an average dc value which is proportional to the duty cycle, as shown in Figure 18.17. The high-frequency switching component can be removed with a low-pass filter. Because the switching frequency is high the low-pass filter can include an inductor, which with a suitable capacitor can result in a very efficient filter. The position of the inductor with respect to the switch affects the output voltage and results in three variations of switching-mode regulator.

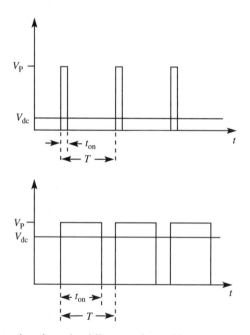

FIGURE 18.17   Average dc voltage for different pulse widths.

Step-Down Configuration

The most straightforward configuration places the filter immediately after the switch, as shown in Figure 18.18.

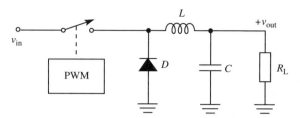

FIGURE 18.18   Simplified representation of step-down switched regulator.

The switch is controlled by the pulse-width modulator. The unregulated input voltage, v_{in}, is greater than the dc output voltage, v_{out}. The low-pass filter is formed by L and C. The diode is necessary to short-circuit the back emf from the inductor caused when the switch is opened and the magnetic field in the inductor collapses. The energy stored in the magnetic field generates a back emf which could damage a semiconductor which replaces the mechanical switch in a practical realization of the circuit shown in Figure 18.18. The output voltage is given by:

$$v_{out} = v_{in}\left(\frac{t_{on}}{T}\right) \tag{18.10}$$

The ratio t_{on}/T is known as the *duty cycle* (D). A practical realization of the circuit is shown in Figure 18.19. The switch is replaced by an npn transistor (a MOSFET could also be used). A Zener diode is used to provide a reference voltage which is compared with a fraction of the output voltage, obtained from the potential divider comprising R_1 and R_2, with the operational amplifier. If the load voltage falls then the width of the pulses applied to the switching transistor is increased in order to increase the dc output. If the output voltage increases then the width of the pulses is decreased to lower the output voltage.

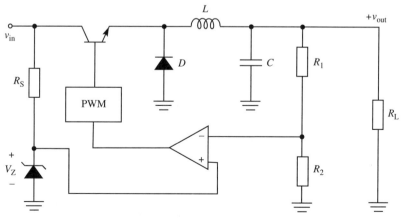

FIGURE 18.19   Practical realization of a step-down switching-mode regulator.

The diode prevents large reverse voltages appearing across the base–emitter junction of the transistor.

The operating frequency is usually in the kilohertz range, but usually less than 100 kHz. In the arrangement described above the output voltage is less than the input voltage. A rearrangement of the inductor and the switch can result in the output voltage being greater than the input.

Step-Up Configuration

Consider the basic configuration shown in Figure 18.20. Consider the instant when the switch closes. The anode of the diode is grounded and it becomes reverse biased by the voltage on the capacitor, v_{out}. Provided that $CR_L \gg t_{on}$ then the drop in v_{out} during the ON period of the switch is very small. At the instant that the switch closes the voltage across the inductor rises to a maximum value of v_{in}. During the ON period the voltage across L falls as the current through the inductor rises, and the magnetic field increases. When the switch opens, the magnetic field in the inductor collapses and the voltage across the inductor reverses so that the voltage applied to the diode is now $v_{in} + V_L$. During the time that the switch is open the diode is forward biased and the capacitor is charged.

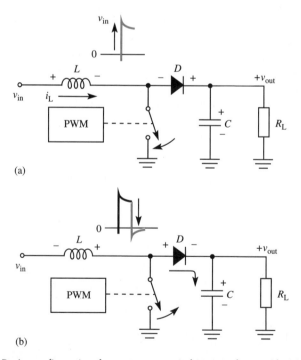

FIGURE 18.20   Basic configuration for a step-up switching regulator with (a) the switch closing and (b) the switch opening.

The shorter the time that the switch remains closed the larger the voltage which remains across L at the time that the switch is due to open, and therefore the larger the voltage applied to the diode ($v_{in} + V_L$). The longer the switch remains closed the smaller the voltage across the inductor and the smaller the voltage applied to the diode. That is, the output voltage is inversely proportional to the ON time and the output voltage can be expressed as:

$$v_{out} = v_{in}\left(\frac{T}{t_{on}}\right) = \frac{v_{in}}{D} \tag{18.11}$$

The schematic of the control circuit for the step-up regulator which provides the correct controlling action is shown in Figure 18.21.

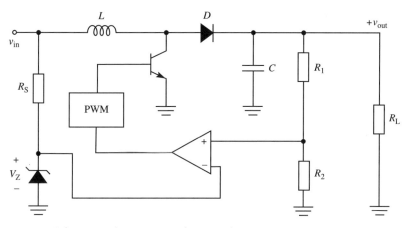

FIGURE 18.21   Schematic of step-up switching regulator.

Notice the polarity of the operational amplifier compared with that for the step-down switching regulator in Figure 18.19.

Negative Voltage Configuration

A negative voltage may be obtained from an unregulated supply with the configuration shown in Figure 18.22.

Consider the instant when the switch closes. The input voltage immediately appears across the inductor and i_L increases building up the magnetic field. During the ON period the voltage across the inductor falls. At the instant when the switch opens the voltage across L reverses and a negative voltage is applied to the cathode of the diode. The diode becomes forward biased and current flows to charge the capacitor so that v_{out} is negative. As for the step-down regulator the output voltage is inversely proportional to the ON time as given by equation 18.11.

A practical realization of the negative voltage regulator is shown in Figure 18.23.

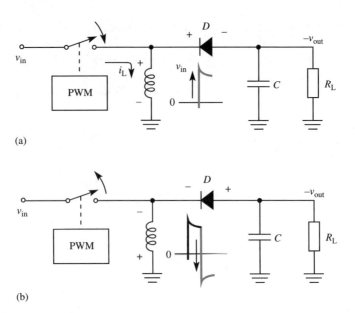

(a)

(b)

FIGURE 18.22   Negative voltage switching regulator.

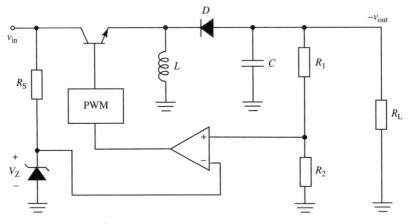

FIGURE 18.23   Negative voltage switching voltage regulator.

Switching regulators are very efficient and in many cases heat sinks are not required, but care must be taken to prevent electromagnetic interference, which can result from the high frequency switching circuitry, from interfering with circuits within the same piece of equipment and other pieces of equipment.

IC Switching Regulators

Switching regulators are now available from a number of manufacturers as an integrated circuit which contains the switching transistor, oscillator components, pulse-width modulator and temperature-compensated voltage reference. The only external components required in many cases are the inductor, smoothing capacitors and the diode. In some cases additional components are required to establish the oscillator frequency. Some examples of IC regulators are listed in Table 18.2.

TABLE 18.2   Switching-mode voltage regulators

Type	Supply voltage	Output voltage	Max output current	Efficiency
LM2574T5	7 V–40 V	5 V	0.5 A	77%
LM2575T5	8 V–40 V	5 V	1.0 A	77%
LM2575T12	15 V–40 V	12 V	1.0 A	77%
LM2575T15	18 V–40 V	15 V	1.0 A	88%
LM2576T5	8 V–40 V	5 V	3.0 A	77%
LM2576T12	18 V–40 V	12 V	3.0 A	77%
LM2577	3.5 V–40 V	0 V–6.0 V	3.0 A	80%

Typical circuits for these regulators are shown in Figure 18.24.

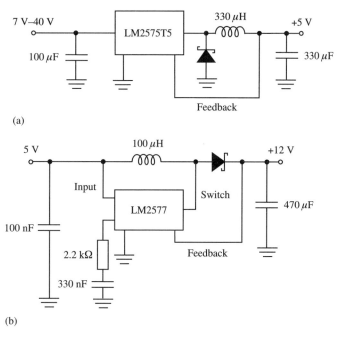

(a)

(b)

FIGURE 18.24   Examples of switching-mode regulator circuits with (a) a fixed voltage step-down regulator and (b) a step-up regulator.

The switching-mode regulator is now used extensively in modern computer equipment and provides a very efficient and compact means of converting ac line voltage to dc. The power units are usually packaged as a separate unit with electromagnetic screening to minimize EMI.

▷ PROBLEMS

18.1 For the circuit shown $\beta = 100$, $V_{BE} = 0.7$ V and v_{in} varies from 10 V to 20 V. Determine the output voltage, the value of R_S if the minimum Zener diode current is 8 mA and the maximum power dissipation in the Zener diode.

[5.4 V, 567 Ω, 120 mW]

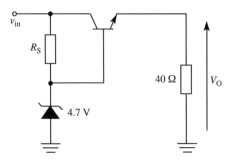

18.2 For the circuit shown $\beta = 70$, $V_{BE} = 0.7$ V and v_{in} varies from 15 V to 30 V. Determine the maximum power dissipation in the transistor and the maximum current supplied by v_{in}.

[3.3 W, 206 mA]

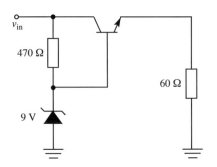

18.3 For the circuit shown v_{in} varies from 18 V to 30 V. Determine the output voltage, the value of R_S if the minimum Zener diode current is 10 mA and the maximum power dissipation in the Zener diode.

[12.2 V, 1.47 kΩ, 60 mW]

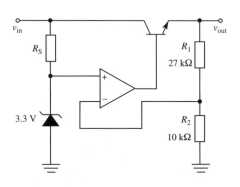

18.4 For the circuit shown in Problem 18.3 determine the maximum power dissipation in the transistor when $R_L = 20\ \Omega$.

[10.8 W]

18.5 Determine the minimum and maximum output voltages as the potentiometer is varied from minimum to maximum value, and determine the power dissipated in the transistor for each setting.

[5.6 V to 29 V, 1.9 W to 3.2 W]

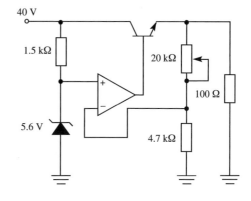

18.6 For the circuit shown in Figure 18.4 $R_1 = 10\ \text{k}\Omega$, $R_2 = 5\ \text{k}\Omega$, $V_Z = 5.6\ \text{V}$ and $R_3 = 1\ \Omega$. Determine the output voltage when $R_L = 100\ \Omega$ and $10\ \Omega$.

[16.8 V, 7 V]

18.7 If for the circuit shown in Figure 18.4 $R_1 = 22\ \text{k}\Omega$, $R_2 = 8\ \text{k}\Omega$, $v_{in} = 40\ \text{V}$ and $V_Z = 6.8\ \text{V}$, determine the value of R_3 to limit the power dissipation in the series transistor to 5 W. Determine the output voltage if $R_L = 20\ \Omega$.

[2 Ω, 7.2 V]

18.8 For the circuit shown in Figure 18.8 $R_1 = R_2 = 5\ \text{k}\Omega$, $R_3 = 1.7\ \Omega$, $R_4 = 560\ \Omega$, $R_5 = 4.7\ \text{k}\Omega$ and $V_Z = 4.7\ \text{V}$. Determine the output voltage, the short-circuit current and the power rating for the series transistor (Q_1).

[9.4 V, 1 A, 20 W]

18.9 For the circuit shown determine the output voltage and the power rating of the 82 Ω series resistor and the transistor.

[9 V, 19.5 W, 3.4 W]

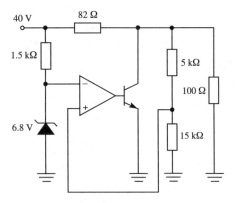

18.10 The mains input to the regulator shown is 110 V at 60 Hz. Determine the load current and the output voltage ripple.

[80 mA, 5.6 mV]

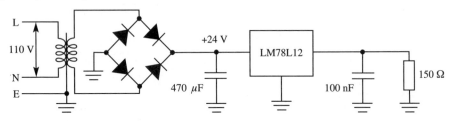

18.11 If the switching frequency is 20 kHz and the ON time is 10 μs, determine the output voltage.

[4.8 V]

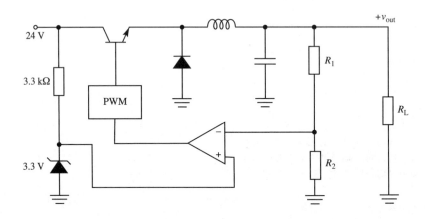

18.12   If the ON time for Problem 18.11 is
increased to 40 μs determine the output
voltage and the value of R_1 if R_2 is 5 kΩ.

   [19.2 V, 24 kΩ]

18.13   For the circuit shown the switching frequency is 40 kHz. Determine the ON time to
produce an output voltage of 30 V.

   [10 μs]

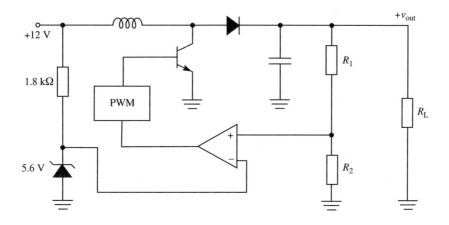

18.14   For the circuit in Problem 18.13
determine the ON and OFF times if
$R_1 = 7.2$ kΩ and $R_2 = 4.3$ kΩ.

   [$t_{on} = 20$ μs, —]

18.15   If the oscillator frequency is 25 kHz determine the duty cycle for the regulator shown.
Explain the purpose of the diode and suggest why a Schottky diode is used.

   [25%]

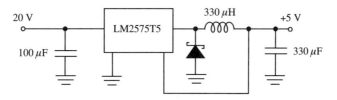

RMS Value of a Triangular Wave and Conduction Angle

RMS Value of a Triangular Wave

A triangular wave is shown in Figure A1.1.

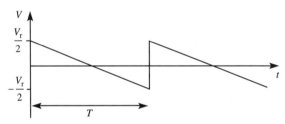

FIGURE A1.1 Triangular wave with peak values of $\pm V_r/2$.

The expression for the voltage is:

$$V = \frac{V_r}{2}\left(1 - \frac{2t}{T}\right)$$

$$= \frac{V_r}{2} \quad \text{when } t = 0$$

$$= 0 \quad \text{when } t = T/2$$

$$= -\frac{V_r}{2} \quad \text{when } t = T$$

The rms value is defined as:

$$V_{rms} = \left(\frac{1}{T}\int_0^T V^2\, dt\right)^{1/2}$$

$$= \left[\frac{1}{T}\int_0^T \frac{V_r^2}{4}\left(1 - \frac{2t}{T}\right)dt\right]^{1/2}$$

$$= \left[\frac{V_r^2}{4T} \int_0^T \left(1 - \frac{4t}{T} + \frac{4t^2}{T^2} \right) dt \right]^{1/2}$$

$$= \left\{ \frac{V_r^2}{4T} \left[t - \frac{2t^2}{T} + \frac{4t^3}{3T^2} \right]_0^T \right\}^{1/2}$$

$$= \left\{ \frac{V_r^2}{4T} \left[t - \frac{2T^2}{T} + \frac{4T^3}{3T^2} \right] \right\}^{1/2}$$

$$= \left\{ \frac{V_r^2}{4T} \left[\frac{T}{3} \right] \right\}^{1/2}$$

$$V_{rms} = \frac{V_r}{2\sqrt{3}}$$

A1.2 ▷ Conduction Angle

The waveform diagram for a half-wave rectifier with a capacitive filter is shown in Figure A1.2. The current surges through the diode are shown shaded.

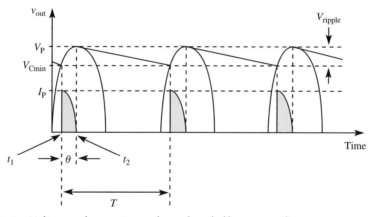

FIGURE A1.2   Voltage and current waveforms for a half-wave rectifier.

The voltage supplied to the diode from the transformer secondary is given by:

$$v = V_P \sin \omega t$$

At t_1 the voltage from the transformer secondary exceeds the voltage across the capacitor and the diode conducts. The voltage at this instant is:

$$v = V_{Cmin} = V_P \sin \omega t_1$$

At t_2 the voltage applied to the diode reaches a peak and the current through the diode ceases. The conduction angle (θ) is:

$$\theta = 90° - \omega t_1 = 90° - \sin^{-1}\left(\frac{V_{Cmin}}{V_P}\right) = \cos^{-1}\left(\frac{V_{Cmin}}{V_P}\right)$$

But $V_{Cmin} = V_P - V_{ripple}$:

$$\therefore \quad \theta = \cos^{-1}\left(\frac{V_P - V_{ripple}}{V_P}\right)$$

Emitter-Coupled Amplifier with Active Loads

An emitter-coupled difference amplifier with active loads is shown in Figure A2.1.

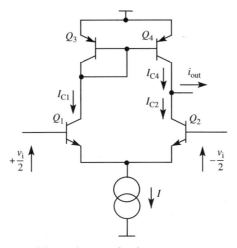

FIGURE A2.1 Difference amplifier with active loads.

Transistors Q_3 and Q_4 form a pnp current mirror which results in:

$$I_{C3} = I_{C1} = I_{C4} = \frac{I}{2}$$

where I is the constant current source for the emitter-coupled pair. If the emitter-coupled pair, Q_1 and Q_2, are assumed to be identical, then when a signal v_i is connected between the bases of Q_1 and Q_2 there will be equal and opposite changes of collector current in these transistors. By symmetry the source can be divided into two equal and opposite parts as $+v_i/2$ and $-v_1/2$. Thus when $+v_i/2$ is applied to the base of Q_1 the collector current increases from I_{C1} to $I_{C1} + i$, while the collector current in Q_2 decreases by the same amount, that is from I_{C2} to $I_{C2} - i$.

Thus:

$$+ \frac{v_i}{2} \quad \text{produces} \quad I_{C1} + i = \frac{I}{2} + i$$

and

$$-\frac{v_i}{2} \quad \text{produces} \quad I_{C2} - i = \frac{I}{2} - i$$

Summing the currents at the output gives:

$$i_{out} = I_{C4} - I_{C2}$$

Because of the current mirror formed by Q_3 and Q_4, $I_{C4} = I_{C1}$ and thus:

$$i_{out} = I_{C1} - I_{C2}$$

$$i_{out} = \left(\frac{I}{2} + i\right) - \left(\frac{I}{2} - i\right) = 2i$$

The transconductance may be defined as:

$$g_m = \frac{di}{dv_i} = \frac{i}{v_i/2}$$

or

$$i = g_m \frac{v_i}{2}$$

and substituting for i

$$i_{out} = g_m v_i$$

The output may be expressed as a voltage as:

$$v_{out} = i_{out} R_o = g_m v_i R_o$$

where R_o is the output resistance, and the gain may be expressed as:

$$A_v = \frac{v_{out}}{v_i} = g_m R_o$$

The output resistance is obtained by looking back into the amplifier at the junction of the two collectors of Q_2 and Q_4. The output resistance of a single transistor can be described in terms of the Early voltage as:

$$R_o = \frac{V_A}{I_C}$$

where V_A is the Early voltage and I_C is the collector current. For the difference amplifier the two outputs R_{oN} and R_{oP} appear in parallel and the output resistance is:

$$R_o = R_{oN} \| R_{oP}$$

Noting that with $v_i = 0$ then $I_{C2} = I_{C4} = I/2$, then:

$$R_o = \frac{2}{I} \frac{V_{AN}V_{AP}}{(V_{AN} + V_{AP})}$$

The transconductance may be obtained for a bipolar transistor as an alternative to the common-emitter current gain by reference to the 'r' parameter equivalent circuit as follows:

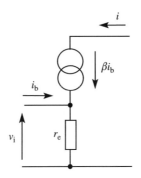

$$g_m = \frac{i}{v_i} = \frac{\beta i_b}{(i_b + \beta i_b)r_e} \cdot \approx \frac{1}{r_e} \quad \text{if } \beta \gg 1$$

and

$$g_m = \frac{1}{0.026/I_E} = \frac{I_E}{0.026} \approx \frac{I_C}{0.026}$$

Thus the voltage gain is:

$$A_v = g_m R_o = \frac{I/2}{0.026} \frac{2}{I} \frac{V_{AN}V_{AP}}{(V_{AN} + V_{AP})} = \frac{1}{V_T} \left(\frac{V_{AN}V_{AP}}{V_{AN} + V_{AP}} \right)$$

where V_T is the thermal voltage (0.026 V at 300 K).

APPENDIX 3

Magnitude and Phase of Low-Pass and High-Pass Filters

The output voltage expressions for low-pass and high-pass filters contain real and imaginary terms, and consequently can be expressed in terms of magnitude and phase.

A3.1 High-Pass Filter

The expression for the output voltage for the high-pass filter is obtained by application of KVL as:

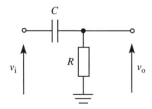

$$v_o = \frac{v_i}{R + (1/j\omega C)} R$$

or

$$\frac{v_o}{v_i} = \frac{j\omega RC}{1 + j\omega RC} \tag{A3.1}$$

Multiplying numerator and denominator by $(1 - j\omega RC)$ to remove the complex term from the denominator gives:

$$\frac{v_o}{v_i} = \frac{j\omega RC(1 - j\omega RC)}{(1 + j\omega RC)(1 - j\omega RC)} = \frac{\omega^2 R^2 C^2 + j\omega RC}{(1 + \omega^2 R^2 C^2)}$$

or

$$\frac{v_o}{v_i} = X + jY \tag{A3.2}$$

where

$$X = \frac{(\omega RC)^2}{1 + (\omega RC)^2} \text{ and } Y = \frac{\omega RC}{1 + (\omega RC)^2}$$

Magnitude

The magnitude of the expression in equation A3.2 is:

$$\left| \frac{v_\text{o}}{v_\text{i}} \right| = \sqrt{X^2 + Y^2}$$

$$\left| \frac{v_\text{o}}{v_\text{i}} \right| = \sqrt{\left(\frac{(\omega RC)^2}{1 + (\omega RC)^2} \right)^2 + \left(\frac{\omega RC}{1 + (\omega RC)^2} \right)^2}$$

and

$$\left| \frac{v_\text{o}}{v_\text{i}} \right| = \frac{\omega RC}{\sqrt{1 + (\omega RC)^2}} \tag{A3.3}$$

Phase

The phase angle is given by:

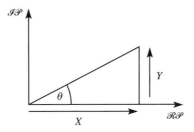

$$\theta = \tan^{-1}\left(\frac{Y}{X} \right)$$

From equation A3.2:

$$\theta = \tan^{-1}\left(\frac{\omega RC}{1 + (\omega RC)^2} \right) \bigg/ \left(\frac{(\omega RC)^2}{1 + (\omega RC)^2} \right)$$

and

$$\theta = \tan^{-1}\left(\frac{1}{\omega RC} \right) = \tan^{-1}\left(\frac{X_\text{C}}{R} \right) \tag{A3.4}$$

A3.2   **Low-Pass Filter**

The expression for the output voltage for the low-pass filter is:

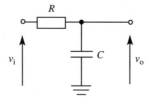

$$v_o = \frac{v_i}{R + (1/j\omega C)} \frac{1}{j\omega C}$$

or

$$\frac{v_o}{v_i} = \frac{1}{1 + j\omega RC} \qquad (A3.5)$$

Multiplying numerator and denominator by $(1 - j\omega RC)$ gives:

$$\frac{v_o}{v_i} = \frac{1 - j\omega RC}{1 + \omega^2 R^2 C^2} = X - jY \qquad (A3.6)$$

where

$$X = \frac{1}{1 + (\omega RC)^2} \text{ and } Y = \frac{\omega RC}{1 + (\omega RC)^2}$$

Magnitude

The magnitude of the expression in equation A3.6 is:

$$\left| \frac{v_o}{v_i} \right| = \sqrt{\left(\frac{1}{1 + (\omega RC)^2} \right)^2 + \left(\frac{\omega RC}{1 + (\omega RC)^2} \right)^2}$$

and

$$\left| \frac{v_o}{v_i} \right| = \frac{1}{\sqrt{1 + (\omega RC)^2}} \qquad (A3.7)$$

Phase

The phase angle is:

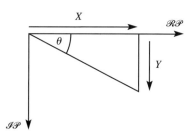

$$\theta = \tan^{-1}\left(\frac{-\omega RC}{1 + (\omega RC)^2}\right)\Bigg/\left(\frac{1}{1 + (\omega RC)^2}\right)$$

and

$$\theta = \tan^{-1}(-\omega RC) = -\tan^{-1}\left(\frac{1}{\omega RC}\right) = -\tan^{-1}\left(\frac{X_C}{R}\right) \tag{A3.8}$$

Hybrid π Equivalent Circuit

The hybrid pi, or hybrid π, equivalent circuit for the bipolar transistor is shown in Figure A4.1.

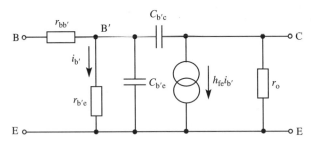

FIGURE A4.1   Hybrid π equivalent circuit for the bipolar transistor.

The circuit is similar to the h parameter circuit with the addition of the junction capacitors $C_{b'c}$ and $C_{b'e}$ and the base resistance $r_{bb'}$. A comparison of the h parameter circuit (Figure 4.14) with that above shows that:

$$h_{ie} = r_{bb'} + r_{b'e} = r_{bb'} + \frac{0.026h_{fe}}{I_E} \tag{A4.1}$$

The approximation that $h_{ie} \approx r_{b'e}$ is generally valid.

The output resistance r_o ($1/h_o$) can often be neglected at high frequencies, since it is much larger than the external load.

The capacitors in the circuit represent the junction capacitance for the base–emitter and base–collector junctions. The base–collector capacitance is simply that of a reverse-biased junction, and is typically $1-10$ pF. For the base–emitter junction there is a junction capacitance resulting from the depletion layer of the base–emitter junction, but there is an additional capacitance which results from the finite amount of time required for the charge carriers injected by the emitter to cross the base and reach the collector. This delay is represented by means of a capacitor, and this component of $C_{b'e}$ varies with the dc emitter current. The total base–emitter capacitance may typically be $50-500$ pF.

An alternative representation for the hybrid π is shown in Figure A4.2, where the current-controlled generator is replaced by a voltage-controlled generator.

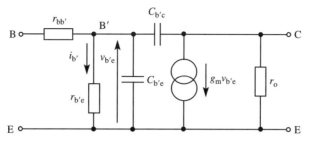

FIGURE A4.2   Hybrid π with voltage-controlled current generator.

From a comparison of the two diagrams it is a simple matter to show that:

$$i_{b'} = \frac{v_{b'e}}{r_{b'e}}$$

and

$$h_{fe}i_{b'} = h_{fe}\frac{v_{b'e}}{r_{b'e}} = \frac{h_{fe}}{0.026h_{fe}/I_E}\,v_{b'e}$$

that is:

$$h_{fe}i_{b'} = g_m v_{b'e} \tag{A4.2}$$

where

$$g_m = \frac{I_E}{0.026}\,\text{S} \tag{A4.3}$$

The parameter g_m is known as the transconductance, and is proportional to the dc emitter current.

Cut-off Frequency

The capacitors, as shown in the hybrid π circuit, affect the high-frequency performance of the transistor, and in particular the common-emitter cut-off frequency, f_β. The cut-off frequency may be defined as the frequency at which the short-circuit common-emitter current gain is reduced by 3 dB. Based on Figure A4.2 the frequency response of the basic transistor may be obtained by placing a small-signal short circuit across the output and observing the variation of the common-emitter current gain, $i_c/i_i|_{v_{ce}=0}$, with frequency. The resultant equivalent circuit for this configuration is shown in Figure A4.3.

With a short circuit at the output r_o disappears and $C_{b'c}$ appears in parallel with $C_{b'e}$. The short-circuit output current is:

$$i_{sc} = -g_m v_{b'e}$$

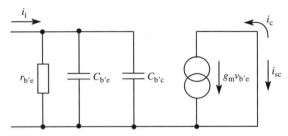

FIGURE A4.3   Simplified equivalent circuit for obtaining f_β.

where

$$v_{b'e} = \frac{i_i r_{b'e}}{1 + j\omega r_{b'e}(C_{b'e} + C_{b'c})}$$

The short-circuit current gain is:

$$\frac{i_{sc}}{i_i} = \frac{-g_m r_{b'e}}{1 + j\omega r_{b'e}(C_{b'e} + C_{b'c})} \tag{A4.4}$$

This equation is similar to that for the low-pass filter, equation A3.5 in Appendix 3, and the 3 dB frequency is defined as the frequency at which the real and imaginary parts of the denominator are equal. Using this fact gives:

$$f_\beta = \frac{1}{2\pi r_{b'e}(C_{b'e} + C_{b'c})} \approx \frac{1}{2\pi r_{b'e} C_{b'e}} \tag{A4.5}$$

At a frequency of f_β the common-emitter current gain is 0.707 of its low-frequency value.

A4.1   Unity Gain Frequency f_t

As well as f_β a further parameter may be defined by noting the frequency at which the magnitude of the current gain is reduced to unity. From equation A4.4 the condition for unity gain is:

$$\left| \frac{i_{sc}}{i_i} \right| = 1 \approx \frac{g_m r_{b'e}}{\sqrt{1 + (\omega r_{b'e} C_{b'e})^2}} \quad \text{since } C_{b'e} \gg C_{b'c}$$

Provided that $\omega r_{b'e} C_{b'e} \gg 1$ then:

$$2\pi f_t r_{b'e} C_{b'e} = g_m r_{b'e}$$

and

$$f_t = \frac{g_m}{2\pi C_{b'e}} \tag{A4.6}$$

From equations A4.5 and A4.4 and an approximation of A4.1, it is seen that:

$$f_t = h_{fe} f_\beta \tag{A4.7}$$

The frequency f_t is often referred to as the gain–bandwidth product of a transistor. It is usually specified in manufacturers' data and it provides a means for calculating $C_{b'e}$ for the hybrid π circuit. From equation A4.7:

$$C_{b'e} = \frac{g_m}{2\pi f_t} \tag{A4.8}$$

Active Filter Derivations

A low-pass active filter is shown in Figure A5.1.

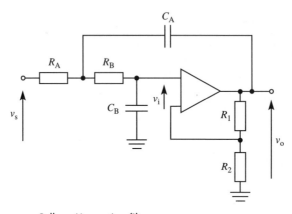

FIGURE A5.1 Low-pass Sallen–Key active filter.

For the analysis of the circuit it is convenient to redraw the circuit, as shown in Figure A5.2, where the amplifier is replaced by a voltage source.

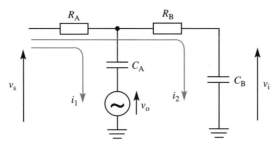

FIGURE A5.2 Equivalent circuit of low-pass filter.

The transfer function required to obtain the frequency response is:

$$H(j\omega) = \frac{v_o}{v_s}$$

Applying KVL for the two mesh currents gives:

$$v_s = i_1 R_A + i_2 R_A + i_1 X_{CA} + v_o \qquad (A5.1)$$

$$v_s = i_1 R_A + i_2 R_A + i_2 R_B + i_2 X_{CB} \qquad (A5.2)$$

Note that:

$$v_i = i_2 X_{CB} \quad \text{and} \quad v_o = A_v v_i$$

$$\therefore \quad i_2 = \frac{v_o}{A_v X_{CB}}$$

where A_v is the gain of the non-inverting amplifier $(1 + R_1/R_2)$.

Substituting for I_2 in equation A5.1 and solving for I_1 gives:

$$i_1 = \frac{v_s A_v X_{CB} - v_o (R_A + A_v X_{CB})}{A_v X_{CB}(R_A + X_{CA})}$$

Substituting for both i_1 and i_2 in equation A5.2 gives, with some manipulation:

$$\frac{v_o}{v_i} = \frac{A_v X_{CA} X_{CB}}{X_{CA} X_{CB} + R_A R_B + R_A X_{CA} + R_B X_{CA} + R_A X_{CB}(1 - A_v)}$$

or

$$\frac{v_o}{v_s} = \frac{A_v}{\omega^2 R_A R_B C_A C_B + \omega[C_B(R_A + R_B) + R_A C_A(1 - A_v)] + 1}$$

To simplify the calculations it is usual to make $R_A = R_B = R$ and $C_A = C_B = C$. Then:

$$H(j\omega) = \frac{A_v}{\omega^2 R^2 C^2 + \omega RC(3 - A_v) + 1}$$

Defining the corner frequency as $\omega_c = 1/RC$ gives:

$$H(j\omega) = \frac{A_v}{(\omega/\omega_c)^2 + (\omega/\omega_c)(1/Q) + 1} \qquad (A5.3)$$

where $Q = 1/(3 - A_v)$ and represents a quality factor.

The expression in equation A5.3 represents a second-order response, as would be expected for a filter which contains two RC networks. The gain can obviously have many different values, but selecting a value of 1.586 results in the expression having the following value:

$$H(j\omega) = \frac{1.586}{(\omega/\omega_c)^2 + 1.414(\omega/\omega_c) + 1} \qquad (A5.4)$$

A plot of this expression normalized to ω_c is shown in Figure A5.3. The particular

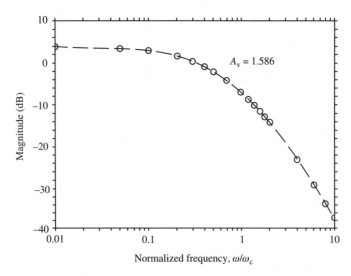

FIGURE A5.3   Second-order low-pass Butterworth response.

polynomial described in the denominator of equation A5.4 is for the class of filters known as Butterworth. Other filter types can be produced by applying different conditions to the value of the RC products and the amplifier gain.

Derivation of Capacitive Transient and Average Value of a Rectified Sine Wave

A6.1 Capacitive Transient

The charge and discharge of a capacitor through a resistor is an important process for waveform generation. The simple relationship for the voltage across a capacitor is:

$$v_c = E[1 - \exp(-t/RC)]$$

This only applies when the initial voltage across the capacitor is zero. For a typical waveform generation circuit, involving dual power supplies, this may not be the case.

The basic circuit to be investigated is shown in Figure A6.1.

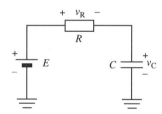

FIGURE A6.1 Circuit for the charging of a capacitor.

Applying KVL gives:

$$E = v_R + v_C$$

since

$$i_C = C \frac{d(v_C)}{dt} \quad \text{then } v_R = i_C R = RC \frac{d(v_C)}{dt}$$

and

$$E = RC \frac{d(v_C)}{dt} + v_C \qquad \text{(A6.1)}$$

Equation A6.1 is a differential equation and is solved by means of integration as follows:

$$RC \frac{d(v_C)}{dt} = E - v_C$$

$$\frac{d(v_C)}{(E - v_C)} = \frac{1}{RC} dt$$

$$\int \frac{d(v_C)}{(E - v_C)} = \frac{1}{RC} \int dt \tag{A6.2}$$

Equation A6.2 is solved by parts to give:

$$\ln(E - v_C) = -\frac{t}{RC} + \text{constant} \tag{A6.3}$$

It is the solution of this equation which establishes the relationship between the voltage on C at any given time and which is affected by the initial voltage on the capacitor, that is when $t = 0$. Let the initial voltage on the capacitor be V_I when $t = 0$. Then:

$$\ln(E - V_I) = -\frac{0}{RC} + \text{constant}$$

and

$$\text{constant} = \ln(E - V_I)$$

Substituting for the constant gives:

$$\ln(E - v_C) = -\frac{t}{RC} + \ln(E - V_I)$$

or

$$\ln\left(\frac{E - v_C}{E - V_I}\right) = -\frac{t}{RC}$$

and

$$\frac{E - v_C}{E - V_I} = \exp\left(-t/RC\right)$$

or

$$v_C = E - (E - V_I)\exp(-t/RC) \text{ for a positive } V_I \tag{A6.4}$$

If the initial voltage is $-V_I$ then by a similar process the voltage is:

$$v_C = E - (E + V_I)\exp(-t/RC) \text{ for a negative } V_I \tag{A6.5}$$

A6.2   Average Value

For a general periodic function $y(t)$, with a period T, the average value Y_{av} is given by:

$$Y_{av} = \frac{1}{T} \int_0^T y(t)\,dt$$

For the half-wave rectified sine wave shown in Figure A6.2(a) for the interval $0 < \omega t < \pi$, $y = Y_m \sin \omega t$; for $\pi < \omega t < 2\pi$, $y = 0$. The period is 2π. The average value is:

$$Y_{av} = \frac{1}{2\pi} \left(\int_0^\pi Y_m \sin \omega t \, d(\omega t) + \int_\pi^{2\pi} 0 \, d(\omega t) \right) = \frac{1}{2\pi} Y_m\{[-\cos \omega t]_0^\pi + 0\}$$

$$Y_{av} = \frac{Y_m}{\pi} \tag{A6.6}$$

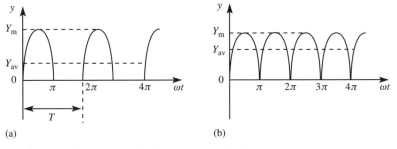

(a)   (b)

FIGURE A6.2   (a) Half-wave- and (b) full-wave-rectified sine waves.

For the full-wave rectified sine wave shown in Figure A6.2(b) for the interval: $0 < \omega t < \pi$, $y = Y_m \sin \omega t$, and the period is π. The average value is:

$$Y_{av} = \frac{1}{\pi} \int_0^\pi Y_m \sin \omega t \, d(\omega t) = \frac{Y_m}{\pi} [-\cos \omega t]_0^\pi \tag{A6.7}$$

$$Y_{av} = \frac{2Y_m}{\pi} \tag{A6.8}$$

Introductory Guide to MicroSim PSpice

MicroSim's evaluation package of PSpice is a fully functioning analog/digital simulator with schematic capture and graphical output. It is restricted to simulating circuits with no more than 20 components. The simulator is primarily for analog circuits, but can also be used for digital and mixed-mode analog/digital. The input data is in the form of a circuit schematic which can contain a wide range of passive and active devices and also a range of behavioural models which can be used to model a subsystem by means of values, tables and polynomial equations. The first step of simulation is to determine the dc operating point of a circuit and then performing an ac or transient analysis. Output is presented as a frequency response for magnitude and phase (small-signal ac analysis), and also as an 'oscilloscope' (transient analysis) for examining waveforms at different nodes in the circuit.

The software operates under Microsoft Windows 3.1 or higher, and this guide refers to PSpice version 6.0, although many of the examples presented in the main text were obtained with version 5.4. For the evaluation package the differences between the two versions are minor. Most of the operational steps to be described below are accessed by means of menus in the usual manner of Windows-based programs, but memorizing a few keyboard commands greatly increases the speed of operation when creating schematics or changing from one operation to another.

Overall Procedure

The process of using PSpice for a simulation involves a number of basic steps as follows:

1 Enter a schematic.
2 Identify the type of analysis required. More than one can be specified.
3 Run PSpice (F11).
4 If there are no problems examine the output with Probe (F12).

If problems are reported at step 3 it may be necessary to examine the text file, which is generated by the simulator, to identify the errors. These are usually corrected by returning to step 1. Most errors are usually concerned with unconnected components, or the incorrect specification of a source, a model or type of analysis.

A7.1   Entering the Schematic

Selecting **Design Center Eval** and **Schematics** from the Windows Program Manager opens a window with a grid of dots. The grid can be turned off if desired, but it does assist in the placing of components. The tool bar contains a number of drop-down menus which are selected by pointing and left clicking with the left-hand mouse button. The most frequently used are **File**, **Edit**, **Draw** and **Analysis**, and the contents of each menu are largely self-explanatory. Most operations are performed with the mouse, with the keyboard being used to enter component values and attributes. Some useful keyboard actions to speed up the operation are as follows:

Ctrl G	Get a component
Ctrl W	add a Wire
Ctrl R	Rotate a component
Ctrl F	Flip a component

The main rules to remember when producing a schematic are:

1 A ground connection must be provided.
2 Every part of the circuit must have a dc path from it to ground.
3 No components or connections must be left floating.

Get a Component

To get a component press Ctrl G, and if it is a resistor, capacitor or inductor, press r, c, or l respectively and press Return. For other components browse through the component libraries. The libraries available in the evaluation version are as follows:

abm.slb	behavioural models
analog.slb	Resistor, capacitor, inductor, controlled sources, transmission line
breakout.slb	Generic diodes and transistors
eval.slb	Named diodes, transistors, op-amps
port.slb	Ground connection, interconnections
source.slb	Current and voltage sources
special.slb	Initial condition flags, parameter boxes, etc.
7400	7400 series components

Components may be rotated or flipped to obtain the correct orientation. Position the component and press the left-hand mouse button to place a component. If several components of the same type have to be placed then continue pressing the left-hand button. Press the right-hand button to complete the operation for either one component or multiple copies of the same component.

NB While it is possible to place components together to form connections it is not advisable because it makes subsequent labelling of nodes impossible. Place components a small distance apart and use wire to connect them.

If a component needs to be repositioned, point and click to highlight the component, press the left-hand button and drag to new position and release the button. Groups of components may be selected by pressing the left-hand button and dragging a box around the components. The group can be moved, copied or deleted.

Make a Connection

To make a connection press Ctrl W. Make connections by left clicking to the end of component connections, or to a previous connection. Right click to complete a connection.

NB Wires which cross are not automatically connected. Click on a wire to make a new connection.

To make more connections press the space bar or double click the right-hand button.

To delete a connection point left click to highlight it and press the Delete key.

Add/Modify Component/Model Values

The default value of components may be altered by pointing at the value and double clicking and entering the value from the keyboard in the menu box displayed. In specifying values, exponential (1.2e3) and prefixes (p, u, m, k, meg) are acceptable.

NB Note use of m for 10^{-3} and meg for 10^6.

For more complex components highlight and double click the component. Each item in the menu box can be given a value, which may be a zero.

For transistors and diodes there are **.model** expressions which may need to be completed for generic models if the default values are unsuitable. Highlight the device and select **Model...** from the **Edit** menu. Note that the schematic must be saved before this can be done. Select **Edit Instance Model...** in the menu box which is displayed. Enter suitable model parameters in the **Edit Model** menu box. For example, for an npn breakout transistor:

```
.model Qbreakn-X NPN
bf = 120
vaf = 80
cje = 10p
cjc = 2p
rb = 80
tf = 0.3ns
```

For a named transistor these values will already exist and should not be changed and the above is only necessary when a suitable named transistor is not available or when a particular parameter value is required to demonstrate some effect.

Label Nodes

It is convenient when using Probe to be able to identify quickly the nodes to investigate. These are labelled in the schematic by highlighting and double left clicking, and then completing the menu box by entering a suitable identity.

Voltage and Current Sources

There are dc sources (ISRC, VSRC) and a number of small-signal sources for providing driving signal (ISIN, VSIN, VPULSE, etc.). The most commonly used are VSRC, VSIN and VPULSE (see the PSpice manual for the other sources).

VSRC　　　　　for a dc source and frequency response
Double click the symbol to open the attribute window
Double click **DC=** and enter a value and press Enter
Double click **AC=** and enter 1 for frequency response and press Enter
Click the OK button

VSIN　　　　　for transient and frequency response.
The attributes to be completed are:

AC=	set equal to 1 for ac frequency response
VOFF=	offset equal to 0 for most cases
VAMP=	set to amplitude for transient analysis
FREQ=	set to value for transient analysis

VPULSE　　　　for transient analysis
The attributes to be completed are:

DC=	default to 0
AC=	default to 0
V1=	first voltage value
V2=	second voltage value
TD=	delay value usually 0

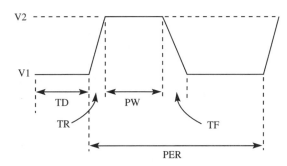

FIGURE A7.1   Illustration of pulse attributes.

TR = risetime
TF = fall-time
PW = pulse width
PER = pulse repetition time

The attributes for a pulse are illustrated in Figure A7.1

A7.2   Setting up the Analysis

The type of analysis is established by selecting **Analysis** from the main menu and then **Setup...** . The main types of analyses are *dc sweep*, *ac frequency response* and *transient*. (See the PSpice manual for other types of operation.)

DC Sweep

This can be used to investigate the effect of changing a dc source in a circuit. It can also be used to investigate the effect of changes of temperature. The name of the source must be entered together with the type of sweep, **linear**, **octave** or **decade**. The start and end values must be specified and the increment, which for the octave and decade is the number of points per octave or decade. Alternatively a list of values can be provided, separated by spaces or commas.

The analysis must be **Enabled** in the main menu box by clicking the appropriate box.

Frequency Response

For small-signal frequency response select **AC Sweep...** . Choose the sweep type, **linear**, **octave** or **decade**, and enter the sweep parameters. Note that for this type of analysis the **AC=** parameter in the source attributes must be set to 1 V. The output voltage is then numerically equal to the gain.

Click the **Enable** box before closing the main **Analysis** menu box.

Transient

The transient or time domain mode is used to examine a waveform at any point in the circuit, much as an oscilloscope would be used in a bench setup. Select **Transient...** and enter **print steps:** and **final time:**. For an input signal frequency of 1 kHz then a print step of 10 µs for 2 ms would produce 200 data points for the graphical output which will cover two cycles of the input waveform. A smaller print step will improve the resolution for a pulse input, or if there is oscillation present, but there will be a bigger data file and it will take longer to produce a trace with Probe.

Transient analysis is the most complex of the circuit simulator modes and for a very non-linear device, for example the switching of an SCR, the simulation time can be many tens of minutes.

Running PSpice and Examining the Output

Start the simulation by pressing function key 11 (F11). If there are no errors then the analysis will proceed. When complete run Probe by pressing function key 12 (F12) to examine the output.

If there are errors then select **Examine Output...** from the **Analysis** menu and examine the text file for an indication of the errors. Go back to the schematic to correct them and rerun PSpice.

Depending on how Probe has been set up it may or may not start automatically at the end of the simulation. The setup for Probe is under the main **Analysis** menu as **Probe Setup...**. Enable the options required.

From the main Probe tool bar select **Trace** for adding traces or frequency graphs. From the **Trace** menu select **Add...** and select the signals to be examined. Any number of signals can be added to the list. When the list is complete select **OK** or press Enter. Mathematical operators can be used to observe the difference between two signals, the product or the ratio. For example:

To plot the difference of two voltages enter
$$v(node1) - v(node2)$$
To plot the gain in dB enter
$$dB(v(node\_name))$$
To plot the instantaneous power in a 1.5 kΩ resistor enter
$$I(R5)^*I(R5)^*1.5e3$$

To make measurements on traces select **Tools** and **Cursor**. There are two cursors: **A1** which is controlled by the left mouse button and **A2** which is controlled by the right mouse button. If there are more than two traces then click the appropriate button on the trace symbol beneath the x axis. The cursor is positioned by clicking at the position on the graph where a measurement is required. The co-ordinates are presented in the **Probe Cursor** box. Incremental movements can be made by pressing the arrow keys on the keyboard for **A1** and Shift plus the arrow keys for cursor **A2**.

If the parameters being investigated have widely different scales, for example mA and volts, then separate plots can be made by selecting from **Plot** the **Add Y Axis** or **Add Plot**. In the former an additional y axis is added, while for the latter a separate plot is formed in a new box. The 'active' box or axis is identified by **SEL≫** on the y axis. The position of this can be changed by left clicking in either box or axis.

Parameter Values

For some simulations it is of interest to investigate the effect of changing model attributes, or the value of one or more components, and to observe the effect of these changes with Probe. This can be achieved by replacing the numerical value with a parameter name, for example the value of a bias resistor may be changed or the value of the common-emitter gain may be altered. This is done as follows:

1   Replace the numerical value of a resistor with **{Rbias}** on the schematic, or **{beta}**

in the **.model** expression for the transistor. The name is not important, but it must be enclosed in curly brackets.

2   From **special.slb** library get **param**.

3   Make **NAME1** attribute **Rbias** and **NAME2** attribute **beta** if the two parameters are to be investigated at the same time. Enter an initial value in **VALUE1** and **VALUE2**. If nothing more is done then the simulation will now run with these two values being used.

4   To change *one* of these values automatically during the simulation then from **Analysis** and **Setup...** choose **Parametric...**.

5   For the resistor select **Global Parameter** and for the **Name:** enter Rbias.

6   Enter the values to be examined under **Sweep Type** either as incremented values or as discrete values.

7   Enable the **Parametric** check box and run PSpice.

8   Use Probe to examine the multiple outputs.

For the transistor gain at step 5 enter for the **Name:** beta in place of Rbias and suitable sweep values.

A7.5 Behavioural Models

A very powerful feature of PSpice is the availability of behavioural models. These models are to be found in the **abm.slb** library. These models provide a variety of transfer functions covering such things as band-pass, band-reject, low pass, high pass, summing, integration, multiplying, Laplace and many mathematical functions. Notice that these models have open-circuited terminals. PSpice does not regard this as a connection, and if this does produce an error message, then it can be resolved by connecting a large resistor (1 GΩ) between the terminals, or from the terminal(s) to ground.

Behavioural models can be used to model complex circuits by means of a box with a suitable transfer function. Thus a phase-locked loop can be represented by three voltage-controlled voltage sources: EMULT, ELAPLACE and ETABLE. The advantage of using such a model is increased simulation speed, but also the ability to simulate a complex circuit without the need to design the discrete component form. Also, with the evaluation copy of PSpice, which does not handle more than 20 components, it provides a means for simulating complex circuits. An actual PLL contains far more than 20 components.

Of the many controlled voltage or current sources some of the simpler and more useful ones are those which begin with the letter E. These are voltage-controlled voltage sources which can be used to represent operational amplifier blocks. The simplest are ESUM and EMULT. The first produces an output which is the sum of two input voltages. The output voltage of the second is numerically equal to the product of two input voltages.

EVALUE produces an output which is a function of the input, where the function is determined by the **EXPR** attribute. To generate an ideal full-wave rectifier characteristic the **EXPR** would be ABS(V(%IN+,%IN−)). A voltage-controlled oscillator for the PLL can be obtained with $\sin(2^*pi^*fo^*TIME+k^*V(\%IN+))$, where pi, fo and k are set with a **PARAM** box.

ETABLE allows a transfer function to be defined in a piece-wise linear fashion by means of a table of co-ordinate pairs. These must be in a linear increasing order. For example, setting the **TABLE** attribute to $(-12,2)(0,12)(20,20)$ produces the transfer characteristic shown below in Figure A7.2. Note that outside the ranges indicated that the output holds the first and last value given, that is, -12 and $+20$.

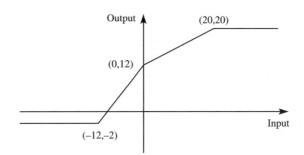

FIGURE A7.2   ETABLE transfer function.

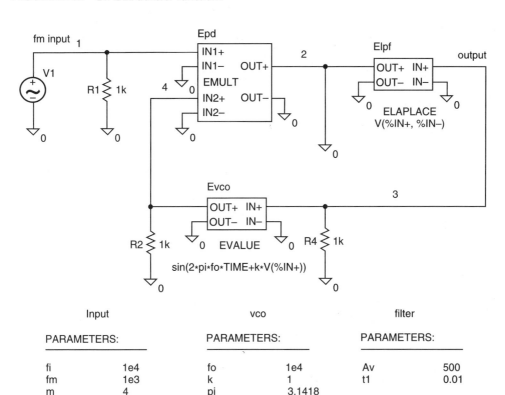

Input		vco		filter	
PARAMETERS:		PARAMETERS:		PARAMETERS:	
fi	1e4	fo	1e4	Av	500
fm	1e3	k	1	t1	0.01
m	4	pi	3.1418		

FIGURE A7.3   Behavioural schematic for a PLL.

ELAPLACE can be used to describe a frequency-dependent function of the form $1/s$. A simple low-pass filter can be described by entering for the **XFORM** attribute $Av/(1+t1^*s)$, where Av is a voltage gain and t1 a time constant which can be specified in the **XFORM** attribute or separately by means of a **PARAM** box.

A complete PLL schematic from PSpice is shown in Figure A7.3.

The different waveforms from the input, after the multiplier and the output of the PLL, are shown in Figure A7.4.

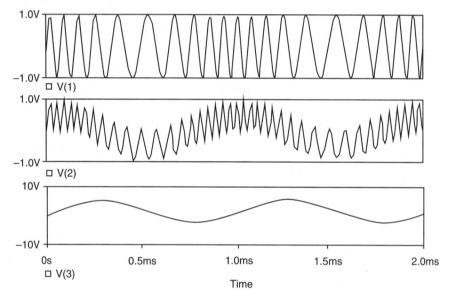

FIGURE A7.4   Waveforms for the PLL.

The input signal to the PLL is a frequency-modulated voltage source VSFFM with the attributes set as follows:

VOFF = 0
VAMPL = 1
FC = {fi}
MOD = {m}
FM = {fm}

Note the use of curly brackets to identify parameters which are specified in the **PARAM** box. This provides a simple means for varying parameters when examining the operation of the PLL under different conditions.

A7.6 ▷ Full Version

The full version of PSpice handles many more components and has far more comprehensives libraries with over 13000 analog and digital parts. It can handle mixed-mode analysis with analog and digital parts in the same circuit. The graphical output by means of Probe allows signals at any point in the circuit to be examined in many different ways. Timing delays and digital timing violations and hazards can be examined in digital simulations.

The evaluation package provides an ideal introduction to circuit simulation and it is hoped that the examples provided in this text will help in the understanding of how analog circuits operate.

A7.7 ▷ Tutorial

This tutorial is based on the first PSpice example in Example 1.3. The schematic is reproduced in Figure A7.5. Use PSpice to simulate the waveforms produced in the 'dimmer' control circuit when the input is 110 V and 60 Hz.

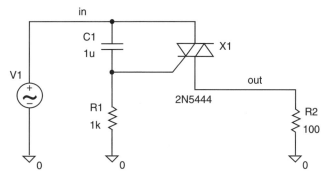

FIGURE A7.5   PSpice schematic of a TRIAC light dimmer.

The purpose of the tutorial is to illustrate how a circuit is entered as a schematic and an analysis performed. It is not intended to cover all aspects of PSpice analysis, nor to provide an explanation of the operation of the TRIAC, but rather to provide some familiarity with the basic operations for using PSpice. Additional features of PSpice are described as required in subsequent examples within the main text.

Creating the Schematic

From the Windows Program Manager select the **Design Centre Eval X.X** icon, where X.X is the current version number. From the **Design Centre** workgroup select the **Schematics** icon. The **Schematics** window is shown in Figure A7.6.

Use the mouse to point and select the menu options on the horizontal menu bar. None of these menus will be used initially, but if a circuit already existed then it could be opened by selecting **Open...** from the **File** menu.

For a new circuit it is possible to start selecting components immediately. Press Ctrl G to open a box for getting a component, as shown in Figure A7.7.

If the part number is known then enter this in the box for **Part:**, else select **Browse...** and the appropriate library, and select the part from the library by left clicking with the mouse on the part number listed in the library, as shown in Figure A7.8.

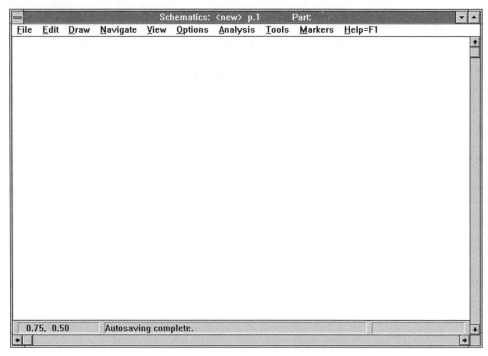

FIGURE A7.6   PSpice schematics workspace.

FIGURE A7.7   Add a part.

For a resistor it is simply a matter of entering **r** and pressing Return or selecting **OK** by left clicking with the mouse. The parts box closes and a resistor symbol appears lying horizontally and attached to the mouse pointer. It is positioned in the work area by moving the mouse pointer. For the required schematic the resistor symbol is vertical. Press Ctrl R to rotate the symbol through 90° steps. Try this.

A component symbol is placed by left clicking. Multiple copies are placed by continuing to move the mouse and left clicking. The placement operation is cancelled by right clicking. Unwanted components may be removed by left clicking the component, when it changes from green to red, and pressing Delete from the keyboard. A group of components may be selected by pressing and holding down the left mouse button and drawing a rectangle around the components to be removed. Place some resistors on the screen and remove them individually and collectively.

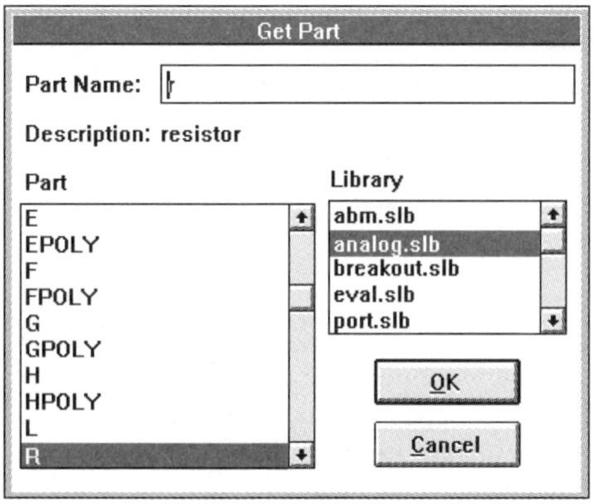

FIGURE A7.8   Part and library selection box.

Start to construct the circuit by placing two resistors, a capacitor – enter **c** in the part box – and the TRIAC by entering **2N5444** in the part box, or by selecting **Browse…** and **eval.slb** from the library, left click on the part 2N5444, and then select **OK**.

For the voltage source select **Browse…** and in the library box use the vertical scroll box to scan through the different libraries to **source.slb**. From the part box use the vertical scroll bar to select **VSIN**.

Notice in the schematic in Figure A7.5 the ground connections. These are important for the success of the simulation. It is not necessary to have ground connections for each grounded component. The grounds could be wired together and one ground connection used. However, it is often simpler to use multiple grounds as shown. Select **Browse…** and the **port.slb** library, and left click on **AGND**.

Adding Wires

The components are wired together by using the mouse to draw horizontal and vertical lines from the end of each component termination. Wiring commences by pressing Ctrl W. The mouse pointer changes from an arrow to a pencil. To start a connection position the tip of the pencil on the end of a component termination and left click. Move the pointer to produce a dotted line showing the position of the connection. Left click on the termination of another component, and right click to complete the connection. If the connection is to continue to another component, then move the pointer to the new position and left click again rather than right click. Connection to another wire is made by left clicking on the wire. A change of direction (a right-angle) is obtained by left clicking at the point where the change is required.

Adding Attributes

Attributes such as component values, amplitude and frequency of the source and component and node (or wire) names are added by double clicking on the component, wire, or the actual default value, for example the default 1 kΩ value of the resistor, or the 1 nF value of the capacitors.

For the schematic in Figure A7.5 the value of R1 is the default value of 1 kΩ, while that of the load is 100 Ω; the capacitor is 1 μF – enter **1u** in the attribute value box. For the source enter the following:

AC = 0
voff = 0
vampl = 155
freq = 60

Further useful attributes are names for connections, and in Figure A7.5 the input connection is identified as **in** and the output as **out**. These are added by double clicking on the wire, and entering the name in the **Label** box.

Selecting the Type of Analysis

From the **Analysis** menu box select **Setup...** to produce the setup box, as shown in Figure A7.9.

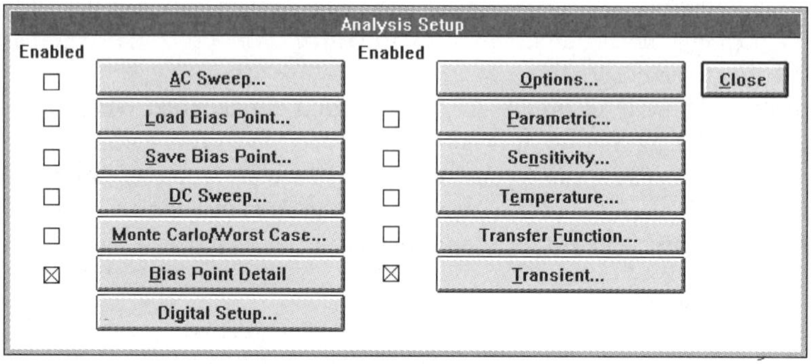

FIGURE A7.9   Analysis setup box.

The different types of analysis are selected by clicking on the appropriate **Enable** box. Notice that **Bias Point Detail** is enabled by default. For the dimmer circuit enable the **Transient** box, and then select the **Transient...** button. This produces the transient setup box, as shown in Figure A7.10.

Enter 200us for the **Print Step:** and 33ms for the **Final Time:**. This corresponds to output print data being produced every 200 μs for a duration of 33 ms, or two complete cycles of the 60 Hz waveform. Click the **OK** button and the **Close** button in the setup box. Remember to click the **Enable** box.

Run PSpice

Either select **Simulate** from the **Analysis** menu, or press F11 to start the simulation. If

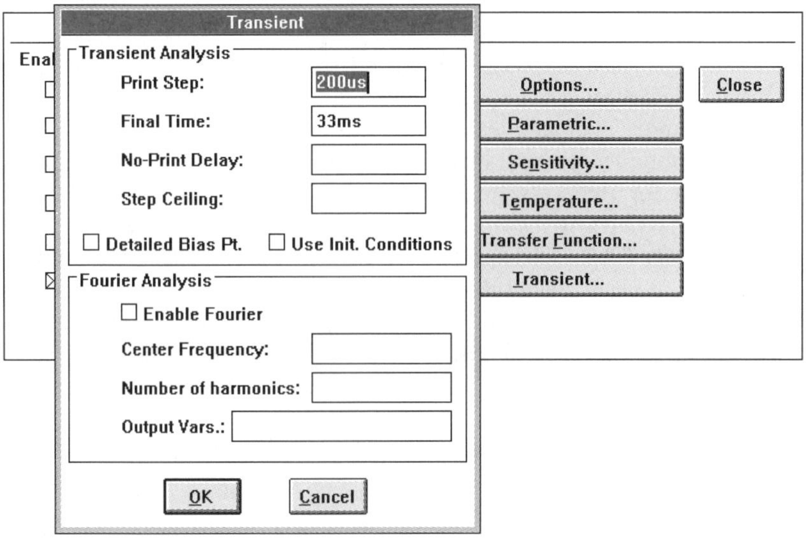

Transient
Enal
Transient Analysis
Print Step:
200us
Final Time:
33ms
No-Print Delay:
Step Ceiling:
Detailed Bias Pt.
Use Init. Conditions
Fourier Analysis
Enable Fourier
Center Frequency:
Number of harmonics:
Output Vars.:
OK
Cancel
Options...
Parametric...
Sensitivity...
Temperature...
Transfer Function...
Transient...
Close

FIGURE A7.10 Transient setup box.

the connections are correct and the correct attributes have been supplied then the simulator runs. With a 386 machine with a maths co-processor or a 486 machine the simulation should take between 2 and 3 minutes.

Run Probe

To examine the output waveform either select **Run Probe** from the **Analysis** menu or press F12. From the Probe menu bar select **Trace** and **Add...** from the menu. From the **Add Traces** box left click on V(in) and V(out) and select **OK** to close the **Add Traces** box and produce the waveform shown in Figure A7.11.

200V
−200V
V(in)
V(out)
0s
10ms
20ms
30ms
40ms
Time

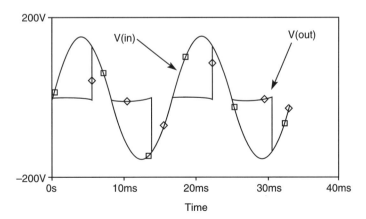

FIGURE A7.11 Input and output waveforms from Probe.

Change the Value of R1

Repeat the simulation for a different value of R1, for example 5 kΩ. To investigate the effect of a number of different values of R1 without having to restart the simulation each time, it is possible to use the parameter feature. In the schematic replace the value of R1 with a name, for example {phase}. Note the use of curly brackets. These are necessary and are the means for the software to identify a parameter. It is also necessary to identify the name separately by means of a parameter box. Use Ctrl G to get a component and select **PARAM** from the **special.slb** library. Place the parameter box anywhere on the schematic. Double click on the parameter box and enter phase for **Name1:** and a start value for **Value1:**, for example 1 kΩ. Next from **Setup...** in the **Analysis** menu enable the box for **Parametric**. Select **Parametric** and select **Global Parameter**, enter phase for **Name:**, select **Value List** for sweep type, and enter 1k, 3k, 5k for **Values:**. Select **OK** and **Close**. Run the simulation. The simulation now runs for each of the resistor values and the three waveforms can be examined in a composite Probe diagram.

PSpice contains many features for examining different aspects of the operation of a circuit, some of which are described in appropriate examples within the main text. A much fuller account of the operation is to be obtained by studying the manuals provided with the full version of the simulator, or purchased separately from MicroSim.

Index